Inner Solar System

Viorel Badescu · Kris Zacny
Editors

Inner Solar System

Prospective Energy and Material Resources

Springer

Editors
Viorel Badescu
Candida Oancea Institute
Polytechnic University of Bucharest
Bucharest
Romania

Kris Zacny
Honeybee Robotics
Pasadena, CA
USA

ISBN 978-3-319-19568-1 ISBN 978-3-319-19569-8 (eBook)
DOI 10.1007/978-3-319-19569-8

Library of Congress Control Number: 2015941111

Springer Cham Heidelberg New York Dordrecht London

Printed on acid-free paper

Springer International Publishing AG Switzerland is part of Springer Science+Business Media (www.springer.com)

This book is dedicated to the past, present, and future explorers of Venus and Mercury.

Foreword

Throughout history, voyages of discovery and exploration have been followed by periods of relative quiet, as people absorb the new findings, but were often "motionless," fearing to follow in the footsteps of these courageous individuals, or simply not knowing what to do. Then came tentative expeditions of assessment of the new lands, reporting back on the nature and beauty of the landscape and its potential. And then, attempts at settlement and living off the land, and finally, exploration for natural resources to support an ever-increasing population and complex infrastructure.

The well-known exploration of the American West comes to mind. Early explorers were followed by Lewis and Clark, sent by insightful leaders to map and assess the potential of the land. Then followed wagon trains, cowboys and ranchers, sodbusters and homesteaders, railroads to connect the country, and finally even greater infrastructure to support an ever-increasing and organized society and the cities they came to live in.

What must it have been like to be Captain James Cook, setting sail on his first voyage of exploration in 1768, a whole new and unknown world awaiting? Imagine yourself being Captain Cook, setting off to the vast Pacific Ocean, not knowing the location and magnitude of landmasses, with only the Sun and stars as your guide. Who could have dreamed that a mere 250 years later, we could visit any place on the planet with Google Earth.

Critical to this rapid evolution on Earth were several factors:

(1) brave explorers venturing into the unknown,
(2) insightful governments funding expeditions of discovery and assessment,
(3) scientists driven by a desire to explore the unknowns of nature,
(4) entrepreneurs anxious to locate and develop potential resources,
(5) technologists at the forefront of innovation developing ways to apply the latest capabilities,
(6) engineers anxious to create new and useful designs and infrastructure, and most importantly,

(7) dreamers, living in the world of the future, unencumbered by the chains of short-term reality.

Today we stand on the verge of the next two centuries of exploration of the inner solar system. Twelve humans have ventured to the surface of the Moon. Completed or underway are missions with heroic names bringing to mind our earlier mythical explanations of nature and exploration history, such as Mercury, Gemini, Apollo, Mariner, Viking, Magellan, Surveyor, and MESSENGER. In a manner analogous to our earlier epiphany that the Earth was indeed round, we now know the basic characteristics and broad histories of the Solar System. In the last 50 years, Solar System objects, planets, satellites, comets, and asteroids have changed from astronomical objects to geological objects. We know where they are, what they are made of, and broadly, how they came to be. Astronomical objects are now planets orbiting other stars, being discovered by the hundreds. Our own Solar System has become the neighborhood, and the Earth, our Home Planet.

But what next? What will we find in our Solar System neighborhood in the *next* century? What will be the infrastructure necessary to take the next steps to explore, exploit, and live on or around these other planetary bodies? We still have many, many unanswered questions, as exemplified by the name of the latest Mars rover, The Mars Science Laboratory "Curiosity." But who will be the scientists, entrepreneurs, technologists, engineers, and most importantly, the dreamers of the future?

The *Inner Solar System: Prospective Energy and Material Resources* brings these individuals together for an exciting glimpse into the future. In this remarkable volume you will find syntheses of the current knowledge of planets and their interiors, presentation of outstanding questions to drive future exploration and exploitation, assessments of new techniques to undertake human and robotic exploration of Venus and Mercury, imaginative engineering and systems approaches to explore through drilling, orbiting, floating in balloons, insightful infrastructures for providing infrastructure and power, as well as shielding from harmful radiation, and finally, business plans and models that begin to explore the reality of how all these inevitable steps might be financed.

But what ultimate motivation for these visions of the future? Many years ago I had a long discussion with Apollo 16 Commander John Young, in which we considered, from my geological perspective, the fact that over 99 % of the species that ever existed on the Earth were now extinct. A man of few, but important words, he thought for a moment and said: "Hmmm… Single-planet species don't survive."

Viorel Badescu and Kris Zacny have gathered the doers and the dreamers to describe their visions for the next steps in the evolution of our species.

March 2015

James W. Head III
Department of Earth
Environmental and Planetary Sciences
Brown University, Providence
Rhode Island, USA

Preface

This is a fourth volume of a Springer book series making an inventory of the material and energy resources of our Solar system. The first three books, referring to resources existent on Mars, Moon, and asteroids were published in 2009, 2012, and 2013, respectively.

This book presents a present-day perspective on the energy and material resources in the inner Solar System for prospective human use. One investigates the advantages and limitations of various systems thought-out for future mankind utilization. The book collects together recent proposals and innovative options and solutions. It is a good starting point for researchers involved in current and impending Mercury- and Venus-related activities.

The book is structured along logical lines of progressive thought and may be conceptually divided into seven sections.

The first section deals with what *we know about Mercury and Venus* and contains four chapters. After the introductory Chap. 1, treating the origins and interiors of the Inner Planets, Chap. 2 shows the Significant Results from MESSENGER and Venus Express Missions. Chapter 3 focus on Mercury while Chap. 4 refers to previous and past missions accessing the Venus lower atmosphere and surface.

The second section of the book deals with *transportation from Earth to Mercury and Venus* and consists of three chapters. Chapter 5 refers to Special orbits for Mercury observation while Chap. 6 examines the low-thrust Earth–Venus trajectories. The estimation of the fuel consumption for space trip to Venus and Mercury is made in Chap. 7.

The third section of the book, dealing with the *drilling techniques,* consists of two chapters. Chapter 8 focus on drilling and sample transfer mechanisms for potential missions to Venus while Chap. 9 treats in detail pneumatic drilling and excavation in support of Venus science and exploration.

The fourth section of the book, referring to *power systems,* consists of three chapters. Chapter 10 describes the past, present, and future power system options for Venus exploration missions. Chapter 11 deals with production of energy for Venus by electron wind generator and Chap. 12 refers to photovoltaic power resources on Mercury and Venus.

The fifth section of the book, referring *exploration and exploitation of Mercury and Venus resources,* consists of five chapters. Chapter 13 describes flight apparatuses and balloons in Venus atmosphere while Chap. 14 refers to conditions on Mercury and Venus that could be of use to engineers who are tasked with designing structures for robotic operations and human habitation on these planets. Chapter 15 treats deployable structures for Venus surface and atmospheric missions and Chap. 16 presents a systems approach to the exploration and resource utilization of Venus and Mercury. Artificial magnetic fields for Venus are proposed in Chap. 17.

The sixth section of the book, dealing with the *management of the Inner Solar System,* consists of Chap. 18, with business concerns and considerations.

The seventh section of the book, dealing with the *resource utilization of Mercury and Venus in the far future,* consists of four chapters. Chapter 19 refers to the Economic Development of Mercury in Comparison with Mars Colonization. Chapter 20 refers to terraforming of Mercury while Chap. 21 proposes a method of terraforming Mercury and Venus. Living on Venus is treated in Chap. 22.

More details about the 22 chapters of the book are given below.

Chapter 1, by Marvin Herndon, shows that the presently popular description of planetary formation in our Solar System, an updated version of the Chamberlin-Moulton planetesimal hypothesis began as assumption based models nearly 50 years ago and has proven inadequate to explain current heat production and magnetic field generation in small bodies, such as Mercury and Jupiter's moons Io and Ganymede. Moreover, within that framework there exists no logical, causally related explanation why the terrestrial planets have such diverse surface dynamics even though they have more-or-less similar bulk chemical compositions. In 2013, the author of this chapter published a new indivisible planetary science paradigm, a wholly self-consistent vision of the nature of matter in the Solar System, and dynamics and energy sources of planets is described here. Massive-core planets formed by condensing and raining-out from within giant gaseous protoplanets at high pressures and high temperatures. Earth's complete condensation included a ~300 Earth-mass gigantic gas/ice shell that compressed the rocky kernel to about 66 % of Earth's present diameter. T-Tauri eruptions stripped the gases away from the inner planets and stripped a portion of Mercury's incompletely condensed protoplanet and transported it to the region between Mars and Jupiter where it fused with in-falling oxidized condensate from the outer regions of the Solar System and formed the parent matter of ordinary chondrite meteorites, the main-Belt asteroids and veneer for the inner planets, especially Mars. In response to decompression-driven planetary volume increases, cracks form to increase surface area and mountain ranges characterized by folding form to accommodate changes in curvature. The differences between the inner planets are primarily the consequence of different degrees of protoplanetary compression: By contrast to Earth, Mercury and presumably Mars experienced no significant protoplanetary compression. The degree of compression for Venus cannot presently be ascertained. The internal composition of Mercury is calculated by analogy with Earth. The rationale is provided for Mars potentially having a greater subsurface water reservoir capacity than previously realized.

Chapter 2, by Sanjay S. Limaye, states that MESSENGER, the second mission launched toward Mercury became the first orbiter around the planet and Venus Express orbiter have collected vast data on the two planets from orbital observations which are shedding light on many questions about them and raising new ones. This chapter presents some significant discoveries from these two missions that are still being made at present and for years to come.

Chapter 3, by Johannes Benkhoff, reviews the current knowledge about Mercury and gives some details about various spacecraft missions. Mercury is the planet of extremes. It is the smallest among the planets of our solar system and the only planet beside Earth with an internally generated magnetic field. Mercury's density is much higher than densities of all the other planets. Because of its close distance to the Sun, Mercury's surface experiences temperatures up to 700 K. In contrast, the interiors of some large craters close to the poles are permanently shadowed and remain as cold that Scientists believe that the influx of ice from in-falling comets and meteorites could be cold-trapped in these craters over billions of years, even until today. On the other hand the thermal and radiation environment of Mercury is extremely aggressive, which makes a spacecraft mission to explore this planet technically very challenging. Two NASA spacecraft have visited Mercury so far. Mariner 10 flew by Mercury three times during the 1970s. The MESSENGER spacecraft flew by Mercury in 2008 and 2009 before inserted into orbit around in March 2011. In short the European Space Agency, ESA, will launch the BepiColombo mission to explore Mercury. BepiColombo a joint project between Europe and Japan. From dedicated orbits two spacecraft will be studying the planet and its environment. All this will hopefully reveal further secrets on the "sunshine planet" Mercury essential to gain further knowledge about the evolution history of our Solar System.

Chapter 4, by Michael J. Amato, states that Venus was a prime target for atmospheric probes and landers, particularly Soviet spacecraft, in the early days of space exploration. More recently, however, this has not been the case, the last in situ missions (Venera 13 and 14) having landed on Venus in 1982. Even with the later orbital missions, Magellan and Venus Express, our understanding of Venus is very incomplete, and future in situ missions would be critical in filling these knowledge gaps. The first Successful probe into the atmosphere of Venus was the Soviet Venera 4 in 1967. This was followed by more Venera probes which survived penetrations successively deeper into the atmosphere and culminated in a series of landers that survived the torrid conditions long enough to send data back from the Venus surface. A set of NASA Pioneer Venus probes were dropped into various parts of the atmosphere in 1978 and returned data as they fell to the surface. These probes and landers returned valuable information on Venus environmental and surface conditions, but were limited by the technological capabilities of the time. The future of Venus in situ missions is driven by a large set of unanswered science questions that demand better instruments, access to new locations, or longer access to the atmosphere and surface than past missions. Much of the technology needed to address many of the highest priority questions already exists. For the first time in decades, there is a possibility the human race will return spacecraft to the atmosphere of Venus. For example, medium-sized Venus probes and landers have been proposed in the past decade in the U.S. and to European Space Agency opportunities. NASA planetary decadal reports recommend a Venus in situ mission for the NASA mid-tier competed mission program. Examples of more capable and complicated missions are in the U.S. NASA decadal survey, and are driven by long-term science needs at Venus. The National Research Council's Planetary Decadal Survey Inner Planets Panel commissioned three Venus in situ mission studies as part of the 2010 survey that led to the "Visions and Voyages for Planetary Science in the Decade 2013–2022." These mission's science, designs, and cost were studied and reviewed and they serve as an excellent measure of possible near term future missions from NASA to the Venus atmosphere.

Chapter 5, by Generoso Aliasi, Giovanni Mengali and Alessandro A. Quarta, deals with an advanced scientific mission concept in which the existence of suitable positions for the observation and the measurement of the Mercury's magnetotail are investigated. The scientific mission is based on the use of artificial equilibrium points in the elliptic three-body system, constituted by the Sun, Mercury, and a spacecraft, which is modeled as a massless point. The spacecraft motion in the Sun–Mercury system is first discussed under the assumption that the propulsion system provides a radial continuous thrust with respect to the Sun. In particular, the spacecraft is assumed to have a generalized sail as its primary propulsion system. A generalized sail models the performance of different types of advanced propulsion systems, including a (photonic) solar sail, an electric solar wind sail and an electric thruster, by simply modifying the value of a thrusting parameter. The location of the artificial equilibrium points is derived, and their stability is also investigated. It is shown that that collinear artificial equilibrium points are always unstable, except for a range of L_2-type points which are placed far away from Mercury. A similar result is obtained for triangular equilibrium points. A control strategy is introduced to maintain the spacecraft in the neighborhood of an artificial equilibrium point. In this context, a simple and effective way to actively control the spacecraft dynamics is by means of a Proportional-Integral-Derivative feedback control law. The latter control law is finally employed in the magnetotail mission scenario, whose fundamental idea is to continuously and slowly displacing the artificial equilibrium point along the Sun–Mercury direction. Numerical simulations show the effectiveness of the proposed mission strategy.

Chapter 6, by Alessandro A. Quarta, Giovanni Mengali and Generoso Aliasi, discusses the simulation results involving the minimum-time trajectories for an Earth-Venus mission transfer using a spacecraft with an electric propulsion system, with both a nuclear and a solar electric power source. The analysis has been performed in a parametric way as a function of some design parameters, such as the available thruster electric power and the initial in-flight mass. Various models have been considered to describe the propulsion system behavior with different levels of approximations and obtain increasingly refined information about the mission performance. Some simplifying assumptions have been introduced in order to make the mathematical problem tractable, to reduce the simulation time and guarantee a thorough parametric investigation of the mission performance. The analysis performed in this chapter is useful to obtain a first order estimate of the mission requirements as a function of the specific thruster characteristics.

Chapter 7, by Alexander A. Bolonkin, researches the most useful space trajectories to Mercury and Venus. It gives the theory of space flights, methods and estimation of fuel consumption for space trips, and trajectories close to optimal (minimum fuel usage). It also provides example computations of fuel consumption for flights to Venus and Mercury. It also gives brief data about former space apparatus and rocket system flight to Venus.

Chapter 8, by Yoseph Bar-Cohen, Xiaoqi Bao, Mircea Badescu, Stewart Sherrit, Hyeong Jae Lee, Kris Zacny, Nishant Kumar, and Erik Mumm describes a device for transferring acquired powder samples. The mechanism uses pneumatic actuation and involves the use of capsules. The inclusion of such a sample transfer system allows the use of science instruments inside the thermally controlled enclosure within the lander. The in situ exploration of our Solar System bodies poses many challenges, and these are unique to each body. Without a doubt, the toughest conditions are encountered on Venus and to some extent on Mercury. Temperatures on both planets are over 400 °C. However, while the temperature on Mercury can drop to −173 °C, which poses its own challenges, the temperature on Venus is constant at 465 °C. This extreme temperature combined with the Venus surface atmospheric pressure of 90 atmospheres (equivalent to pressure at 1 km deep Earth ocean), makes carbon dioxide atmosphere super critical and in turn corrosive. The temperature combined with corrosive atmosphere poses biggest challenge. A key challenged to the in situ exploration is the ability to sample its surface and the subsurface. This chapter covers methods of drilling at high temperature conditions. The methods covered in this chapter include high temperature piezoelectric materials and electromagnetic actuators as well as thermal techniques (thermal-spalling and thermal-melting). Piezoelectric actuated drills with various diameters have been developed to operate at temperatures as high as 500 °C. The use of various piezoelectric materials including LiNbO3 and Bismuth Titanate was tested by fabricating Ultrasonic/Sonic Driller/Corer (USDC) based drills and testing them at high temperatures. One of the drills was designed with a novel method of operating the USDC as a rotary-hammer where the rotation is induced by the vibration of the piezoelectric actuator. The drill driven by Bismuth Titanate was demonstrated to penetrate a brick sample to a depth of 25 mm in 21 min. The chapter also covers more conventional rotary drills powered by two switch reluctance motors (SRM). The two motors were used to rotate the auger with a carbide bit at its end and for advancing the bit into the rock. Tests were conducted at temperatures as high as 460 °C drilling to a depth of 20 cm in approximately 20 min with a maximum power of 45 W.

Chapter 9, by Kris Zacny, Justin Spring, Gale Paulsen, Stephen Ford, Philip Chu, and Steve Kondos, states that Venus is considered to be Earth's sister planet and hence we can learn a lot about Earth by investigating Venus tectonics, volcanism, and atmosphere. The most recent Planetary Science Decadal Survey publication called the Visions and Voyages for Planetary Science in the Decade 2013–2022, recommended Venus In Situ Explorer (VISE) as one of the five candidates for the NASA's New Frontiers mission. The VISE mission was reaffirmed by the Decadal Survey because many questions about Venus cannot be addressed by observations from an orbiter. The VISE mission requires sample acquisition and delivery to remotely located sensors technology. Honeybee Robotics, funded by the NASA's Small Business Innovation Research investigated methods of excavation (Drill or a Trencher), Sample Delivery (Pneumatic Suction vs Blower vs 3 DOF Robotic Arm) and Sampler Deployment (Vertical Z-stage, Spring Loaded Arm, 3 DOF Robotic Arm), in support of the technology development for the VISE mission. This chapter reports on test results and makes recommendations with respect to sample acquisition and delivery for Venus surface missions.

Chapter 10, by Simon Fraser, states that Venus, the planet so close to Earth in terms of distance and size, provides an equally fascinating and challenging environment for atmospheric and surface exploration missions. Venus is actually quite the opposite of Mars, the second one of Earth's two sister planets. Mars has got a very thin atmosphere with low surface temperatures. Venus, on the other hand, has got a very thick atmosphere with pressures exceeding 90 bars and surface temperatures exceeding 450 °C. Exploration missions therefore face completely different ambient conditions than on Moon and Mars; this is reflected in surface lander design and operation profiles. This chapter summarizes the history of Venus exploration missions, and has a special focus on power system solutions applied with the different mission elements. In this, the power system options can be broken down into two sub-categories: first, power systems applied en route to Venus and/or in orbit around Venus; secondly, power systems for descent stages designed to enter the atmosphere, and in some cases even to (soft-) land on the surface of Venus. Selected exploration missions, their scientific payload and their power systems are described in more detail to outline how the challenges of Venus exploration have been addressed in the past.

Chapter 11, by Alexander A. Bolonkin, offers a new method of getting electric energy from wind. A special device injects electrons into the atmosphere. Wind picks up the electrons and moves them in the direction of the wind which is also against the direction of electric field. At some distance from injector, a unique grid acquires the electrons, thus charging and producing electricity. This method does not require, as does other wind energy devices, strong columns, wind turbines, or electric generators. This proposed wind installation is cheap. The area of wind braking may be large and produces a great deal of energy. The author applies this method to Venus's atmosphere, which has higher density and temperature than the Earth's atmosphere.

Chapter 12, by T.E Girish and S. Aranya, addressed the severe challenges for photovoltaic power generation on Mercury and Venus. This includes high daytime temperatures, hazardous radiation environments and long night periods. Some possibilities for utilizing photovoltaic (PV) power resources on these inner planets are also discussed. For short duration Venus missions lasting 80–90 days daytime PV power generation near the top of Venus atmosphere (50–70 km) is preferable using Gr III–V solar cells mounted on balloons. A pilot lander mission to polar regions of Mercury which are significantly illuminated during daytime is suggested where photovoltaic power generation is with triple junction solar cells is relevant during periods of calm space weather conditions such as sunspot minima.

Chapter 13, by Alexander A. Bolonkin, gathered most recent information about Venus and Venus' atmosphere and offers a new tethered aircraft for directed operations within Venus' atmosphere. This airplane can fly for very long time periods. The author has also developed the theory of estimation and computation of the flight of the tethered airplane, capturing energy in Venus' atmosphere and has worked out an example of the estimation and computation for the proposed balloon, tethered airplane.

Chapter 14, by Sushruth Kamath, Jullian Rivera, Michael Garcia, and Haym Benaroya provided a summary of currently known environmental conditions on Mercury, Venus, and Titan that could be of use to engineers who are tasked with designing structures for robotic operations and human habitation on these respective bodies in our Solar System. Recommendations are made for scientific explorations to collect key data necessary for the design and sustenance of engineering systems. A brief overview is also included on the kinds of criteria that are used in the design of hypothetical lunar habitats. Similar criteria will hold for habitats on any extraterrestrial body.

Chapter 15, by Kürsad Ozdemir, documents and assesses the array of deployable components of the Surface and Atmospheric missions to planet Venus. Designed to be transported in a compact layout which is dictated by the conditions of interplanetary space flight, planetary mission hardware systems traditionally include expandable elements. Unfolded, extended, or inflated these units of the exploration architectures deploy to their final operational form upon arrival to their mission environments. Venus, our planet's "hellish twin," with her extreme conditions in terms of temperature, pressure and acidity, is just another stage of operations for deployable structures. The atmosphere is monolithically dense, surface temperatures can reach as high as 465 °C, the atmosphere hosts large amounts of sulfuric acid. Let alone operating deployable mechanisms, even highly robust and rigid structures are challenged by the mentioned environment, resulting in the short life-spans of the surface hardware of the past Venus missions. Yet, instrument arms, buoyancy envelopes, rover sails, and wheels do not cease to deploy, defying the hellish environment of Venus. Ranging from folding instrument arms to inflatable metal-skin balloons bellows, this article presents an overview to the existing and upcoming deployable system concepts and designs of the Venus surface and atmospheric missions. Figures, tables, and trade-off charts provide the readers a comparative insight into the nature of these precious pieces of the planetary exploration clockwork.

Chapter 16, by Dragos Alexandru Paun, states that a total of 38 missions have been launched having Venus as their destination and they have contributed greatly to the knowledge of the planet's atmosphere and surface. The purpose of the present chapter is that of providing an overview of the Venusian importance, its potential resources and of the essential elements required for a successful global exploration of Venus as well as to underline the particular technologies that require further development. A short description is given of previously used architectures of essential subsystems and finally, a short overview of Mercury exploration systems is included.

Chapter 17, by Alexander A. Bolonkin, states that the Earth has a sufficiently powerful magnetic field which protects the people and organic nature of the planet from solar and cosmos high energy particles. But some planets, including Venus, which may in future be populated by humanity, do not have this natural protection and, therefore, will need to have such protection if humans are ever to use them in the same way we utilize Earth. Some scientist believe that Venus, in the past, had more suitable magnetic field—that is, suitable for people—atmosphere pressure and temperature than now. Over the geological timescale, it was lost (and continues to our time, to be lost); atmospheric gases are lost to interplanetary space because Venus does not have the necessarily stabilizing magnetic field. The author estimates the possibility to create an artificial magnetic field around Venus. He shows it is possible if the cost of the equipment and other materials launched from Earth will be significantly decreased.

Chapter 18, by Mike H. Ryan and Ida Kutschera, states that the possibility for commercial activity in space is no longer a question. Various companies regularly engage in commercial operations in space including launch service, satellite systems, and even supply missions to the International Space Station. What remains a question is how extensive commercial operations might become in the future and where they might be possible. Among the more difficult propositions is whether the inner solar system planets of Venus and Mercury might have business potential. The brief answer is perhaps. The reality is that the long-term prospects of using the resources of either planet commercially are quite speculative but not zero. This chapter discusses in detail some of the possible scenarios under which commercial use of Mercury, Venus, and their surrounding space might provide business opportunities. Although the time frames involved are exceptionally long, they are not beyond the realm of possibility especially if new transportation systems were to be developed over the next decade or two. And, as development of business opportunities on Earth has shown, it may never be too soon to consider what the future might bring or to plan for those opportunities.

Chapter 19, by Alexander A. Bolonkin, states that at present-time the main attention of scientists attends the future economic development on Moon and Mars. Mercury has very high temperature at surface (423 °C). The author shows the high temperature of the Mercury surface is not a barrier for human life activity on Mercury. People can easily be protected from the harsh environment by thin-film enclosures with variable reflectivity. Thus, Mercury can be made to have excellent conditions for substantial human life work, agriculture, and solar energy utilization. There, conditions may be better than on Mars or on the Moon.

Chapter 20, by Kenneth Roy, asks about the future role of the planet Mercury, as humanity establishes a space faring civilization and begins to explore and exploit the resources of the solar system. Can it be made habitable by human beings and possibly for most Earth-life currently living on Earth? Mercury is a small planet but its surface area is still about half the total land area of Earth.

Terraforming is a process of planetary engineering directed at an extraterrestrial planetary environment to alter it in such a way to make it fully habitable for human beings. Can Mercury be terraformed? This question is important apart from this one small planet for if Mercury can be terraformed then many other marginal planets orbiting our sun and other stars can also be made into living worlds—and Earth life could have many future homes. This chapter looks at the prospects of terraforming Mercury and concludes that it is possible, given sufficient resources and time. Based on our understanding of science and technology today, this chapter examines how Mercury could be terraformed and what would it take in the form of material and energy resources to do so. What might a terraformed Mercury look like? The ultimate fate of Mercury is left to a future space faring civilization that will be far richer, having advanced energy and propulsion technologies far beyond what is available today, and whose citizens are, hopefully, somewhat wiser than humans today.

Chapter 21, by Alexander A. Bolonkin, states that current physics assumes that vacuum can produce energy and Universes. The basis of any Universe is energy. The author assumes: energy may be positive or negative. Positive energy produces our positive matter, negative energy produces negative matter. Using this effect the author offers terraforming Mercury and Venus, making them suitable for humanity. Negative matter repels our (positive) matter. Using this effect, the author offers a space propulsion system which allows reaching a speed close to the light speed and to enable massive retrieval of extraterrestrial materials to construct various technical works. Such plan may be the best method for human colonization of Mercury and Venus. Concept of negative energy allows solving many, very important macro-problems that humanity faces e.g.: production of any artificial material or food.

Chapter 22, by Magnus Larsson and Alex Kaiser, starts by saying that No one can tell what lies ahead. In the future, a natural or man-made planetary-scale disaster may or may not force us to abandon Earth. Following such a potential forced departure, space colonisation might be our only chance of survival as a species. Prospective heavenly bodies that might support human life include the Moon, Mars, the asteroid belt—and Venus, our acid cloud-veiled sister planet with the hellishly hot surface. As Mark Twain would have it, the secret of getting ahead is getting started, and so we examine the latter option. The Cloud Ten space architecture proposal investigates the unique opportunities offered by a future Venus habitat. Tracing the footsteps of precedent schemes for "floating cities"—lightweight balloon-like structures that soar at the cloud tops some 50 km into the Venusian atmosphere, where the environment is reasonably Earth-like and relatively benign—we envisage the launching of a series of initial capsule structures that contain everything needed to grow—in situ, in space—the materials needed to create the skeleton frames and membrane skins for expandable/deployable cellular structures that increase in size through an adaptation of the famous sequential geometries called the Jitterbug Transformation (invented by neo-futuristic architect and systems theorist Richard Buckminster Fuller). This new paradigm of growing construction materials in outer space is positioned as a an enhanced version of

ISRU (in situ resource utilization), and given its own acronym, ISMG (in situ materials generation). Using a unique combination of swarm intelligence, stygmergy, and evolutionary logics, the cells are then controlled by algorithms connected to live data feeds and allowed to self organise and attach to each other in order to create a full-sized habitat high above the surface of Venus. To produce an outline for how the schedule for a construction of Cloud Ten might be constituted, the process is subdivided into ten phases, from Preparation to Permanent living. In a final coda, Cloud Ten is shown to be the best of paradoxes: a cellular habitat at the intersection of many conflicting aspects—an interstellar lightweight megastructure, an enormous city made from grass and bacteria, volumetrically expanded in an exponential fashion from within, folded into shape through a geometric dance in space, self-assembled as an incredibly intelligent swarm, soaring high in the acid clouds above the most fiendish planet we know—that represents a single instant that coalesces varying strands of an mindboggingly intricate process, while at the same time having the potential to become a true sanctuary in a post-terrestrial universe, offering the clean air and safe environment that we struggle to find on Earth.

The book allows the reader to acquire a clear understanding of the scientific fundamentals behind specific technologies to be used in the inner Solar System region in the future. The principal audience consists of researchers (engineers, physicists) involved or interested in space exploration in general and in Mercury and Venus exploration in special. Also, the book may be useful for industry developers interested in joining national or international space programs. Finally, it may be used for undergraduate, postgraduate, and doctoral teaching in faculties of engineering and natural sciences.

Viorel Badescu
Kris Zacny

Acknowledgments

A critical part of writing any book is the review process, and the authors and editors are very much obliged to the following researchers who patiently helped them read through subsequent chapters and who made valuable suggestions:

Prof. Giulio Avanzini (University of Salento, Lecce, Italy), Prof. Viorel Badescu (Polytechnic University of Bucharest, Romania), Dr. Nanan Balan (University of Sheffield, UK), Prof. Haim Baruh (Rutgers University, Piscataway, New Jersey, US), Prof. Leonhard Bernold (Universidad Tecnica Federico, Santa Maria Valparaiso, Chile), Mr. Richard Cathcart (Geographos, Burbank, California, USA), Prof. Christian Circi (University of Rome, Italy), Prof. George Cooper (University of California, Berkeley, USA), Dr. David Fields (Ultimax Group, Knoxville, Tennessee, USA), Dr. Joseph Friedlander (Shavei Shomron, Israel), Dr. Lori Glaze (NASA GSFC, Greenbelt, Maryland, USA), Prof. Viktor Hacker (Graz University of Technology, Graz, Austria), Dr. Sandra Haeuplik-Meusburger (Institute for Architecture and Design, Vienna, Austria), Dr. Jeannette Heiligers (University of Strathclyde, Glasgow, UK), Dr. Dario Izzo (ESA/ESTEC, Noordwijk, Netherlands), Dr. Natasha Johnson (NASA GSFC, Greenbelt, Maryland, USA), Dr. Vikram Kumar (Anna University, Chennai, India), Dr. Mark Krinker (City University of New York, New York, USA), Dr. Jonathan McAuliffe (ESAC, Madrid, Spain), Dr. Alessandra Migliorini (IAPS-INAF, Rome, Italy), Dr. Masato Nakamura (SAS/JAXA, Yokohama, Kanagawa, Japan), Prof. Shmuel Neumann, (Michlala Jerusalem College, Israel), Dr. Adriana Ocampo (NASA HQ, Washington DC, USA), Prof. Oleg Pensky (National Research Perm University, Russia), Dr. Stephen Ransom (SR Consulting, Vienna, Austria), Prof. Radu Rugescu (University Politehnica Bucharest, Romania), Dr. David Rothery (Open University, Milton Keynes, UK), Dr. Max Schautz (ESA), Dr. W. Scott Sherman (Texas A&M University, Corpus Christi, TX, USA), Dr. Paul van Susante (Michigan Technological University, Houghton, MI, USA), Dr. Peter Wurz (University of Bern, Switzerland), Dr. Kris Zacny (Honeybee Robotics, Pasadena, CA, USA).

Thanks are addressed to Alexandra Rzepiejewska (Honeybee Robotics) for her help with proofreading.

The editors, furthermore, owe a debt of gratitude to all authors. Collaborating with these stimulating colleagues has been a privilege and a very satisfying experience.

Contents

Contributors

Generoso Aliasi University of Pisa, Pisa, Italy

Michael Amato NASA-Goddard Space Flight Center, Greenbelt, USA

S. Aranya University College, Trivandrum, India

Mircea Badescu Jet Propulsion Laboratory (JPL)/California Institute of Technology, Pasadena, USA

Xiaoqi Bao Jet Propulsion Laboratory (JPL)/California Institute of Technology, Pasadena, USA

Yoseph Bar-Cohen Jet Propulsion Laboratory (JPL)/California Institute of Technology, Pasadena, USA

Haym Benaroya Rutgers University, Piscataway, NJ, USA

Johannes Benkhoff ESA/ESTEC Science Support Office, Noordwijk, The Netherlands

Alexander A. Bolonkin C&R, Brooklyn, NY, USA

Philip Chu Honeybee Robotics, Pasadena, CA, USA

Stephen Ford Honeybee Robotics, Pasadena, CA, USA

Simon D. Fraser Mechanical Engineering Consultant, Graz, Austria

Michael Garcia Rutgers University, Piscataway, NJ, USA

T.E. Girish University College, Trivandrum, India

J. Marvin Herndon Transdyne Corporation, San Diego, USA

Alex Kaiser Ordinary Ltd., London, UK

Sushruth Kamath Rutgers University, Piscataway, NJ, USA

Steve Kondos Honeybee Robotics, Pasadena, CA, USA

Nishant Kumar Honeybee Robotics, Pasadena, CA, USA

Ida Kutschera Bellarmine University, Louisville, Kentucky, USA

Magnus Larsson Ordinary Ltd., London, UK

Hyeong Jae Lee Jet Propulsion Laboratory (JPL)/California Institute of Technology, Pasadena, USA

Sanjay S. Limaye University of Wisconsin, Madison, USA

Giovanni Mengali University of Pisa, Pisa, Italy

Erik Mumm Honeybee Robotics, Pasadena, CA, USA

K. Ozdemir Yeditepe University, Istanbul, Turkey

Gale Paulsen Honeybee Robotics, Pasadena, CA, USA

Dragoş Alexandru Păun Airbus Defence and Space, Bremen, Germany

Alessandro A. Quarta University of Pisa, Pisa, Italy

Jullian Rivera Rutgers University, Piscataway, NJ, USA

Kenneth Roy Tennessee Valley Interstellar Workshop, Chattanooga, TN, USA

Mike H. Ryan Bellarmine University, Louisville, Kentucky, USA

Stewart Sherrit Jet Propulsion Laboratory (JPL)/California Institute of Technology, Pasadena, USA

Justin Spring Honeybee Robotics, Pasadena, CA, USA

David Williams NASA-Goddard Space Flight Center, Greenbelt, USA

Kris Zacny Honeybee Robotics, Pasadena, CA, USA

Abstract

The Earth has limited material and energy resources while these resources in space are virtually unlimited. Further development of humanity will require going beyond our planet and exploring of extraterrestrial resources and sources of unlimited power.

Thus far, all missions to Venus and Mercury have been motivated by scientific exploration. However, given recent advancements in various space technologies, mining Venus and Mercury for their resources is becoming more feasible.

This book investigates Venus and Mercury prospective energy and material resources. It is a collection of topics related to exploration and utilization of these bodies. It presents past and future technologies and solutions to old problems that could become a reality in our lifetime. The book therefore is a great source of condensed information for specialists interested in current and impending Venus- and Mercury-related activities and a good starting point for space researchers, inventors, technologists, and potential investors.

Written for researchers, engineers, and businessmen interested in Venus and Mercury exploration and exploitation.

Keywords Venus · Mercury · Exploitation · Energy sources · Space resources · Material resources · In situ resource utilization · Mining

Chapter 1
Inner Planets: Origins, Interiors, Commonality and Differences

J. Marvin Herndon

1.1 Introduction

There are many among us who remember a time when the surface features of the inner planets, also called terrestrial planets, were known only from blurry, earthbound-telescopic images. In barely more than four decades, that changed dramatically. Now, data and images from orbiting spacecraft and landers have revealed new and unanticipated features of the inner planets that, to use the words of Galileo (1623), “ought to have opened the mind’s eye much room for admirable speculation”. While space-exploration technology has burgeoned over the past four decades, curiously, understanding the myriad observations and data has posed a serious challenge for planetary investigators. Instead of understanding observations from a framework of discoveries that are securely anchored to the properties and behavior of matter, planetary investigators became accustomed to make computational models based upon assumptions, and to ignore advances that contradict ‘consensus favored’ models, especially, a model of the internal composition of Earth that had its origin circa 1940 and the so-called ‘standard model of solar system formation’ that dates from the 1960s. Decades of model proliferation (Raymond et al. 2014) from these beginnings has led to, again the words of Galileo, “dark and confused labyrinths”, virtually devoid of logical, causally-related understanding.

Generally, science progresses by replacing less-precise understanding with more-precise understanding. Much of the prevailing confusion in planetary and astrophysical science stems from building upon old ideas, as if they were correct

J. Marvin Herndon (✉)
Transdyne Corporation, San Diego, USA
e-mail: mherndon@san.rr.com

V. Badescu and K. Zacny (eds.), *Inner Solar System*,
DOI 10.1007/978-3-319-19569-8_1

ideas, and never reconsidering the validity of those ideas in light of subsequent discoveries (Herndon 2010). Traveling backward in time through the scientific literature, questioning the validity of long-held fundamental assumptions, and making appropriate corrections led me to a different, logical, causally-related understanding of formation of the inner planets: Mercury, Venus, Earth and Mars (Fig. 1.1), including the main-Belt Asteroids. Recently, I published this understanding in a scientific article entitled "New Indivisible Planetary Science Paradigm" (Herndon 2013). This chapter is derived from and amplifies that article.

Fig. 1.1 Comparison of relative sizes of the inner (terrestrial) planets. From left to right, Mercury, Venus, Earth, and Mars

The popular view of planet formation for decades has been the idea that dust, condensed at low pressures and low temperatures, gathered into balls, then into grains, then rocks, and finally into planetisimals that gathered to form planets. But there are problems with this scheme. Instead, I submit, the massive-core planets formed by condensing and raining-out from within giant gaseous protoplanets at high pressures and high temperatures. Earth's complete condensation included a Jupiter-like stage when its gigantic gas/ice shell, about 300 Earth-masses, compressed the rocky kernel to about 66 % of Earth's present diameter. Super-eruptions from the Sun as it ignited stripped the gases away from the inner planets and stripped a portion of Mercury's incompletely condensed protoplanet, and transported it to the region between Mars and Jupiter where it fused with in-falling oxidized condensate from the outer regions of the solar system and formed the parent matter of ordinary chondrite meteorites, the main-Belt asteroids, and the veneer that added to the outer layers of the inner planets, especially Mars. Here I describe the basis for that new understanding of the origin of the inner planets, the nature and commonality of their internal structures and magnetic field generation, and why their surface dynamics are so different from one another.

1.2 Internal Composition of Earth

Interpretations of other planets, their internal compositions and magnetic field origination, are strongly influenced by interpretations of our own, better-studied planet. Part of the present confusion in planetary science stems from interpretations based upon a flawed Earth-model that has been built upon by specialists since 1940. It is therefore important to present in some detail the corrected basis for understanding Earth's composition in a logical causally-related manner.

Wiechert (1897) observed that Earth's mean density (5.5 g/cm^3), measured by Cavendish (1798), is too great for the planet to be composed solely of non-metallic rocks and in 1897 postulated the existence of a dense central dense core similar to meteorites composed of nickel-iron metal. Oldham (1906) discovered that earthquake waves travel faster with depth, but at 2900 km they slow markedly upon entering a different material which he postulated is the core. The core was subsequently inferred from earthquake waves to be liquid. In 1936, analysis of records of a surprisingly large earthquake near New Zealand led Lehmann (1936) to discover the Earth's almost-Moon-sized inner core.

The physical structure of Earth's interior is deducible from seismic observations, but the chemical composition of these inaccessible regions must be inferred from meteorites. Abundant data indicate that Earth, our Moon, meteorites and other solar system bodies formed about 4,600 million years ago from "primordial" matter of uniform composition (Dalrymple 1991). That primordial composition is seen today in the outer part (photosphere) of the Sun and in the non-gaseous elements of chondrite-meteorites. The importance of chondrites is that their non-gaseous elements never appreciably separated from one another as they did in other, more changed meteorites. Consequently, chondrites are appropriately accepted to resemble Earth's bulk chemical composition. But, there are three different types of chondrites characterized by strikingly different mineral composition. In my view the incorrect choice of chondrite led to an erroneous assumption of Earth's internal mineral composition. This has long confused geophysicists' ideas about Earth's origin and dynamics which, consequently, led to confusion about planetary interiors.

The three groups of chondrite-meteorites, *carbonaceous*, *enstatite*, and *ordinary* differ markedly in oxygen content and, hence, in mineral composition. Although Lehmann (1936) correctly deduced the presence of Earth's inner core, Birch (1940), on the basis of geophysical understanding in the late '30s and early '40s, incorrectly postulated its composition. The Earth was assumed to be derived from ordinary chondrite material. In the metal of ordinary chondrites, nickel invariably occurs alloyed with metallic iron. And, the solar abundances of elements heavier than nickel and iron, even together, are insufficient to form such a massive inner core. These considerations led Birch (1940) to postulate, analogous to an ice-cube in a freezing glass of ice-water, that the inner core is composed of nickel-iron metal in the process of freezing from the liquid nickel-iron core. The explanation proffered was one of changed physical state, not chemical difference.

By 1940, Bullen (1938, 1940) had recognized a seismic discontinuity in the mantle, an interface where earthquake waves change speed and direction, at a depth of about 660 km, thus separating the mantle into two major parts, upper and lower (Fig. 1.2). Additional seismic discontinuities were later discovered in the upper mantle. Bullen (1949) subsequently discovered a zone of seismic "roughness", called D″, located between the core and the seismically-featureless lower mantle (Bina 1991; Manglik 2010; Vidale and Benz 1993). Generally, seismic discontinuities within the Earth's mantle, including at D″, have been ascribed to physical changes in a medium of uniform composition, *i.e.*, pressure-induced changes in crystal structure, rather than boundaries between layers having different chemical compositions; physics without chemistry.

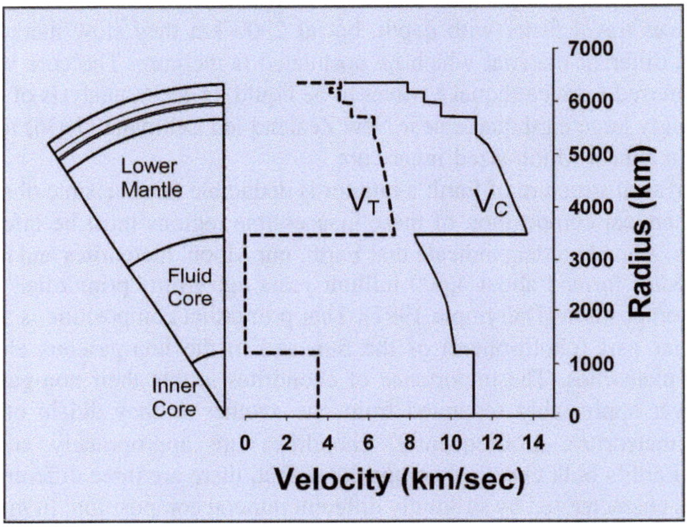

Fig. 1.2 Seismic discontinuities, *i.e.*, abrupt changes in earthquake wave velocities, indicate structures within the Earth. V_T signifies transverse waves; V_C compression waves

Nearly four decades after Birch's ideas were entrenched in the literature of geophysics, I studied enstatite chondrites. Combined with two 1960s discoveries, my investigations generated a fundamentally different view of Earth's inner core composition. What was the news? (1) Elemental silicon occurs in the metallic iron of enstatite chondrites (Ringwood 1961), and (2) perryite, nickel-silicide, Ni_2Si, is present in many enstatite-chondrite meteorites (Ramdohr 1964; Reed 1968). I realized that in Earth's core, silicon in chemical combination with nickel would have settled by gravity to the center and, in principle, formed a mass virtually identical to the relative mass deduced for the inner core. When my nickel-silicide-inner-core concept was accepted for publication (Herndon 1979), Lehmann commented: "I admire the precision of your reasoning based upon available information, and I congratulate you on the highly important result you have obtained" (Herndon 2010).

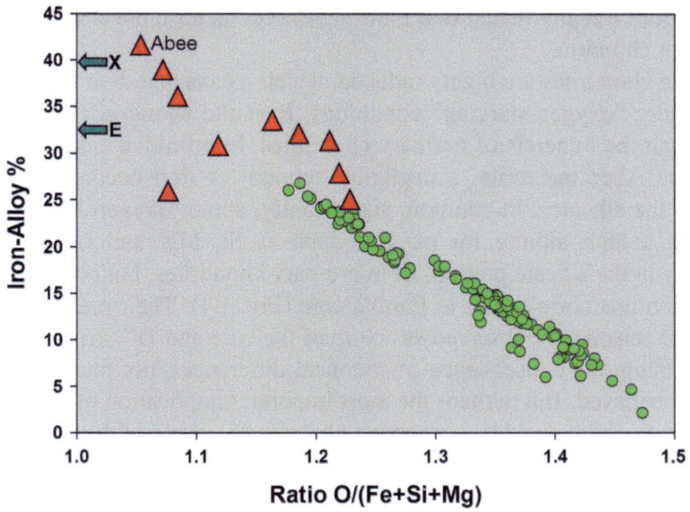

Fig. 1.3 Evidence that Earth resembles an enstatite chondrite. The percent alloy (iron metal plus iron sulfide) of 157 ordinary (circles) and 9 enstatite chondrite meteorites (triangles) plotted against oxygen content. The core percent of the whole-Earth, "arrow E", and of the core-plus-lower mantle, "arrow X", shows that Earth is like an Abee-type enstatite chondrite and unlike an ordinary chondrite

Next, I showed conclusively that, if the Earth resembles a chondrite meteorite as widely believed for good reasons, then our planet is in the main like an enstatite chondrite, not an ordinary, chondrite (Fig. 1.3) and, moreover, the relative masses of inner parts of Earth, derived from seismic data, exactly match the corresponding, chemically-identified, relative masses of enstatite-chondrite-components, observed by microscopic examination (Table 1.1) (Herndon 1980, 1993, 2011). Aside from

Table 1.1 Fundamental mass ratio comparison between the endo-Earth (lower mantle plus core) and the Abee enstatite chondrite. Above a depth of 660 km seismic data indicate layers suggestive of veneer, possibly formed by the late addition of more oxidized chondrite and cometary matter, whose compositions cannot be specified with certainty at this time

Fundamental earth ratio	Earth ratio value	Abee ratio value
Lower mantle mass to total core mass	1.49	1.43
Inner core mass to total core mass	0.052	theoretical 0.052 if Ni_3Si 0.057 if Ni_2Si
Inner core mass to lower mantle + total core mass D″ mass to total core mass	0.021 0.09***	0.021 0.11*
ULVZ** of D″ CaS mass to total core mass	0.012****	0.012*

* = avg. of Abee, Indarch, and Adhi-Kot enstatite chondrites
D″ is the "seismically rough" region between the fluid core and lower mantle
** ULVZ is the "Ultra Low Velocity Zone" of D″
*** calculated assuming average thickness of 200 km
**** calculated assuming average thickness of 28 km
**** calculated assuming average thickness of 28 km
data from (Dziewonski and Anderson 1981; Keil 1968; Kennet et al. 1995)

core size, does it really matter that Earth resembles an enstatite chondrite instead of an ordinary chondrite?

Enstatite chondrites are highly reduced, which means that their minerals initially formed under "oxygen-starving" conditions. Enstatite chondrites are more highly reduced than the minerals of ordinary chondrites. In primitive enstatite chondrites, such as the Abee meteorite, a negligible amount of iron occurs combined with oxygen in the silicates. In addition, significantly, some 'oxygen loving' elements, those with a high affinity for oxygen, such as Si, Mg, and Ca, do not occur exclusively in the silicate portion, as in ordinary chondrites, but occur in part in the alloy portion that corresponds to Earth's core (Fig. 1.4). Their presence in the core leads to the seismically observed structure of the core and D'' structure of the core (Fig. 1.4) being the consequence of chemical differences, not phase differences as previously believed. But perhaps the most important implication of my discovery of the mass ratio relationships connecting the internal parts of Earth with parts of primitive enstatite chondrites (Table 1.1) is in revealing the nature of Earth's formation, which is strikingly unlike the so-called 'standard model of solar system formation'.

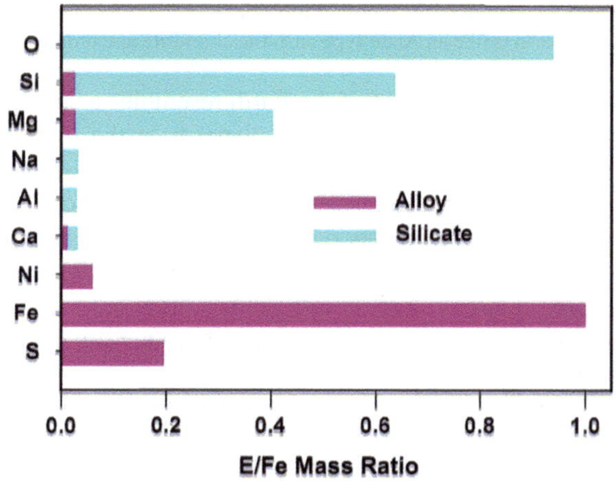

Fig. 1.4 Relative abundances of the major and minor elements in the Abee enstatite chondrite, normalized to iron, showing their relative amounts in the alloy and silicate portions. Note that calcium (Ca), magnesium (Mg), and silicon (Si), normally lithophile elements, occur in part in the alloy portion

1.3 Popular Problematic Version of Planet Formation

Modern ideas on the origin of the Sun and the planets can be traced to the late 18th Century hypothesis of Immanuel Kant, which was later modified by Pierre-Simon de Laplace. Early in the 20th Century, Laplace's nebula hypothesis was replaced with the Chamberlin-Moulton hypothesis which held that a passing star pulled matter from the Sun which condensed into large protoplanets and small planetesimals. Although the passing star idea fell out of favor, the nomenclature of protoplanets and planetesimals remained. Generally, concepts of planetary formation fall into one of two categories that involve either (1) condensation at high pressures, hundreds to thousands of times greater than the pressure of Earth's present atmosphere at the surface; or (2) condensation at very low pressures.

Eucken (1944) considered the thermodynamics of Earth condensing and raining-out within a giant gaseous protoplanet at pressures of 100–1000 atm. There was discussion of planetary formation at such pressures in the 1950s and early 1960s (Bainbridge 1962; Kuiper 1951a, b), but that abruptly changed with the publication by Cameron (1963) of a model of solar system formation from a primordial gas of solar composition at low pressure, circa 10^{-4} atm. Cameron's low pressure model became the basis for (1) condensation models that (wrongly) purported to produce minerals characteristic of ordinary chondrites as the equilibrium condensate from that medium (Grossman 1972; Larimer and Anders 1970) and (2) planetary formation models based upon the Chamberlin-Moulton planetesimal hypothesis. The idea was that dust would condense from the gas at this very low pressure. Dust grains would collide with other grains, sticking together to become progressively larger grains, then pebbles, then rocks, then planetesimals and finally planets (Goldrich and Ward 1973; Wetherill 1980).

Since the 1960s, the planetary science community has almost unanimously concurred that Earth formed from primordial matter that condensed at a very low pressure, circa 10^{-4} atm. (Grossman 1972; Larimer 1973). However, as I discovered, there is an inherent flaw in that concept (Herndon 1978, 2006b, 2009). The inner planets all have massive cores, as known from their high relative densities. I was able to show by thermodynamic calculations that the condensate of primordial matter at those very low pressures would be oxidized, like the Orgueil C1/CI meteorite wherein virtually all elements are combined with oxygen. In such low pressure, low temperature condensate, there would be essentially no iron metal for the massive cores of the inner planets, a contradiction to the observation of massive-core planets. Those who make planet-formation models based upon assumptions, however, seem not to have noticed.

The planetesimal hypothesis, *i.e.*, the 'standard model of solar system formation', not only stands in contradiction of observed planetary bulk-density, but as well necessitates the following additional *ad hoc* hypotheses: (1) Hypothesis of a radial solar system temperature gradient during planetary formation, an assumed warm inner region delineated by a hypothetical 'frost line' between Mars and Jupiter; ice/gas condensation is assumed to occur only beyond that frost line, and; (2) Hypothesis of whole-planet melting, *i.e.*, the 'magma ocean', to account for core formation from essentially undifferentiated material.

The so-called 'standard model of solar system formation' is so deeply ingrained in planetologists' thinking that it was applied without question to other planetary systems. So how were exoplanet gas giants explained that were observed as close (or closer) to their star than Earth is from the Sun? Rather than questioning the validity of that model, another *ad hoc* assumption was added, namely, the idea of planet migration. Close-to-star gas giants were assumed to have formed at Jupiter-like distances and then migrated inward to reside in their present orbits. Then, planet migration migrated to models of our own solar system: "In our solar system, the "Grand Tack" model has Jupiter migrating closer to the Sun, approaching 1.5 astronomical units (AU) from the Sun before the presence of Saturn caused Jupiter to reverse direction back out to its current location at 5.2 AU" (Chambers and Mitton 2013).

1.4 Primary Mode of Planet Formation

The above described popular version of planetary formation consists of an assemblage of assumption-based hypotheses that lack substantive connection with one another; they do not, explain, for example, in a logical, causally-related manner the highly reduced state of planetary interiors, the origin of main-Belt asteroids, Mercury's relatively large core mass, or the reason that Earth's surface behavior is quite different than other planets. In the popular version, profound confusion exists because planetary scientists are trying to make advances within an incorrect mode of planet formation. Planet formation, I submit, takes place primarily by raining-out from within a giant gaseous protoplanet, a circumstance that can be deduced from thermodynamic calculations and the mass ratio relationships shown in Table 1.1. Secondarily, only during the final stages, planet formation occurs by the planetesimal mode, adding veneer to the outer layers with in-falling more oxidized matter.

As scientific knowledge burgeons, specialization becomes an inevitable, but limiting activity. In complex processes, such as biological speciation or planet formation, one may advance by envisioning the subject in its entirety, not looking through a narrow aperture as a specialist. So, how did I determine the manner of planetary formation without making models based upon arbitrary assumptions? Here I explain.

In circa 1940, the primitive enstatite chondrites were not considered representative of Earth's composition because of their rarity and their highly reduced mineral assemblage. For decades, geoscientists were puzzled as to what circumstance in nature could produce such oxygen-starved matter. Larimer (1975) considered the possibility that carbon was more abundant than oxygen in the primordial gas in the region of enstatite chondrite formation, but for that there is neither corroborating evidence nor even a plausible mechanism. Herndon and Suess (1976), on the other hand, showed from thermodynamic considerations that such highly reduced condensate may be expected by condensation from solar matter at pressures >1 atm., provided the condensate was promptly removed/isolated from further reaction with the gases at lower temperatures.

In primordial matter of solar composition, there is a relationship between condensation pressure, condensation temperature, and the state of oxidation of the condensate. Ideally, when the partial pressure of a particular substance in the gas exceeds the vapor pressure of that condensed substance, the substance will begin to condense. In a gas of solar composition, the partial pressure of a substance is directly proportional to the total gas pressure, so at higher pressures substances condense at higher temperatures. The degree of oxidation of the condensate, on the other hand, is determined by the gas phase reaction:

$$H_2 + 1/2O_2 \leftrightarrow H_2O$$

which is a function of temperature but essentially independent of pressure. As I discovered, that reaction leads to an oxidized condensate at low temperatures and to a highly reduced condensate at high temperatures, provided the condensate is isolated from further reaction with the gas (Herndon 2006b; Herndon and Suess 1976).

Thermodynamic considerations led Eucken (1944) to conceive of Earth formation from within a giant, gaseous protoplanet when molten iron rained out to form the core, followed by the condensation of the silicate-rock mantle. By similar, extended calculations I verified Eucken's results and deduced that oxygen-starved, highly reduced matter characteristic of enstatite chondrites and by inference the Earth's interior, condensed at high temperatures and high pressures from primordial solar system gas under circumstances that isolated the condensate from further reaction with the gas at low temperatures (Herndon 2006b; Herndon and Suess 1976). Note that the Earth was not formed from enstatite chondrite meteorites. Rather, the matter that formed primitive Abee-like enstatite chondrites formed under similar conditions as planetary interiors. Perhaps the Abee enstatite chondrite formed from the debris of inner planet formation.

Figure 1.5 shows that, at total pressures ≥1 atm., iron metal ideally begins to condense from primordial solar matter as a liquid under highly reducing conditions. Liquid iron, under such conditions, is capable of dissolving significant amounts of hydrogen as well as a portion of oxygen-loving elements such as Ca, Mg, Si, and U. The composition and structure of the Earth's core (Fig. 1.6) can be understood from the metallurgical behavior of an iron alloy of this composition initially with all of the core-elements fully dissolved at some high temperature.

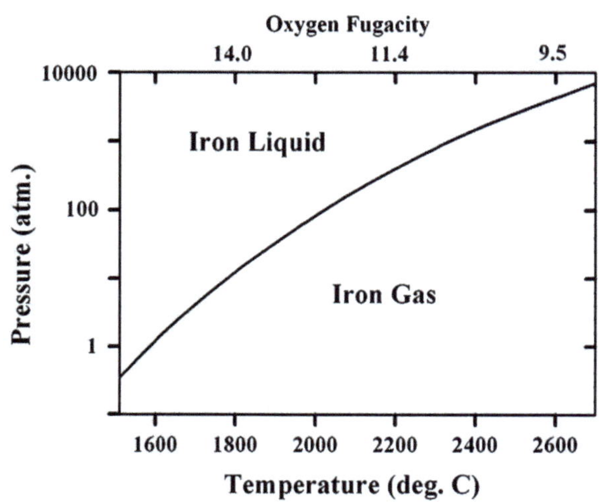

Fig. 1.5 The curve in this figure shows the temperatures and total pressures in a cooling atmosphere of solar composition at which liquid iron will ideally begin to condense. The pressure-independent oxygen fugacity is shown on the upper abscissa

Elements with a high affinity for oxygen are generally incompatible in an iron alloy. So, when thermodynamically feasible those elements escaped from the liquid alloy. Calcium and magnesium formed CaS and MgS, respectively, which floated to the top of the core and formed the region referred to as D″. Silicon combined with nickel, presumably as Ni_3Si, and formed the inner core. The trace element uranium precipitated, presumably as US, and, through one or more steps, settled at the center of the Earth where it engaged in self-sustaining nuclear fission chain reactions (Herndon 1993, 1994, 1996, 2003, 2006b; Hollenbach and Herndon 2001).

The gaseous portion of primordial solar system matter, as is the Sun's photosphere today, was about 300 times as massive as all of its rock-plus-metal forming elements. I posited Earth's complete condensation formed a gas-giant planet virtually identical in mass to Jupiter (Herndon 2005, 2009, 2012b). Giant gaseous planets of Jupiter size are observed in other planetary systems as close or closer to their star than Earth is to the Sun (Seager and Deming 2010).

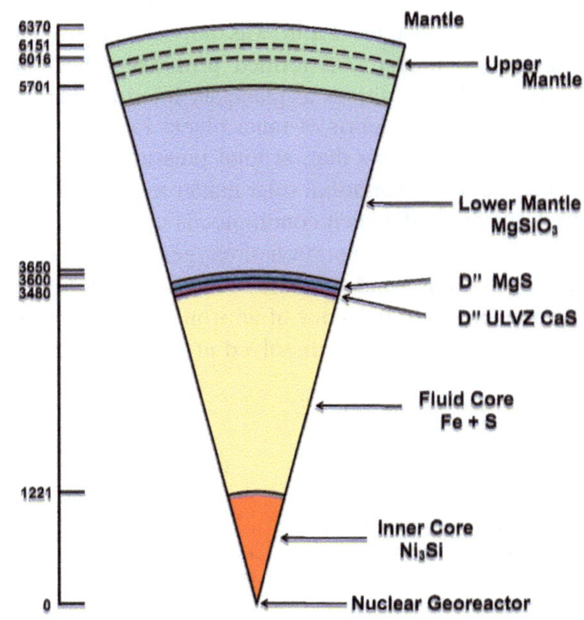

Fig. 1.6 Chemical compositions of the major parts of the Earth, inferred from the Abee enstatite chondrite (see Table 1.1). The upper mantle, above the lower mantle, has seismically-resolved layers whose chemical compositions are not yet known. Radial distance scale in km

Of the eight planets in the Solar system, the outer four (Jupiter, Saturn, Uranus, and Neptune) are gas-giants, whereas the inner four are rocky (Mercury, Venus, Earth, and Mars), without primary atmospheres. But the inner planets originated from giant gaseous protoplanets and their massive, primordial gases. How were the gases lost?

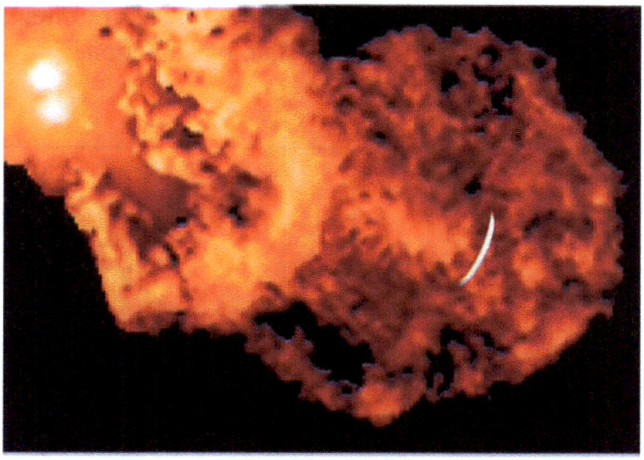

Fig. 1.7 Hubble Space Telescope image of binary star XZ-Tauri in 2000 showing a T-Tauri phase outburst. The white crescent label shows the position of the leading edge of that plume in 1995, indicating a leading-edge advance of 130 A.U. in 5 years. T-Tauri eruptions are observed in newly formed stars. Such eruptions from our nearly-formed Sun, I submit, stripped the primordial gases from the inner four planets of our solar system

A brief period of violent activity, the T-Tauri phase, occurs during the early stages of star formation with grand eruptions and super-intense "solar-wind". The Hubble Space Telescope image of an erupting binary T-Tauri star is seen here in Fig. 1.7. The white crescent shows the leading edge of the plume from a five-year earlier observation. The plume edge moved 130AU, a distance 130 times that from the Sun to Earth, in just 5 years. A T-Tauri outburst by our young Sun, I posit, stripped gas from the inner four planets. A rocky Earth, compressed by the weight of primordial gases, remained. Eventually Earth began to decompress driven primarily by the stored energy of protoplanetary compression. The consequences of Earth's formation in this manner provide rich new ways to interpret planetary data, especially when viewed in the broader context of solar system processes responsible for the diversity of planet-forming matter.

1.5 Matter of the Asteroid Belt, Mercury, and Ordinary Chondrites

The near-constancy in isotopic compositions of most of the elements of the Earth, the Moon, and the meteorites indicates formation from primordial matter of common origin (Herndon 1978). Exceptions do occur and are important cosmochemical tracers, for example, oxygen and, in refractory inclusions of carbonaceous chondrites, magnesium, silicon, calcium, and titanium. Primordial elemental composition is yet evident to a great extent in the photosphere of the Sun and, for the less volatile, rock-forming elements in chondrite meteorites where many elements have not been separated from one another to within a factor of two. But there is complexity: rather than just one type of chondrite, there are three, with each type characterized by its own strikingly unique state of oxidation.

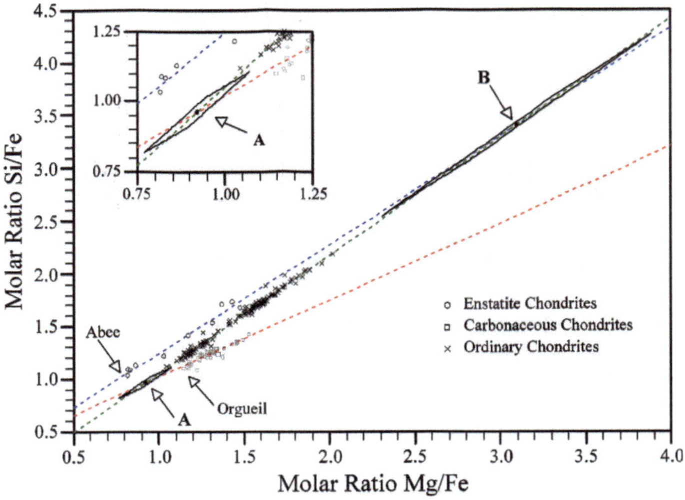

Fig. 1.8 Molar (atom) ratios of Mg/Fe and Si/Fe from analytical data on 10 enstatite chondrites, 39 carbonaceous chondrites, and 157 ordinary chondrites. Data from (Baedecker and Wasson 1975; Jarosewich 1990; Wiik 1969). Members of each chondrite class data set scatter about a unique, linear regression line. Upper line, enstatite chondrites; lower line carbonaceous chondrites, and; intersecting line, ordinary chondrites. The locations of the volatile-rich Orgueil carbonaceous chondrite and the volatile-rich Abee enstatite chondrite are indicated. Line intersections A and B are designated, respectively, *primitive* and *planetary* components. Error estimates of points A and B are indicated by solid-line parallelograms formed from the intersections of the standard errors of the respective linear regression lines. Inset shows in expanded detail the standard error parallelogram of point A

Understanding the nature of the processes that yielded those three distinct types of matter from one common progenitor forms the basis for understanding much about planetary formation, their compositions, and the processes they manifest, including magnetic field production.

Ordinary chondrites possess the common characteristic of being markedly depleted in refractory siderophile elements such as iridium and osmium. The degree of iridium and osmium depletion in each ordinary chondrite correlates with the relative proportion of its *planetary* component (Herndon 2007a). One can therefore conclude that the *planetary* component originated from a single large reservoir, characterized by a depletion in iridium and in osmium. From the inferred composition of the *planetary* component indicated in Fig. 1.8, I suggested the partially-differentiated *planetary* component might be comprised of matter stripped from the protoplanet of incompletely-formed Mercury, presumably by the T-Tauri outbursts during thermonuclear ignition of the Sun. In the region between Mars and Jupiter, the ejected Mercury-component fused with in-falling Orgueil-like matter that had condensed at low pressures and low temperatures in the far reaches of the Solar system and/or in interstellar space. That fused binary combination of materials became the parent matter of ordinary chondrites and asteroids of that region.

The molar (atom) Mg/Fe = 3.1 deduced for the *planetary* component indicates the stripping of Mercury's protoplanetary gases took place during the time when Mercury was only partially formed. The idea of heterogeneous protoplanetary differentiation/accretion is not new. Eucken (1944) first suggested Earth's core formation as a consequence of successive condensation on the basis of relative volatility from a hot, gaseous protoplanet, with iron metal raining out at the center. The approximately seven-fold greater depletion within the *planetary* component of refractory iron-loving, siderophile, elements (iridium and osmium) than other more volatile siderophile elements (nickel, cobalt, and gold) indicates that planetary-scale differentiation and/or accretion progressed in a heterogeneous manner. The first liquid iron to condense and rain-out preferentially scavenged the refractory siderophile elements from the hot gaseous protoplanet.

I estimated the original total mass of ordinary chondrite matter present in the solar system as a function of the core mass of Mercury (Herndon 2007a). For a core mass equal to 75 % of Mercury's present mass, the calculated original total ordinary chondrite mass amounts to 1.83×10^{24} kg, about 5.5 times the mass of Mercury. That amount of mass is insufficient to have formed a planet as massive as the Earth, but may have contributed significantly to the formation of Mars, as well as adding a veneer to other planets, including Earth. Presently, only about 0.1 % of that mass remains in the asteroid belt.

1.6 Interiors of the Inner Planets

During the formation of the solar system only three processes were primarily responsible for the diversity of matter in the solar system and were directly responsible for planetary internal compositions and structures (Herndon 2006b). These are: (i) High pressure, high temperature condensation from primordial matter associated with planetary formation by raining-out from the interiors of giant-gaseous protoplanets; (ii) Low pressure, low temperature condensation from primordial matter in the remote reaches of the solar system and/or in the interstellar medium associated with comets; and, (iii) Stripping of the primordial volatile components from the inner portion of the solar system by super-intense T-Tauri phase outbursts during the thermonuclear ignition of the Sun. The internal composition of massive-core planets derives from (i) above, and leads to a simple commonality of highly reduced internal planetary compositions. The outer portions of the inner planets, however, appear in varying degree to be 'painted' by an additional veneer of more-oxidized matter derived from (ii) and (iii) above. That veneer is the main source of oxidized iron, which is necessary for sustaining life as we know it.

As noted in Sect. 1.5, Mercury appears to have been stripped of its primordial gases before it had completely condensed. The low FeO content of surface silicates indicates minimal 'painting' of additional veneer by more-oxidized matter (Blewitt et al. 2009). So, even lacking seismic data, it is possible to estimate the Mercury's internal composition by analogy with the inner 82 % of Earth (Fig. 1.6), which is seismically well-characterized.

The compositions of the interior parts of Mercury, calculated according to the mass ratio relationships presented in Table 1.1, are shown in Fig. 1.9. Mercury's $MgSiO_3$ mantle mass is taken as the difference between planet mass and calculated core mass. Only 9 elements account for about 98 % of the mass of a chondrite meteorite and the planet Mercury. Of the major and minor elements comprising Mercury's core, depicted in Fig. 1.9, only aluminum and sodium, which have a high affinity for oxygen, are not represented. Presumably all aluminum and most, if not all, sodium occurs in Mercury's mantle/crust. Possibly a minor amount sodium might occur in Mercury's core as $NaCrS_2$ (Okada and Keil 1982). In the extreme case, if all of the trace element Cr formed $NaCrS_2$, a maximum of 18 % of Mercury's sodium might occur as $NaCrS_2$.

As with Earth, the composition and structure of the Mercury's core (Fig. 1.9) can be understood from the metallurgical behavior of an iron alloy initially with all of the core-elements fully dissolved at some high temperature. Upon cooling sufficiently, calcium and magnesium formed CaS and MgS, respectively, which floated to the top of the Mercurian core and formed the region analogous to Earth's D″. Silicon combined with nickel, presumably as Ni_3Si, and formed the inner core. The trace element uranium precipitated, presumably as US, and through one or more steps settled at Mercury's center where it inevitably engaged in self-sustaining nuclear fission chain reactions (Herndon 2009).

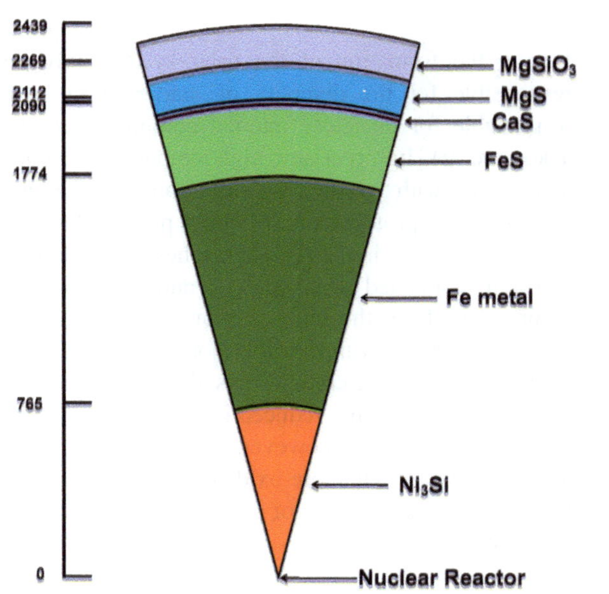

Fig. 1.9 Internal structure of Mercury calculated from the mass ratio relationships of Earth shown in Table 1.1. Mercury's core is assumed to be fully solidified. The initial location of the planetocentric 'georeactor' is indicated. Radial distance scale in km

Venus' mass is 81.5 % of Earth's mass. That is about the same mass as the endo-Earth (lower mantle plus core) which is shown in Table 1.1 to be of similar composition to the enstatite chondrites like the Abee meteorite. Without moment of inertia and seismic data, little more can be said about the internal composition of Venus, except that the interior must be highly reduced, indicated by its massive core, which is a consequence of its formation by high pressure, high temperature condensation from the interior of a giant gaseous protoplanet. The degree of 'painting' with more oxidized matter is unknown, but suspected to be less than that of Earth. Mars, on the other hand, may possess a small highly reduced interior region and a relative large veneer of more oxidized minerals.

1.7 Commonality of Nuclear Fission Heat and Magnetic Field Generation

Internally generated, currently active magnetic fields have been detected in six planets (Mercury, Earth, Jupiter, Saturn, Uranus and Neptune) and in one satellite (Jupiter's moon Ganymede). Magnetized surface areas of Mars and the Moon indicate the former existence of internally generated magnetic fields in those bodies. Furthermore, Jupiter, Saturn and Neptune radiate about twice as much energy as each receives from the Sun. Energy from nuclear fission chain reactions provides logical and causally related explanations (Herndon 2009).

The condensate from within a giant gaseous protoplanet resembles an enstatite chondrite; thermodynamic condensation considerations are similar (Eucken 1944; Herndon 2006b; Herndon and Suess 1976). As described in Sect. 1.4, the interior of Earth, below 660 km, resembles an enstatite chondrite (Table 1.1). Thus, one may reasonably conclude that the Earth formed by raining-out from within a giant gaseous protoplanet and that the interiors of other planets are similar to Earth's interior, which means their interiors are highly reduced like the Abee enstatite chondrite. In the Abee meteorite, uranium occurs in the non-oxide part that corresponds to the Earth's core.

In cores of planets, density is a function of atomic number and atomic mass. Uranium, being the densest substance tends ultimately to accumulate at the planets center. Applying Fermi's nuclear reactor theory, I demonstrated the feasibility of planetocentric nuclear fission reactors as energy sources for Jupiter, Saturn, and Neptune (Herndon 1992, 1994) and for Earth as the energy source for the geomagnetic field (Herndon 1993, 1994, 1996). Numerical simulations subsequently made at Oak Ridge National Laboratory verified those calculations and demonstrated that the georeactor could function over the entire age of the Earth as a fast neutron breeder reactor (Herndon 2003; Hollenbach and Herndon 2001). Moreover, the calculations showed that helium would be produced in precisely the range of isotopic compositions observed exiting Earth (Rao 2002).

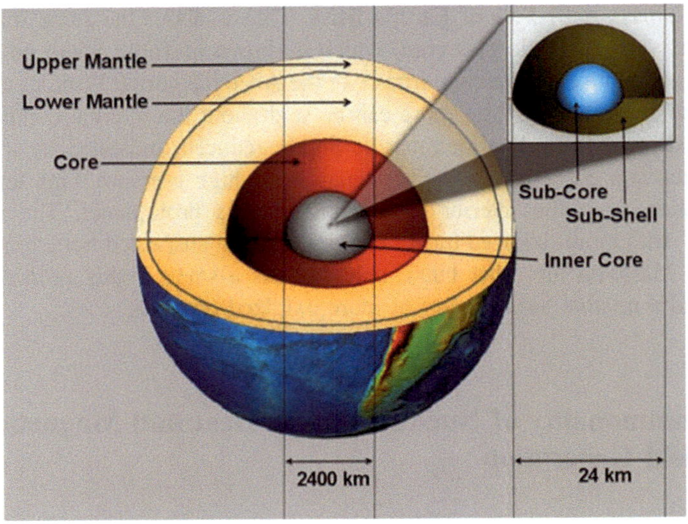

Fig. 1.10 Earth's nuclear fission georeactor (inset) shown in relation to the major parts of Earth. The georeactor at the center is one ten-millionth the mass of Earth's fluid core. The georeactor sub-shell, I posit, is a liquid or a slurry and is situated between the nuclear-fission heat source and inner-core heat sink, assuring stable convection, necessary for sustained geomagnetic field production by convection-driven dynamo action in the georeactor sub-shell (Herndon 1996, 2007b, 2009)

The georeactor is a two-part assemblage, as illustrated in Fig. 1.10, consisting of a fissioning nuclear sub-core surrounded by a sub-shell of radioactive waste products, presumably a liquid or slurry. The ~24 km diameter assemblage is too small to be presently resolved from seismic data. Oceanic basalt helium data, however, provide strong evidence for the georeactor's existence (Herndon 2003; Rao 2002) and antineutrino (sometimes called geoneutrino) measurements have not refuted that (Bellini et al. 2010; Gando et al. 2011). To date, detectors at Kamioka, Japan and at Gran Sasso, Italy have detected antineutrinos coming from within the Earth. After years of data-taking, an upper limit on the georeactor nuclear fission contribution was determined to be either 26 % (Kamioka, Japan) (Gando et al. 2011) or 15 %

Table 1.2 Geoneutrino (antineutrino) determinations of radiogenic heat production (Bellini et al. 2010; Gando et al. 2011) shown for comparison with Earth's heat loss to space (Pollack et al. 1993). See original report for discussion and error estimates

Heat flux (TW)	Source
44.2	Global heat loss to space
20.0	Neutrino contribution from ^{238}U, ^{232}Th, and georeactor fission
5.2	Georeactor KamLAND data
3.0	Georeactor Borexino data
4.0	^{40}K theoretical
20.2	Loss to space minus radiogenic

(Gran Sasso, Italy) (Bellini et al. 2010) of the total energy output of uranium and thorium, estimated from deep-Earth antineutrino measurements (Table 1.2). The actual total georeactor contribution may be somewhat greater, though, as some georeactor energy comes from natural decay as well as from nuclear fission.

Before the Mariner 10 flybys in 1974 and 1975, in light of predictions of early core-solidification (Solomon 1976), there was essentially no expectation that Mercury possesses a currently generated magnetic field. That changed. The MESSENGER observations confirmed the existence of an actively generated, albeit very weak, global magnetic field centered close to the spin axis (Anderson et al. 2011). Efforts to explain Mercury's magnetic field generation within the problematic planetary science paradigm have proven to be challenging. This is why: Popular cosmochemical models fashioned on the idea that the internal composition of Mercury resembles an ordinary chondrite do not predict a substantial source of heat in Mercury's core. Without such a heat source, the core would solidify within about one billion years thus rendering assumes core-convection impossible (Solomon 1976).

Elsasser (1939, 1946, 1950) first published his idea that the geomagnetic field is produced by convective motions in the Earth's fluid, electrically conducting core, interacting with rotation-produced Coriolis forces, creating a dynamo mechanism, a magnetic amplifier. Elsasser's convection-driven dynamo mechanism seemed to explain so well the generation of the geomagnetic field that for decades geophysicists believed convection in the Earth's fluid core 'must' exist. Later, when it was discovered that many planets had internally generated magnetic fields, they were assumed, by analogy to Earth, to have convecting fluid iron alloy cores. But there is a problem, not with Elsasser's idea of a convection-driven dynamo, but with its location; as I discovered, convection is physically impossible in the Earth's fluid core and, presumably, as well in the cores of the various planets (Herndon 2009, 2011).

Why no Earth-core convection? The core is bottom-heavy, being approximately 23 % denser at the bottom than at the top. The small decrease in density at the bottom due to thermal expansion is insufficient to overcome such a great density gradient. Moreover, for sustained convection the core-top must be maintained at a lower temperature than the core-bottom which is impossible because the Earth's core is wrapped in the mantle, a 2900 km thick thermally insulating blanket that has considerably lower thermal conductivity and heat capacity than the core.

In the popular problematic planetary science paradigm another problem is evident: There is no basis for the existence of a central heat source to drive the assumed planetary-core convection. But, in the "New Indivisible Planetary Science Paradigm" I published (Herndon 2013) and base this chapter on, all of those problems are moot.

I have suggested that the geomagnetic field is produced by Elsasser's convection-driven dynamo operating within the georeactor's radioactive waste sub-shell (Herndon 2007b). Unlike the Earth's core, sustained convection appears quite feasible in the georeactor sub-shell. The top of the georeactor sub-shell is in contact with the inner core, a massive heat sink, which is in contact with the fluid core, another massive heat sink. Heat brought from the nuclear sub-core to the top of the georeactor sub-shell by convection is efficiently removed by these massive heat sinks thus maintaining the sub-shell adverse temperature gradient. Moreover, the sub-shell is not bottom heavy. Further, decay of neutron-rich radioactive waste in the sub-shell provides electrons that might provide the seed magnetic fields for amplification.

Location of the source of the geomagnetic field in the georeactor sub-shell has implications for magnetic field reversals. The mass of the georeactor is only one ten-millionth the mass of the fluid core. High-intensity changing outbursts of solar wind, through the intermediary of the geomagnetic field, will induce electric currents into the georeactor, causing ohmic heating, which in extreme cases, might disrupt dynamo-convection and lead to a magnetic reversal. Massive trauma to the Earth might also disrupt sub-shell convection and lead to a magnetic reversal.

Among massive-core planets and large moons, there is a commonality of formation by condensing and raining-out of a gas of solar composition at high temperatures and high pressures, which leads to a commonality of internal compositions and highly reduced states of oxidation, which in turn leads to a commonality of georeactor-like planetocentric nuclear fission reactors. In each case the central nuclear reactor is about one ten-millionth as massive as the planet's core and its operation does not depend upon the physical state of the core. That small mass means that major impacts could in principle offset the nuclear core from the planets center which, for example, might explain why Mercury's magnetic field is offset ~484 km north of center (Anderson et al. 2011).

Venus currently has no internally generated magnetic field. Four potential explanations are: (1) Venus' rotation rate may be too slow; (2) Venus currently may be experiencing interrupted sub-shell convection such as might occur during a magnetic reversal; (3) Fuel breeding reactions at some point may have been insufficient for continued reactor operation, or; (4) Venus' 'georeactor' may have consumed all of its fissionable fuel. In light of helium evidence portending the eventual demise of Earth's georeactor (Herndon 2003), the fourth explanation seems most reasonable.

1.8 Why the Inner Planets Have Different Surface Dynamics

Earth's surface is markedly different from that of the other inner planets in two pronounced ways: (1) About 41 % of Earth's surface area is comprised of continental rock such as granite with the balance being ocean floor basalt, and; (2) Like stitching on a baseball, a series of mid-ocean ridges encircles the Earth from which basalt extrudes, creeps across the ocean basins, and disappears into trenches. As disclosed in *Whole-Earth Decompression Dynamics (WEDD)*, these are consequences of Earth's early formation as a Jupiter-like gas giant with the rocky portion initially compressed to about 66 % of present diameter by about 300 Earth-masses of primordial gases and ices (Herndon 2005).

Surface differences among the inner planets, I posit, are the consequence of circumstances that prevented the rocky kernels of other inner planets from being fully compressed by condensed gigantic gas/ice shells. As described in Sect. 1.5, stripping of Mercury's protoplanetary gases is inferred to have taken place during the time when Mercury was only partially formed (Herndon 2007a). One might speculate from relative density that the rocky kernel of Venus was fully formed, but the extent of its compression may differ from that of Earth due to the prevailing thermal environment and/or relative time of the Sun's T-Tauri outbursts. Eventually, the degree of compression experienced should be able to be estimated by understanding Venetian surface geology. Mars may be a special circumstance, having a relatively small, highly reduced kernel surrounded by a relatively large shell of ordinary chondrite matter; additional information is needed to be more precise.

Earth's crust is markedly different from that of the other inner planets in harboring a geothermal gradient, a phrase which describes the observed rise in temperature with increasing depth within the crust. Like Earth's two-component crust, the otherwise inexplicable geothermal gradient is understandable as a consequence of our planet's early formation as a Jupiter-like gas giant.

Since 1939, scientists have been measuring the heat flowing out of continental-rock (Benfield 1939; Bullard 1939) and, since 1952, heat flowing out of ocean floor basalt (Revelle and Maxwell 1952). Continental-rock contains much more of the long-lived radioactive nuclides than does ocean floor basalt. So, when the first heat flow measurements were reported on continental-rock, the heat was assumed to arise from radioactive decay. But later, ocean floor heat flow measurements, determined far from mid-ocean ridges, showed more heat flowing out of the ocean floor basalt than out of continental-rock measured away from heat-producing areas (Blackwell 1971; Stein and Stein 1992). This seemingly paradoxical result, I posit, arises from a previously unanticipated mode of heat transport that emplaces heat at the base of the crust. I call this mode of heat transport *Mantle Decompression Thermal Tsunami* (Herndon 2006a).

Heat generated deep within the Earth may enhance mantle decompression by replacing the lost heat of protoplanetary compression. The resulting decompression, beginning within the mantle, will tend to propagate throughout the mantle, like a tsunami, until it reaches the impediment posed by the base of the crust. There, crustal rigidity opposes continued decompression; pressure builds and compresses matter at the mantle-crust-interface resulting in compression heating. This compression heating, I submit, is the source of heat that produces the geothermal gradient.

Earth's geothermal gradient serves as a barrier that limits the downward migration of water. The 'geothermal gradient' is minimal or non-existent for inner planets that lack the compression-stage characterized by an early, massive, fully condensed shell of primordial gases and ices. Mars appears to have lacked an early massive shell of compressive condensed gases. Without subsequent decompression of the Martian kernel, there is no basis to assume the existence of a 'geothermal gradient'; there is no thermal barrier to the downward percolation of water. The absence of such a thermal barrier suggests that Mars may have a much greater subsurface water reservoir potential than previously realized.

In the popular, problematic planetary science scheme, internal planetary heat is produced through the decay of long-lived radionuclides, the only non-hypothetical heat source, although for moons sometimes tidal friction is also included. In my new indivisible planetary science paradigm, the following two important energy sources are added: (1) Stored energy of protoplanetary compression which, in the case of Earth, is the principle driving-energy for decompression and for heat emplacement at the base of the crust by *Mantle Decompression Thermal Tsunami*, and; (2) Planetocentric 'georeactor' nuclear fission energy.

During Earth's early formation as a Jupiter-like gas giant, the weight of ~300 Earth-masses of gas and ice compressed the rocky kernel to approximately 66 % of present diameter. Because of rheology and crustal rigidity, the protoplanetary energy of compression was locked-in when the T-Tauri outbursts stripped away the massive gas/ice layer leaving behind a compressed kernel whose crust consisted entirely of continental rock (Fig. 1.11). Over time, internal pressures began to build and eventually the crust began to crack.

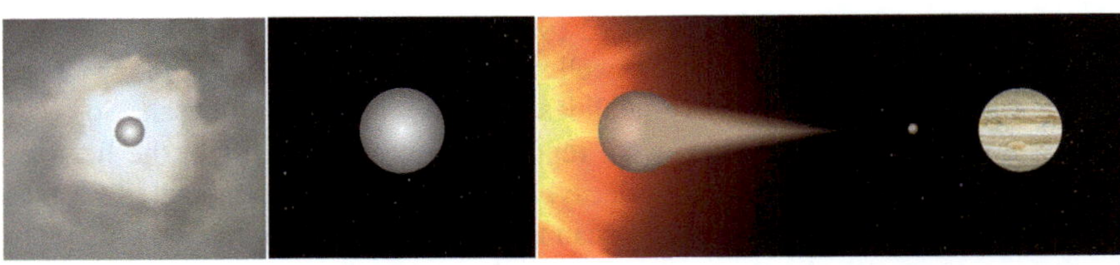

Fig. 1.11 Whole-Earth Decompression Dynamics formation of Earth. From left to right, same scale: (1) Earth condensed at the center of its giant gaseous protoplanet; (2) Earth, a fully condensed a gas-giant; (3) Earth's primordial gases stripped away by the Sun's T-Tauri solar eruptions; (4) Earth at the onset of the Hadean eon, compressed to 66 % of present diameter; (5) Jupiter for size comparison

The *Whole-Earth Decompression Dynamics* basis of Earth's surface dynamics is this: To accommodate decompression-driven increases in volume in planetary volume, Earth's surface responds in two fundamentally different ways; by crack formation to increase surface area (Fig. 1.12) and by the formation of mountain chains characterized by folding to accommodate changes in curvature (Fig. 1.13).

Cracks form to increase the surface area required as a consequence of planetary-volume increases. *Primary* cracks are underlain by heat sources and are capable of basalt extrusion, for example, mid-ocean ridges; *secondary* cracks are those without heat sources, for example, submarine trenches, and which become the ultimate repositories for basalt extruded by *primary* cracks. In addition to crack formation, decompression-increased planetary volume necessitates adjustments in surface curvature. Decompression-driven increases in volume result in a misfit of the continental rock surface formed earlier at a smaller Earth-diameter. This misfit results in 'excess' surface material confined within continent margins, which adjusts to the new surface curvature by buckling, breaking and falling over upon itself producing fold-mountain chains (Herndon 2012c).

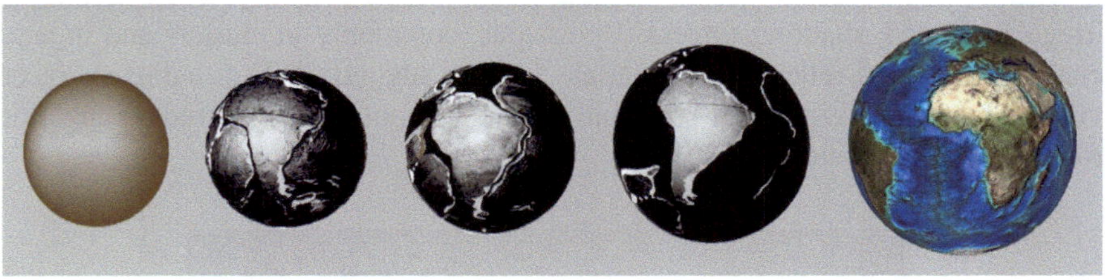

Fig. 1.12 Schematic representation of the decompression of Earth (WEDD) from Hadean to present. From left to right, same scale: (1) Earth after T-Tauri removal of gases, 66 % of present Earth diameter, fully covered with continental-rock crust; (2), (3), and (4) Formation of primary and secondary decompression cracks that progressively fractured the continental crust and opened ocean basins. Timescale not precisely established. (5) Holocene Earth. The geology of Earth, according to WEDD, is principally determined by Earth's decompression: Surface crack formation to accommodate increased planetary volume, and mountain formation to accommodate changes in surface curvature (see Fig. 1.13)

Fig. 1.13 Demonstration illustrating the formation of fold-mountains as a consequence of Earth's early formation as a Jupiter-like gas giant. On the left, two balls representing the relative proportions of 'present' Earth (large), and 'ancient' Earth (small) before decompression. In the center, a spherical section, representing a continent, cut from 'ancient' Earth and placed on the 'present' Earth, showing: (1) the curvature of the 'ancient continent' does not match the curvature of the 'present' Earth and (2) the 'ancient continent' has 'extra' surface area confined within its fixed perimeter. On the right, tucks remove 'extra' surface area and illustrate the process of fold-mountain formation that is necessary for the 'ancient' continent to conform to the curvature of the 'present' Earth. Unlike the ball-material, rock is brittle so tucks in the Earth's crust would break and fall over upon themselves producing fold-mountains. From (Herndon 2012c)

Crack formation and the production of mountains characterized by folding, are both consequences of protoplanetary compression: These are pronounced processes on Earth, are irrelevant on Mercury, may have some relevance to Venus, but have little relevance to Mars. Planetocentric 'georeactor' nuclear fission, on the other hand, has relevance to virtually all planets, generating their magnetic fields and providing heat for volcanic activity.

1.9 Evidence from Mercury's Surface

One of the most important Project MESSENGER discoveries were images from the spacecraft that revealed '... an unusual landform on Mercury, characterized by irregular shaped, shallow, rimless depressions, commonly in clusters and in association with high-reflectance material ... and suggest that it indicates recent volatile-related activity' (Fig. 1.14) and which have not been observed on any other rocky planet (Blewett et al. 2011).

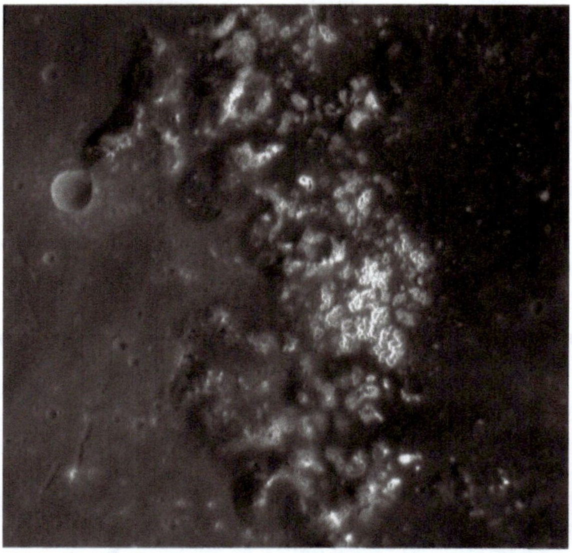

Fig. 1.14 NASA MESSENGER image, taken with the Narrow Angle Camera, shows an area of hollows on the floor of Raditladi basin on Mercury. Surface hollows were first discovered on Mercury during MESSENGER's orbital mission and have not been seen on the Moon or on any other rocky planetary bodies. These bright, shallow depressions appear to have been formed by disgorged volatile material(s) from within the planet

But the planetary investigators were unable to describe a scientific basis for the source of those volatiles or to suggest identification of the 'high-reflectance material'. I posited that during formation, condensing and raining-out as a liquid at high pressures and high temperatures from within a giant gaseous protoplanet, Mercury's iron alloy core dissolved copious amounts of hydrogen, one or more Mercury-volumes at STP. Hydrogen is quite soluble in liquid iron, but much less soluble in solid iron. I suggested that dissolved hydrogen from Mercury's core, released during core-solidification and escaping at the surface, produced hydrogen geysers that were responsible for forming those 'unusual landform on Mercury', sometimes referred to as pits or hollows, and for forming the associated 'high-reflectance material', bright spots, which I suggested is iron metal reduced from an exhaled iron compound, probably iron sulfide, by the escaping hydrogen (Herndon 2012a).

So, here is a test: Verifying that the 'high-reflectance material' is indeed metallic iron will not only provide strong evidence for Mercury's hydrogen geysers, but more generally will provide evidence that planetary interiors rained-out by condensing at high pressures and high temperatures within giant gaseous protoplanets. The high reflectance metallic iron can be distinguished by its low-nickel content from meteoritic metallic iron.

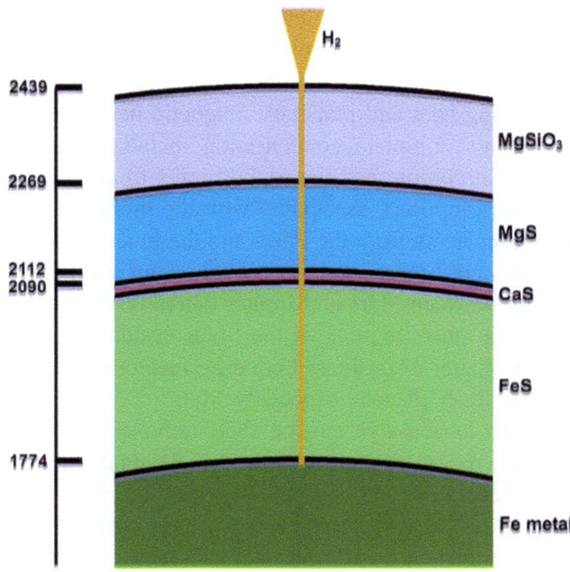

Fig. 1.15 Schematic illustration of the source and path of hydrogen which is exhausted as hydrogen geysers and forms hollows (pits) on Mercury's surface. Radial distance scale in km

One of the surprising early discoveries of the Project MESSENGER mission was abundant sulfur on Mercury's surface (Nittler et al. 2011). That observation is understandable as a consequence of hydrogen geysers. Figure 1.15 is a schematic representation of the path taken by exsolved hydrogen. Note the exiting hydrogen gas traverses regions of various sulfide compositions: iron sulfide (FeS), calcium sulfide (CaS), and magnesium sulfide (MgS). The exiting hydrogen, I submit, may scavenge sulfides from these layers and deposit them on Mercury's surface and perhaps may even emplace some in Mercury's exosphere.

Mercury is about 6 % as massive as Earth. In the 1970s, this tiny planet's core, based upon heat-flow calculations, was thought to have solidified within the first billion years after formation (Solomon 1976). But that was before my demonstration of the feasibility of planetocentric nuclear fission reactors (Herndon 1993, 2007b, 2009) whose energy production considerably delayed solidification. Later, upon subsequent cooling, iron metal began to precipitate from Mercury's iron-sulfur alloy fluid core; the endpoint of core solidification is depicted in Fig. 1.15. Core solidification with its concomitant release of dissolved hydrogen provides explanations for Mercurian surface phenomena.

1.10 Summary

Massive-core planets formed by condensing and raining-out from within giant gaseous protoplanets at high pressures and high temperatures, accumulating heterogeneously on the basis of volatility with liquid core-formation preceding mantle-formation; the interior states of oxidation resemble that of the Abee enstatite chondrite. Core-composition was established during condensation based upon the relative solubilities of elements, including uranium, in liquid iron in equilibrium with an atmosphere of solar composition at high pressures and high temperatures. Uranium settled to the central region and formed planetary nuclear fission reactors, producing heat and planetary magnetic fields.

Earth's complete condensation included a ~300 Earth-mass gigantic gas/ice shell that compressed the rocky kernel to about 66 % of Earth's present diameter. T-Tauri eruptions, associated with the thermonuclear ignition of the Sun, stripped the gases away from the Earth and the inner planets. The T-Tauri outbursts stripped a portion of Mercury's incompletely condensed protoplanet and transported it to the region between Mars and Jupiter where it fused with in-falling oxidized condensate from the outer regions of the Solar system and/or interstellar space, forming the parent matter of ordinary chondrite meteorites, the main-Belt asteroids, and veneer for the inner planets, especially Mars.

With its massive gas/ice shell removed, pressure began to build in the compressed rocky kernel of Earth and eventually the rigid crust began to crack. The major energy source for planetary decompression and for heat emplacement at the base of the crust is stored energy of protoplanetary compression is the stored energy of protoplanetary compression. In response to decompression-driven volume increases, cracks form to increase surface area and fold-mountain ranges form to accommodate changes in curvature.

One of the most profound mysteries of modern planetary science is this: As the terrestrial planets are more-or-less of common chondritic composition, how does one account for the marked differences in their surface dynamics? Differences among the inner planets are principally due to the degree of compression experienced. Planetocentric georeactor nuclear fission, responsible for magnetic field generation and concomitant heat production, is applicable to compressed and non-compressed planets and large moons.

The internal composition of Mercury is calculated based upon an analogy with the deep-Earth mass ratio relationships. The origin and implication of Mercurian hydrogen geysers is described. Besides Earth, only Venus appears to have sustained protoplanetary compression; the degree of which might eventually be estimated from understanding Venetian surface geology. A basis is provided for understanding that Mars essentially lacks a 'geothermal gradient' which implies potentially greater subsurface water reservoir capacity than previously expected.

References

Anderson BJ et al (2011) The global magnetic field of Mercury from MESSENGER orbital observations. Science 333:1859–1862

Baedecker PA, Wasson JT (1975) Elemental fractionations among enstatite chondrites. Geochim Cosmochim Acta 39:735–765

Bainbridge J (1962) Gas imperfections and physical conditions in gaseous spheres of lunar mass. Astrophys J 136:202–210

Bellini G et al (2010) Observation of geo-neutrinos. Phys Lett B 687:299–304

Benfield AF (1939) Terrestrial heat flow in Great Britain. Proc R Soc Lond Ser A 173:428–450

Bina CR (1991) Mantle Discontinuities. Rev Geophys Supplement:783–793

Birch F (1940) The transformation of iron at high pressures, and the problem of the earth's magnetism. Am J Sci 238:192–211

Blackwell DD (1971) The thermal structure of continental crust. In: Heacock JG (ed) The structure and physical properties of the earth's crust, geophysical monograph 14. American Geophysical Union, Washington, pp 169–184

Blewett DT, Chabot NL, Denevi BW, Ernst CM, Head JW, Izinberg NR, Murchie SL, Solomon SC, Nittler LR, McCoy TJ, Xiao ZZ, Baker DMH, Fassett CI, Braden SE, Oberst J, Scholten F, Preusker F, Hurwitz DM (2011) Hollows on Mercury: MESSENGER evidence for geologically recent volatile-related activity. Science 333:1859–1859

Blewitt DT, Robinson MS, Denevi BW et al (2009) Multispectral images of Mercury from the first MESSENGER flyby: analysis of global and regional color trends. Earth Planet Sci Lett 285:272–282

Bullard EC (1939) Heat flow in South Africa. Proc R Soc Lond Ser A 173:474–502

Bullen KE (1938) Note on the density and pressure inside the Earth. Trans Roy Soc New Zealand 67:122

Bullen KE (1940) The problem of the Earth's density variation. Bull Seis Soc Amer 30:235–250

Bullen KE (1949) Compressibility-pressure hypothesis and the Earth's interior. Mon Not R Astron Soc Geophys Suppl 5:355–398

Cameron AGW (1963) Formation of the solar nebula. Icarus 1:339–342

Cavendish H (1798) Experiments to determine the density of Earth. Phil Trans Roy Soc Lond 88:469–479

Chambers J, Mitton J (2013) From dust to life: the origin and evolution of our solar system. Princeton University Press

Dalrymple GB (1991) The age of the earth. Stanford University Press, Stanford

Dziewonski AM, Anderson DA (1981) Preliminary reference Earth model. Phys Earth Planet Inter 25:297–356

Elsasser WM (1939) On the origin of the Earth's magnetic field. Phys Rev 55:489–498

Elsasser WM (1946) Induction effects in terrestrial magnetism. Phys Rev 69:106–116

Elsasser WM (1950) The Earth's interior and geomagnetism. Revs Mod Phys 22:1–35

Eucken A (1944) Physikalisch-chemische Betrachtungen ueber die frueheste Entwicklungsgeschichte der Erde. Nachr Akad Wiss Goettingen, Math-Kl:1–25

Galilei G (1623) The Assayer. In: Drake S (ed) Discoveries and opinions of Galileo. Doubleday, New York

Gando A, Gando Y, Ichimira K, Ikedia H, Inoue K, Kibe Y, Kishimoto Y, Koga M, Minekawa Y, Mitsui T, Morikawa T, Nagai N, Nakajima K, Nakamura K, Narita K, Shimizu I, Shimizu Y, Shirai J, Suekane F, Suzuki A, Takahashi H, Takahashi N, Takemoto Y, Tamae K, Watanabe H, Xu BD, Yabumoto H, Yoshida H, Yoshida S, Enomoto S, Kuzlov A, Murayama H, Grant C, Keefer G, Piepke A, Banls TI, Bloxham T, Detwiler JA, Freedman SJ, Fujikawa BK, Han K, Kadel R, O'Donnell TO, Steiner HM, Dwyer DA, McKeown RD, Zhang C, Berger BE, Lane CE, Maricic J, Miletic T, Batygov M, Learned JG, Matsuno S, Sakai M, Horton-Smith GA, Downum KE, Gratta G, Tolich K, Efremenko Y, Perevozchikov O, Karwowski HJ,

Markoff DM, Tornow W, Heeger KM, Decowski MP (2011) Partial radiogenic heat model for Earth revealed by geoneutrino measurements. Nature Geosci 4:647–651
Goldrich P, Ward WR (1973) The formation of planetesimals. Astrophys J 183(3):1051–1061
Grossman L (1972) Condensation in the primitive solar nebula. Geochim Cosmochim Acta 36:597–619
Herndon JM (1978) Reevaporation of condensed matter during the formation of the solar system. Proc R Soc Lond A363:283–288
Herndon JM (1979) The nickel silicide inner core of the Earth. Proc R Soc Lond A368:495–500
Herndon JM (1980) The chemical composition of the interior shells of the Earth. Proc R Soc Lond A372:149–154
Herndon JM (1992) Nuclear fission reactors as energy sources for the giant outer planets. Naturwissenschaften 79:7–14
Herndon JM (1993) Feasibility of a nuclear fission reactor at the center of the Earth as the energy source for the geomagnetic field. J Geomag Geoelectr 45:423–437
Herndon JM (1994) Planetary and protostellar nuclear fission: implications for planetary change, stellar ignition and dark matter. Proc R Soc Lond A455:453–461
Herndon JM (1996) Sub-structure of the inner core of the earth. Proc Nat Acad Sci USA 93:646–648
Herndon JM (2003) Nuclear georeactor origin of oceanic basalt $^{3}He/^{4}He$, evidence, and implications. Proc Nat Acad Sci USA 100(6):3047–3050
Herndon JM (2005) Whole-Earth decompression dynamics. Curr Sci 89(10):1937–1941
Herndon JM (2006a) Energy for geodynamics: mantle decompression thermal tsunami. Curr Sci 90(12):1605–1606
Herndon JM (2006b) Solar System processes underlying planetary formation, geodynamics, and the georeactor. Earth Moon Planet 99(1):53–99
Herndon JM (2007a) Discovery of fundamental mass ratio relationships of whole-rock chondritic major elements: implications on ordinary chondrite formation and on planet Mercury's composition. Curr Sci 93(3):394–398
Herndon JM (2007b) Nuclear georeactor generation of the earth's geomagnetic field. Curr Sci 93 (11):1485–1487
Herndon JM (2009) Nature of planetary matter and magnetic field generation in the solar system. Curr Sci 96(8):1033–1039
Herndon JM (2010) Inseparability of science history and discovery. Hist Geo Space Sci 1:25–41
Herndon JM (2011) Geodynamic basis of heat transport in the earth. Curr Sci 101(11):1440–1450
Herndon JM (2012a) Hydrogen geysers: explanation for observed evidence of geologically recent volatile-related activity on Mercury's surface. Curr Sci 103(4):361–361
Herndon JM (2012b) Indivisible Earth: consequences of earth's early formation as a Jupiter-like gas giant. In: Margulis L (ed)
Herndon JM (2012c) Origin of mountains and primary initiation of submarine canyons: the consequences of Earth's early formation as a Jupiter-like gas giant. Curr Sci 102 (10):1370–1372
Herndon JM (2013) New indivisible planetary science paradigm. Curr Sci 105(4):450–460
Herndon JM, Suess HE (1976) Can enstatite meteorites form from a nebula of solar composition? Geochim Cosmochim Acta 40:395–399
Hollenbach DF, Herndon JM (2001) Deep-earth reactor: nuclear fission, helium, and the geomagnetic field. Proc Nat Acad Sci USA 98(20):11085–11090
Jarosewich E (1990) Chemical analyses of meteorites: a compilation of stony and iron meteorite analyses. Meteoritics 25:323–337
Keil K (1968) Mineralogical and chemical relationships among enstatite chondrites. J Geophys Res 73(22):6945–6976
Kennet BLN, Engdahl ER, Buland R (1995) Constraints on seismic velocities in the earth from travel times Geophys J Int 122:108–124
Kuiper GP (1951a) On the evolution of the protoplanets. Proc Nat Acad Sci USA 37:383–393
Kuiper GP (1951b) On the origin of the Solar System. Proc Nat Acad Sci USA 37:1–14

Larimer JW (1973) Chemistry of the solar nebula. Space Sci Rev 15(1):103–119
Larimer JW (1975) The effect of C/O ratio on the condensation of planetary material. Geochim Cosmochim Acta 39(3):389–392
Larimer JW, Anders E (1970) Chemical fractionations in meteorites-III. Major element fractionations in chondrites. Geochim Cosmochim Acta 34:367–387
Lehmann I (1936) P'. Publ Int Geod Geophys Union, Assoc Seismol, Ser A, Trav Sci 14:87–115
Manglik A (2010) New insights into core-mantle boundary region and implications for Earth's internal processes. Curr Sci 99(12):1733–1738
Nittler LR et al (2011) The major element composition of Mercury's surface from MESSENGER X-ray spectrometry. Sci 333:1847–1850
Okada A, Keil K (1982) Caswellsilverite, $NaCrS_2$: a new mineral in the Norton County enstatite achondrite. Am Min 67:132–136
Oldham RD (1906) The constitution of the interior of the earth as revealed by earthquakes. Q T Geol Soc Lond 62:456–476
Pollack HN, Hurter SJ, Johnson JR (1993) Heat flow from the Earth's interior: analysis of the global data set. Rev Geophys 31(3):267–280
Ramdohr P (1964) Einiges ueber Opakerze im Achondriten und Enstatitachondriten. Abh D Akad Wiss Ber Kl Chem Geol Biol 5:1–20
Rao KR (2002) Nuclear reactor at the core of the Earth!—a solution to the riddles of relative abundances of helium isotopes and geomagnetic field variability. Curr Sci 82(2):126–127
Raymond SN, Kokubo E, Morbidelli A, Morishima R, Walsh KJ (2014) Terrestrial planet formation at home and abroad. In: Beuther H, Klessen R, Dullemond C, Henning Th (eds) Protostars and planets VI. University of Arizona Press, Tucson. http://arxiv.org/pdf/1312.1689v1311.pdf
Reed SJB (1968) Perryite in the Kota-Kota and South Oman enstatite chondrites. Mineral Mag 36:850–854
Revelle R, Maxwell AE (1952) Heat flow through the floor of the eastern North Pacific Ocean. Nature 170:199–200
Ringwood AE (1961) Silicon in the metal of enstatite chondrites and some geochemical implications. Geochim Cosmochim Acta 25:1–13
Seager S, Deming D (2010) Exoplanet atmospheres. Ann Rev Astron Astrophys 48:631–672
Solomon SC (1976) Some aspects of core formation in Mercury. Icarus 28(4):509–521
Stein C, Stein S (1992) A model for the global variation in oceanic depth and heat flow with lithospheric age. Nature 359:123–129
Vidale JE, Benz HM (1993) Seismological mapping of the fine structure near the base of the Earth's mantle. Nature 361:529–532
Wetherill GW (1980) Formation of the terrestrial planets. Ann Rev Astron Astrophys 18:77–113
Wiechert E (1897) Ueber die Massenverteilung im Inneren der Erde. Nachr K Ges Wiss Goettingen, Math-Kl:221–243
Wiik HB (1969) On regular discontinuities in the composition of meteorites. Commentationes Phys Math 34:135–145

Chapter 2
Mercury and Venus: Significant Results from MESSENGER and Venus Express Missions

Sanjay S. Limaye

2.1 Introduction

Understanding how the solar system evolved has been one of the driving reasons for exploring the solar system until recently. Now the focus includes planets around other stars and particularly terrestrial planets and habitability. Mercury and Venus are two extreme members in our solar system but are poorly understood. Our closest planetary neighbors, with their proximity to the sun made it challenging to learn much about them from telescopes, as they are accessible for only a short time before sunrise or after sunset for large telescopes (due to scattered light and sensitive detectors). Venus with it global cloud cover made it impossible to learn about its surface, until new advances in radar and microwave techniques. Only partly surveyed by Mariner 10 from three fly-bys during 1974–1976, Mercury remained enigmatic until MESSENGER. By contrast, Venus has been explored from fly-by spacecraft, orbiters, entry probes and landers and even balloons, yet the major science questions have only become sharper. MESSENGER and Venus Express, the two current spacecraft visitors from Earth to the innermost, entered the final phase of their mission lives in summer 2014 as the fuel required for orbit maintenance was depleted. Orbiting Venus since 15 April 2006, Venus Express conducted an aerobraking experiment in June 2014. It collected its last observations on 27 November 2014 when the fuel was exhausted during orbit raise maneuvers, and the spacecraft entered the atmosphere on 18 January 2015. Over more than eight years of observing Venus from its 24 h, polar elliptic orbit, it collected a large amount of data from its operating instruments which have provided new insights into the atmosphere of Venus and to a limited extent, its surface. The MESSENGER spacecraft also observed Venus on its third fly-by of Venus which also yielded some new results.

S.S. Limaye (✉)
University of Wisconsin, Madison, USA
e-mail: sanjay.limaye@ssec.wisc.edu

V. Badescu and K. Zacny (eds.), *Inner Solar System*,
DOI 10.1007/978-3-319-19569-8_2

Mercury, the primary target of the MESSENGER spacecraft since it entered into orbit around Mercury on 18 March, 2011 now awaits its next visitor, BepiColombo, a joint mission of the European Space Agency (ESA) and the Japan Aerospace Exploration Agency (JAXA) to be launched in July 2016. MESSENGER made three fly-bys of Mercury on 14 January 2008, 6 October 2008, and 29 September 2009 and has provided a deeper understanding of the planet which was explored previously by Mariner 10 in 1970s.

The larger goal behind exploring the two innermost planets is the same – how did the terrestrial planets evolve? The differences between the two innermost planets—Venus with its thick atmosphere and Mercury without any significant atmosphere—are starker than the differences between the two outer members – Earth and Mars. The impact histories of the four planets are different. It is believed that the presence of the Moon and impacts affected the spin states of Earth and Mars but the evolution of the spin states of Mercury and Venus are not well understood.

Basic orbital information about the two innermost planets have been known for a few long time, but their rotation rates were measured only in the 1960s with the advent of radar. Details of their spin state – polar wander, changes in rotation rate are not yet known. Whether or not the two planets had moons at some point is also a question that is being considered. Mercury's iron rich composition is a challenge to understand in the context of the other terrestrial planets (Asphaug and Reufer 2014). However, Venus' backward spin raises questions about its orbital evolution. The cloud-shrouded surface of Venus could be penetrated only by radar to measure the rotation rate, but precise recognition from optical telescopes of surface features on Mercury; despite being devoid of cloud cover had been hampered by its proximity to the Sun. Knowledge about Mercury before the observations from MESSENGER has been summarized by Solomon (2003) and about Venus by Taylor (2006).

While Mercury at first glance from Mariner 10 pictures resembles the Moon with its crater-filled surface, they are quite different, except for being comparable in size and devoid of significant atmospheres. Even from Mariner 10 images it was apparent that the volcanic plains on Mercury are brighter than those on the Moon, suggesting different composition and weathering, and the impact history surmised from crater size distribution is different for the two objects. The craters are both volcanic and impact, and both surfaces show the presence of lobate scarps that have been used to infer global contraction. Interpretation of topography from Mariner 10 images suggested that Mercury has contracted, and recent estimates imply even greater contraction. Recent Lunar Reconnaissance Orbiter (LRO) images also reveal a small contraction of the Moon – only about 0.1 km (Watters et al. 2010), compared to about 8 km for Mercury (Byrne et al. 2014), as inferred from MESSENGER images. Surface compositions are also different – despite being very dense with a massive iron core, Mercury has less iron and more sulfur on the surface as compared to the moon, which may explain differences in the appearance of the volcanic plains.

Mercury is smaller in equatorial diameter (3,397 km) than Mars (4,880 km), and both are smaller than Venus (6,052 km) and Earth 6,378 (km). Mercury and Venus have the most eccentric and the most circular orbits around the Sun (e = 0.2056 and e = 0.0068 respectively), but also have the smallest inclination (0.0° and 2.6° respectively), whereas Earth and Mars have intermediate eccentricities and moderate inclinations. Thus, Mercury and Venus have small seasonal variations – on Mercury the insolation is modulated only by the changing distance from the Sun over its year, whereas Earth and Mars show significant insolation variation over their circuits around the Sun. The combination of the spin rate and the orbital period give Mercury the longest solar day – 175.94 Earth days, followed by Venus at 116.74, with Mars at only slightly longer than Earth at 1.029 Earth days. This results in the solar system's largest range of surface temperatures on Mercury, from about 100 K to 700 K. Due to its thick atmosphere, temperatures on Venus range from ~100 K in the upper atmosphere (inferred from SOIR experiment data on Venus Express; Mahieux et al. 2012) to ~750 K at the surface (VeGa 2 lander, Linkin et al. 1986).

The two smaller inner planets also have the lowest surface pressures – negligibly small on Mercury and about 6.8 mb on Mars, compared to 1025 mb on average on Earth at sea level, and about 93,000 mb on Venus at the mean radius. Compared to Mars, Venus has low topographic relief: about 80 % of the surface lies within ±1 km of the mean surface (6051.78 km radius) and about 51 % within ±0.5 km. Thus the surface temperature and pressure variations due to topography are small. The temperature decreases with altitude at approximately 8 K/km near the surface and may have superadiabatic lapse rates if the VeGa 2 temperature profile (Linkin et al. 1986) is representative of the global conditions, or at least adiabatic (Seiff 1987).

The high pressure and high temperature conditions in the lower atmosphere of Venus imply that the ideal gas law is not valid there (Staley 1970). Carbon dioxide (96.5 % atmospheric abundance) and nitrogen (3.5 %) are supercritical under these conditions; however it is not known how their mixture behaves. It is only recently that the supercritical gases are being investigated theoretically (Bolmatov et al. 2014), but more work is needed for mixtures. How the properties of gases and mixtures of gases change when they approach supercritical conditions is not very well known yet. While the belief was that they change slowly, computer simulations suggest that the change is abrupt and experimental validation is needed. At the same time, Venera probes did not appear to reveal any effect of the supercritical conditions, so some investigation is needed to verify the effect.

Venus has no internal measureable magnetic field, while Mars had an internal magnetic field in the past that disappeared some time ago, perhaps because its core solidified and the dynamo stopped. Mercury and Earth are known to have liquid outer cores, but whether Venus has a liquid outer core is not yet known. The absence of a liquid core is suggested only by the absence of a magnetic field, but the field could be absent if convection is suppressed. The case of the missing magnetic field of Venus is still a puzzle.

Both the Venus Express and MESSENGER missions had set science goals that led to their competitive selection for implementation. Those goals have been met, but the larger questions still await further analysis and interpretation with more observations to answer the new questions arising from the new results. A summary of the results from MESSENGER and Venus Express is presented below.

2.2 Mercury: Discoveries from MESSENGER

MESSENGER mission intended to answer the following six questions:

- Why is Mercury so dense?
- What is the geologic history of Mercury?
- What is the nature of Mercury's magnetic field?
- What is the structure of Mercury's core?
- What are the unusual materials at Mercury's poles?
- What volatiles are important at Mercury?

Major discoveries by MESSENGER include confirmation of a large iron core, presence of water ice and organic material in shadowed craters, presence of sulfur on the surface and an asymmetric magnetic field. Presence of ice was suspected from radar observations of the planet more than two decades ago that showed high reflectivity regions. MESSENGER observations from three different MESSENGER experiments confirmed the presence of water ice. The neutron experiments showed a hydrogen-rich layer in the bright radar locations, consistent with pure water ice. The laser altimeter carried by MESSENGER showed irregular dark and bright deposits in the craters near the north pole. The following subsections discuss these discoveries in further detail.

2.2.1 Surface

High spatial resolution imaging from narrow (monochrome) and wide angle (eleven narrow band filters) cameras of the MESSENGER Dual Imaging System (MDIS, Hawkins et al. 2007) has produced near global coverage of the Mercury. The MESSENGER Laser Altimeter (MLA, Cavanaugh et al. 2007) has yielded global elevation maps or the Northern hemisphere. Due to the eccentric orbit with periapsis latitude near the north pole, the range to the surface is too large for observations in the Southern hemisphere. These new datasets have revealed new insights into the processes modifying the planet's surface, including tectonics, volcanism, impact cratering, and volatile accumulation.

Prior to MESSENGER observations, the role of volcanism on Mercury was uncertain. The smooth plains within craters and between craters could be attributed to both eruptions and ejecta from impacts, as distinguishing features corresponding to each process were not discerned at the resolution of Mariner 10 images (Solomon 2003). Infrared (MASCS), gamma ray (GRNS) and X-ray (XRS) spectrometers also provide composition information key to understanding the surface evolution and the Laser Altimeter (MLA) has provided information about topographic relief.

Figure 2.1 shows a natural color global composite of many MDIS images. An enhanced view is shown in the bottom panel of this figure. The enhanced color was produced by using images from the color base map imaging campaign during

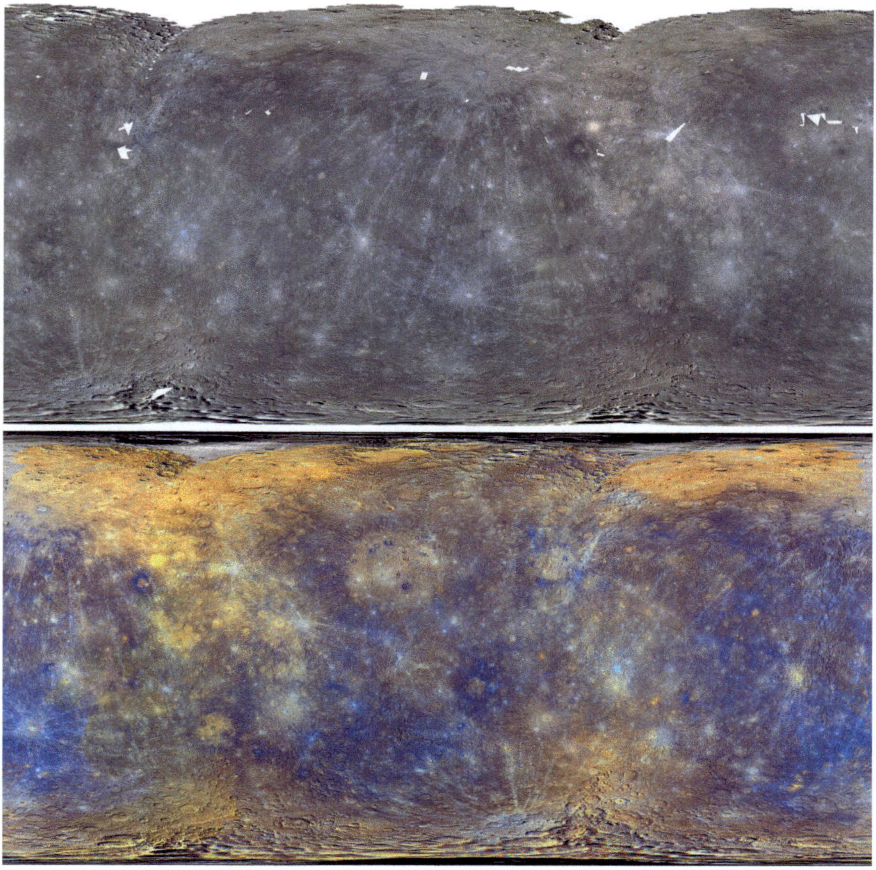

Fig. 2.1 The color global mosaic is comprised of 8 narrow-band color filters of the MDIS Wide Angle Camera (WAC), while the 3-color northern hemisphere mosaic provides higher resolution but only in 3 of the narrow-band filters. Both products below are the latest as of March 7, 2014 (top). Placing the 1000-nm, 750-nm, and 430-nm filters in the red, green, and blue channels, respectively, produced the enhanced-color mosaics available below (bottom). Credit: NASA/Johns Hopkins University Applied Physics Laboratory/Arizona State University/Carnegie Institution of Washington. http://messenger.jhuapl.edu/gallery/sciencePhotos/image.php?image_id=234

MESSENGER's primary mission. These colors are not what Mercury would look like to the human eye, but rather the colors highlight the chemical, mineralogical, and physical differences between the rocks that make up Mercury's surface. This specific color combination places the second principal component in the red channel, the first principal component in the green channel, and the ratio of the 430 nm/1000 nm filters in the blue channel. The color differences suggest compositional differences on the surface and that some of the material may have been exposed due to impact processes. The deposits due to impacts can be seen in all parts of the crater – rim, central peak and the (Ernst et al. 2010).

a. Volcanism and lava flows

Mariner 10 images of Mercury's surface showed the presence of smooth plains, but their volcanic origins were not unanimously accepted. Many scientists interpreted them to be due to fluidized ejecta from impacts (Byrne et al. 2013; Wilhelms 1976; Oberbeck et al. 1977), while others suggested otherwise based on embayment relations between the smooth plains, topography and estimates of ejecta from impacts (Murray et al. 1975; Strom et al. 1975; Robinson and Lucey 1997). One argument offered was based on studies of lunar volcanism: plains thought to be of volcanic origins from ground-based telescope images were later found to be due to impacts (Wilhelms 1976). During its first fly-by, MESSENGER images provided enough evidence to confirm the volcanic origins of the smooth plains. Subsequent fly-bys and orbital imaging data have now provided plenty of evidence of widespread surface volcanism and past large-scale lava flows extending over long distances. Unlike Venus, Mercury's volcanic features are not very diverse; generally smooth plains, while some interesting new details are being discovered from MESSENGER from the altimeter data and the images.

Figure 2.2 shows an example of a volcanic feature, possibly a shield volcano. Lava flows can be seen in many of the MESSENGER images (e.g., Fig. 2.3) and familiar features such as channel islands are also visible (Figs. 2.4 and 2.5). Eruptive or explosive volcanism is also evident in many parts (Fig. 2.6). Other surface features such as hollows are unique to Mercury (Fig. 2.7), as well as ridges and troughs (Fig. 2.8), which may have been altered or filled in by volcanic activity.

Images at high northern latitudes show lava flows with islands in the plains between craters as well as narrow, sinuous channels, similar to those seen on Venus, Mars, and the Moon (Leverington 2004) and consistent with formation by low viscosity lava flows (Byrne et al. 2013).

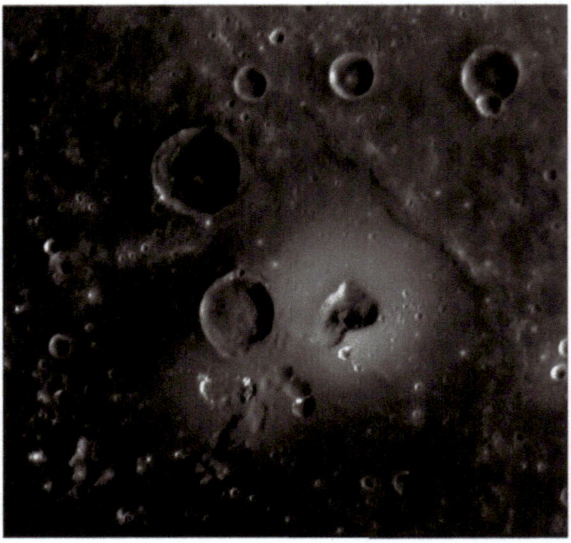

Fig. 2.2 The kidney-shaped depression is believed to be the volcanic vent of a shield-type volcano, similar to the volcanoes of Hawaii. The bright material around the vent is interpreted as material that fell after eruptions. *Photo credit: Science/AAAS*

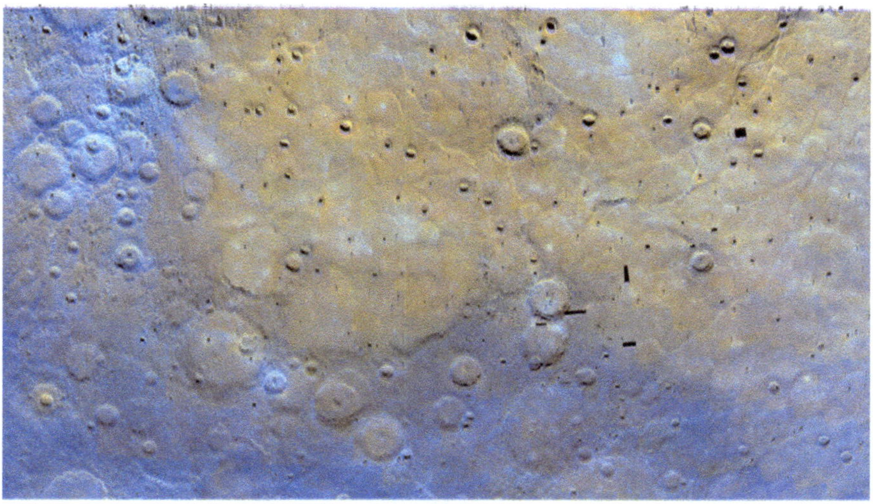

Fig. 2.3 Northern plains in enhanced color. The northern plains are seen here to be distinctive in color and thus composition from the surrounding terrain. The mosaicked images are shown with the 1000, 750, and 430 nm images in red, green, and blue, respectively. The scene is centered at 73 °N, 300 °E. Colors are based on a statistical method that highlights differences among the eight color filters, making variations in color and composition easier to discern. Credit: NASA/The Johns Hopkins University Applied Physics Laboratory/Carnegie Institution of Washington

Fig. 2.4 Flat-floored channels near the northern smooth plains of Mercury (top), contain streamlined islands that appear to have been sculpted by hot lava as it flowed around them. A larger view of the islands is shown in the bottom image. Credit: NASA/JHU - APL/Carnegie Inst. of Washington

There are other differences also – on Mars the islands in the channel tend to be tear drop shaped which gives a clue to the flow direction. On Mercury the islands do not show such shapes, and the flow direction is not indicated by the shape but can be inferred from the laser altimeter data (Fig. 2.5). The source of the lava is also

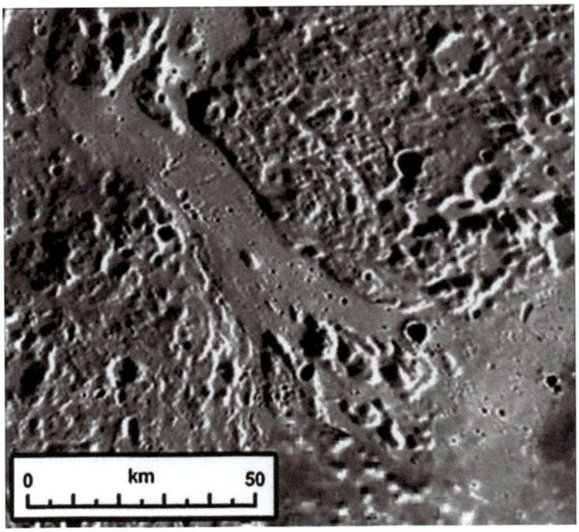

Fig. 2.5 A larger view of the islands is shown in the bottom image. Image Credit: NASA/JHU - APL/Carnegie Inst. of Washington

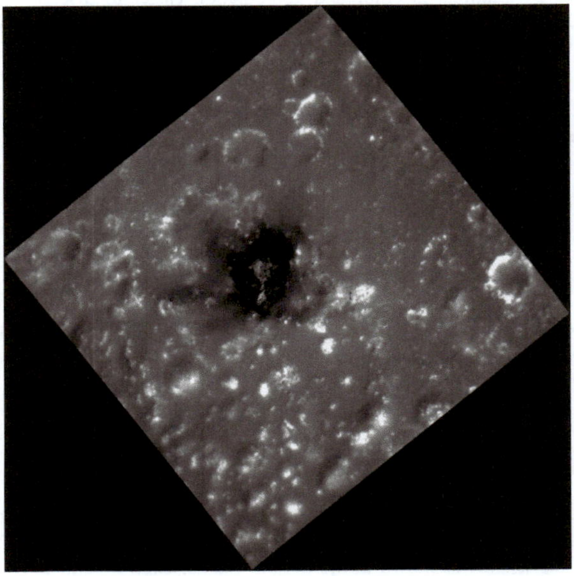

Fig. 2.6 This high-resolution image shows a region of the southern rim of the large Caloris basin. In the center is an irregularly shaped depression believed to be a pyroclastic volcanic vent. In this previously posted image, you can see this feature in the upper right as having a reddish color with a dark center. The composition of the dark material is not known. Most pyroclastic vents do not have this dark material, but other features do show small outcrops of it, such as Berkel and Seuss. Date acquired: May 05, 2013, *Image Credit: NASA/Johns Hopkins University Applied Physics Laboratory*

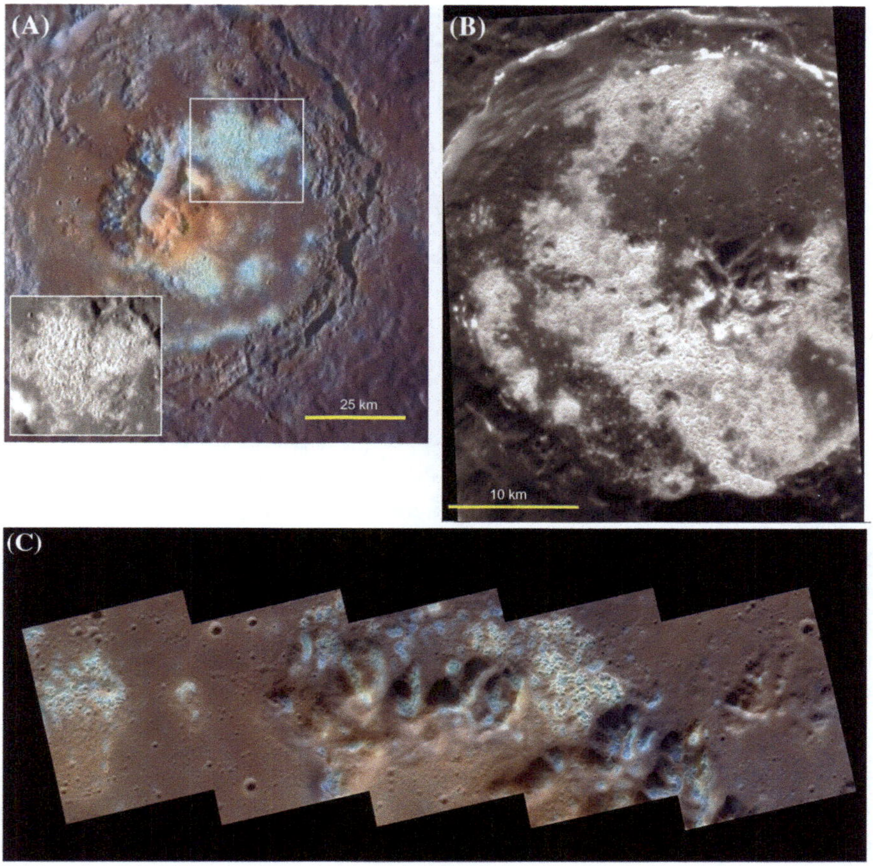

Fig. 2.7 (*A*) Tyagaraja crater, 97 km in diameter. Bright areas with blue color and etched texture correspond to a high density of hollows (inset). The central pit surrounded by reddish material is probably a pyroclastic vent. From monochrome image EN0212327089 M, 111 m/pixel, with enhanced color from the eight-filter set EW0217266882I (34). (*B*) Sander crater, 47 km in diameter. A high density of hollows occurs in the high-reflectance portions of the crater floor; others are found on the northern crater rim and wall terraces. Stereo analysis (35) indicates that any elevation differences between the high-reflectance and darker areas of the floor are ~20 m. Image EN0218289182 M, 30 m/pixel. (C). Hollows inside the Raditladi impact basin. Credit: NASA/Johns Hopkins University Applied Physics Laboratory/Carnegie Institution of Washington

not generally obvious. The other surprising indication is that the surface lava is not rich in Iron and Magnesium, but more sulfur rich, which is indicated by the color differences in the surface images.

b. Eruptive/Explosive Volcanism: Pyroclastic Deposits

The presence of pyroclasts implies presence of volatiles in the magma, its viscosity, ejection velocity and temperature, so the confirmation of such deposits first

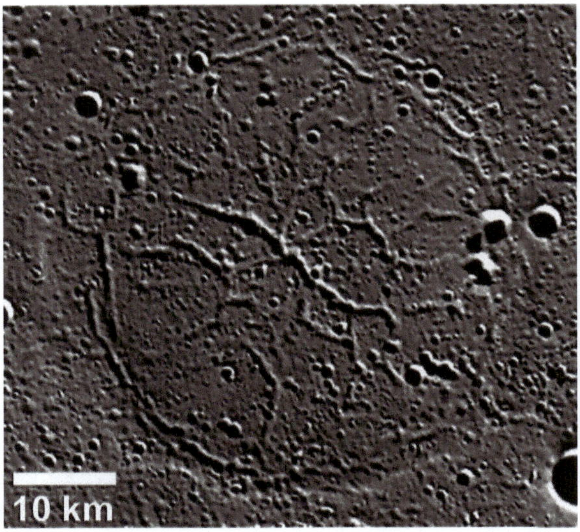

Fig. 2.8 This image shows large graben in a ghost crater on the floor of the Goethe impact basin on Mercury. The extensional troughs or graben are up to 2 km wide in this about 47-km-diameter ghost crater. Ridge and trough systems unique to Mercury have been observed in MESSENGER imagery. These features form in relation to "ghost craters," impact craters that are filled and buried by volcanic deposits, but whose outline is revealed by ridges that form over the crater rim. Credit: NASA/APL

suspected from Mariner 10 images indicates that Mercury is not as depleted in volatiles as planetary formation theories had suggested. Further, the differences in color of such deposits and surrounding area from other regions of Mercury suggest compositional differences that provide clues about the volatiles and thermal history of the planet (Nittler et al. 2014; Kerber et al. 2011). Pyroclastic deposits detected from MESSENGER imaging data appear within impact craters (Fig. 2.5), suggesting that the eruptions occurred after the impacts as otherwise the impacts would have destroyed the deposits (Goudge et al. 2014). Fifty-one pyroclastic deposits have been identified from the near-global coverage of MESSENGER images from orbit after the initial tally from fly-by images (Kerber et al. 2009, 2011). Such features are also found on the Moon. However, after accounting for the different surface gravity, the distribution of deposit radius suggests that the eruption velocity on Mercury is likely much higher (Kerber et al. 2011).

The deposits are volatile-rich as determined from the multispectral images from the MESSENGER Dual Imaging System (MDIS) analyzed by Kerber et al. (2009, 2011). This detection is significant due to the indication of presence of volatiles on Mercury. Being so close to the Sun, it would appear that Mercury would be volatile-depleted (Wetherill 1994) as it is believed to have formed in the hot part of the solar nebula. Subsequent impacts and other heating events would have further driven away the volatiles, and hence the detection of larger-than-expected deposits of volatiles will required a fresh look at the formation of the planet and its evolutionary history.

c. Hollows: Rimless Depressions

One of the strangest features discovered on Mercury are the occurrences in many parts of hollows or pits (Figs. 2.6 and 2.7). Hollows were found in the polar regions of Mars covered with carbon dioxide, but on Mercury these features are found on the rocky surface. High resolution images from MESSENGER revealed many irregular, shallow and rimless depressions for the first time (Blewett et al. 2011). Their age is relatively young as apparent from the fact that such hollows appear within impact craters, thus confirming that Mercury is not a static relic orbiting the Sun, but their origin is not entirely clear.

d. Ridges and Troughs

Some of the most unusual formations seen on Mercury are ridges and troughs, not seen on any other moon or planet. Mercury (Watters et al. 2012). Ridges are believed to be formed when impact craters are covered in volcanic ash deposits towards the rim, creating "ghost craters". Troughs and ridges are believed to be formed due to extensional fault forming process and contraction resulting from cooling of the interior.

e. Impact Craters

Not all craters on Mercury are volcanic as can be expected. Figure 2.9 shows one of the high resolution images of the Cunningham impact crater. The central peak is seen in detail but the rays resulting from the falling debris are outside the image area.

f. Polar Ice Deposits

It was suggested decades ago (Watson et al. 1961) that permanently shadowed regions in polar craters on the moon may harbor water ice. In 1991, radar images of the polar regions of Mercury from Arecibo showed several radar bright regions (Slade et al. 1992) which were co-located with craters, and it was shown that water ice could survive for a long time on the surface of Mercury in shadowed regions (Vasavada et al. 1999). MESSENGER was equipped with instruments capable of detecting the presence of water ice and observations confirmed that water ice is indeed present in the permanently shadowed regions of polar craters. The laser altimeter observations recorded the presence of bright and darker deposits in the 1032 nm echoes from the shadowed regions from the poleward facing walls in the same locations as the radar echoes (Chabot et al. 2014). Paige et al. (2013) calculated the expected temperatures in the shadowed regions. The results show that the bright echoes (Fig. 2.10) are consistent with water ice and the darker regions with a coating of complex organic material on the ice, with impacts of comets and asteroids laden with organic compounds as the likely source (Lucey 2013).

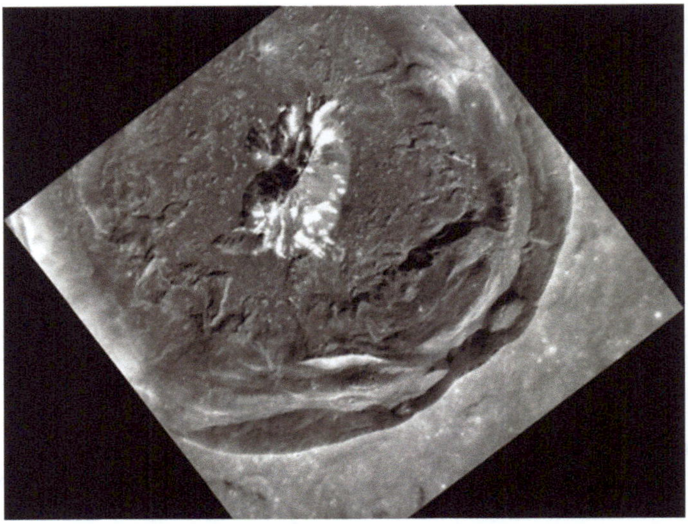

Fig. 2.9 This high-resolution view of Cunningham crater (~200 m/pixel) was recently acquired by MESSENGER. What cannot be seen in this image, which shows striking details of the crater's interior, is the extensive set of rays associated with Cunningham. The bright rays of Cunningham indicate that the crater is relatively young, having formed on Mercury likely within the last billion years. In this view, the preserved terraces of the crater walls, the well-defined central peak, and the limited number of overlying small craters are also all signs of Cunningham's relative youth

Fig. 2.10 Perspective view of Mercury's north polar region with the radar-bright regions shown in yellow. Credit: NASA/Johns Hopkins University Applied Physics Laboratory/Carnegie Institution of Washington

The composition of the polar deposits was confirmed indirectly from the detection of neutrons created when cosmic rays strike the planet (Lawrence et al. 2013). The inference is based on the process as follows. The neutrons travel through the polar deposits and surface material, escaping into space where they are detected by the Neutron Spectrometer. The presence of hydrogen atoms within the ice in the polar deposits stop the neutrons from escaping into space, thus a drop in the rate of neutrons detected is believed to be due to the presence of hydrogen in the form of polar ice thus, a drop in the neutron detection rate can be related to the abundance of hydrogen and hence water ice in the polar deposits (Fig. 2.11).

Dark patches have also been observed in some of these deposits and are believed to be composed of organic compounds (Fig. 2.11). Figure 2.12 shows the calculated maximum temperatures reached in the northern polar regions of Mercury (Paige et al. 2013). The detailed thermal map of Mercury, and others that display additional thermal parameters, are the first to have been calculated from measurements of Mercury's topography by the MESSENGER MLA instrument. Mercury displays the most extreme range of surface temperatures of any body in the solar system. Regions that receive direct sunlight at the equator reach maximum temperatures of 700 K (800 °F), whereas regions in permanent shadow in high-latitude craters can drop below 50 K (−370 °F).

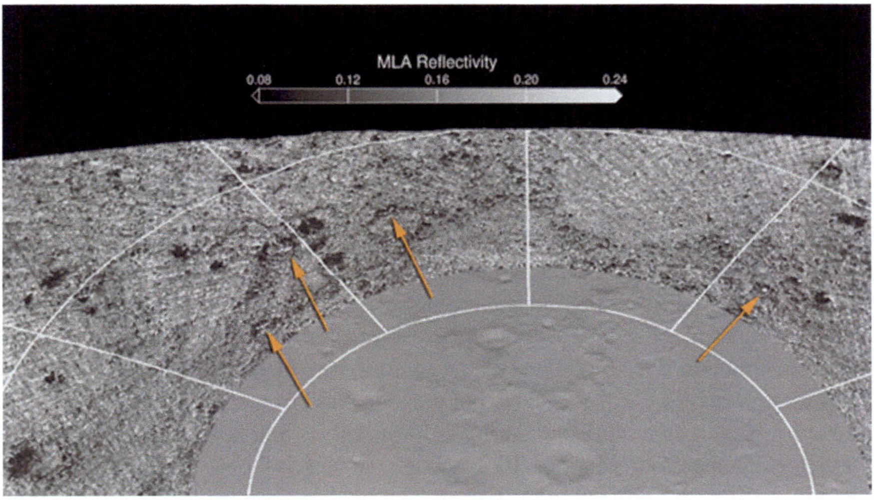

Fig. 2.11 Map of reflectivity determined from the Laser Altimeter data showing isolated areas of brighter and darker reflectance in areas of permanently shadow. Many of the brighter areas detected by MLA (as indicated by the arrows) are in unusually cold regions where surface water ice is predicted, as shown in Fig. 2.2. In some cases, the dark regions are somewhat larger than the areas predicted to have thermally stable water ice. *Credit: NASA/UCLA/Johns Hopkins University Applied Physics Laboratory/Carnegie Institution of Washington*

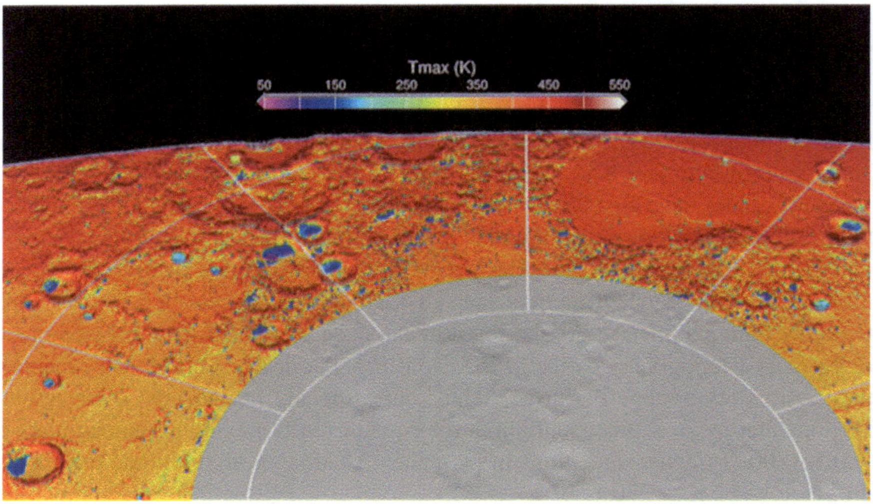

Fig. 2.12 Map of the maximum surface temperature reached over a two-year period over the north polar region of Mercury. In this view from above Mercury's north pole, there are numerous craters with poleward-facing slopes on which the annual maximum temperature is less than 100 K (−280 °F). At these temperatures, water ice is thermally stable over billion-year timescales. Credit: *David A. Paige, MESSENGER Participating Scientist,* University of California, Los Angeles, California. *Credit: NASA/UCLA/Johns Hopkins University Applied Physics Laboratory/Carnegie Institution of Washington*

2.2.2 Mercury's Interior Structure and Gravity Field

a. Contraction of Mercury

Mercury is the only planet for which the contraction has been confirmed observationally. It is believed that all planets were very warm in their formative stages and have since cooled. With cooling, a decrease in size can be expected. Such global contraction due to cooling of the interior was first proposed as a mountain building process for Earth's surface, but the idea was criticized early (Dutton 1874). The solid planet, as it cools and shrinks, forms ridges and scarps, and by studying these formations from Mariner 10 coverage, it was inferred that the planet had shrunk in radius by an amount between 1 and 3 km, smaller than expected from predictions of Mercury's thermal history (Byrne et al. 2014). With global coverage from MESSENGER observations, a more comprehensive survey of ridges and scarps yields a higher value – about 5–10 km, more consistent with the thermal models. This is useful for understanding the history of the planet in terms of heat loss, tectonic and volcanic activity, and its metallic core.

b. Gravity Field and Internal Structure

Indications that Mercury has a molten core came from radar observations of the planet from Earth, prior to MESSENGER spacecraft's arrival at Mercury (Margot et al. 2007). By tracking the MESSENGER spacecraft accurately from the ground

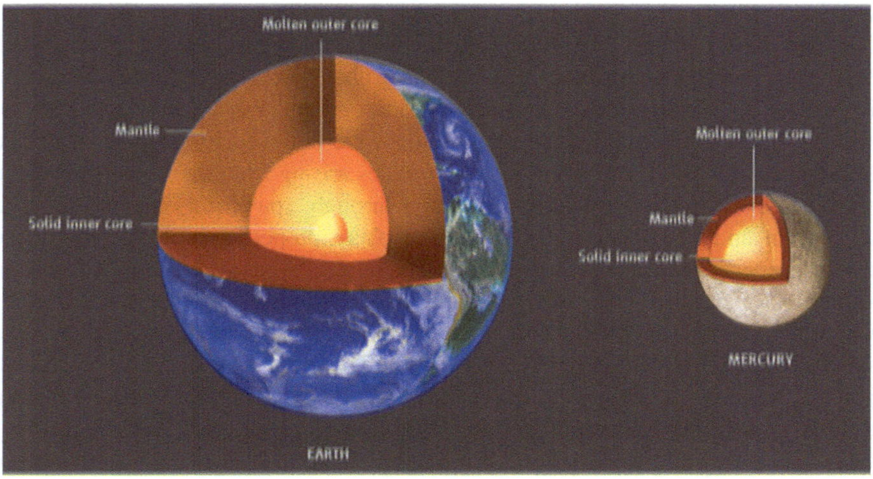

Fig. 2.13 The solid inner core and fluid outer core of Earth are shown to approximate scale. Mercury's outer core is now known to be fluid, but its radius and the nature and radius of any solid inner core remain to be determined. (Credit: NASA and The Johns Hopkins Applied Physics Laboratory)

stations, the northern hemisphere gravity field has been obtained, which reveals several mass concentrations. Another inference is that the low latitude crust is thicker than at higher latitudes. According to Margot et al. (2007), a model for Mercury's radial density distribution consistent with these results includes a solid silicate crust and mantle overlying a solid iron-sulfide layer, an iron-rich liquid outer core, and perhaps a solid inner core.

c. Interior

The discovery of a relatively weak magnetic field around Mercury by Mariner 10 led to the suggestion that the field was created by the dynamo effect from a molten core. Yet Mercury's small size should favor cooling and solidification of a liquid core. Thus the inference of a liquid core by monitoring the variations in the spin rate of the planet by bouncing radar signals was a surprise (Margot et al. 2007). These observations also pinned down the obliquity of Mercury to a high accuracy (2.11 ± 0.1 arc min) and found large amplitude librations in longitude (35.8 ± 2 arc s) which indicate that the core is at least partially molten and decoupled from the mantle (Fig. 2.13).

2.3 Exosphere of Mercury

Mercury's atmosphere was detected for the first time after hydrogen, helium, and atomic oxygen were detected on the dayside observations from ultraviolet spectrometer observations from Mariner 10. Presence of sodium, potassium, and calcium was known before MESSENGER from ground based observations (Killeen et al. 2007) in the exosphere, and MESSENGER found magnesium. What characterizes Mercury's atmosphere, or more accurately exosphere (because of the low densities) is that the collisional process are not with other atoms or molecules, but with the surface of the planet. Hydrogen and helium were believed to originate from the impinging solar wind, and other detected species arising from the interaction of the solar wind with the surface of Mercury through sputtering, photon induced desorption and vaporization. Thus the exosphere properties are affected by the orbital properties and also Mercury's magnetic field. The exosphere thus shows large variations in density with solar longitude and planet latitude. The polar regions which receive little incident sunlight act as cold traps for material that reaches the surface (Killen et al. 2007). Figure 2.14 shows a schematic view of the processes that play a role in maintaining the exosphere – (i) the incident sunlight stimulating desorption and thermal evaporation of volatiles from the surface (low energy process), (ii) sputtering (high energy process), and (iii) release of material from surface through meteoroid impact. The atoms can get ionized and interact with the magnetic field and the solar wind and radiation pressure while some, ejected with lower energies can fall back on the planet at a different location, some carried to the dark polar regions.

MESSENGER detected magnesium in the tenuous atmosphere for the first time, suggesting strongly that the metal is present in the crust since the atmosphere

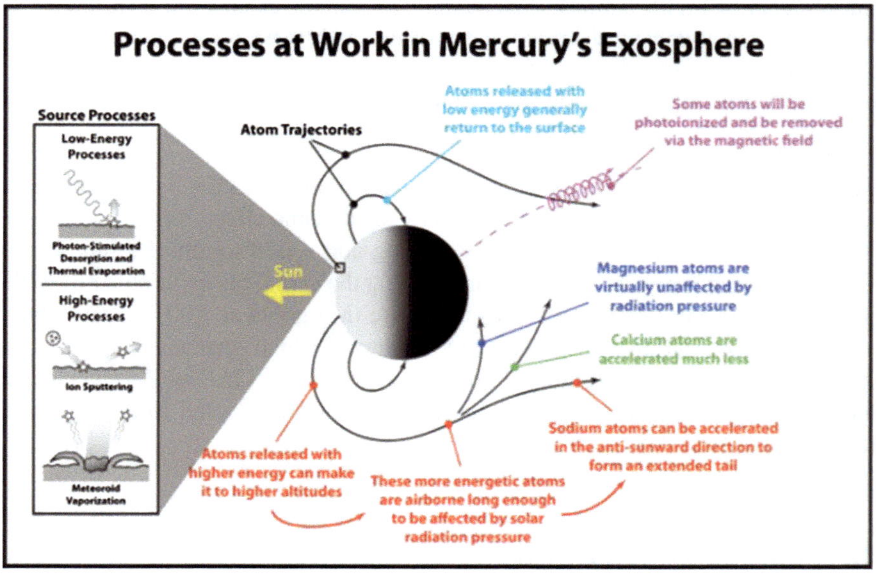

Fig. 2.14 Schematic illustration of the processes that generate and maintain Mercury's exosphere. Determining the relative importance of these processes through observations at different times and positions will yield insight into the interaction between Mercury's surface and the surrounding space environment

survives as atoms are ejected from the surface by the impacting solar wind, solar radiation and micrometeoroids. Prior observations had detected the presence of hydrogen, helium, sodium, calcium and potassium. One of the surprising discoveries from MESSENGER is a difference between the distributions of calcium and magnesium. Calcium abundance peaked near the equator, while magnesium was observed to be distributed more uniformly. Further, more calcium was detected near sunrise than at sunset, but not in sodium or magnesium amounts. Figure 2.15 shows the sodium atmosphere ("tail") seen from MESSENGER during the second and the third fly-bys and results of modeling of the processes involved that determine the abundance. As in previous flybys, the distinct north and south enhancements in the emission that result from material being sputtered from the surface at high latitudes on the dayside are seen.

The mission has succeeded in obtaining valuable observations to address them. A fuller understanding will follow when the implications of the observations and the new questions raised are considered, and after new data from the BepiColombo mission.

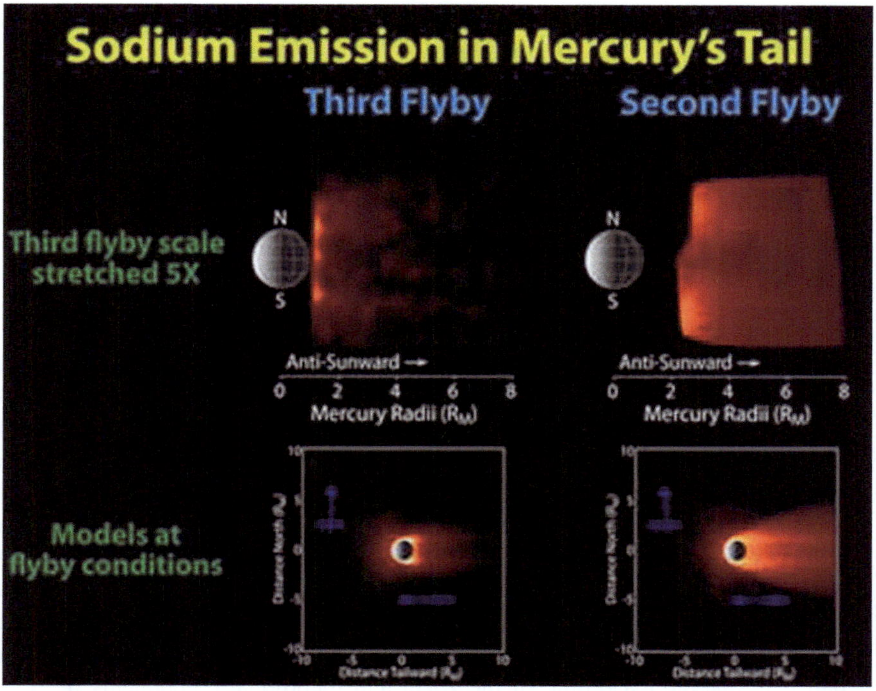

Fig. 2.15 Comparison of the neutral sodium observed during the second and third Mercury flybys to models. The top left and right panels show the same images as in Image 2.2, but the color scale for the third flyby has been stretched to show the distribution of sodium more clearly. The lower two panels show Monte Carlo models of the sodium abundance in Mercury's exosphere for conditions similar to those during the two flybys. Credit: http://www.nasa.gov/mission_pages/messenger/media/flyby20091029.html

2.4 Venus: Results from Venus Express

Operating with most instruments initially designed for Mars, ESA's Venus Express has made many significant discoveries about Venus – its exosphere, atmosphere and surface. Its objectives were to study (Svedhem et al. 2009):

- Atmospheric structure
- Atmospheric dynamics,
- Atmospheric composition and chemistry
- Cloud layer and hazes
- Energy balance and greenhouse effect,
- Plasma environment and escape processes, and
- Surface properties and geology.

The mission formally ended on 18 January 2015 after making many discoveries. Among the most interesting discoveries made by Venus Express, the possible change in the rotation rate of Venus is perhaps most significant, followed by long term changes in the abundance of sulfur dioxide gas above the clouds, the presence of electrical activity, and the presence of a relatively narrow layer of increasing temperature in the exosphere, similar to Earth's stratosphere. Also significant are the detection of hydroxyl molecules escaping Venus, and detection of trace species such as ozone, isotopologues and temporal and spatial changes in the amounts of sulfur dioxide above the clouds, and carbon monoxide. Some of the significant discoveries from Venus Express are summarized below.

2.4.1 Atmospheric Circulation

One of the early discoveries from Venus Express from the insertion orbit was a spectacular view of the hemispheric vortex over the southern pole from VIRTIS. Imaging in near infrared (night side) and in reflected sunlight (dayside) from high above the southern pole, the single image captured the vortex and the dipole structure of its inner core, also seen in Pioneer Venus OIR observations (Taylor et al. 1980). Previously Suomi and Limaye (1978) argued that the hemispheric vortex on Venus was dynamically similar to a tropical cyclone. Limaye et al. (2009) showed that the dipole feature could be simulated numerically from a cyclone circulation model and the result of a dynamical instability. The vortex was also investigated from VIRTIS (Garate-Lopez et al. 2013; Luz et al. 2011) who also obtained the latitudinal profile of the zonal and meridional components of cloud features at latitudes near the pole. The structure of motion of the vortex in the infrared observations was also seen to be asymmetric about the pole (Fig. 2.16).

Long-term changes in the measured cloud motions measured from the Venus Monitoring Camera were reported by Khatuntsev et al. (2013) and also by Kouyama et al. (2013). Periodicities ranging from 4.1 to 5.0 days were found in the

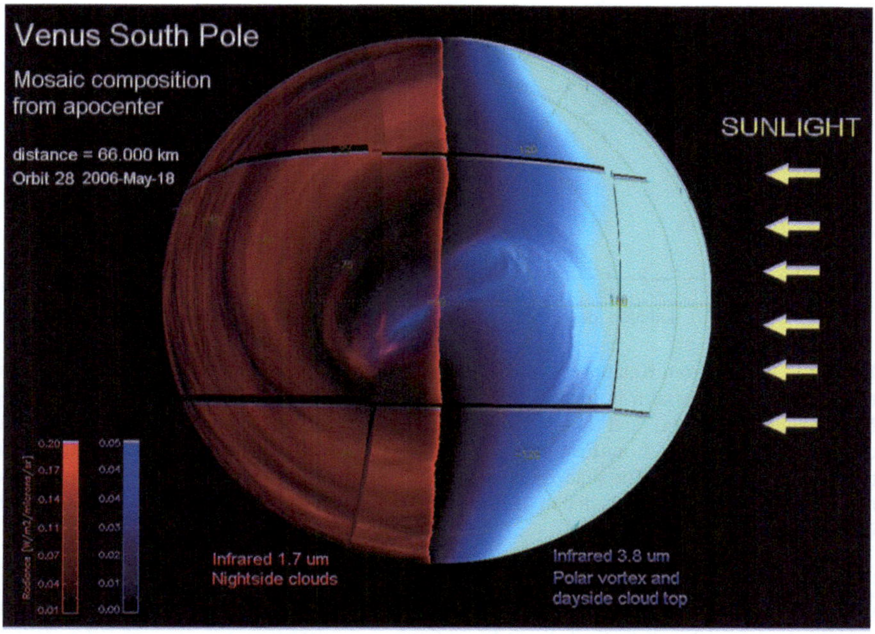

Fig. 2.16 A composite view of the day and night side of Venus from the VIRTIS instrument on Venus Express (ESA Image SEM49273R8F). This global view of Venus is a mosaic of several images taken by the Visible and Infrared Thermal Imaging Spectrometer (VIRTIS) on board ESA's Venus Express on 18 May 2007, at a distance of about 66 000 km from the planet. The images were obtained at 1.7 μ (left) and 3.8 μ (right) wavelengths. The wavelength used to obtain the left-hemisphere composite (1.7 μ) provides a dramatic global view of the night-side clouds in the lower atmosphere (approximately 45 km); while the wavelength used to obtain the right-hemisphere composite (3.8 μ) provides a view of the day-side cloud top

average cloud motions over during 2006–2012 resulting in an average value of 4.83 days. The long-term (Venus Express) average profile of zonal speed is comparable to the older, Mariner 10 results (Limaye and Suomi 1981) obtained from the short-term coverage of the fly-by observations, and both show the presence of a mid-latitude jet. The presence of such a jet is consistent with the hemispheric vortex. The Pioneer Venus average zonal flow did not show such a jet, but those results suffered from low image resolution and much longer time intervals between images (Limaye 2007). Further, these results are representative of the day-side as no nightside measurements are possible at the same vertical level from available data. The mean meridional flow over the Venus Express observations is also similar to Mariner 10, Pioneer and Galileo results – a weak poleward flow, peaking at about the same latitude (Fig. 2.17).

Over the duration of Venus Express observations, the average day-side cloud motions at low latitudes show a gradual increase in speed over time, but its validity remains uncertain. The chief reason may be due to the variable cloud morphology captured in the VMC images. The ability to measure cloud motions reliably

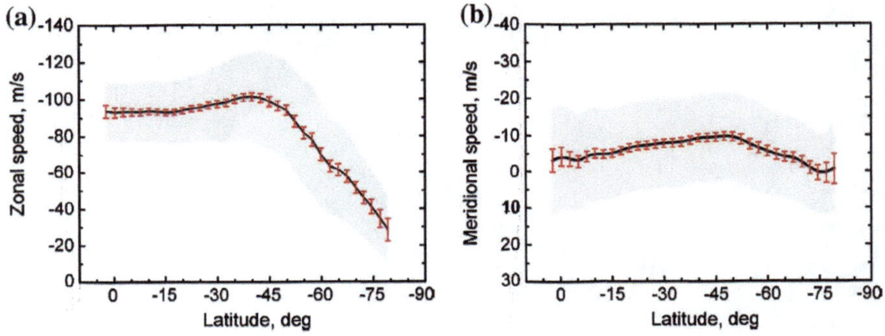

Fig. 2.17 Results from Khatuntsev et al. (2013) of average zonal (East-West) flow inferred from movements of clouds seen in images taken through the ultraviolet filter (365 nm) of the Venus Monitoring Camera (a) and North-South component (b). The negative sign for the zonal component implies flow from East to West (morning to evening direction) and poleward flow in the southern hemisphere. The shaded region indicates the standard deviation in one degree latitude bins. Some of the variability is due to the long-term changes in the measured cloud motions which is apparent in Fig. 2.18

depends on the visibility of discrete cloud features, and the morphology often makes it very difficult to find discrete features over the planet in the sun-lit portions of the planet, as the amount of haze that suppresses image cloud contrast increases towards the morning and evening terminators. Further, there is preliminary evidence that the representative cloud level that the measured motions correspond to may be variable over time by as much as ~1 km or more. Analysis of thermal structure data (Piccialli et al. 2012) and VIRTIS observations (Garate-Lopez et al. 2013) show that the vertical shear can be significant near the cloud top level, so the apparent trend in the temporal evolution may not be real (Fig. 2.18).

2.4.2 Rotation Rate of Venus – Length of Day Variations

By exploiting the fact that the temperature near the surface of Venus decreases with altitude, the Visible InfraRed Imaging Spectrometer (VIRTIS) on board Venus Express is able to relate the observed brightness temperature corresponding to the emitted radiation to topography. It was discovered that the rotation rate of Venus may be variable (Mueller et al. 2012) as the measured locations of some topographic features were different from expected values. The topography of Venus was obtained by the Magellan mission (Ford and Pettengill 1992). Mueller et al. (2012) discovered that the topographic features seen in the VIRTIS data are displaced somewhat from their corresponding topographic locations on Venus in the Magellan SAR data, which could result from a different orientation model, systematic offsets or a change in the rotation rate. Ground based radar estimates of the spin rate from data obtained during 2006–2009 are consistent (Schubert 2010) with

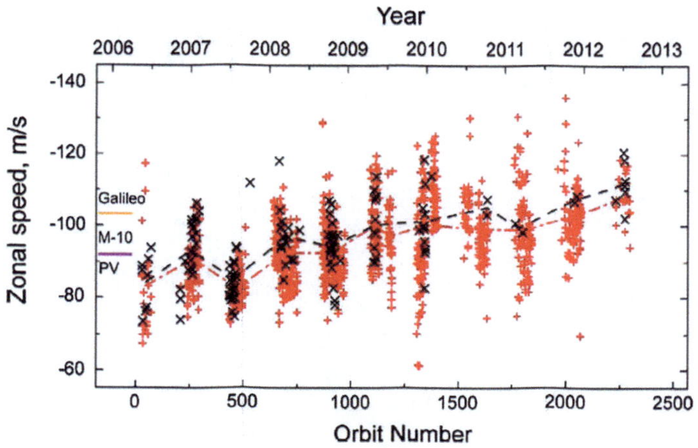

Fig. 2.18 Variations of the mean zonal wind speed at 20° ± 2.5 °S over the mission time. Symbols show orbital averages derived by manual ("x" and black dashed line) and digital ("+" and red dotted line) methods. The results from the Mariner 10 (92 m/s), Pioneer Venus (91.8 ± 3 m/s) and Galileo (103 m/s) missions for the same latitude zone are presented at the left edge of the plot for comparison. (For interpretation of the references to color in this figure legend, the reader is referred to Khatuntsev et al. 2013)

the 243.023 Earth days rotation rate estimated from VIRTIS data. It is also possible that Venus may exhibit measuring wobble arising out of planet's interior structure. A change or a modulation in the rotation rate may also be expected through the exchange of angular momentum between the atmosphere and the solid planet (Schubert 1983).

2.4.3 Upper Atmosphere

2.4.3.1 Detection of Ozone in the Atmosphere of Venus

Although the same photochemistry is possible on Venus and Earth, the composition of the respective atmospheres leads to different processes. The lack of a large stratosphere with increasing temperature was an early clue to the lack of an ozone-rich layer on Venus. Nevertheless, the discovery of ozone on the night side in a 5–10 km thick layer between 90–120 km above the mean surface from the SPICAV instrument on Venus Express came as a surprise (Montmessin et al. 2011).

Ozone is believed to form by the release of an oxygen atom from photo-dissociation of carbon dioxide, combining with an oxygen molecule. Molecular oxygen is formed on the night side resulting from recombination of oxygen atoms transported from the day-side, as evidenced by the airglow (Gerard et al. 2008; Migliorini et al. 2011; Soret et al. 2014). The amount of ozone is ~1000 times

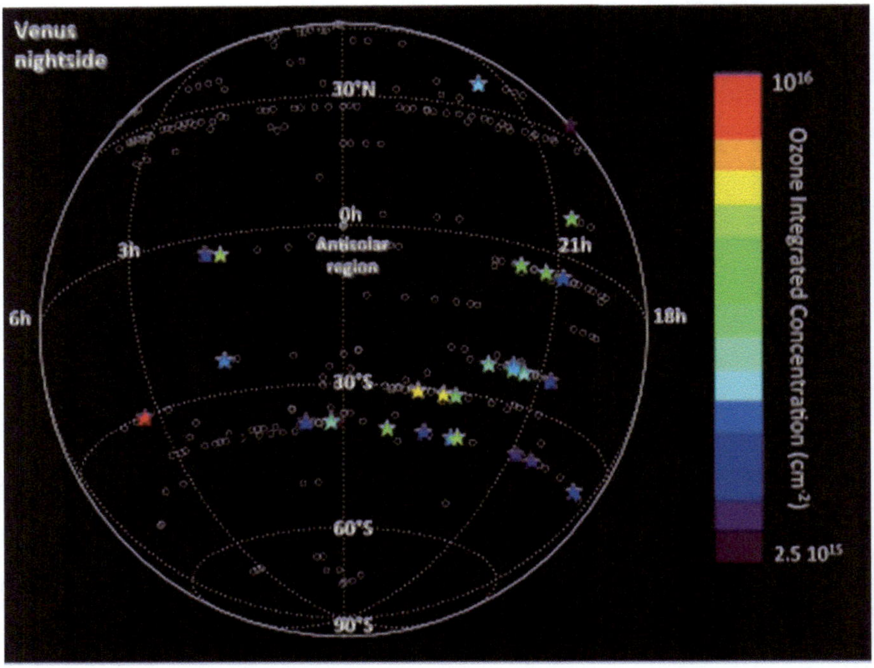

Fig. 2.19 SPICAV measurements of ozone in Venus atmosphere on the night side between 90–120 km altitude. Little ozone is detected at the midnight location at equator, a surprise

less than found in the stratosphere on Earth between about 9–15 km altitude (but ambient pressures are comparable for the ozone layer on Earth and Venus). It is not yet known whether ozone is catalytically destroyed near terminator by the same process that occurs in the polar stratospheric clouds on Earth. SPICAV observations do not provide good spatial coverage as they rely on stellar occultation, so more uniform coverage is needed to reveal the global pattern of ozone abundance in the atmosphere of Venus (Fig. 2.19).

2.4.3.2 Upper Atmosphere Cold Layer and Temperature Inversion

While it has been known for some time that the highest atmospheric temperature observed on any planet in our solar system is found on Venus near the surface, the detection of the coldest atmospheric temperature, about 100 K, also on Venus near the 125 km level, was a new discovery (Fig. 2.20).

Inferred from data collected by the Solar Occultation Infrared Radiometer instrument near the terminator in polar latitudes on Venus, the coldest layer may suggest the presence of frozen carbon dioxide particles (Mahieux et al. 2012). The atmospheric range of temperatures on Venus, from ~100 K to 750 K is comparable to the day-night difference in surface temperatures on the atmosphereless Mercury.

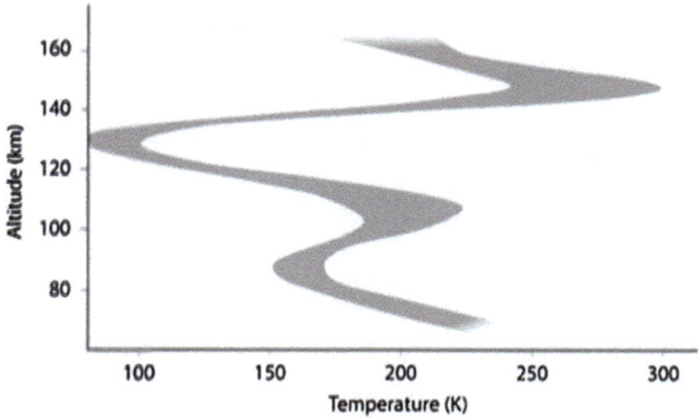

Fig. 2.20 The temperature profile along the terminator for altitudes of 70–160 km above the surface of Venus. The values were derived from the volume density of carbon dioxide molecules measured during solar occultation experiments by Venus Express' SOIR instrument. The shaded curve shows the average range of values calculated from 59 measurements taken along the terminator from 88 °N to 77 °S, during different orbits between 2006 and 2011

But the other surprising features are two layers of increasing temperature with altitude – between about 90–100 km and between 130–150 km. The reason for the warm layer is not yet known.

2.4.4 Possible Recent Volcanism?

The extent of current volcanism is still unknown, but several different clues keep coming. One of the more compelling clues is a rise and decay in the abundance of sulfur dioxide measured above the cloud tops from SOIR/SPICAV experiment (Marcq et al. 2013). While a near-exponential decrease was measured from Pioneer Venus ultraviolet spectrometer (Esposito et al. 1988), a rise from the low levels was observed for the first time from Venus Express observations, peaking in 2007 and decreasing again. Esposito (1984) had proposed explosive volcanism injected large amounts of sulfur dioxide gas punching through the thick cloud layer, and a similar situation is suggested by the Venus Express observations. Dynamical transport by the global atmospheric circulation has also been suggested, but it is not understood how the circulation would lead to the observed rise and decay. One difficulty with both the Venus Express and the older Pioneer Venus observations is the lack of global coverage. Thus, it is plausible that the observed rise and fall is not globally representative, which does not negate the argument for episodic volcanism.

The rise and fall of sulfur dioxide in the upper atmosphere of Venus over the last 40 years, expressed in units of parts per billion by volume (ppbv) is shown in Fig. 2.21. The dataset on the left is mostly from NASA's Pioneer Venus orbiter data (Esposito et al. 1988), which orbited Venus from 1978 to 1992. The dataset on the right shows results from the SOIR instrument observation at morning at evening terminators after Venus Express entered into orbit in 2006. A clear rise in the

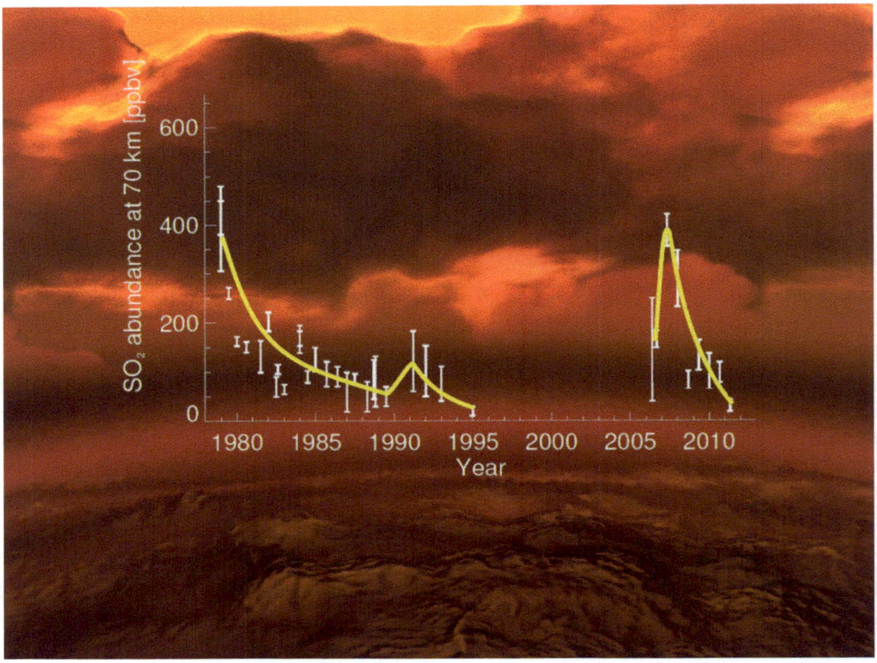

Fig. 2.21 A clear rise in the concentration of sulfur dioxide (SO_2) concentration was observed at the start of the mission, with a subsequent decrease. The increase in sulfur dioxide can be interpreted either as evidence for volcanic activity or for decadal-scale variations in the circulation of Venus' vast atmosphere. The data are superimposed on an artist impression of Venus, depicting a volcanic terrain surrounded by a thick, noxious atmosphere. Credit: ESA graphics (http://sci.esa.int/venus-express/51187-rise-and-fall-of-sulphur-dioxide/)

concentration of sulfur dioxide (SO2) concentration was observed at the start of the mission, with a subsequent decrease.

Bondarenko et al. (2010) detected excess microwave emissions in 1993 in the Magellan radiometry data and concluded that this implied increased sub-surface temperatures due to emplacement of magma. Smrekar et al. (2010) also reported high emissivity locations coinciding with some potential hotspots in low latitude regions of Venus. An emplacement age of <2.5 Ma was argued based on the inference of limited surface weathering. Increased, transient high surface temperatures have also been inferred in some Venus Monitoring Camera near infrared images (Shalygin et al. 2014; Bazilevskiy et al. 2014).

While doubts can be raised about the validity of such claims of near-surface geologic activity due to and observational constraints and data processing issues, the diversity of different clues is intriguing. Current volcanism may explain many aspects of the dynamical behavior of the Venus atmosphere and confirm some of the assumptions about Venus. Nevertheless, a strong confirmation of an active current or recent volcanic eruption that would remove any doubts and remains a future observational prize.

2.5 Summary

MESSENGER and Venus Express missions have exceeded expectations about new knowledge gained about Mercury and Venus. Due to the orbital constraints posed by the elliptic orbit, MESSENGER data have poorer coverage or resolution in the southern hemisphere. At the end of the nominal mission the orbit was trimmed from a twelve hour period to one with an eight hour one, improving imaging ground resolution and also increasing coverage from the laser altimeter. In the near future, ESA/JAXA's BepiColombo mission will add significantly to the knowledge about Mercury.

JAXA's Akatsuki mission is expected to enter into Venus orbit during its second attempt in late November/early December 2015. Designed as a climate observing mission (Venus Climate Orbiter), it carries mostly cameras capable of taking pictures from ultraviolet to the thermal infrared wavelengths. The spacecraft is expected to be placed in a 7–10 day near equatorial orbit. The observations will perform a systematic investigation of the superrotation of the atmosphere. Many questions about Venus will remain unaddressed after this mission and the scientific community is eagerly awaiting a more comprehensive investigation of the planet regarding its surface, atmosphere and magnetosphere.

Acknowledgments I thank Ms. Rosalyn Pertzborn for editing the manuscript. This work was supported partially by NASA Grant NNX09AE85G, and by Space Science and Engineering Center, University of Wisconsin, Madison. Comments by two anonymous referees and Dr. A.B.S. Limaye were useful in improving the manuscript.

References

Asphaug E, Reufer A (2014) Mercury and other iron-rich planetary bodies as relics of inefficient accretion. Nat Geosci. doi:10.1038/ngeo2189

Bazilevskiy A, Ignatiev N, Markiewicz W, Head J, Titov D, Shalygin EV (2014) Volcanism of Venus: insights from the VMC data analysis. In: 40th COSPAR scientific assembly. Held 2–10 August 2014, in Moscow, Russia, Abstract B0.7-1-14

Blewett DT, Chabot NL, Denevi BW, Ernst CM, Head JW, Izenberg NR, Murchie SL, Solomon SC, Nittler LR, McCoy TJ, Xiao Z, Baker DMH, Fassett CI, Braden SE, Oberst J, Scholten F, Preusker F, Hurwitz DM (2011) Hollows on Mercury: MESSENGER evidence for geologically recent volatile-related activity. Science 333(6051):1856–1859. doi:10.1126/science.1211681

Bolmatov D, Zav'yalov D, Gao M, Zhernenkov M (2014) Structural evolution of supercritical CO2 across the Frenkel line, Bo. J Phys Chem Lett 5(16):2785–2790. doi:10.1021/jz5012127

Bondarenko NV, Head JW, Ivanov MA (2010) Present-day volcanism on Venus: evidence from microwave radiometry. Geophys Res Lett 37:L23202. doi:10.1029/2010GL045233

Byrne PK, Klimczak C, Williams DA, Hurwitz DM, Solomon SC, Head JW, Preusker F, Oberst J (2013) An assemblage of lava flow features on Mercury. J Gephys Res. doi:10.1002/jgre.20052

Byrne PK, Klimczak C, Şengör AMC, Solomon SC, Watters TR, SA Hauck II (2014) Mercury's global contraction much greater than earlier estimates. Nat Geosci 7:301–307 doi:10.1038/ngeo2097

Cavanaugh JF, Smith JC, Sun XI, Bartels AE, Ramos-Izquierdo L, Krebs DJ, McGarry JF, Trunzo R, Novo-Gradac AM, Britt JL, Karsh J, Katz RB, Lukemire AT, Szymkiewicz R, Berry DL, Swinski JP, Neumann GA, Zuber MT, Smith DE (2007) The Mercury laser altimeter instrument for the MESSENGER mission. Space Sci Rev 131(1):451–479. doi:10.1007/s11214-007-9273-4

Chabot et al. (2014) Images of surface volatiles in Mercury's polar craters acquired by the MESSENGER spacecraft. Geology 42(10)

Dutton CE (1874) A criticism of the contractional hypothesis. Am J Sci 8:113–123

Ernst CM, Murchie SL, Barnouin OS, Robinson MS, Denevi BW, Blewett DT, Head JW, Izenberg NR, Solomon SC, Head JW (2010) Exposure of spectrally distinct material by impact craters on Mercury: implications for global stratigraphy. Icarus 209:210–223

Esposito LW (1984) Sulfur dioxide - Episodic injection shows evidence for active Venus volcanism, Science. Science 223:1072–1074. (ISSN 0036-8075) doi:10.1126/science.223.4640.1072

Esposito LW, Copley M, Eckert MR, Gates L, Stewart AIF, Worden H (1988) Sulfur dioxide at the Venus cloud tops, 1978–1988. J Geophys Res 93:5267–5276

Ford PG, Pettengill GH (1992) Venus topography and kilometer-scale slopes. J Geophys Res 97:13103–13114

Garate-Lopez I, Hueso R, Sánchez-Lavega A, Peralta J, Piccioni G, Drossart P (2013) A chaotic long-lived vortex at the southern pole of Venus. Nat Geosci 6:254–257. doi:10.1038/ngeo1764

Gérard J-C, Cox C, Saglam A, Bertaux J-L, Villard E, Nehmé C (2008) Limb observations of the ultraviolet nitric oxide nightglow with SPICAV on board Venus Express. J Geophys Res Planet 113(E5). doi:10.1029/2008JE003078

Goudge TA, Head JW, Kerber L, Blewett DT, Denevi BW, Domingue DL, Gillis-Davis JJ, Gwinner K, Helbert J, Holsclaw GM, Izenberg NR, Klima RL, McClintock WE, Murchie SL, Neumann GA, Smith DE, Strom RG, Xiao Z, Zuber MT, Solomon SC (2014) Global inventory and characterization of pyroclastic deposits on Mercury: new insights into pyroclastic activity from MESSENGER orbital data. J Geophys Res doi:10.1002/2013JE004480 (Article first published online: 28 MAR 2014)

Hawkins III SE et al (2007) The Mercury dual imaging system on the MESSENGER spacecraft. Space Sci Rev 131:247–338

Kerber L, Head JW, Solomon SC, Murchie SL, Blewett DT, Wilson L (2009) Explosive volcanic eruptions on Mercury: Eruption conditions, magma volatile content, and implications for interior volatile abundances. Earth Planet Sci Lett 285(3–4):263–271. doi:10.1016/j.epsl.2009.04.037

Kerber L, Head JW, Blewett DT, Solomon SC, Wilson L, Murchie S, Robinson MS, Denevi BW, Domingue DL (2011) The global distribution of pyroclastic deposits on Mercury: the view from MESSENGER flybys 1–3. Planet Space Sci 59:1895–1909

Killen R, Cremonese G, Lammer H et al (2007) Processes that promote and deplete the exosphere of Mercury. Space Sci Rev 132(2–4):433–509. doi:10.1007/s11214-007-9232-0

Khatuntsev IV, Patsaeva MV, Titov DV, Ignatiev NI, Turin AV, Limaye SS, Markiewicz WJ, Almeida M, Roatsch TH, Moissl R (2013) Cloud level winds from the Venus Express Monitoring Camera imaging. Icarus 226(1):140–158. http://dx.doi.org/10.1016/j.icarus.2013.05.018

Kouyama T, Imamura T, Nakamura M, Satoh T, Futaana Y (2013) Long-term variation in the cloud-tracked zonal velocities at the cloud top of Venus deduced from Venus Express VMC images. J Geophys Res Planet 118:37–46. doi: 10.1029/2011JE004013

Lawrence DJ, Feldman WC, Goldsten JO, Maurice S, Peplowski PN, Anderson BJ, Bazell D, McNutt RL, Nittler LR, Prettyman TH, Rodgers DJ, Solomon SC, Weider SZ (2013) Evidence for water ice near Mercury's North pole from MESSENGER neutron spectrometer measurements. Science 292–296 (Published online 29 November 2012)

Leverington DW (2004) Volcanic rilles, streamlined islands, and the origin of outflow channels on Mars. J Geophys Res 109:E10011. doi:10.1029/2004JE002311

Limaye SS (2007) Venus atmospheric circulation: known and unknown. J Geophys Res 112: E04S09. doi:10.1029/2006JE002814

Limaye SS, Suomi VE (1981) Cloud motions on Venus: global structure and organization. J Atmos Sci 38:1220–1235
Limaye SS, Kossin JP, Rozoff C, Piccioni G, Titov DV, Markiewicz WJ (2009) Vortex circulation on Venus: dynamical similarities with terrestrial hurricanes. Geophys Res Lett 36(4):L04204. doi:10.1029/2008GL036093
Linkin VM, Blamont J, Lipatov A, Devyatkin SI, D'yachkov AV, Ignatova B1, Kerzhanovich VV, Malique C, Stadnyk V1 et al (1986) Vertical thermal structure of the Venus atmosphere from temperature and pressure measurements. Pistma V Astron 12(2):100–105
Lucey PG (2013) A wet and volatile Mercury. Science 282–283. doi:10.1126/science.1232556
Luz D et al (2011) Venus's southern polar vortex reveals precessing circulation. Science 332:577–580
Mahieux et al (2012) J Geophys Res Planet 117:E07001. doi:10.1029/2012JE004058
Margot JL, Peale SJ, Jurgens RF, Slade MA, Holin IV (2007) Large longitude libration of Mercury reveals a molten core. Science 316:710–714
Marcq E, Bertaux J-L, Montmessin F, Belyaev D (2013) Variations of sulphur dioxide at the cloud top of Venus's dynamic atmosphere. Nat Geosci 6:25–28. doi:10.1038/ngeo1650
Migliorini A, Altieri F, Zasova L, Piccioni G, Bellucci G, Cardesın Moinelo A, Drossart P, D'Aversa E, CarrozzoFG, Gondet B, Bibring J-P (2011) Oxygen airglow emission on Venus and Mars as seen by VIRTIS/VEX and OMEGA/MEX imaging spectrometers. Planet Space Sci 59:981–987
Montmessin F, Bertaux J-L, Lefèvre F, Marcq E, Belyaev D, Gérard J-C, Korablev O, Fedorova A, Sarago V, Vandaele AC (2011) A layer of ozone detected in the nightside upper atmosphere of Venus. Icarus 216(I):82–85. http://dx.doi.org/10.1016/j.icarus.2011.08.010
Mueller N, Helbert J, Erard S, Piccioni G, Drossart P (2012) Rotation period of Venus estimated from Venus Express VIRTIS images and Magellan altimetry. Icarus 217(21):474–483
Murray BC, Strom RG, Trask NJ, Gault DE (1975) Surface history of Mercury: implications for terrestrial planets. J Geophys Res 80:2508–2514
Nittler LR, Weider SZ, Starr RD, Chabot N, Denevi BW, Ernst CM, Goudge TA, Head JW, Helbert J, Klima RL, McCoy TJ, Solomon SC (2014), Sulfur-depleted composition of Mercury's largest pyroclastic deposit: implications for explosive volcanism and surface reflectance on the innermost planet. Presented at the 45th Lunar and planetary science conference, held 17–21 March, 2014 at The Woodlands, Texas. LPI Contribution No. 1777, p 1391
Oberbeck VR, Quaide WL, Arvidson RE, Aggarwal HR (1977) Comparative studies of lunar, Martian and Mercurian craters and plains. J Geophys Res 82. doi:10.1029/JB082i011p01681
Paige, DA et al (2013) Thermal stability of volatiles in the north polar region of Mercury. Science 339:300–303. doi:10.1126/science.1231106
Piccialli AS, Tellmann DV, Titov SS, Limaye IV, Khatuntsev M, Pätzold B, Häusler Titov DV (2012) Dynamical properties of the Venus mesosphere from the radio-occultation experiment VeRa onboard Venus Express. Icarus 217(2):669–681. http://dx.doi.org/10.1016/j.icarus.2011.07.016 (Advances in Venus Science)
Robinson MS, Lucey PG (1997) Recalibrated Mariner 10 color mosaics: implications for mercurian volcanism. Science 275:197–200
Schubert G (1983) General circulation and the dynamical state of the Venus atmosphere. In: Venus, Hunten D et al (eds) University of Arizona Press, vol 1143, p 1983
Schubert G (2010) Venus rotation, Paper presented at the VEXAG. In: International Venus Workshop, 2010. venus.wisc.edu/venus-workshop-submission/files/schubert_gerald-2.pdf
Seiff A (1987) Further information on structure of the atmosphere of Venus derived from VeGa balloon and lander mission. Adv Space Res 7(12):323–328. doi:10.1016/0273-1177(87)90239-0
Shalygin EV, Markiewicz WJ, Basilevsky AT, Titov DV, Ignatiev NI, Head JW Bright transient spots in ganiki chasma, Venus. Presented at the 45th Lunar and planetary science conference, held 17–21 March, 2014 at The Woodlands, Texas. LPI Contribution No. 1777, p 2556
Slade MA, Butler BJ, Muhleman DO (1992) Mercury radar imaging: evidence for polar ice. Science 258:635–640

Smith DE, Zuber MT, Phillips RJ, Solomon SC, Hauck II SA, Lemoine FG, Mazarico E, Neumann GA, Peale SJ, Margot J-L, Johnson CL, Torrence MH, Perry ME, Rowlands DD, Goossens S, Head JW, Taylor AH (2012) Gravity field and internal structure of Mercury from MESSENGER. Science 336(6078):214–217. doi:10.1126/science.1218809

Smrekar SE, Stofan ER, Mueller N, Treiman A, Elkins-Tanton AL, Helbert J, Piccioni G, Drossart P (2010) Recent hotspot volcanism on Venus from VIRTIS emissivity data. Science 328:605–608

Solomon SC (2003) Mercury: the enigmatic innermost planet. Earth Planet Sci Lett 216:441–455

Soret L, J-C Gérard, Piccioni G, Drossart P (2014) Time variations of $O_2(a^1\Delta)$ nightglow spots on the Venus nightside and dynamics of the upper mesosphere. Icarus 237:306–314. doi:10.1016/j.icarus.2014.03.034

Staley D (1970) The adiabatic lapse rate in the Venus atmosphere. J Atmos Sci 27:219–223

Strom RG, Trask JJ, Guest JE (1975) Tectonism and volcanism on Mercury. J Geophys Res 80:2478–2507

Suomi VE, Limaye SS (1978) Further evidence of vortex circulation on Venus. Science 201:1009–1011

Svedhem H, Titov D, Taylor F, Witasse O (2009) Venus Express mission. J Geophys Res 114: E00B33. doi:10.1029/2008JE003290

Taylor FW (2006) Venus before Venus express. Planet Space Sci 2006:1249–1262

Taylor FW, Beer R, Chahine MT, Diner DJ, Elson LS, Haskins RD, McCleese DJ, Martonchik JV, Reichley PE, Bradley SP, Delderfield J, Schofield JT, Farmer CB, Froidevaux L, Leung J, Coffey MT, Gille JC (1980) Structure and meteorology of the middle atmosphere of Venus Infrared remote sensing from the Pioneer orbiter. J Geophys Res 85:7963–8006. http://dx.doi.org/10.1029/JA085iA13p07963

Vasavada AR, Paige DA, Wood SE (1999) Near-surface temperatures on mercury and the moon and the stability of polar ice deposits. Icarus 141(1999):179–193

Watson KB, Murray BC, Brown H (1961) The behavior of volatiles on the Lunar surface. J Geophys Res 66(9):3033–3045

Watters TR, Robinson MS, Beyer RA, Banks ME, Bell III JF, Pritchard ME, Hiesinger H, van der Bogert CH, Thomas PC, Turtle EP, Williams NR (2010) Evidence of recent thrust faulting on the moon revealed by the Lunar Reconnaissance Orbiter Camera. Science 329(5994):936–940. doi:10.1126/science.1189590

Watters TR, Solomon SC, Klimczak C, Freed AM, Head JW, Ernst CM, Blair DM, Goudge RA, Byrne PK (2012) Extension and contraction within volcanically buried impact craters and basis on Mercury. Geology 40:1123–1126. doi:10.1130/G33725.1

Wetherill G (1994) Provenance of the terrestrial planets. Geochim Comochim Acta 58:4513–4520

Wilhelms DE (1976) Mercurian volcanism questioned. Icarus 28:551–558

Chapter 3
Mercury the Sunshine Planet

Johannes Benkhoff

3.1 Introduction

Mercury is the smallest among the planets of our solar system. Its diameter is only about 2/5 of that of the Earth, about 4880 km. Even some moons of Jupiter and Saturn, Ganymede and Titan, are bigger. Mercury is the planet closest to the Sun and named after the roman God of dealers and thieves. Because Mercury moves in an orbit around the Sun, which lies within Earth's orbit it only appears in Earth's sky in the morning or the evening. In the early days the Greeks were mislead by these observations and assumed two different planets and called them Hermes and Apollo.

Although Mercury can appear as a very bright object when viewed from Earth, its proximity to the Sun makes it more difficult to see than Venus. When the first astronomical telescopes appeared in the 17th century Mercury became an attractive object to observe. Already at that time it was noted that Mercury's orbit slightly deviates from what is to be expected from classical Newtonian mechanics [Mercury's orbit shows an "anomalous" perihelion shift of about to 43 arc-seconds per century, which is about 120 km/year. As a result Mercury orbits on a rosette-like pattern around the Sun]. Trying to explain this mysterious behavior, astronomers postulated a hypothetical planet, Vulcain, which by its gravitation could have caused these deviations. However, the postulated planet was never found and today scientists know that they have to take Einstein's theory of general relativity into account because of the tremendous forces of gravity reigning in Mercury's quarters close to the Sun.

A further peculiarity of Mercury's orbit is that at certain longitudes a hypothetically assumed observer on its surface would see the Sun rise and immediately

J. Benkhoff (✉)
ESA/ESTEC Science Support Office, Noordwijk, The Netherlands
e-mail: johannes.benkhoff@esa.int

V. Badescu and K. Zacny (eds.), *Inner Solar System*,
DOI 10.1007/978-3-319-19569-8_3

set again before rising again and travelling westwards. At sunset the Sun rises briefly before setting once more. These observations are due Mercury's highly elliptical orbit and its 3:2 spin/orbit resonance as a result of tidal forces of the Sun acting on Mercury, meaning that whilst Mercury orbits the Sun twice, it rotates around its own rotational axis exactly three times. Days on Mercury are quite long compared to Earth. Mercury's rotation period is about 58.6 Earth days, while its orbital period is about 88 Earth days. A solar day on Mercury last two orbits (Mercury years), about 176 Earth days. Mercury's highly elliptical orbit ranges between 0.3 and 0.47 astronomical units or 46 million km and 69.8 million km, respectively. [An astronomical unit AU is a measure of distance in Astronomy. 1 AU is equal to the mean distance between Earth and Sun of 149 597 870.7 km]. The mean distance from the Sun is 57.91 million km. Because of its close distance to the Sun (3 times closer than Earth) Mercury's surface experiences temperatures ranging from 100 K (−173 °C) at the night side of the planet up to 700 K (427 °C) during the day at some equatorial regions. Mercury's axis of rotation is oriented nearly perpendicular to the planet's orbit, so that in the Polar Regions sunlight strikes the surface at a constant grazing angle. As a consequence the interiors of large craters at the poles are permanently shadowed and remain perpetually cold, below 65 K (−210 °C). Radar images of the Polar Regions, first obtained in 1991, show that the large crater interiors are highly reflective at radar wavelengths (Harmon et al. 2001). The most common material that could explain this behavior is - water ice. Unthinkable but true for the hottest planet closest to the Sun. However, influx of ice from infalling comets and meteorites could be cold-trapped in Mercury's polar caps over billions of years, or water vapor might originate from the planetary interior and be frozen out at the poles.

Understanding the sunshine planet Mercury, the planet of extremes, is crucial to develop a better understanding of the early processes in the inner solar system, of how our Earth was formed, how it evolved, and how it interacts with the Sun. Some of the important questions one needs to address to understand Mercury are about its evolution history, its high density, its inner structure and composition, the nature of its magnetic field, its environment, and its surface composition.

3.2 Exploration History

Mercury is difficult to observe from the Earth, due to its close proximity to the very bright Sun. For an in-depth study of the planet and its environment, it is therefore necessary to operate a spacecraft equipped with scientific instrumentation around the planet. On the other hand the thermal and radiation environment close to the Sun and close to the hottest planet in the solar system is extremely aggressive, which makes such a mission technically very challenging

Two spacecraft have visited Mercury so far. In 1974 and 1975 Mariner 10 flew by Mercury three times (Strom and Sprague 2003) and in 2004 the MESSENGER spacecraft was launched (Solomon et al. 2007; McNutt et al. 2006). MESSENGER flew by Mercury three times in 2008 and 2009 before inserted into orbit around Mercury in March 2011.

During the fly-bys of the Mariner 10 spacecraft only 45 % of the surface of Mercury was photographed. The surface showed many craters and its age seemed to be comparable to that of surface areas of the Moon. Unlike on Earth and Venus, very few features on Mercury are clearly related to tectonic forces that reshaped the surface. Chaotic terrain on the side of the planet opposite to the largest crater may have formed when the planet's shape focused seismic energy from the impact into concentrated regions. Mercury showed features, lobate scarps, which are thought to have formed by compression.

One of the most unexpected discoveries by Mariner 10 was the measurement that Mercury has a global magnetic field like the Earth (Ness et al. 1975). The accepted model of magnetic field generation valid for the Earth is that swirling motions in portions of the liquid outer core generate the magnetic field. Thus, to drive these motions a liquid core is essential. However, because Mercury is so small the core should have cooled and solidified a long time ago, within the first billion years after its formation. In fact, cooling and contraction of the core may have been the driving force behind the global wrinkling that formed Mercury's lobate scarps. But how could a cooled, solid core generate a magnetic field? One possibility is that Mercury's core is not yet completely frozen due to dissolved low melting-point elements like sulfur or oxygen.

Those early Mariner 10 observations, and subsequent ground-based astronomical discoveries, raised a number of basic questions about Mercury that could not be addressed without further spacecraft exploration. In 1999, the US National Aeronautics and Space Agency (NASA) selected the MErcury Surface, Space ENvironment, GEochemistry, and Ranging (MESSENGER) mission, which in March 2011 became the first spacecraft to orbit the innermost planet (Solomon et al. 2007, 2008).

3.3 Formation

Mercury can teach us some important lessons on the formation of our Solar System. The terrestrial planets consist of a dense iron-rich core surrounded by a rocky mantle of magnesium- and iron-bearing silicates (rock). The core is assumed to be initially fluid, but becomes solid when its temperature falls below the so-called 'Liquidus Temperature' (broken dashed line in Fig. 3.1). The upper topmost layer of the mantle and the crust compose is the lithosphere. A schematic view of the inner structure of Mercury is given in Fig. 3.1. It is assumed that the temperature in the interior is high, around 1900 K (T_{ic}) at the inner core – outer core boundary and decreases with increasing radius R. The total density of Mercury is a composite of the high density of the iron-rich core, the density of the silicate-rich mantle and the density of the crust.

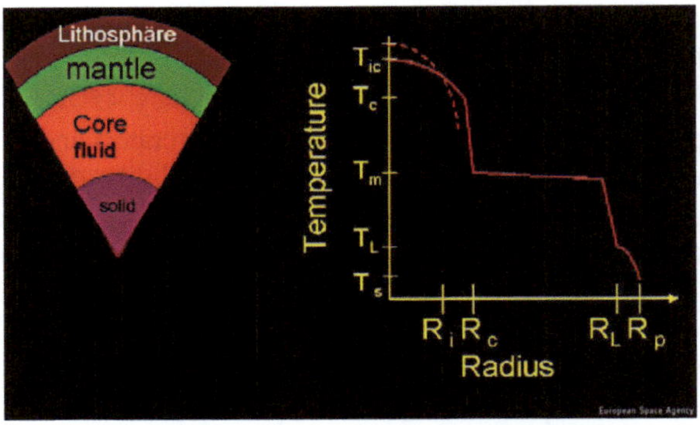

Fig. 3.1 Schematic cut of the inner structure of Mercury and a model temperature profile based on assumptions of the inner structure and its physical and chemical behavior as a function of radius. Subscripts indicate the boundaries of the different layers (*c* core, *i* inner core, *m* convecting mantle, *l* lithosphere, *p* planet)

Mercury's uncompressed density (density without compaction of its interior by the planet's own gravity) is 5300 kg/m^3, much higher than densities of all the other terrestrial planets. Mercury's density implies that 65 % of the planet is the metal-rich core - twice as much as Earth.

There are numerous formation models addressing the Mercury's unusual high density, indicative of a large metallic core and the apparent low iron oxide, FeO, surface abundance. The models fall in several classes. One class focuses on the enrichment of metals, either by differential accumulation of metal and silicate or by blasting away the silicate rich crust by a giant impact. Weidenschilling (1978) presented the idea of aerodynamic fractionation during the accretion phase, which would separate metal from silicate and increase the abundance of metal in the formation region of Mercury. In the giant impact model (Benz et al. 1988; Cameron et al. 1988; Wetherill 1988) the silicate mantle is stripped off the planet by a giant impact after its differentiation. In this model the impact body must have about 20 % of the size of the planet and depending on impact geometry the predicted impact velocity ranges from 20 km/s to 35 km/s. Both models can account for the high metal/silicate ratio, but they put no real constraints on the FeO abundance or the surface composition.

Evaporation models are another class of models. Based on models by Cameron (1985), during the early T-Tauri phase, the formation region of Mercury might have experienced temperatures up to 2500–3500 K. Recent models however do not show such high temperatures at this heliocentric distance range. If we still assume this high temperature and further assume that Mercury was formed very early, the composition of planet might have been affected. According to Cameron (1985) two main effects should occur. Most of the silicate would be vaporized and lost, accounting for the high metal to silicate ratio. In this model Mercury had originally ~2.25 times the mass observed now. The second effect is an enrichment of refractory elements. Fegley and Cameron (1987) modelled several cases assuming different silicate magmas as starting conditions. One of their models with a low FeO abundance would predict very low silicon dioxide and high magnesium and no Potassium. This is in contradiction to recent measurements from MESSENGER indicating that Mercury contains much more volatile material than expected by Fegley and Cameron (1987) (Fig. 3.2). The Potassium/Thorium ratio indicates the volatile material content of a planet. Scientists expected that this ratio to be similar to the one of the Moon. However these new results point in another direction and some formation models may need to be re-written.

For example the class of refractory condensation models, based on the predicted temperature gradient in the solar nebula by theoretical models. If these gradients were steep enough the composition of the solar nebula would vary as a function of temperature and distance (Lewis 1972, 1974). Mercury would have been formed from high-temperature condensates, like metallic iron and FeO-free silicates. Since iron condenses at higher temperatures than magnesium silicates, this would also explain the large metallic core. It would, however, require a very rapid accretion, because the condensation temperatures are very close. This class of models is little considered today, because newer models of the early solar nebula indicate a widespread radial mixing of planetesimals, essentially erasing the compositional differences. If Mercury had been formed according to the refractory model, the surface would be rich in aluminum, calcium, and magnesium and show no FeO, sodium or potassium.

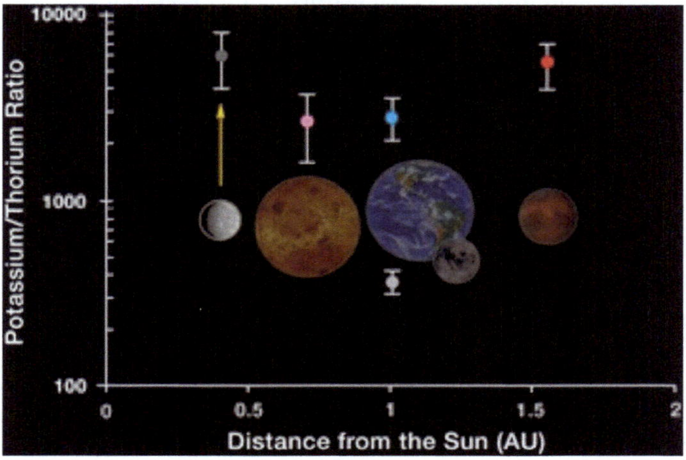

Fig. 3.2 The figure shows the Potassium/Thorium ratio as a function of the distance from the Sun. This ratio indicates the volatile material content of a planet. Scientists believe that this ratio increases with distance from the Sun and assume a value much lower than measured (Credit: NASA/The Johns Hopkins University Applied Physics Laboratory/Carnegie Institution of Washington)

Variations of the refractory model are refractory-volatile models taking into account the mixing throughout the inner solar system. Goettel (1988) and Morgan and Anders (1980) performed calculations using mixtures of refractory and volatile end-members to form the planet. In these models the abundance of refractory elements on Mercury are modest, but larger than on Earth (Taylor and McLennan 1985). Low abundances of sodium and potassium are expected, in agreement with the presence of these elements in Mercury's exosphere. In the Morgan and Anders (1980) model the normative mineralogy of the surface would show plagioclase and high-calcium pyroxene.

Another class of formation models, the chondritic models, based on the long discussed notion that chondrites were the building blocks for the planets. [Chondrites are stony (non-metallic) meteorites that have not been much modified due to melting or differentiation. Scientists believe that they are from primitive asteroids formed during the early history of solar system. They are the most common type of meteorite that falls onto Earth]. Enstatite chondrites have been proposed as possible analogues for Mercury (Wasson 1988), based on the idea that these chondrites formed also close to the Sun.

Recently Asphaug and Reufer (2014) proposed a new formation hypothesis involving hit-and-run collisions, where Mercury loses mayor parts of its mantle by impacting a larger proto planet (e.g. Venus) but failing to accrete. This seems to be a possible process occurring in about one out of 10 impacts. However, Asphaug and Reufer have not only shown that in rare cases a proto-planet can survive hit-and-run collisions, but also that survivors of multiple hit-and-run events exists. Already one or two of these hit-and-run collisions could occur in the early phase of planet formation can explain Mercury's high density and its very thin mantle. In addition volatiles could also be retained.

3.4 Surface Properties

Numerous craters of all sizes cover Mercury's surface. During the time of the late heavy bombardment (LHB) between 4.1 and 3.8 billion years ago, the terrestrial planets experienced intense and frequent impacts of smaller solar system bodies (asteroids or comets), which created craters on their surfaces. Erupting lava filled the surrounding craters and created large plains with smooth surfaces. At the end of the LHB, the impact rate decreased significantly. As a result heavily cratered surfaces are assumed to be old while regions exhibiting only few craters seemed to be younger.

Images taken by the MESSENGER spacecraft show that huge areas of Mercury's surface were volcanically emplaced (Head et al. 2008). Rimless depressions similar to what one observes on the Moon provided are among other strong evidence for this assumption.

Mercury also shows tectonic structures interpreted to be the result of planetary cooling and contraction. However, the amount of global contraction inferred from spacecraft images has been far lower than that predicted by models of the thermal evolution of the planet's interior. Latest results from orbital observations acquired by the MESSENGER spacecraft (Byrne et al. 2014) show that Mercury has contracted radially by as much as 7 km. These findings provide a key constraint to Mercury's thermal history, its bulk silicate abundances of heat-producing elements, mantle convection and the structure of its large metallic core.

The Moon has often been used for comparison with Mercury. Like lunar regolith, Mercury's surface should be modified by so called "space weathering" micro-meteoritic bombardment and solar wind irradiation, most probably to a larger extent than on the Moon (Cintala 1992). The expected optical effects of lunar-like space weathering are darkening of the surface and a reduction of the spectral contrast. However, recent results from MESSENGER indicate significant differences of the average surface composition of Mercury compared to lunar highlands or lunar soils (Fig. 3.3).

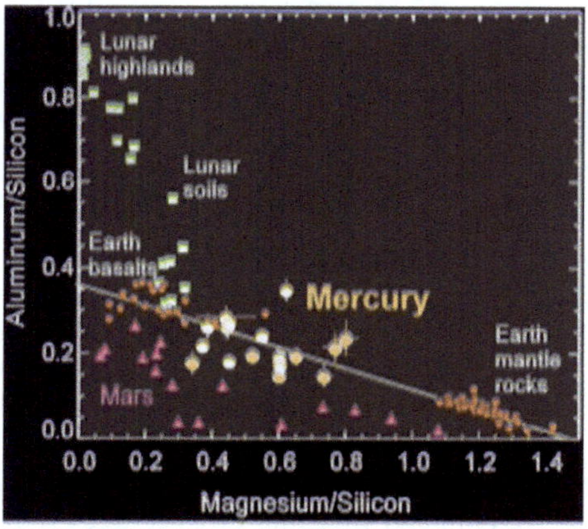

Fig. 3.3 The figure shows the Aluminum to Silicon ratio as a function of the Magnesium to Silicon ratio measured by MESSENGER (yellow dots) and indicate that Mercury's surface is much different from the lunar highlands and Lunar soils (green squares) Credit: NASA/The Johns Hopkins University Applied Physics Laboratory/Carnegie Institution of Washington

Although all observations indicate a very low iron oxide content on the Mercurian surface, the question remains whether the interior is similarly depleted. The composition of volcanic deposits can be used to probe composition of the planetary interior. Many smooth plains on Mercury origin from lava flows. Smooth plains in those regions have lower concentration of opaque minerals than surrounding highlands. This suggests that the smooth plains are partial melts of the interior samples.

High-resolution images of Mercury's surface from orbit reveal that many bright deposits within impact craters exhibit fresh-appearing, irregular, shallow, rimless depressions. The depressions, or 'hollows' (Fig. 3.4), range from tens of meters to a few kilometers across, and many have high-reflectance interiors and halos (Blewett et al. 2011).

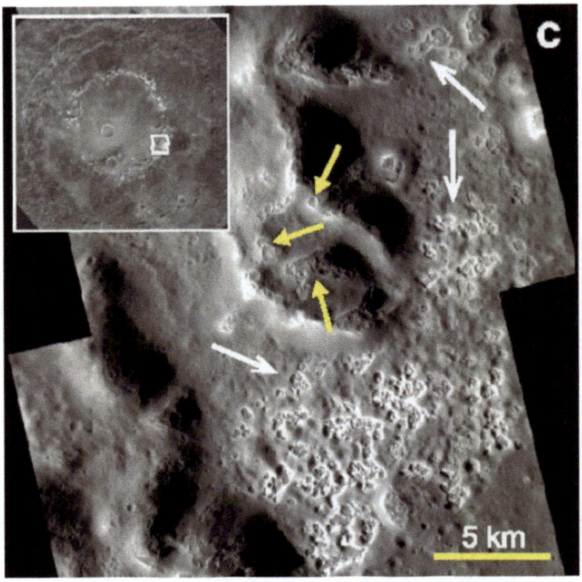

Fig. 3.4 Hollows on Mercury. The features are characterized by irregular rimless depressions and associated high-reflectance materials. The image shows a portion of the peak ring of the 265-km-diameter Raditladi basin (inset). White arrows indicate hollows on the basin floor; yellow arrows show those on the peak mountains. Images EN0220979987M and -9993 M, 17 m/pixel taken by MESSENGER dual imaging system (MDIS) and taken from Blewett et al. 2011)

'Hollows' appear to form by the loss of a moderately-volatile substance from the planet's surface and it is believed that this process has continued until relatively recently and may be on-going. Hypotheses to explain the volatile-loss have included sublimation and space weathering, and it has been suggested that hollow-forming volatiles are endogenic and are exposed at the surface during impact cratering. However, hollows occur at the surface of thick flood lavas where a lower-reflectance substrate has been exhumed from beneath them (Thomas et al. 2014).

3.5 Mercury Environment

Mercury is the smallest planet with an internal magnetic field. Being closest to the Sun and lacking a collisional atmosphere, its surface experiences the most extreme temperature variations when compared with the other planets. One of the mainstream research trends in magnetospheric physics today is directed towards the fundamental understanding of the physical processes in the Universe. Here, Mercury is a key natural laboratory for cosmic plasma physics research. Mercury is also a key laboratory for surface bounded exospheres. Surface bounded exospheres are seen throughout the solar system (e.g., Moon (Earth system), Galilean satellites of Jupiter) and would be ubiquitous in the systems of exo-planets.

3.5.1 Mercury's Exosphere

The atmosphere of Mercury is very tenuous, with a pressure of a fraction of pico-bar, and is non-collisional so that atmospheric neutral particles move on ballistic paths. Therefore the atmosphere is called an exosphere. From Mariner 10 UV and Earth- based optical spectroscopy measurements, six elements have been identified: Calcium (Ca), Sodium (Na), Potassium (K), Hydrogen (H), Helium (He) and Oxygen (O). Measurements of MESSENGER confirmed these findings (Solomon et al. 2008; Zurbuchen et al. 2008). Other species are expected such as H_2, OH (possibly released by impacting bodies) and noble gases (both non-radiogenic Neon (Ne) and radiogenic Argon and Xenon (^{40}Ar, ^{129}Xe)).

The dynamics of Mercury's exosphere are complex due to the interplay of several factors, such as solar wind, solar radiation, atmosphere, magnetosphere, planetary magnetic field and the rocky surface of the planet. This leads to high temporal and spatial variations in the exosphere, characterized by global asymmetries between day and night side and as well as between the northern and southern hemispheres. Although the source of the exosphere lies in Mercury's surface layers, the drivers are varying solar wind and magnetospheric activity, and especially vast coronal mass ejections from the Sun. Many details about the dynamics, structure, and composition of the Hermean exosphere can be found in Killen et al. (2007) and Milillo et al. (2010).

There are many source processes acting on Mercury's surface that can cause the ejection of neutrals and ions from the surface, such as thermal desorption (TD), photon stimulated desorption (PSD), ion sputtering (either solar wind or magnetospheric ions) and micrometeoroid impact vaporization (Killen and Ip 1999). The estimates of yields for various source processes are made difficult by our limited knowledge of the regolith and its absorption processes. As neutrals are on ballistic orbits, one expects that on time scales of a day, atmospheric species are absorbed by the surface during night and released at sunrise or during energetic particle precipitation. In addition due to the irregular nature of the surface material, absorption/desorption may also happen on longer time scales than expected. Impacting meteorites can provide an additional source for the atmosphere. Volatiles continuously diffusing from the deep crust are another additional source.

3.5.2 Mercury's Magnetosphere

The physics of the Hermean magnetosphere has been considered based on what is known about Earth's magnetosphere (e.g., Slavin 2004; Baumjohann et al. 2006; Fujimoto et al. 2007) and recent results from MESSENGER.

The overall size of the Hermean magnetosphere is not very large compared to the Larmor radius of a solar wind proton but seems to be large enough to contain its own dynamics. [The Larmor radius is the radius of the circular motion of a charged particle in the presence of a uniform magnetic field].

The two major factors that determine the shape and dimensions of a planet's magnetosphere are the ram pressure of the solar wind, which varies with distance from the Sun, and the magnitude of the planet's intrinsic magnetic field. Since Mercury is closer to the Sun and has a weaker magnetic field than Earth, one can easily infer that the size of its magnetosphere is smaller.

Absence of the atmosphere makes the planet's surface act as a source of plasma in the magnetosphere, and so the role of heavy ions can be substantial in the magnetospheric dynamics. The discovery of Mercury's planetary magnetic field (Ness et al. 1975) and confirmation by MESSENGER measurements in 2008 (Slavin et al. 2008) encouraged the magnetospheric physicists to start developing ideas in the new parameter regime that had not been visited.

As Mercury has only a tenuous atmosphere and no ionosphere one expects a weak planetary plasma population in the magnetosphere. This, however, may not necessarily be so. There is an argument claiming that contribution by the heavy ions of planetary origin can be rather significant, so significant in fact that the mass density in the inner-magnetosphere is larger than that in the solar wind (Fujimoto et al. 2007). The interaction of the solar wind with this heavily loaded magnetosphere could give rise to interesting magnetospheric dynamics.

3.6 Future Spacecraft Missions to Mercury

In 2017 the European Space Agency, ESA, will launch its BepiColombo mission. BepiColombo is an interdisciplinary mission to explore Mercury, and is a joint project between ESA and the Japanese Aerospace Exploration Agency (JAXA). From dedicated orbits two spacecraft, the JAXA provided Mercury Magnetosphere Orbiter (MMO) and the European Mercury Planetary Orbiter (MPO) will be studying the planet and its environment. The mission will address a comprehensive set of scientific questions to study the planet, its evolution and its surrounding environment. A suite of state-of-art scientific instruments, flying on the two spacecraft, allow a wide range of scientific questions to be addressed that will provide clues on the origin and formation of terrestrial planets and help to answer fundamental questions like: "How do Earth-like planets form and evolve in the Universe?" The MPO will focus on a global characterization of the planet itself, while the second spacecraft, the MMO, will study the environment around the planet including the planet's exosphere and magnetosphere. Upon arrival in the first half of 2024 after a cruise phase of about 7.5 years, the Solar Electric Propulsion Module will be jettisoned and chemical propulsion will be used to inject both spacecraft into their dedicated polar orbits. The MMO will be released first, after which an additional thrust phase will guide the MPO into its final orbit. Both orbits are elliptical with eccentricity and inclination optimized for the study of Mercury (MPO orbit: 480 × 1,500 km) and its magnetosphere (MMO orbit: 590 × 11,639 km). The base-lined lifetime of the MPO and MMO in Mercury orbit is one Earth year (about four Mercury years, or two Mercury solar days). A mission extension by another Earth year is optional.

The BepiColombo MPO (Fig. 3.5) accommodates eleven scientific instruments. It is designed to take scientific measurements in all parts of the orbit throughout the Mercury year, requiring that most of the apertures of the remote sensing instruments are continuously nadir pointing. As a consequence, five out of six spacecraft faces may be illuminated by the Sun at some point. This leaves only one spacecraft side for a radiator to dump excess heat into space and to avoid solar exposure of the radiator. The heat load is tremendous: at the perihelion sub-solar point the MPO gets 14 kW/m^2 from the Sun and 6 kW/m^2 from Mercury.

Fig. 3.5 The BepiColombo MPO and MMO in Mercury orbit (artist view, from Benkhoff et al. 2010)

The BepiColombo MMO (Fig. 3.5) is a spin-stabilized spacecraft. The MMO is optimized for in situ plasma and electromagnetic fields and waves measurements in Mercury orbit. The nominal spin rate is 15 rpm (spin period of 4 s) due to the scientific requirements. The spin axis is pointed nearly perpendicular to the Mercury orbital plane. The MMO contains four deployment units of electric probe antennas for Plasma Wave Instrument sensors (for more details on the BepiColombo mission see Benkhoff et al. 2010).

The mission has been named in honor of the Italian mathematician Giuseppe (Bepi) Colombo (1920–1984), who made many contributions to planetary research, celestial mechanics, including the development of new space flight concepts. He is well known for explaining why Mercury rotates three times about its axis while it completes two orbits around the Sun. He also pointed out to NASA how a modification to the trajectory of Mariner 10 would permit multiple flybys of Mercury rather than only the one originally planned

The dual spacecraft mission BepiColombo will provide a rare opportunity to collect multi-point measurements in a planetary environment. This will be particularly important at Mercury because of short temporal and spatial scales in the Mercury's environment. The orbits of MPO and MMO will be selected so as to allow close encounters of the two spacecraft throughout the mission and to study the planet and its environment in the most optimal way. This will allow gaining scientifically valuable information in an environment where both spatial and temporal scales can be very short. BepiColombo will follow up on MESSENGER results. It will provide high resolution, stereo coverage of the whole planet, very high accurate position measurements, and a comprehensive characterization of the planet's surface. In addition BepiColombo will study physical and chemical processes to better characterize the planet's environment. All this will hopefully reveal further secrets on the "sunshine planet" Mercury essential to gain further knowledge about the evolution history of our Solar System.

References

Anderson BJ, Acuna MH, Korth H, Purucker ME, Johnson CL, Slavin JA, Solomon SC, McNutt RL Jr (2008) The structure of Mercury's magnetic field from MESSENGER's first flyby. Science 321:82–85

Asphaug E, Reufer A (2014) Mercury and other iron-rich planetary bodies as relics of inefficient accretion. Nat Geosci 7:564–568

Balogh A, Grard R, Solomon SC, Schulz R, Langevin Y, Kasaba Y, Fujimoto M (2007) Missions to Mercury. Space Sci Rev 132:611–645

Baumjohann W, Matsuoka A, Glassmeier CH, Russell CT, Nagai T, Hoshino M, Nakagawa T, Balogh A, Slavin JA, Nakamura R, Magnes W (2006) The magnetosphere of Mercury and its solar wind environment: open issues and scientific questions. Adv Space Res 38:604

Benkhoff J, van Casteren J, Hayakawa H, Fujimoto M, Laakso H, Novara M, Ferri P, Middleton HR, Ziethe R (2010) BepiColombo—comprehensive exploration of Mercury: mission overview and science goals. Planet Space Sci 58:2–20

Benz W, Slattery WL, Cameron AGW (1988) Collisional stripping of Mercury's mantle. Icarus 74:516–528

Blewett DT, Chabot NL, Denevi BW, Ernst CM, Head JW, Izenberg NR, Murchie SL, Solomon SC, Nittler LR, McCoy TJ, Xiao Z, Baker DMH, Fassett CI, Braden SE, Oberst J, Scholten F, Preusker F, Hurwitz DM (2011) Hollows on Mercury: MESSENGER evidence for geologically recent volatile-related activity

Byrne PK, Klimczak Celengor AM, Solomon SC, Watters TR, Hauck SA (2014) Mercury's global contraction much greater than earlier estimates. Nat Geosci 7:301–307

Cameron AGW (1985) The partial volatilization of Mercury. Icarus 64:285–294

Cameron AGW, Benz W, Fegley B Jr, Slattery W (1988) The strange density of Mercury—theoretical considerations. Mercury (A89-43751:19-91). University of Arizona Press, Tucson, AZ, pp 692–708

Cintala MJ (1992) Impact-induced thermal effects in the lunar and Mercurian regoliths. J Geophys Res 97:947–973

Fegley B, Cameron AGW (1987) A vaporization model for iron/silicate fractionation in the Mercury protoplanet. Earth Planet Sci Lett 3–4(82):207–222

Fujimoto M, Baumjohann W, Kabin K, Nakamura R, Slavin JA, Terada N, Zelenyi L (2007) Hermean Magnetosphere-solar wind interaction. Space Sci Rev Issue 2–4(132):529–550

Goettel KA (1988) Present bounds on the bulk composition of Mercury: implications for planetary formation processes. In: Vilas F, Chapman CR, Matthews MS (eds) Mercury University of Arizona Press, Tucson, pp 613–621

Grard R, Balogh A (2001) Returns to Mercury: science and mission objectives. Planet Space Sci 49:1395–1407

Harmon JK, Perillat PJ, Slade MA (2001) High-resolution radar imaging of Mercury's North pole. Icarus 149:1–15

Head JW, Murchie SL, Prockter LM, Robinson MS, Solomon SC, Strom RG, Chapman CR, Watters TR, McClintock WE, Blewett DT, Gillis-Davis JJ (2008) Volcanism on Mercury: evidence from the first MESSENGER flyby. Science 321:69–72

Killen RM, Ip WH (1999) The surface-bounded atmosphere of Mercury and the Moon. Rev Geophys 37:361

Killen RM, Cremonese G, Lammer H, Orsini S, Potter AE, Sprague AL, Wurz P, Khodachenko M, Lichtenegger HIM, Milillo A, Mura A (2007) Processes that promote and deplete the exosphere of Mercury. Space Sci Rev 132:433–509

Lewis JS (1972) Metal/silicate fractionation in the solar system. Earth Planet Sci Lett 15:286–290

Lewis JS (1974) Chemistry of the planets. Ann Rev Phys Chem 24:339–351

McNutt RL, Solomon SC, Gold RE, Leary JC, the MESSENGER Team (2006) The MESSENGER mission to Mercury: development history and early mission status. Adv Space Res 38:564–571

Morgan JW, Anders E (1980) Chemical composition of Earth, Venus, and Mercury. Proc Natl Acad Sci 77:6973–6977

Milillo A, Fujimoto M, Kallio E, Kameda S, Leblanc F, Narita Y, Cremonese G, Laakso H, Laurenza M, Massetti S, McKenna-Lawlor S, Mura A, Nakamura R, Omura Y, Rothery DA, Seki K, Storini M, Wurz P, Baumjohann W, Bunce E, Kasaba Y, Helbert J, Sprague A and Hermean Environment WG (2010) The BepiColombo mission: an outstanding tool for investigating the Hermean environment. Planet Space Sci 58:40–60

Ness NF, Behannon KW, Lepping RP, Whang YC (1975) Magnetic field of Mercury confirmed. Nature 255:204

Slavin JA (2004) Mercury's magnetosphere. Adv Space Res 33:1859–1874

Slavin JA, Acuña MH, B. Anderson BJ, Baker DN, Benna M, Gloeckler G, Gold RE, Ho GC, Killen RM, Korth H, Krimigis SM, McNutt RM Jr, Nittler LR, Raines JM, Schriver D, Solomon SC, Starr RD, Trávnícek P, Zurbuchen TH (2008) Mercury's magnetosphere after MESSENGER's first flyby. Science 321:85–89

Solomon SC, McNutt RL, Gold RE, Domingue DL (2007) MESSENGER mission overview. Space Sci Rev 131:3–39

Solomon SC, McNutt RL Jr, Watters TR, Lawrence DJ, Feldman WC, Head JW, Krimigis SM, Murchie SL, Phillips RJ, Slavin JA, Zuber MT (2008) Return to Mercury: a global perspective on MESSENGER's first Mercury flyby. Science 321:59–62

Strom RG, Sprague AL (2003) Exploring Mercury: the iron planet. Springer, Verlag, Heiderberg, New York. ISBN 1852337311

Taylor SR, McLennan S (1985) The continental crust: its composition and evolution. Blackwell, Malden

Thomas RJ, Rothery DA, Conway SJ, Anand M (2014) Hollows on Mercury: materials and mechanisms involved in their formation. Icarus 229:221–235

Wasson JT (1988) The building stones of the planets. In: Vilas F, Chapman CR, Matthews MS (eds) Mercury. University of Arizona Press, Tucson, pp 622–650

Weidenschilling SJ (1978) Iron/silicate fractionation and the origin of Mercury. Icarus 35:99–111
Wetherill GW (1988) Accumulation of Mercury from planetesimals. In: Vilas F, Chapman CR, Matthews MS (eds) Mercury. University of Arizona Press, Tucson, pp 670–691
Wurz P (2012) Mysterious Mercury. SPATIUM published by the Association Pro ISSI No. 29
Zurbuchen TH, Raines JM, Gloeckler G, Krimigis SM, Slavin JA, Koehn PL, Killen RM, Sprague AL, McNutt RL Jr, Solomon SC (2008) MESSENGER observations of the composition of Mercury's ionized exosphere and plasma environment. Science 321:90–92

Chapter 4
Accessing the Venus Lower Atmosphere and Surface- from Venera and Pioneer Venus to VISE and VITaL

Michael Amato and David Williams

4.1 Introduction

Access to the chemical constituents, and dynamics of the atmosphere and surface of Venus has long been a primary goal and challenge of robotic exploration. After early missions and observations dashed the hopes that Venus was a habitable place, the challenge of understanding Venus has driven robotic exploration into the Venus atmosphere and towards the surface. A variety of science and technology goals resulted in a fascinating set of robotic mission designs and approaches. These past missions and the resulting data combine with new desires and science to drive the Venus mission technologies and missions of the future. Contemplation of humankind's further exploration of Venus requires us to look at the approaches of both past in situ missions and proposed future missions.

4.2 Why Venus and Why Enter the Venus Atmosphere?

Venus is the closest planet in location and size to Earth, understanding today's Venus informs us of the Venus of the past. This is important because Venus' history is linked to Earth's past and future. Comparing Earth to our closest neighbor and its sibling as it grew up, can help us understand our past and future. Explaining Venus' runaway greenhouse of today can help understand Earth and predict its future. Despite its closeness in size and distance to Earth, Venus is one of the most poorly understood planets. Is it active volcanically and/or tectonically? Where did the

M. Amato (✉) · D. Williams
NASA-Goddard Space Flight Center, Greenbelt, USA
e-mail: michael.j.amato@nasa.gov

D. Williams
e-mail: dave.williams@nasa.gov

V. Badescu and K. Zacny (eds.), *Inner Solar System*,
DOI 10.1007/978-3-319-19569-8_4

water go? What is the history of its surface and atmosphere? How does its surface interact with its atmosphere? Are some surface terrains older and/or more chemically evolved? These are all critical questions. Most of these questions require in situ missions that access the atmosphere and surface directly.

4.3 Why Are Venus in situ Missions Challenging

Entering any atmosphere from space requires an entry and descent system to allow a mission's experiments to survive the energy and heat encountered. Any entry and descent system that encounters Venus has to be designed carefully with margins that allow for variances and unknowns. Missions to Venus can challenge flight path design and flight dynamics designers. The trip or cruise from Earth to Venus requires careful consideration of entry system margins, payload designs and requirements. Other considerations include keeping any primary spacecraft close enough to access the data sent back from the entry probe or lander.

Entry - Heat shields are used to absorb the heat as one enters and slows down. Thermal protection material selection, material maturity, entry angles, dimensions, entry speeds, and test plans must be designed carefully to reduce risk. We don't know as much about the Venus atmosphere as we do Earth's or even Mars'. Release of the heat shield and deployment of parachutes involve proven but still complex systems that must work.

Descent - Once a missions' probe or lander enters the atmosphere, the decent speeds, stability, parachute deployment and performance are all impacted by the harshness and unknowns of the very environment the mission is there to learn about. The Venus atmosphere is dense and caustic. At lower levels it is characterized by extreme heat and crushing pressures (~460 C, 92 atmospheres). The super-critical fluid CO2, thick sulfuric acid clouds, extreme temperatures and pressures at lower altitudes challenge most thermal and mechanical designs for landers and probes. This limits survivability and drives mass and complexity up.

Near surface and surface – The extreme heat and pressure drive impressive pressure vessel, electrical, thermal protection and material designs. Even so, surviving more than an hour or two near the surface or at the surface is challenged by the physics of the tremendous heat. Returning the precious data requires careful control of the transmission system and either the position of the primary spacecraft above as it flies by or tenuous, power limited, lower bandwidth direct to Earth transmissions.

4.4 Venus in situ Missions of the Past

It is useful and necessary to consider the missions that have already gone to Venus and entered its atmosphere for perspective on what has been done. An impressive set of missions were attempted or accomplished in past decades. Each had its successes, failures and limitations. Some missions used a more brute-force design approach to the pressures and heat, as technologies of the day limited options. A top level description of each past Venus in situ mission gives needed perspective.

The Soviet Union made exploration of Venus, particularly exploration of the Venusian atmosphere and surface, high priorities of its space program. Because of the orbital configuration of Earth and Venus, optimal launch windows occur approximately every 19 months. The Soviet Union would typically build spacecraft in pairs to take advantage of each launch opportunity. This was not always obvious at the time because unsuccessful launches or missions that failed to leave Earth orbit for Venus were not given a "Venera" designation or publicly acknowledged as Venus spacecraft.

The first Venus flyby, Venera 1, occurred in 1961, although contact had been lost a few months earlier. In 1965, Venera 3 was launched with the objective of landing on the surface of Venus, but again contact was lost before the spacecraft reached its destination. It is believed that thermal control problems resulted in overheating of the spacecraft and failure of the communications system. Venera 3 ended up impacting the Venus surface on 1 March 1966 without returning any data, becoming the first probe to reach the surface of another planet.

For the next Venus launch opportunity, in June 1967, a pair of identical probes were built. These were larger, at 383 kg, basically spherical pressure vessels (see Fig. 4.1), and had improved thermal control. They were equipped with thermometers, barometers, descent radio altimeters, and gas analyzers. The first, Venera 4, was launched on 12 June and headed towards Venus as planned. The second probe was launched on 17 June, but the upper stage failed to fire and the spacecraft remained stranded in Earth orbit, and therefore was named Cosmos 167. Meanwhile, Venera 4 successfully completed a midcourse correction on 29 July and reached Venus on 18 October 1967. The descent probe was released from the spacecraft bus 45,000 km out and entered the nightside Venus atmosphere (200 km altitude) at 04:34 UT. Initial aerobraking involved an ablative heatshield. At about 52 km altitude, at a velocity of 1032 km/hr, a 2.2 square-meter drogue chute was deployed, followed by a 55 square-meter main chute. As the probe slowly descended through the atmosphere under the parachute, the temperature and pressure rose rapidly. After 93 min of descent, transmission ended. Presumably the sealed cabin broke open due to the high temperature and pressure, measured at 277 C and 20 bar. 23 sets of readings were returned during the descent, and initially there was some speculation that the probe had actually transmitted from the surface, but further study indicated it had failed at an altitude of about 22 to 27 km over the Eisila region.

Venera 4 was the first probe to transmit data from the atmosphere of another planet, and provided the greatest scientific return of any Soviet Venus mission up to that time, but the designers were determined to reach the surface of Venus with a working probe. With this in mind, the probe design for the next launch opportunity, January of 1969, involved a greater temperature and pressure tolerance with smaller parachutes to get it through the atmosphere faster. The probe mass was 405 kg, the parachute area was 15 m^2. Probe experiments included thermometers, a barometer,

Fig. 4.1 The Venera 4 descent probe (Courtesy Brian Harvey)

gas analyzers, an ionization densitometer, and photoelectric sensors. Again, a pair of identical probes were built. In this case both launched successfully to Venus, Venera 5 on 5 January 1969 and Venera 6 five days later. Venera 5 reached Venus on 16 May and was released from the bus 37,000 km from the planet. It entered the nightside atmosphere at 06:01 UT at 11.18 km/sec, deploying the parachute and initiating transmissions when the speed slowed to 210 m/s. Signals were returned from each instrument every 45 s over the next 53 min until the probe succumbed to the temperature and pressure, measured at over 320 C and 27 bar. The altitude was estimated to be roughly 16 km over the region east of Navka Planitia. The Venera 6 probe reached Venus a day later and was released from the bus 25,000 km out. It entered the nightside atmosphere at 06:05 UT on 17 May 1969, deploying its parachute and returning readings for 51 min before ceasing operations due to pressure and temperature effects, also east of Navka Planitia. The measured pressure was similar to that on Venera 5, but the estimated altitude was 10 to 12 km.

The August 1970 launch window saw further modifications to the probe design. To withstand higher pressures the pressure vessel was made from a single spherical shell of titanium, with no seams or welds. Even though the only instruments it carried were a thermometer and pressure meter, the probe was more massive, at 490 kg. It used an even smaller parachute (2.5 m^2), and was lined inside with shock-absorbing material to help survive the landing. This was clearly built as a lander to reach the surface of Venus. Two identical lander probes were built, the first, Venera 7, launching on 17 August 1970 and successfully leaving Earth parking orbit to Venus. The second probe, launched on 22 August, went into Earth parking orbit, but the escape stage failed during firing. The spacecraft was left in an elliptical geocentric orbit, and was designated Cosmos 359.

After two mid-course corrections, the Venera 7 probe was allowed to cool to −8 °C before atmospheric entry. It separated from the bus and entered the nightside atmosphere of Venus on 15 December 1970 at 04:58:44 UT. After aerodynamic braking, the top hatch was blown and the parachute system was deployed at about 60 km altitude. The capsule antenna was extended, and signal return commenced. Six minutes later the parachute ripped, and then collapsed, leaving the probe to fall towards the surface for another 29 min. The probe impacted the Venus surface at 05:34:10 UT at about 17 m/s and the signals weakened, reached full strength for about one second, and then seemingly ceased. Later analysis of the recorded radio signals revealed that the probe had survived the impact and continued transmitting a weak signal for another 23 min. It is believed that the spacecraft may have bounced upon impact and come to rest on its side, so the antenna was not pointed towards Earth. The pressure sensor had failed during the descent, but the temperature sensor showed a steady reading of 475 C at the surface. A pressure of 92 bar and a wind speed of 2.5 m/s were extrapolated from other measurements. The landing point was 5 °S, 351 °E. Venera 7 was the first spacecraft to return signals after landing on another planet.

A pair of spacecraft were built for the March 1972 launch window based on the success of the Venera 7 probe. Now that the surface pressure was known, a lighter lander could be built. (Venera 7 was designed to withstand up to 180 bar pressures.) With this savings in mass, more thermal protection and scientific instrumentation could be added. A stronger parachute was employed, but otherwise the design was the same. In addition to the temperature and pressure sensors, these landers carried an anemometer, photometer, gamma-ray spectrometer, gas analyzer, and altimeter. Venera 8 launched successfully on 27 March 1972, but its companion, launched into Earth parking orbit four days later, fell victim to a failure of the upper stage and was designated Cosmos 482.

Before reaching Venus the interior of Venera 8 was cooled to −15 °C. It separated from the bus on 22 July 1972 and entered the dayside atmosphere at 08:37 UT. Descent speed was reduced from 11 km/sec at entry to about 250 m/s at 67 km altitude by aerobraking. The parachute opened in reefed mode at an altitude of 60 km, and a refrigeration system was used to cool the interior components. Instruments were turned on at 50 km. At 30 km altitude the parachute was fully opened. A decrease in illumination was noted at 35 to 30 km altitude and wind speeds of less than 1 km/s were measured below 10 km. Venera 8 landed at 09:32 UT at 10 °S, 335 °E. It continued to send back data for 63 min after landing, including some first simple measurements of Venus' surface regolith, before failing due to the harsh surface conditions.

The June 1975 launch opportunity saw a pair of much larger and more robust Venera spacecraft headed to Venus (Fig. 4.2). Each mission included both an orbiter and lander. Experience with the dense Venus atmosphere convinced the designers that a parachute was not necessary to slow the descent through the lower atmosphere, so the only braking below 50 km altitude was provided by atmospheric drag on a titanium disc mounted near the top of the craft. The lander portion of the spacecraft had a mass of 660 kg and stood about 2 m high (see Fig. 4.6). The pressure vessel was designed to withstand temperatures up to 2000 °C. It was mounted by struts on a crushable toroidal landing platform and protected by an aluminum heat shield. A cylindrical tower on top held the helical antenna, which transmitted to the orbiter. A panoramic imaging system was mounted 90 cm above the base. The lander also carried a thermometer, barometer, anemometer, mass spectrometer, photometers, nephelometer, gamma-ray spectrometer, radiation densitometer, and accelerometers.

Fig. 4.2 The Venera 9 lander and orbiter (Courtesy Brian Harvey)

Venera 9 launched on 8 June 1975 and Venera 10 six days later. The Venera 9 lander separated from the orbiter on 20 October and entered the atmosphere on 22 October. After jettisoning the parachute at 50 km altitude, it landed at 05:13 UT at 31.7 N, 291 E near Beta Regio. The lander had been pre-cooled before landing to −10 C and was able to operate on the surface for 53 min. The imaging system could only complete a 180° panorama instead of the planned 360° image because one of the two covers failed to eject. These were the first close-ups of Venus and the first images returned from the surface of another planet. Venera 10 separated from the orbiter on 23 October, entered the atmosphere on 25 October, dropped the parachute at 49 km altitude, and landed at 02:17 UT at 16 N, 291 E, also near Beta Regio, but some 2,000 km south of the Venera 9 landing site. Venera 10 operated for at least 65 min after landing, until the orbiter was out of range to act as a relay. As with Venera 9, one of the covers failed to release, and only a 180° panorama could be taken. Both landers found pressures of about 90 bar, temperatures of about 465 C, and light levels comparable to Earth on an overcast day. The Venera 9 and 10 orbiters also carried a suite of instruments which returned data on the atmosphere and space environment (Fig. 4.3).

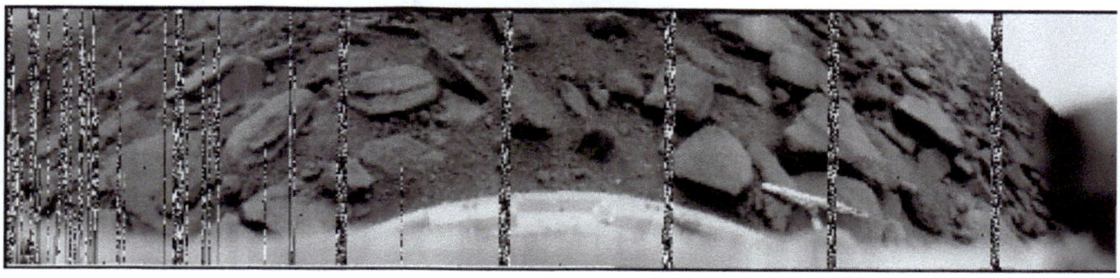

Fig. 4.3 First image of the surface of Venus, from Venera 9 on 22 October 1975 (Courtesy Don Mitchell)

The first U.S. missions targeted towards a detailed study of Venus were the Pioneer Venus missions launched at the 1978 opportunity. These were the Pioneer Venus Orbiter and the Pioneer Venus Multiprobe.

The Pioneer Venus Multiprobe mission comprised a bus, one large probe, and three small probes with a total mass of 875 kg. The Multiprobe Bus was a 2.5 meter diameter cylinder on top of which were mounted the Large Probe in the center and the three Small Probes spaced 120° apart around the outside. A solar array around the outside of the cylinder provided power. The bus itself had no heat shield or parachute and was only designed to survive entry into the upper atmosphere. However, a neutral mass spectrometer and an ion mass spectrometer on the bus afforded the only measurements of the upper atmosphere, as the other probes did not begin making measurements until lower altitudes. A picture of the spacecraft and probes is shown in Fig. 4.4.

The large probe comprised a spherical pressure vessel, a forward aeroshell heat shield, and an aft cover. The probe was 1.5 m in diameter and had a mass of 315 kg. The pressure vessel was built of three machined titanium parts: an aft hemisphere, a

Fig. 4.4 Pioneer Venus spacecraft and probes assembled

flat central ring section, and a forward cap. Each section was flanged and bolted together and sealed with O-rings and graphoil gaskets, which kept the internal nitrogen pressure during descent at 143 kPa. The interior of the shell was lined with a 2.5 cm thick Kapton blanket. The blunt cone-shaped aeroshell had an ablative carbon phenolic coating as a heat-shield. There were no thrusters on any of the probes, once released they were on ballistic trajectories.

The Large Probe pressure vessel carried a neutral mass spectrometer, a solar flux radiometer, an atmospheric structure experiment, a nephelometer, a cloud particle size spectrometer, a gas chromatograph, an infrared radiometer, and radio science experiments. None of the probes carried imaging devices.

The three small probes were identical to each other. Each consisted of a spherical pressure vessel surrounded by a permanently affixed forward conical heat shield and an afterbody. Each probe had a mass of 90 kg and a diameter of 0.8 m. The pressure vessel held all the scientific instruments and spacecraft systems. It was made of two flanged titanium hemispheres joined by bolts with O-rings and graphoil flat gaskets between. The vessel walls were lined on the inside with Kapton blankets and the interior was filled with xenon at a pressure of 102 kPa. The aeroshell was a 45° blunt cone made of titanium which used a bonded carbon phenolic ablative coating as a heat shield. The aeroshell was permanently attached to the pressure vessel, as was the fiberglass honeycomb afterbody. The small probes

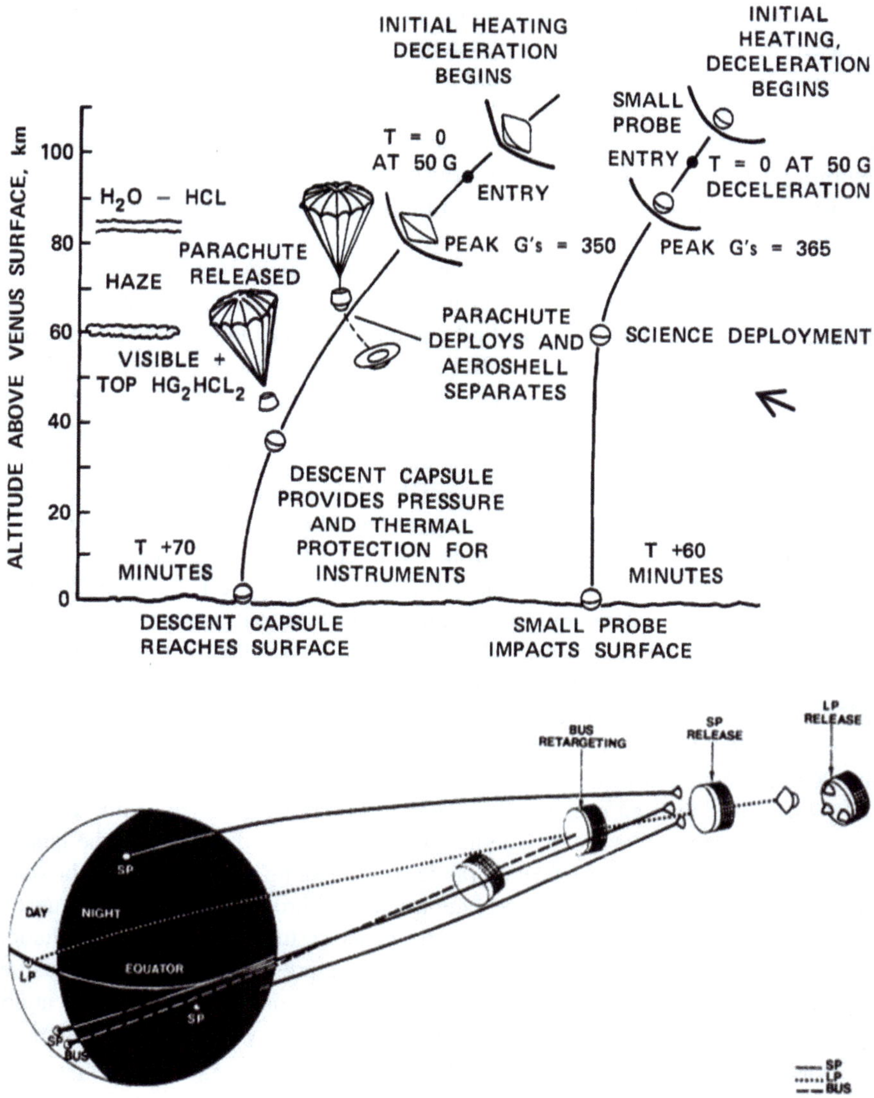

Fig. 4.5 Pioneer Venus large and small probe entry locations and descent profile – Ref – Pioneer Venus project Plan – Charles Hall et al., NASA ARC modified by M. Amato NASA GSFC

did not have parachutes. Each carried an atmospheric structure experiment, a nephelometer, a net-flux radiometer, and radio science experiments. The small probes were named based on their planned atmospheric entry points: The North Probe, Day Probe, and Night Probe. Figure 4.5 above shows the Pioneer Venus large and small probe entry locations and descent profile.

The Pioneer Venus Multiprobe mission launched on 8 August 1978. The large probe was released on 16 November 1978 by a pyrotechnic spring-separation system. On 20 November, while spinning at approximately 48 rpm, the probe bus released the small probes by opening clamps and letting them slide radially off the platform. The bus then used its forward thrusters to slow slightly behind the probes. The large probe entered the dayside Venus atmosphere on 9 December at 18:45 UT. The heat shield was jettisoned at 67 km altitude, at which time all instruments were returning data, and the parachute was released at 47 km. After a total descent time of just under an hour the probe impacted the surface at 19:40 UT at 4.4 N, 304 E, ending transmissions.

The North Probe entered the nightside atmosphere at 18:50 UT. After a 53 min descent, the probe touched down on the surface (59.3 N, 4.8 E) at 19:43 UT. Signals ended at this time. The Day Probe plunged into the atmosphere two minutes after the North Probe. After 56 min, the probe touched down on the surface (31.3 S, 317 E) at 19:48 UT. The probe survived the landing and continued to transmit for another 67 min. The Night Probe entered the atmosphere at 18:56 UT. Fifty-six minutes later the probe touched down on the surface (28.7 S, 56.7 E) at 19:52 UT. The probe transmitted for 2 more seconds after impact, then fell silent. The transmission rate of the small probes had to be reduced to 16 bits/s below about 30 km altitude due to interference from the thick atmosphere, but no data were lost. The probe bus reached the dayside atmosphere at 20:22 UT at 37.9 S, 290.9 E. It returned signals until reaching an altitude of 110 km one minute later. All five probes returned valuable information on the Venus atmosphere, returning day- and nightside temperature, pressure, and wind profiles and detailed information on the makeup and structure of the clouds.

For this 1978 launch opportunity, the successful landings of Venera 9 and 10 led the Soviet Union to use the same basic design, the main difference being use of a flyby bus rather than an orbiter as a communications relay. Once again, a pair of identical spacecraft were flown. The instrumentation carried on each of the landers was more ambitious, including a panoramic color imaging system. Among the other instruments on board was a surface sampler, a gas chromatograph to measure the composition of the Venus atmosphere, instruments to study scattered solar radiation and soil composition, a soil penetrator, temperature, pressure, and wind sensors, an accelerometer, and a device named Groza which was designed to measure atmospheric electrical discharges. Figure 4.6 shows the basic Venera lander design.

Venera 11 launched on 9 September 1978, Venera 12 followed on 14 September. Despite launching later, Venera 12 actually reached Venus first, falling through the atmosphere for one hour and landing on the surface at 03:30 UT on 21 December 1978. From the landing site at 7 S, 294 E, Venera 12 transmitted to the flyby platform for 110 min until it was out of range. Unfortunately, the lens cover failed to eject, so no images were returned. Additionally, the soil sample was not placed correctly into the chamber, so the planned analysis could not be done. Four days later the Venera 11 lander separated from its flight platform and, after a one hour descent through the atmosphere, landed (at 7–8 m/s) at 03:24 UT on 25 December. From the landing site at 14 S, 299 E, the lander transmitted for 95 min after touchdown, until the flyby relay flew out of range. Venera 11 had the same problems with the lens cover and soil sample as Venera 12, so no pictures or soil

Fig. 4.6 Basic Venera lander design showing (top to bottom) helical antenna, disk aerobrake, pressure vessel, landing platform. (Courtesy Brian Harvey)

analyses were returned. Results reported included evidence of lightning and thunder, sulfur and chlorine in the cloud layers, a high Ar36/Ar40 ratio (relative to Earth), and the discovery of carbon monoxide at low altitudes.

The same basic lander design, with improvements in high-temperature technologies and more sophisticated instruments, was flown at the next launch window. Venera 13 and 14 were again an identical pair of spacecraft, carrying instruments to take chemical and isotopic measurements, monitor the spectrum of scattered sunlight, and record electric discharges during the descent phase through the Venusian atmosphere. Each spacecraft utilized a camera system (with red, green, blue, and clear filters), an X-ray fluorescence spectrometer, a screw drill and surface sampler, a dynamic penetrometer, an acoustic detector, and a seismometer to conduct investigations on the surface.

Venera 13 launched on 30 October 1981 and reached Venus four months later, entering the atmosphere on 1 March 1982. The parachute was released at an altitude of 47 km and the disk aerobrake slowed the spacecraft the rest of the way through the thick Venus atmosphere. After a one hour descent, Venera 13 landed at 03:57 UT at 7.5 S, 303 E, just east of the eastern extension of the Phoebe Regio highlands. After landing the imaging panorama sequence was started and a mechanical drilling arm reached to the surface and obtained a sample, which was deposited in a

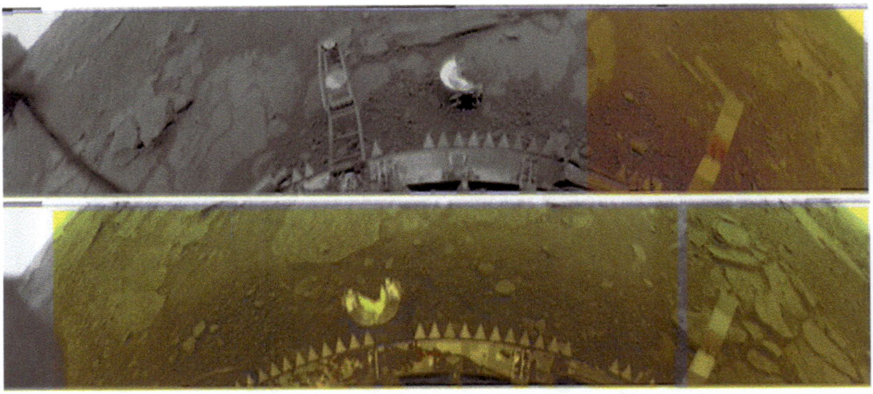

Fig. 4.7 Venera 13 panorama of the Venus surface taken 1 March 1982 (Courtesy Don Mitchell)

hermetically sealed chamber for analysis. The lander survived for 127 min, transmitting data to the bus, which acted as a data relay as it flew by Venus. Venera 13 returned the first color images of the surface of Venus, revealing an orange-brown flat bedrock surface covered with loose regolith and small flat thin angular rocks (Fig. 4.7). The composition of the sample determined by the X-ray flourescence spectrometer showed it was similar to terrestrial leucitic basalt. The acoustic detector returned the sounds of the spacecraft operations and the background 0.5 m/sec wind.

Venera 14 launched on 4 November 1981. An error in the initial midcourse correction necessitated two additional corrections before Venus was reached, but the corrections were successful and the lander entered the Venus atmosphere on 5 March 1982. With the same descent profile as Venera 13, Venera 14 landed at 07:00 UT near the eastern flank of Phoebe Regio at 13.3 S, 310 E. The imaging panorama was started and a sample collected and analyzed. The lander transmitted for 57 min. A panoramic image was completed (Fig. 4.8) and the sample was

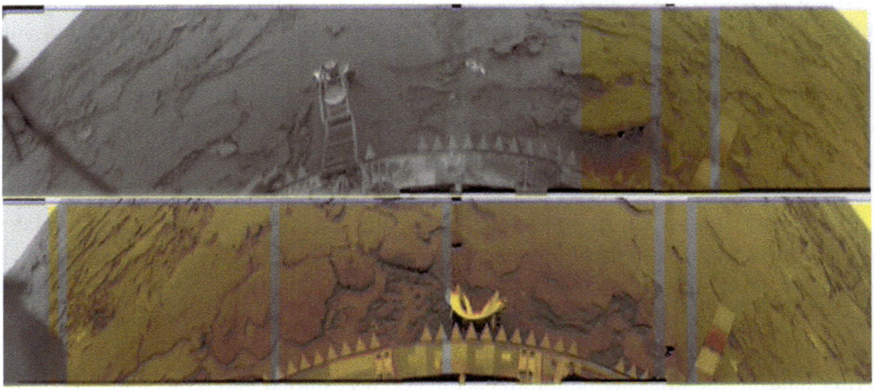

Fig. 4.8 Venera 14 panorama of the Venus surface taken 5 March 1982 (Courtesy Don Mitchell)

shown to be similar to terrestrial oceanic tholeiitic basalts. The seismometer detected two events, although they may have been associated with lander operations. The nephelometer identified three distinct cloud layers and the acoustic detector, as on Venera 13, returned sounds of the spacecraft operations and background wind.

A pair of orbiters, but no probes or landers, were launched at the next opportunity in 1983. For the December 1984 launch window a pair of possibly the most ambitious missions ever designed for Venus were launched. Vega 1 and 2 not only carried landers and balloon aerostats, but also had flyby buses that were targeted for Comet Halley after passing Venus. The landers were identical to the earlier Venera design. Figure 4.9 shows a Vega lander. In addition to atmospheric temperature and pressure measuring instruments, the lander carried a UV spectrometer, a hygrometer, an aerosol analyzer, a particle size spectrometer/nephelometer and other instruments for determination of the chemical composition of the condensed phase: a gas-phase chromatograph; an X-ray spectrometer; and a mass spectrograph. A drilling device, a gamma ray spectrometer and an X-ray fluorescence spectrometer were on the lander to study the surface composition. The UV spectrometer, mass spectrograph, and pressure- and temperature-measuring instruments were developed in cooperation with French investigators.

The Vega balloon probe comprised a 3.4 meter diameter balloon and a gondola, suspended below the balloon by a 13-meter long tether strap (see Fig. 4.10). The total mass of the deployed balloon probe was 21.5 kg. The balloon, gondola, parachute, ballast, tanks of helium, and timing electronics and pyrotechnic release devices with a total mass of 120 kg were stored in a toroidal compartment in the upper heat-protection aeroshell.

The balloon was made of a Teflon cloth matrix coated with Teflon film and filled with helium to 30 mbar overpressure. The diffusion of helium from the balloon was slow enough that the balloon would outlast the probe battery lifetime, losing less than 5 % of its helium and 500 meters of altitude. The balloon itself was transparent to the downlink radio frequency used. The gondola carried a nephelometer, a meteorology package, and a radio transmitter for data relay and tracking.

Vega 1 was launched on 15 December 1984 and Vega 2 on 21 December. Vega 1 reached Venus and entered the nightside atmosphere on 11 June 1985. After aerobraking, the upper aeroshell was released and the parachute was deployed at about 64 km altitude. The balloon package was pulled by parachute out of the aeroshell compartment and inflated at about 02:00 UT. The parachute and inflation system were jettisoned, and then the ballast was dropped at 50 km altitude. The balloon floated back up to its stable height between 53 and 54 km, in the middle layer of the Venus cloud system, with a pressure of 535 mbar and temperature of 27 to 42 C (300–315 K). The balloon drifted westward in the zonal wind with an average speed of about 69 m/s at a nearly constant latitude of about 8 N. The probe crossed the terminator from night to day at 12:20 UT on 12 June after traversing 8500 km. The probe continued to operate in the daytime until the final transmission was received at 00:38 UT on 13 June from 8.1 N, 68.8 E after a total traverse distance of 11,600 km over 46.5 h.

Fig. 4.9 Vega lander being lowered into the aeroshell capsule (Courtesy Brian Harvey)

After release of the aeroshell and balloon, the lander released the parachute at 47 km. At an altitude of 18 km a mechanical shock of unknown origin (possibly due to a jammed valve in an upper compartment suddenly releasing) triggered a ground-contact accelerometer which caused early deployment of the soil drill. The drill was rendered unusable at landing due to the premature deployment. The lander touched down at 03:03 UT on 11 June 1985 at 7.5 N, 177.7 E, just north of eastern Aphrodite Terra. It transmitted data back to the flyby relay for 56 min.

Fig. 4.10 The Vega balloon gondola (Courtesy Brian Harvey)

Vega 2 entered the nightside atmosphere on 15 June 1985 and ran through the same descent profile as Vega 1. The balloon was deployed and floated to its stable altitude between 53 and 54 km at about 02:00 UT. The balloon drifted westward in the zonal wind with an average speed of about 66 m/s at a nearly constant latitude of 7.5 S. The probe crossed the terminator from night to day at 9:10 UT on 16 June after traversing 7400 km. The probe continued to operate in the daytime until the batteries ran out. Final transmission was received at 00:38 UT on 17 June from 7.5 S, 76.3 E after a total traverse distance of 11,100 km over 46.5 h. It is not known how much further either balloon travelled after the final communication.

The Vega 2 lander touched down at 03:01 UT on 15 June 1985 at 6.5 S, 181.1 E, in eastern Aphrodite Terra. The measured pressure at the landing site was 91 atm and the temperature was 736 K. In this case, the drill worked successfully. The surface sample was found to be similar to terrestrial basalt, but also contained an anorthosite-troctolite, similar to lunar highland rocks. This was thought to be the oldest area sampled by the Venera landers.

Table 4.1 shows a summary of past Venus In situ missions as well as some future missions discussed.

Table 4.1 Table of past and potential future Venus in situ missions. (M. Amato, D. Williams, older data modified from Pioneer Venus, NASA SP-461, R.O. Fimmel, L. Colin, and E. Burgess, eds., 1983)

Name	Mission approach	Organization	Launch date	Venus entry or approach date	Landing site latitude	Landing site longitude	Landing site solar angle	Measurements
Venera 4	Descent module and flyby vehicle	USSR Lavochkin	12 Jun 1967	18 Oct 1967	19	38	−20	Temperature, pressure, density, wind velocity; CO2, N2, H2O content; ion number density in the ionosphere, magnetic field
Venera 5	Descent module and flyby vehicle	USSR Lavochkin	5 Jan 1969	16 May 1969	−3	18	−27	Temperature, pressure, wind velocity; CO2, N2, H2O content
Venera 6	Descent module and flyby vehicle	USSR Lavochkin	10 Jan 1969	17 May 1969	−5	23	−25	Same plasma measurements as Venera 4
Venera 7	Descent module (soft landing)	USSR Lavochkin	17 Jul 1970	15 Dec 1970	−5	351	−27	Temperature
Venera 8	Descent module (soft landing)	USSR Lavochkin	26 Mar 1972	22 Jul 1972	−10	335	+5	Temperature, pressure, solar scattered radiation, wind velocity
Venera 9	Descent module (soft landing) and orbiter	USSR Lavochkin	8 Jun 1975	22 Oct 1975	32	291	+54	Temperature, pressure, wind velocity; CO2, N2, H2O content, solar scattered radiation, clouds, panoramic survey of surface
Venera 10	Descent module (soft landing) and orbiter	USSR Lavochkin	14 Jun 1975	25 Oct 1975	16	291	+62	Same as Venera 9
Pioneer Venus 2	Descent modules	USA NASA	8 Aug 1978	9 Dec 1978	Multiple probes			Temperature, pressure, atmospheric structure and composition, clouds, solar flux
Venera 11	Descent module (soft landing) and flyby vehicle	USSR Lavochkin	9 Sep 1978	25 Dec 1978	−14	299	+73	Temperature, pressure, wind velocity, composition, solar scattered radiation spectrum, clouds, thunderstorm activity

(continued)

Table 4.1 (continued)

Name	Mission approach	Organization	Launch date	Venus entry or approach date	Landing site latitude	Landing site longitude	Landing site solar angle	Measurements
Venera 12	Descent module (soft landing) and flyby vehicle	USSR Lavochkin	12 Sep 1978	21 Dec 1978	−7	294	+70	Same as Venera 11; gas chromatograph, particle composition of clouds
VEGA 1	Descent module (soft landing), balloon, and flyby vehicle	USSR Lavochkin	15 Dec 1984	11 Jun 1985	8	178		Descent module: Temperature, pressure, atmospheric composition Balloon: Clouds, winds, meteorology
VEGA 2	Descent module (soft landing), balloon, and flyby vehicle	USSR Lavochkin	21 Dec 1984	15 Jun 1985	−8	76		Descent module: Temperature, pressure, atmospheric composition, surface composition Balloon: Clouds, winds, meteorology
Possible Future Missions								
VITaL	Rough terrain lander	USA/NASA	TBD	TBD	Ovda Regio	Ovda Regio	TBD	Elemental chemistry and mineralogy at the surface, images surface morphology and texture, in situ measurements of noble and trace gases in the atmosphere
VCM	Deployed balloon gondola, orbiter	USA/NASA	TBD	TBD	TBD – atmosphere above 55 km	TBD – atmosphere above 55 km	TBD	Atmospheric motions, radiation balance, cloud composition and chemistry, elemental and isotopic measurements
VIsE	TBD probe, lander orbiter, etc.	USA/NASA	TBD	TBD	TBD	TBD	TBD	NASA New Frontiers physics and chemistry of Venus' atmosphere and crust, characterizing detailed composition of the lower atmosphere, measuring the elemental and mineralogical composition of surface materials

(continued)

Table 4.1 (continued)

Name	Mission approach	Organization	Launch date	Venus entry or approach date	Landing site latitude	Landing site longitude	Landing site solar angle	Measurements
Venus Flagship Design Reference Mission	Orbiter, two balloons, two landers	USA/NASA	TBD	TBD	TBD	TBD	TBD	Extensive list covering many measurements with ~25 instruments between orbiter, balloons and landers covering three broad themes – what can the greenhouse tell us about climate change? How active is Venus? When and where did the water go?
VENERA-D	Lander, surface station, and orbiter	USSR	TBD	TBD	Possibly older tesserae	Possibly older tesserae	TBD	Structure and chemical composition of the atmosphere, nature of the super rotation, radiative balance, ionosphere, electrical activity, magnetosphere, escape rate. Surface composition, mineralogy, geochemistry, volcanic and seismic activity
VME	Mobile lander using inflatable metallic bellows	US/NASA	TBD	TBD	Venus highlands	Venus highlands	TBD	Surface composition and mineralogy in two locations, constraints on the origin of crustal material, the history of water, and the variability of the surface composition

4.5 The Future of Venus in Situ Missions

Many science questions remain after earlier in situ missions. The future of Venus in situ missions is driven by a large set of unanswered science questions that demand better instruments, access to new locations or longer access to the atmosphere and surface than past missions. The technology needed to address many of the highest priority questions exists. For the first time in decades there is a possibility the human race will return to the atmosphere of Venus. For example, medium sized, lower cost in situ Venus missions have been proposed in the past decade in the U.S. by NASA GSFC and NASA JPL. The European Space Agency has also received some proposals. Recent planetary decadal reports recommend a Venus in situ mission for the NASA mid-tier competed mission program. The medium or small sized missions can fit within more limited cost and mass parameters. Medium sized in situ missions of the type that can fit into competed opportunities have two top level approaches.

One approach is an entry system that delivers a then-to-be deployed balloon, glider airplane or other method of staying in the atmosphere for many days or even weeks. These missions avoid the complication of handling the worst of the environmental problems and have time to pursue desired science in the upper atmosphere. A balloon or airplane deployment adds a different set of technical and risk challenges such as the additional deployment and survival of the balloon and glider. Balloon missions attempt to access and study issues such as composition and dynamics of the upper atmosphere over longer time periods. The designs require entry and descent with a challenging deployment scenario. The solutions to these challenges have been studied. Mass spectrometers, IR instruments and other instruments measure noble gasses, chemistry and thermal environment over time until the batteries run out. An orbiter makes regular contact to download the data from the flying or floating platform. The environment at these altitudes, typically 50 to 70 km, can be less challenging.

Avoiding the challenges of the lower atmosphere, however, also avoids the critical science needed near the surface. So another approach is a deeper atmosphere probe or mini lander concept that gets to the less accessed lower atmosphere and approaches or lands on the surface. Access to the lower atmosphere adds environmental challenges which necessitates a measurement approach to enable all of the science to be achieved within limited survival times. Probe designs often leverage previous probe missions such as Pioneer Venus discussed earlier. This approach enables bringing better measurements to the hard-to-access near surface atmosphere driven by science questions regarding surface composition and the interaction zone between the surface and atmosphere. The trade for this approach that gives access to the highly desired science is the challenge of the thermal and pressure environment. Pressures are over 92 atm depending on surface elevation. The temperature exceeds 460 C (900 F). The thick atmosphere enables slower final descent speeds. The design solutions often involve simple but heavy designs. Some approaches for probes and short lived landers have already been prototyped and tested.

Examples of more capable and complicated missions are in the U.S. NASA decadal survey, and are driven by long term science needs at Venus. The National Research Council's Planetary Decadal Survey Inner Planets Panel commissioned three Venus in situ mission studies as part of the 2010 survey that led to the "Visions and Voyages for the decades from 2013–2022" (V&V). V&V lays out the science goals and possible mission approaches for the next decade. These mission's science, designs, and cost were studied and reviewed and they serve as an excellent measure of possible near term future missions from NASA to the Venus atmosphere.

4.6 Potential Future Venus in-situ Missions

The Venus Intrepid Tessera Lander (**VITaL**) was one of the three Venus in situ missions studied for the decadal panel. A science team and Goddard Space Flight Center (GSFC) led the mission concept study with substantial contributions from other NASA centers and organizations like JPL and LaRC. VITaL's mission concept achieves key surface chemistry and mineralogy measurements in a Venus tessera target area. The study baseline target is a key tessera region within Ovda Regio. From the report, VITaL would accomplish a landing in one of the mountainous tesserae regions of Venus on a NASA New Frontiers-like budget. 'The ability to characterize the surface composition and mineralogy within the unexplored Venus highlands would provide essential new constraints on the origin of crustal material and the history of water in Venus 'past'. VITaL also would provide 'new high spatial resolution images of the surface at visible and/or near infrared (NIR) wavelengths from three vantage points: on descent (nadir view), and two from the surface (panoramic view and contextual images of the linear surface chemistry survey)'. These data would provide insight into the processes that have contributed to the evolution of the surface of Venus. The science objectives are achieved by a flexible payload that measures elemental chemistry and mineralogy at the surface as well as images surface morphology and texture on descent and after landing. The instruments also make in situ measurements of noble and trace gases in the atmosphere, measure other attributes of the atmosphere and attempt to detect a potential crustal dipole magnetic field. Table 4.2 shows VITAL's top level science requirements.

The carrier spacecraft delivers VITaL to Venus after an initial Venus flyby, which helps meet landing condition requirements. After release from the carrier, the VITaL probe enters the atmosphere, descends on a parachute, and then free-falls to the surface, descending for almost an hour. The instruments take measurements on descent through the atmosphere and at the surface. The lander does science on the surface for an additional 2 h. VITaL transmits data to the flyby carrier spacecraft throughout the 3 h descent and surface operations.

Table 4.2 VITaL's top level science requirements – (VITaL fact sheet)

Mission Driving Science Objectives	Measurement	Instrument	Functional Requirement
Characterize chemistry and mineralogy of the surface	Major, trace elements, mineralogy, NIR spectroscopy	Raman/LIBS (1.0 micron) descent imager below 1 km, Raman/LIBS context camera	Access to tessera terrain >25 in situ sample measurements, sample context images
Place constraints on the size and temporal extent of a possible ocean in Venus's past	Measure D/H ratio in atmospheric water, mineralogy and major element chemistry of surface rocks	NMS; TLS; Raman/LIBS	In situ sampling of the upper and lower (<16 km) atmosphere. Access to and measurement of tessera terrain.
Characterize the morphology and relative stratigraphy of surface units.	Visible and NIR observations of multiple surface units at cm to m scale spatial resolution.	NIR (1.0 micron) descent imager and surface panoramic camera with ~5 filters from 550–1000 nm.	Position of cameras to image the surface, while accommodating expect slopes, platform stability from clear images.

The VITAL lander design accommodates typical kilometer scale slopes for many of the target tessera landing regions of <30° and local hazards such as ~1 m size boulders. The mission has a primary and back up design approach that enable landing in these potentially rougher areas. The baseline is a low center of gravity Ring Lander. The ring design is stable on lander scale slopes of up to 60°, and can handle an additional 12° of dynamic motion upon landing. The back up design is a cage lander design that uses gravity to orient the science payload after landing and can also flip. Both lander designs can be flown in an aeroshell and entry system that has heritage from past missions, keeping cost and risk down (see Fig. 4.11). Crushable materials and active hazard avoidance may also be utilized. The lander thermal design uses thermal insulation and phase change material to allow the lander electronics and instruments to work for at least 2 h at the Venus surface. All of the imaging and surface chemistry occurs in less than that time. The straw man lander payload includes a Raman/Laser Induced Breakdown Spectroscopy (LIBS) system, a mass spectrometer, and IR and visible cameras.

The Venus Climate Mission (**VCM**) is another of the three Venus in situ mission concepts developed for the NRC Planetary Decadal Survey. The mission's goal is to study the Venus climate within a New Frontiers cost range. From the report, 'The Venus Climate Mission (VCM) undertakes a thorough examination of the radiation balance, atmospheric motions, cloud physics, and atmospheric chemistry and composition of Venus. VCM will be the first ever truly 3-dimensional (and to a large extent 4-dimensional, including many measurements of temporal changes) characterization of the Venus atmosphere.' VCM will measure elemental and isotopic features to answer the origin and evolution of the atmosphere and extreme greenhouse climate questions. It will gather evidence for the existence, nature and timing of the suspected ancient radical global change from habitable, Earthlike

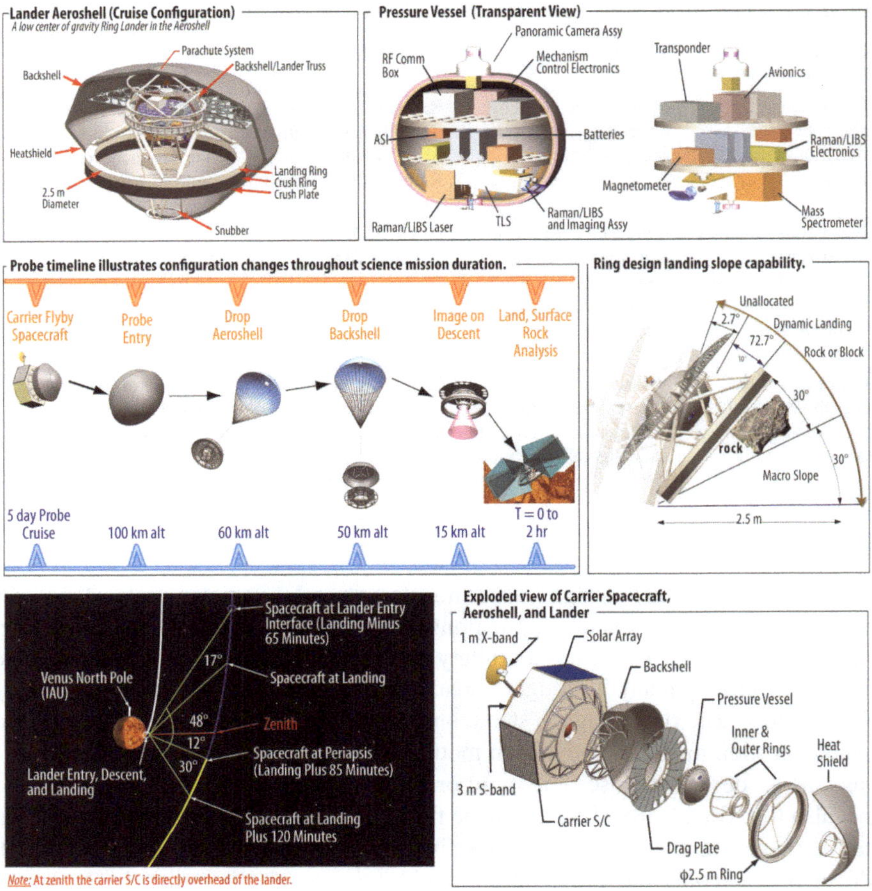

Fig. 4.11 VITaL designs – Lander, pressure vessel, timeline, assembly (VITaL report)

conditions to the current hostile runaway greenhouse climate, with important implications for understanding climate stability and predicting the long term fate of Earth's climate under a future warming Sun.'

Understanding of the current state and evolution of the strong CO_2 greenhouse climate on Venus should enable fundamental advances in our understanding of, and ability to model, climate and global change on Earth-like planets. VCM will answer many still open questions will make measurements helping answer issues on the origin, and history of the atmosphere and the resulting extreme greenhouse climate. Figure 4.12 shows the resulting carrier spacecraft and balloon system features.

VCM would cruise or carrier spacecraft delivers VCMs entry system to Venus about a year later depending on the launch opportunity. The carrier spacecraft performs a divert maneuver after entry system release then performs the Venus orbit insertion maneuver 2 h prior to the entry system's contact with the top of the

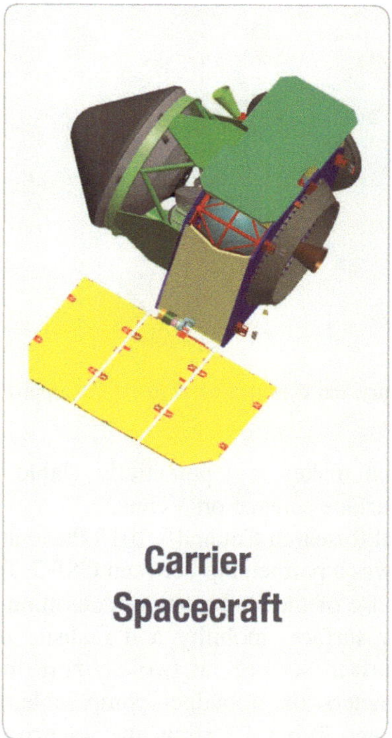

Gondola/Balloon System

Function: 21 day science campaign at 55.5 km float altitude
Power: Lithium-thionyl chloride ($Li\text{-}SOCl_2$) primary battery
Telecom: Two way S-band (plus Doppler) to Carrier Spacecraft; one way S-band from Mini-Probe and Drop Sondes
Science Data Return: 135 Mb from Gondola science + 7 Mb from Probe & Sondes science
Balloon Design: 8.1 m diameter helium filled balloon; teflon coated for sulfuric acid resistance; Vectran fabric plus Mylar film construction; metalized for low solar heating
Inflation System Design: 4 x 0.5 m dia. titanium tanks; pipes; valves

Fig. 4.12 VCM resulting carrier spacecraft and balloon system features (VCM report)

atmosphere (125 km). The entry system aeroshell is jettisoned and the parachute is deployed. The entry system contains the balloon and gondola system and the balloon inflation system. After slowing of the Gondola-Balloon System, balloon inflation and deployment occurs. During balloon inflation, the Mini-Probe is released. After inflation, the empty helium tanks are jettisoned and the Gondola-Balloon System rises to a 56 km nominal float altitude. Science measurements are conducted over a 20-day campaign as the Gondola/Balloon drifts. The two simple Drop Sondes are released from the Gondola-Balloon System during the 20 days and measure the pressure and temperature structure from the clouds towards the surface. The Drop Sondes are tracked by the Gondola-Balloon system to obtain deep atmosphere wind measurements. They will descend into the atmosphere collecting data and transmitting the data back to the Gondola which then transmits the data back to the orbiting spacecraft. Figure 4.13 shows VCM characteristics, layout and entry and deployment steps.

The Venus Mobile Explorer mission (**VME**) is an example of the lengths possibly needed to tackle one of the toughest problems in planetary science. It tries to do the difficult job of surviving longer periods of time on the surface and adds mobility in an attempt to achieve the most difficult Venus science. The VME design tries to solve these problems with a design that is achievable, has reasonable risk

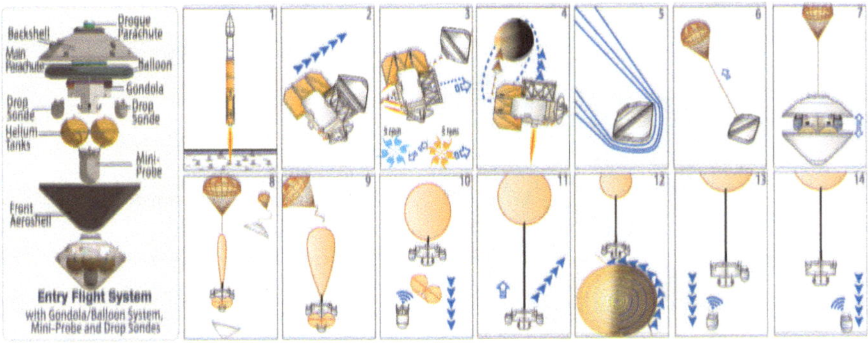

Fig. 4.13 VCM entry system layout and entry and deployment steps (VCM report)

and has enough detail to cost, which makes it a potentially viable long term approach to longer duration mobile surface science on Venus.

VME was also a part of the National Research Council's 2010 Planetary Decadal Survey Inner Planets Panel work and was a partnership between GSFC, JPL, LaRC, ARC and other institutions. The purpose of the study was to determine whether a Venus mission with surface, or near surface, mobility and realistic operational lifetime could achieve the needed surface science 'at two or more independent locations separated by several kilometers on a budget comparable to a New Frontiers cost envelope. Figure 4.14 and Table 4.3 show the science goal table. 'The VME mission concept affords unique science opportunities and vantage points not previously attainable at Venus. The ability to characterize the surface composition and mineralogy in two locations within the Venus highlands (or volcanic regions) will provide essential new constraints on the origin of crustal material, the history of water in Venus' past, and the variability of the surface composition within the unexplored Venusian highlands.' The mission concept will provide new high resolution views of the surface, along a multi-km profile, in the near infrared allowing an unprecedented look at the past evolution of the surface.

As with the future missions described above, VME is designed to be launched on an Atlas V 551 or equivalent. The carrier spacecraft carries the VME entry system and probe to Venus. After release from the carrier, the VME probe enters the atmosphere, deploys and descends on a parachute, and then free-falls to the surface. Atmospheric science is conducted on the descent. The probe then lands at the first site. As the science data is done at the first site, the metallic bellows are filled with helium. Once filled, the probe rises, leaving the helium pressure tank behind. To achieve the desired science, the VME surface package then floats to a new landing site. Driven by the ambient winds, the probe floats with the bellows close to the surface for 3 to 4 h, taking science measurements and images as it traverses over the surface. The probe will travel 8–16 km. At this point the metallic bellows are released and the probe free falls back to the surface at a second surface site, where the surface science measurements are performed again. The total mission time in the Venus atmosphere is 6 h, including descent, landed operations and traverse.

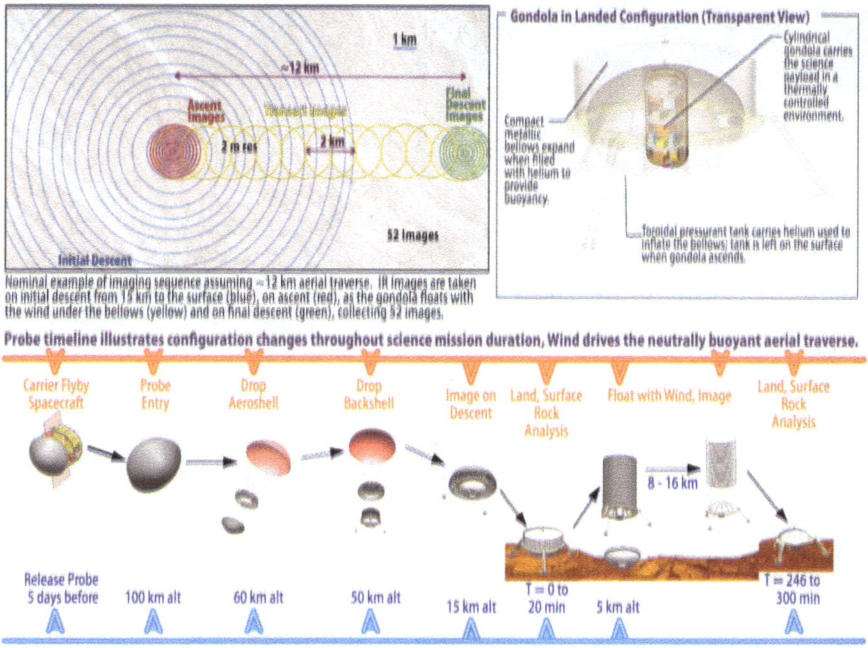

Fig. 4.14 VME design characteristics. The Gondola in landed position, probe timeline, imaging (VME report)

Table 4.3 VME science Goals (VME report)

Mission Driving Science Objectives	Measurement	Instrument	Functional Requirement
Determine the origin and evolution of the Venus atmosphere, and rates of exchange of key chemical species between the surface and atmosphere	In situ measurements of Noble gas isotopes, trace gas mixing ratios and trace gas isotopic ratios	Neutral Mass Spectrometer (NMS) combined with Tunable Laser Spectrometer (TSL)	In situ sampling of the atmosphere as functions of altitude and time
Characterize fundamental geologic units in terms of major rock forming elements, minerals in which those elements are sited, and isotopes	Identify mineralogy and elemental chemistry of surface rocks in 2 locations separated by >8 km	Laser Raman/Laser Induced Breakdown Spectrometer (LIBS)	Land in 2 locations, ~2 m path-length for compositional observation; stable platform for measurement duration
Characterize the geomorphology and relative stratigraphy of major surface units	Airborne near IR imaging along a transect ~8 km in length, at <5 m spatial resolution	Near infrared (~1.1 micron) imager (FOV TBD, and SNR > 100)	Near-surface aerial mobility; >45° solar incidence, contiguous images of the surface during aerial traverse; 5 h near surface operational lifetime

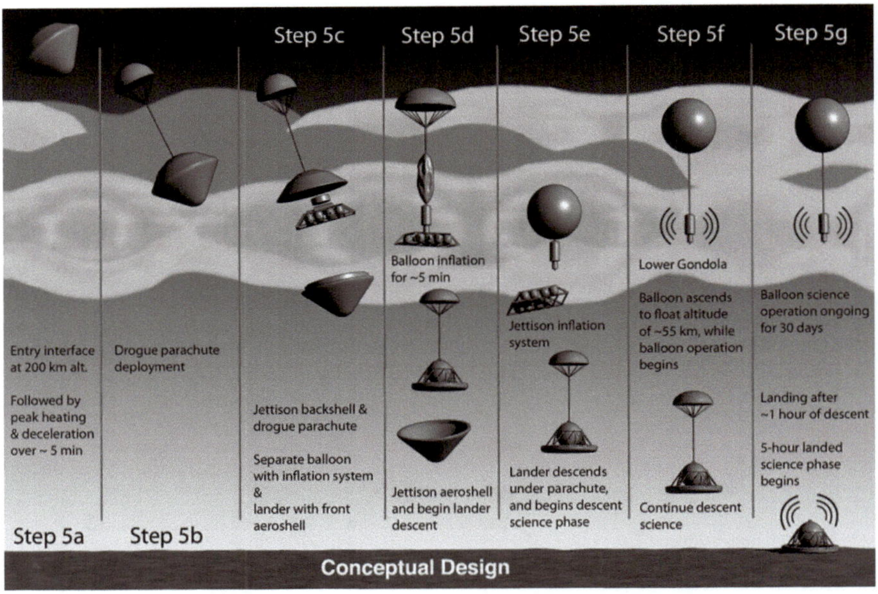

Fig. 4.15 Venus Flagship Design Reference Mission (**VFDRM**) in situ elements summary. (Venus Flagship Mission Study, April 2009, NASA, J. Hall, M. Bullock, et al.)

The VME probe transmits data to the flyby carrier spacecraft continuously during this time. Figure 4.15 shows VME basic design characteristics.

The science payload conducts in situ measurements of noble and trace gases in the atmosphere, and measures elemental chemistry and mineralogy at two surface locations. The instruments image the surface during descent and along surface during traverse. During this path between the two landing sites, the instruments also measure the atmosphere, and look for the crustal dipole magnetic field. The design is a ‘cylindrical probe/gondola that houses the science payload in a thermally controlled environment. The lower volume design ‘surrounds the gondola with a toroidal pressure tank capped with the bellows, enabling the entire lander system to fit in an aeroshell with heritage geometry.’ The thermal approach uses specialized heat pipes and phase change material to allow the probe to operate over 5 h at and near the Venus surface.

In 2008 and 2009 NASA studied a mission with a team that included members from U.S., France, Germany, Japan, the Netherlands and Russia. The mission was driven by Venus science goals and resulted in the Venus Flagship Reference Mission (**VFDRM**). The mission was a NASA flagship mission with cost estimate in the $2.7B to $4B range at the time. The mission attacked three major science themes. What does the Venusian greenhouse tell us about climate change? How active is Venus? When and where did the water go? The mission design has two launches for two spacecraft. The first spacecraft delivers two landers and the two balloons to Venus on a Type-IV trajectory. The second spacecraft would be an orbiter on a Type-II trajectory to Venus. The orbiter would arrive at Venus first, with the landers and balloons arriving 3.5 months later. The orbiters first duty was to be the collect the data from the landers and balloons as they did their in situ measurements. Once the landers and balloons missions were complete, the orbiter would begin a 2 year orbiting science phase with its remote observation instrument suite. The straw man or notional instrument suite was extensive for such a large and ambitious mission, The orbiter has seven instruments included a radar, magnetometer and mass spectrometer. The balloons have five instruments including a gas chromatograph/mass spectrometer and visible/near infrared (IR) camera. The two landers each have different instrument suites that include atmospheric structure instrument, gas chromatograph/mass spectrometer, drill, net flux radiometer, magnetometer and visible near IR cameras.

The two balloons and instrument suites would operate for up to 1 month at an altitude of ~55 km, attempting to circumnavigate the planet several times, moving from mid-latitudes towards the polar Regions. The two landers and their instrument suites would last up to 5 h after a 1 h descent, and likely require some high temperature components, a robust thermal system design, and sample systems advancements. A summary of the in situ elements is shown in Fig. 4.15.

In the United States., the NASA New Frontiers program, which compete mid-size planetary missions, has a list of allowed high priority targets that include the Venus in Situ Explorer (**ViSE**). Teams and organizations can design the mission they think does the best job at attacking many of the goals listed for that mission. The primary goals include: studying the physics and chemistry of Venus' atmosphere and crust, and characterizing variables that cannot be measured from orbit, including the detailed composition of the lower atmosphere, and measuring the elemental and mineralogical composition of surface materials. Versions of the mission architecture have been proposed to the program and might consist of a probe or lander that acquires atmospheric measurements during descent and at the surface, as well as other elements.

The Russian Space Agency along with some partners are in the preliminary planning stages of a potential future Venus in situ mission, currently called **Venera-D**. The very early mission concepts are ambitious and include a lander, orbiter and a station on the surface. The goals of the mission concentrate on multiple areas. The mission hopes to measure parts of the structure and chemical composition of the atmosphere, which may include isotopic ratios. It hopes to investigate the nature of the super rotation, radiative balance and greenhouse drivers. In the upper atmosphere it hopes to investigate additional items like ionosphere, electrical activity, magnetosphere, and escape rate. The near surface goals include the study of surface composition, mineralogy, geochemistry, interaction between the surface and atmosphere, and search for volcanic and seismic activity; (Ref 1 - Venera D web site and paper). Figure 4.16 shows the spacecraft and mission design elements.

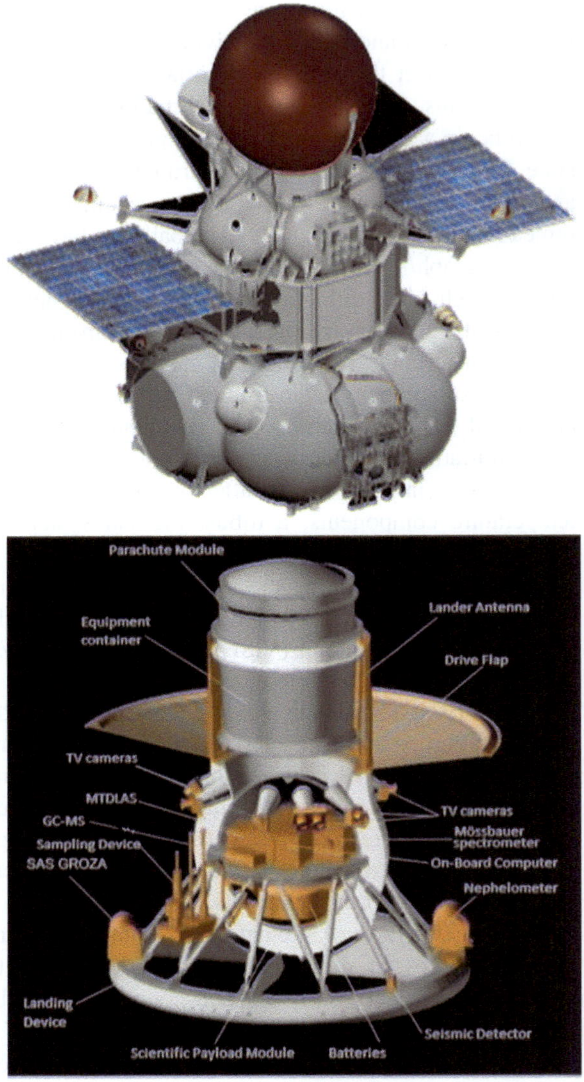

Fig. 4.16 Venera-D spacecraft, lander and instruments (Ref http://venera-d.cosmos.ru/index.php?id=658&L=2 – Russian Federal Space Program)

The large potential suite of potential instruments being considered might attempt to measure light and noble gases and their isotopes; composition of clouds; elemental and mineralogical composition of the surface; and radioactive elements. The envisioned set of possible instruments on the lander includes: gas chromatograph/mass spectrometer, nephelometer and the particle sizes spectrometer, optical package, active gamma spectrometer, Tunable Diode Laser Spectrometer, and camera suites. On the orbiter the envisioned instrument suite under consideration incudes several spectrometers, monitoring camera and a plasma package.

The mission design baselines a one day polar orbit for the orbiter. The lander might target tesserae, where material may still be found that pre-date the volcanic plains.

4.7 Summary

Resolving unanswered questions that require access to composition and dynamics of the atmosphere and surface of Venus has long been a primary goal and challenge of robotic exploration. A variety of science and technology goals and the resulting mission designs that have been applied to past missions. Early exploration were great for testing approaches still used as the basis for future mission concepts. Despite these missions, major questions still remain to complete the data sets begun by Venera, Vega and Pioneer Venus. Venus is the closest planet in location and size to Earth, understanding today's Venus informs us of the Venus and Earth of the past and future. We will continue to aspire to access the atmosphere and surface of Venus to answer these important questions. The variety of future in situ missions being developed on shows the possibilities. Some leverage past success, some attempt new and compelling design approaches and science. Venus in situ missions have challenges we are up to solving.

References

Introduction Section

2013–2022 Committee on the Planetary Science Decadal Survey Space Studies Board, Division on Engineering and Physical Sciences, NRC, *Vision and Voyages for Planetary Science in the Decadal* National Academies Press, Washington, D.C., 2011

Past Missions Section

Chertok B (2009) Rockets and people, NASA history series, vol 3, NASA-SP-2009-4110

Fimmel RO, Colin L, Burgess E (eds) (1983) Pioneer Venus, NASA-SP-461

Hall Charles et al Pioneer Venus project plan, NASA ARC modified by Michael Amato NASA GSFC

Harvey Brian (2007) Russian planetary exploration - history, development, legacy, and prospects. Springer/Praxis, Chichester

Hunten DM, Colin L, Donahue TM, Moroz VI (eds) 1983 Venus. University of Arizona Press, Tucson, AZ, USA

Siddiqi AA (2002) Deep space chronicle, monographs in aerospace history, No. 24, NASA SP-2002-4524

Images used with permission of Brian Harvey and Don Mitchell, copyrighted, all rights reserved

Future Missions Section

Dr. Martha S. Gilmore, Wesleyan University, Dr. Lori S. Glaze, NASA's Charles L. Baker, Michael Amato Gabriel Karpati, Michael Adams, NASA's Goddard Space Flight Center, George J. Tahu, NASA Headquarters, et al. - March 19, 2010 - VITAL Information, images and quotes from '*Venus Intrepid Tessara Lander Mission Concept Study Report to the NRC*

Decadal Survey Inner Planets Panel - Venus Intrepid Tessera Lander', NASA-GSFC • NASA-ARC, NASA LaRC

Dr. David Grinspoon Denver Museum of Nature and Science, Michael Amato, Michael Adams, NASA's Goddard Space Flight Center, Tibor Balint, NASA Jet Propulsion Laboratory, George J. Tahu, NASA Headquarters, et al. NASA GSFC, NASA JPL, NASA ARC, NASA LaRC- June 2010 - VCM information, quotes and images from *'Venus Climate Mission Mission Concept Study Report to the NRC Decadal Survey Inner Planets Panel'*

Dr. Lori S. Glaze, NASA's Charles L. Baker, Michael Amato, Gabriel Karpati, Michael Adams, NASA's Goddard Space Flight Center, George J. Tahu, NASA Headquarters, et al. NASA GSFC, NASA JPL, NASA ARC, NASA LaRC - December 19, 2009 - VME information, quotes and images from '*Venus Mobile Explorer' Mission Concept Report to the NRC Decadal Survey Inner Planets Panel'*

Venus Exploration Analysis Group (VEXAG) Venus Exploration Themes, May 2014

J. Hall, M. Bullock, et al, April 2009 - *NASA Venus Flagship Mission Study, April 2009,* -

VENERA-D information and images from VENERA –D web site – http://venera-d.cosmos.ru/index.php?id=658&L=2 - Russian Federal Space Program

Chapter 5
Special Orbits for Mercury Observation

Generoso Aliasi, Giovanni Mengali and Alessandro A. Quarta

5.1 Introduction

Planetary observations of Mercury started in the mid 70s with the Mariner 10 mission. The three flybys of Mariner 10 with Mercury in 1974 and 1975 discovered the existence of a strong Hermean magnetic field and an active magnetosphere surrounding the planet (Ness et al. 1974, 1975). These flybys were too brief to provide a comprehensive understanding of the physical mechanisms behind the structure and evolution of the magnetic field and the magnetosphere. However, the mission achievements had important consequences for understanding the origin, evolution, and composition of the planet, as they were used to develop scientific questions and proposals for future missions (Baumjohann et al. 2006; Slavin et al. 2007).

NASA's MESSENGER and ESA's BepiColombo missions originate from Mariner 10's achievements, and, while MESSENGER (launched in August 2004) is enriching the global knowledge of Mercury, BepiColombo (whose launch is planned in July 2016) will further enhance it with two Mercury orbiters, aimed at obtaining a comprehensive knowledge of various characteristics of the planet. Its mission objectives involve the investigation about the origin, evolution and composition of the closest planet to the Sun and the interaction among magnetosphere, exosphere, solar wind, and interplanetary magnetic field at Mercury's heliocentric orbit. However, the two spacecraft will only partially observe an important region

G. Aliasi (✉) · G. Mengali · A.A. Quarta
University of Pisa, Pisa, Italy
e-mail: g.aliasi@dia.unipi.it

G. Mengali
e-mail: g.mengali@ing.unipi.it

A.A. Quarta
e-mail: a.quarta@ing.unipi.it

V. Badescu and K. Zacny (eds.), *Inner Solar System*,
DOI 10.1007/978-3-319-19569-8_5

of the Hermean magnetosphere, i.e. its magnetotail, which represents a unique environment where theories about magnetospheric physics around rocky planets can be tested (Baumjohann et al. 2006). Therefore, the relevance of Mercury's magnetotail requires the definition of new scientific missions, specifically devoted to its observation and study (Circi 2004).

In this chapter an advanced mission concept is proposed, based on the use of Artificial Equilibrium Points (AEPs) in the elliptic three-body system, constituted by the Sun, Mercury, and the spacecraft, to provide a family of positions for the magnetotail's observation and measurement. An enhancement of such a mission concept is also suggested, which exploits the proximity of the spacecraft to the Sun for providing in situ measurements of the solar wind plasma and coronal mass ejection, observing the heliosphere, and verifying general relativity with improved accuracy.

The chapter is organized as follows. The spacecraft motion in the Sun-Mercury system is briefly discussed under the assumption that the propulsion system provides a radial continuous thrust with respect to the Sun. A dimensionless vectorial equation of motion is then derived that provides the location of the AEPs in a suitable reference frame. The stability and control of the AEPs is also discussed, to better characterize the dynamical behavior of a spacecraft in the neighborhood of the AEPs and to find a control strategy for maintaining a prescribed position. Finally, an observational mission inside Mercury's magnetotail and its improvement are proposed, by exploiting the advantages provided by the AEPs.

5.2 The Elliptic Restricted Sun-Mercury-Spacecraft Problem

The restricted three-body problem for a spacecraft in the Sun-Mercury system is the description of the motion of a massless point under the gravitational effects of both Sun and Mercury (of masses $m_{\odot}$ and m_M, respectively) and the acceleration provided by the spacecraft's propulsion system, if available. Because Mercury covers an elliptic orbit around the Sun, the problem is called elliptic, and is referred to as the Elliptic Restricted Three-Body Problem (ER3BP), see Fig. 5.1.

A growing amount of work about the ER3BP exists for spacecraft with continuous-thrust propulsion system. Biggs et al. (2009) discussed the existence of periodic orbits in the solar sail ER3BP, while Baoyin and McInnes (2006) investigated the generation of AEPs by means of solar sails. A comprehensive survey about this topic can be found in a review by McKay et al. (2011).

5.2.1 *Equations of Motion*

The motion of the (massless) spacecraft is conveniently described within a non-uniformly rotating (synodic) reference frame $\mathcal{T}(C; x, y, z)$, with origin at the

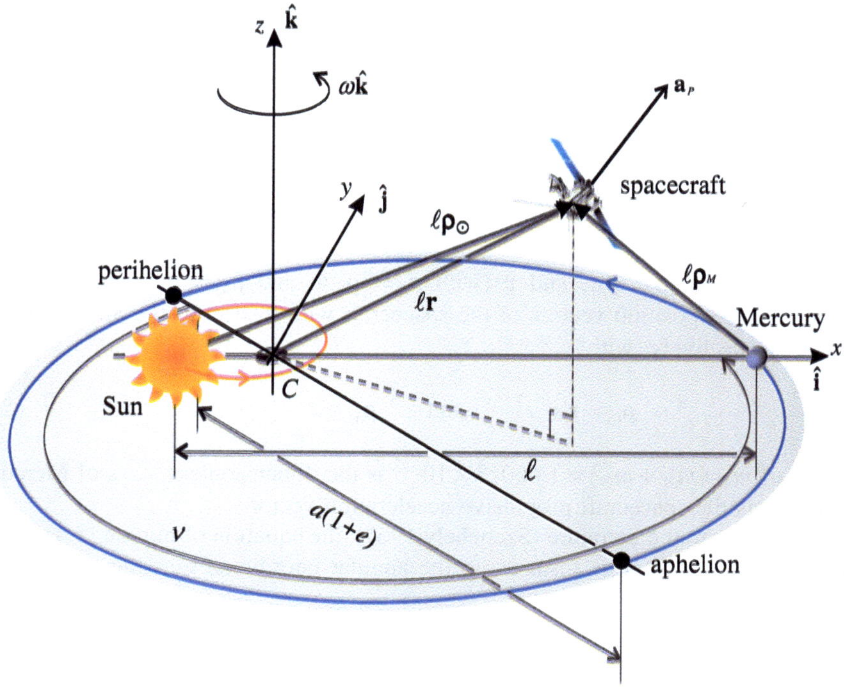

Fig. 5.1 Schematic view of the elliptic restricted three-body problem in the Sun-Mercury system

center-of-mass C of the system and with unit vectors $\hat{\mathbf{i}}$, $\hat{\mathbf{j}}$, and $\hat{\mathbf{k}}$. The z-axis of the synodic reference frame is in the direction of the angular momentum of the Sun-Mercury system, while the x-axis points toward Mercury at any time instant, and the y-axis defines a right-handed frame, see Fig. 5.1.

The time-dependent distance ℓ between Sun and Mercury is (Szebehely 1967):

$$\ell = a\left(1 - e^2\right) g \tag{5.1}$$

with

$$g \triangleq \frac{1}{1 + e\cos\nu} \tag{5.2}$$

where ν is the true anomaly, $a = 0.3871$ au is the semi-major axis and $e = 0.2056$ is the eccentricity of Mercury's heliocentric orbit. The angular velocity of the synodic frame with respect to an inertial reference frame is $\boldsymbol{\omega} = \omega\,\hat{\mathbf{k}}$, with

$$\omega = \frac{\sqrt{a\left(1 - e^2\right) G\left(m_\odot + m_M\right)}}{\ell^2} \tag{5.3}$$

where G is the universal gravitational constant.

The spacecraft vectorial equation of motion in the synodic reference frame is written as (Battin 1987)

$$\frac{\mathrm{d}^2(\ell \mathbf{r})}{\mathrm{d}t^2} + 2\boldsymbol{\omega} \times \frac{\mathrm{d}(\ell \mathbf{r})}{\mathrm{d}t} + \ell \frac{\mathrm{d}\boldsymbol{\omega}}{\mathrm{d}t} \times \mathbf{r} + \ell \boldsymbol{\omega} \times (\boldsymbol{\omega} \times \mathbf{r}) = -\frac{G m_\odot}{\ell^2 \rho_\odot^3} \boldsymbol{\rho}_\odot + \\ -\frac{G m_M}{\ell^2 \rho_M^3} \boldsymbol{\rho}_M + \mathbf{a}_P. \tag{5.4}$$

In Eq. (5.4), $\boldsymbol{\rho}_\odot$, $\boldsymbol{\rho}_M$, and $\mathbf{r}$ (with $\rho_\odot \triangleq \|\boldsymbol{\rho}_\odot\|$, and $\rho_M \triangleq \|\boldsymbol{\rho}_M\|$) represent the dimensionless position vectors of the spacecraft with respect to the Sun, Mercury, and C, respectively, with

$$\boldsymbol{\rho}_\odot = \mathbf{r} + \mu \hat{\mathbf{i}} \qquad \text{and} \qquad \boldsymbol{\rho}_M = \mathbf{r} - (1-\mu)\hat{\mathbf{i}} \tag{5.5}$$

where $\mu \triangleq m_M / (m_M + m_\odot) \approx 1.66013 \times 10^{-7}$ is the dimensionless mass of Mercury, while $\mathbf{a}_P$ is the spacecraft propulsive acceleration vector.

According to the literature (Szebehely 1967), the equation of motion (5.4) can be written in dimensionless form using the angular coordinate ν as the independent variable:

$$\mathbf{r}'' + 2\hat{\mathbf{k}} \times \mathbf{r}' = g\left[-\frac{1-\mu}{\rho_\odot^3}\boldsymbol{\rho}_\odot - \frac{\mu}{\rho_M^3}\boldsymbol{\rho}_M + \frac{\ell^2 \mathbf{a}_P}{G(m_\odot + m_P)} - \hat{\mathbf{k}} \times (\hat{\mathbf{k}} \times \mathbf{r}) + \\ - e\cos\nu (\mathbf{r} \cdot \hat{\mathbf{k}})\hat{\mathbf{k}}\right]. \tag{5.6}$$

It is now necessary to point out the meaning of the dimensionless vectorial Eq. (5.6). In fact, using the variable distance ℓ as the unit of distance, Eq. (5.6) describes the motion of the spacecraft in a suitable rotating coordinate system $\mathcal{T}_P(C; x/\ell, y/\ell, z/\ell)$, which is pulsating synchronously to the distance ℓ. In particular, within this pulsating coordinate system, the positions of the Sun and Mercury are fixed, whereas they actually oscillate (i.e., pulsate) along the x-axis when seen by an observer in the rotating reference frame $\mathcal{T}(C; x, y, z)$.

5.2.2 The Propulsive Acceleration Model

The spacecraft is assumed to have a generalized sail as its primary propulsion system. The generalized sail concept has been introduced by Aliasi et al. (2011, 2012) as a convenient mathematical model for describing the propulsive acceleration of a spacecraft whose propulsion system provides a continuous and purely radial thrust with respect to a reference celestial body, see Fig. 5.2.

The main characteristics of a generalized sail model can be summarized as follows. Assume that the magnitude of the propulsive acceleration $\mathbf{a}_P$ of the spacecraft can be written as the modulus of the product of two functions, the first

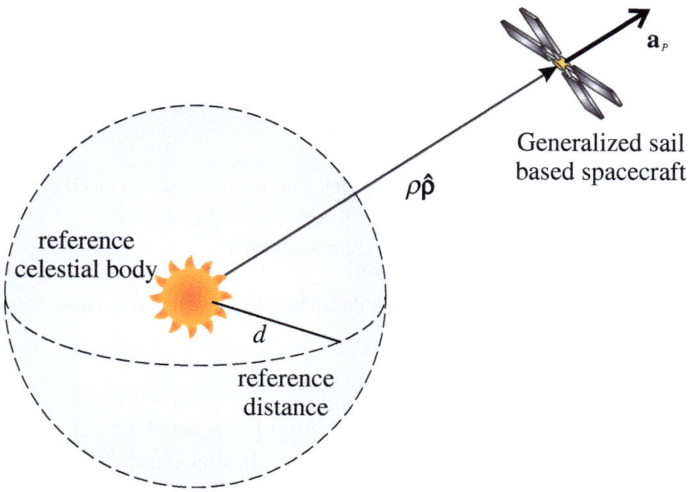

Fig. 5.2 Generalized sail concept

one, called a_c, being at most time-dependent, and the other varying with the distance ρ from a reference celestial body with a power law variation. The purely radial propulsive acceleration is therefore written as

$$\mathbf{a}_P = a_c \left(\frac{d}{\rho}\right)^{\eta} \hat{\boldsymbol{\rho}} \tag{5.7}$$

where d is a reference distance at which the so-called characteristic acceleration a_c is measured, η is a coefficient depending on the propulsion system type, while $\hat{\boldsymbol{\rho}}$ is the unit vector, oriented in the radial outward direction with respect to the celestial body.

The spacecraft characteristic acceleration a_c is a dimensional (scalar) parameter that quantifies the performance of the propulsion system. In this chapter, a_c may be negative for some type of propulsion system, thus indicating a propulsive acceleration directed toward the reference celestial body. When a mission analysis is performed within the Solar System, the reference distance is usually set equal to $d_\star = 1$ au, roughly corresponding to the Earth's heliocentric orbit semimajor axis. This is useful because the acceleration provided by the propulsion system is more easily measured in the proximity of the Earth. The characteristic acceleration varies with the distance d according to the following relationship:

$$a_c = a_c^\star \left(\frac{d_\star}{d}\right)^{\eta} \tag{5.8}$$

where $a_c^\star$ is the value of the characteristic acceleration at 1 au from the Sun.

Table 5.1 Summary of propulsion systems included in the generalized sail model

η	Propulsion system	Notes and references
0	M2P2 Electric thruster	Electric power independent of Sun-spacecraft distance; Winglee et al. (2000)
1	E-sail	Janhunen (2010)
4/3	MagSail	Plasma fluid model; Zubrin and Andrews (1991)
2	MagSail Electric thruster Solar sail	Particle model; Zubrin and Andrews (1991) Ideal solar array; Sauer (1978) McInnes (1999)
η	Electric thruster	Electric power depending on the solar array characteristics; Sauer (1978)

Any thruster capable of generating a propulsive acceleration in the form of Eq. (5.7) will be referred to as a generalized sail. In this sense, Eq. (5.7) models the performance of different types of propulsion systems, either currently available or under development, by simply modifying the value of the parameter η as is shown in Table 5.1.

For example, $\eta = 2$ corresponds to the acceleration provided by a solar sail (McInnes 1999), whereas the cases $\eta = 4/3$ or $\eta = 2$ (depending on the size of the magnetosphere surrounding the spacecraft) describe the acceleration of a magnetic sail (MagSail) based spacecraft (Zubrin and Andrews 1991; Quarta et al. 2013), while $\eta = 1$ corresponds to an electric solar wind sail (E-sail) propulsion system (Janhunen 2010). Furthermore, the case $\eta = 0$ is representative of a propulsive acceleration that is independent of the distance from the reference body (the Sun in this scenario). Such a situation is consistent with either a mini-magnetospheric plasma thruster (M2P2, see Winglee et al. 2000) or an electric thruster, when the power source is independent of the Sun-spacecraft distance, as, for example, in a nuclear powered system when the power subsystem is represented by a radio-isotope thermoelectric generator (Hunt 1993; Lyngvi et al. 2007).

An electric thruster can also be characterized by a coefficient $\eta \neq 0$. In fact, Jahn (1968) suggests that the propulsive acceleration of a spacecraft equipped with an electric thruster of constant specific impulse is proportional to the input power P of the power processing unit, with a coefficient of proportionality depending on time through the variable spacecraft mass and the variable thruster efficiency.

For a nuclear-powered spacecraft, the input power is roughly constant in both time and space, and Eq. (5.7) describes the propulsive acceleration of such a spacecraft with $\eta = 0$. In case of solar-powered spacecraft, when the electric power is supplied by photovoltaic arrays, the maximum input power P is a function of the distance from the Sun (Sauer 1978; Rayman and Williams 2002), but it also depends on the flight time due to solar cells degradation (Richardson and Warren 1971; Bourke and Sauer 1972). Under the assumption that the time degradation of solar cells is negligible, Sauer (1978) shows that the maximum input power P can be described using a rational function of ρ in an interplanetary (robotic) mission of a spacecraft with a solar electric propulsion system. To a first order approximation the maximum input power can be modeled as $P = P_r/\rho^\eta$, where P_r is a reference value that usually coincides with the maximum power generated at a given reference distance d from the Sun, while η depends on the numerical coefficients that model the solar cells performance variation with the Sun-spacecraft distance. In the ideal case, when the dependence of P on ρ is due only to the radial variation of the solar intensity, the exponent becomes $\eta = 2$, and an inverse square law of P with the Sun-spacecraft distance is obtained. In general, the value of η depends on the solar array characteristics and on the range of distances from the Sun. For example, assuming $\eta \approx 0.9$ in the range $\rho \in [0.5,\ 1.15]$ au or $\eta \approx 1.6$ when $\rho \in [0.9,\ 1.5]$ au, the power model approximates the results of Sauer (1978) with a percentage error less than 6 %.

Paralleling the concept of the solar sail lightness number β (McInnes 1999), which is usually used in the context of solar sail based mission design, it is now useful to introduce the modified lightness number B, defined as

$$B \triangleq \frac{a_c^\star\, d_\star^\eta\, \ell^{2-\eta}}{G m_\odot} \tag{5.9}$$

Using Eq. (5.7) with the Sun as the reference celestial body (i.e., $\rho \triangleq \ell \rho_\odot$) and $d^\star$ as the reference distance, the equation of motion (5.6) for a generalized sail-based spacecraft is written as

$$\mathbf{r}'' + 2\hat{\mathbf{k}} \times \mathbf{r}' = g\left[-\frac{1-\mu}{\rho_\odot^3}\boldsymbol{\rho}_\odot - \frac{\mu}{\rho_M^3}\boldsymbol{\rho}_M + B\frac{1-\mu}{\rho_\odot^{\eta+1}}\boldsymbol{\rho}_\odot - \hat{\mathbf{k}} \times (\hat{\mathbf{k}} \times \mathbf{r}) + \right. \\ \left. - e\cos\nu(\mathbf{r}\cdot\hat{\mathbf{k}})\hat{\mathbf{k}}\right]. \tag{5.10}$$

5.3 Artificial Equilibrium Points

The AEPs are the equilibrium solutions of Eq. (5.10) when $B \neq 0$, that is, when the propulsion system is switched on. By enforcing the stationary conditions (Szebehely 1967)

$$\mathbf{r}' = 0, \quad \mathbf{r}'' = 0 \tag{5.11}$$

the position of AEPs are found as the solutions of the vectorial equation

$$-\frac{1-\mu}{\rho_\odot^3}\boldsymbol{\rho}_\odot - \frac{\mu}{\rho_M^3}\boldsymbol{\rho}_M + B\frac{1-\mu}{\rho_\odot^{\eta+1}}\boldsymbol{\rho}_\odot - \hat{\mathbf{k}} \times (\hat{\mathbf{k}} \times \mathbf{r}) - e\cos\nu(\mathbf{r}\cdot\hat{\mathbf{k}})\hat{\mathbf{k}} = 0. \tag{5.12}$$

Note that Eq. (5.12) in the case $B=0$ (corresponding to a spacecraft without any propulsive acceleration) gives the locations of the five classical Lagrangian equilibrium points.

An analysis of Eq. (5.12) reveals that AEPs are possible only in the x-y plane (when $\mathbf{r}\cdot\hat{\mathbf{k}}=0$), even though an out-of-plane AEP may be maintained by means of a feedback control strategy (Baoyin and McInnes 2006) or by means of a low-thrust feed-forward control system (Macdonald et al. 2011). The same Eq. (5.12) also provides the locus of AEPs in the pulsating rotating reference frame $\mathcal{T}_P(C;\, x/\ell,\, y/\ell,\, z/\ell)$, see Fig. 5.3, as well as the constant value of B_0, i.e. the modified lightness number necessary to maintain the equilibrium position $\mathbf{r}_0$ for a given value of the parameter η (Aliasi et al. 2012).

The locus of AEPs, whose shape is independent of η, consists of two families of points, referred to as collinear and triangular points. The collinear points belong to the x-axis of the pulsating rotating frame and are divided into three sub-families: the L_1-type points between the Sun and Mercury, the L_2-type points in opposition to the Sun with respect to Mercury, and the L_3-type points in opposition to Mercury with respect to the Sun. The family of the triangular points is, instead, formed by the circle centered at Mercury and with radius equal to the Sun-Mercury distance in the pulsating reference frame. Note that the exact locations and the number of equilibrium points in the previous families depend on the values of both B and η.

Before proceeding further, it is worth noting that an AEP corresponds to a spacecraft position $\mathbf{r}_0$ fixed with respect to the pulsating rotating reference frame $\mathcal{T}_P(C;\, x/\ell,\, y/\ell,\, z/\ell)$. Such a position, however, when seen by an observer in the rotating frame $\mathcal{T}(C;\, x,\, y,\, z)$, oscillates back and forth along the segment described by the vector $\ell\mathbf{r}_0$. Moreover, Eq. (5.9) states that $a_c^\star$ must be varied periodically and synchronously to ℓ to maintain a constant value of B. Consequently, a propulsive acceleration modulation is required to maintain an equilibrium position, and the maximum value of $a_c^\star$ is

$$a_{c_{\max}}^\star = B_0\,\frac{G m_\odot}{d_\star^\eta}\,\max_{\nu}\left\{\ell^{\eta-2}\right\} \tag{5.13}$$

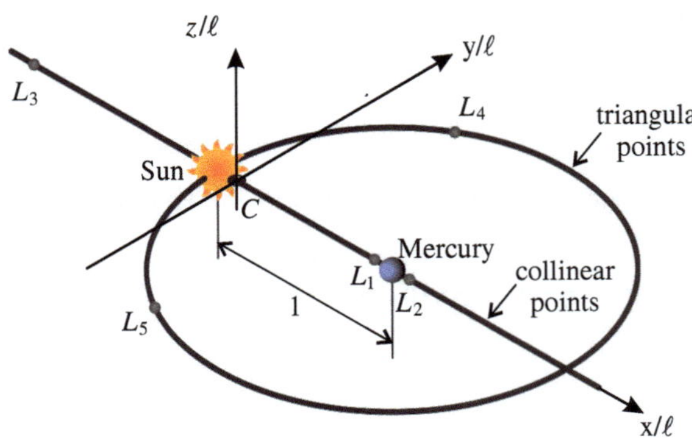

Fig. 5.3 Sketch of locus of artificial equilibrium points

with a maximum variation w.r.t. this value of

$$\frac{\Delta a_c^\star}{a_{c_{\max}}^\star} = 1 - \frac{\min_\nu\{\ell^{\eta-2}\}}{\max_\nu\{\ell^{\eta-2}\}} = 1 - \left[\frac{1-e}{1+e}\right]^{|2-\eta|} \approx 1 - 0.6589^{|2-\eta|} \tag{5.14}$$

where

$$\min_\nu\{\ell^{\eta-2}\} = \begin{cases} [a(1+e)]^{\eta-2} & \text{if} \quad \eta < 2 \\ 1 & \text{if} \quad \eta = 2 \\ [a(1-e)]^{\eta-2} & \text{if} \quad \eta > 2 \end{cases} \tag{5.15}$$

$$\max_\nu\{\ell^{\eta-2}\} = \begin{cases} [a(1-e)]^{\eta-2} & \text{if} \quad \eta < 2 \\ 1 & \text{if} \quad \eta = 2 \\ [a(1+e)]^{\eta-2} & \text{if} \quad \eta > 2 \end{cases} \tag{5.16}$$

Such a modulation represents a difficult task to achieve in order to maintain an AEP. Different modulation strategies can be used, depending on the propulsion system type. For example, in case of an E-sail based mission, a variation of the voltage of the tethers would permit the required modulation to be obtained. When an electric thruster is used, the modulation is given by the mass flow rate variation of the propellant. Finally, note that, for what concerns the ideal solar sail propulsion system case, no modulation is required as $\eta = 2$ in Eq. (5.14), whereas electrochromic material panels (Aliasi et al. 2013) could be used in case of a non ideal solar sail.

In the following a more comprehensive analysis of the two families of AEPs is given. However, only the case $\eta \leq 2$ will be considered, as only this case corresponds to current and near-future propulsion systems modeled by Eq. (5.7), see also Table 5.1.

5.3.1 Collinear Points

The collinear points are aligned with the x-axis, therefore their positions are identified by either $\rho_{\odot_0} = \rho_{\odot_0}(\hat{\rho}_{\odot_0} \cdot \hat{\mathbf{i}})\hat{\mathbf{i}}$ or $\boldsymbol{\rho}_{M_0} = \rho_{M_0}(\hat{\boldsymbol{\rho}}_{M_0} \cdot \hat{\mathbf{i}})\hat{\mathbf{i}}$, with

$$\rho_{M_0} = \left|\rho_{\odot_0}(\hat{\boldsymbol{\rho}}_{\odot_0} \cdot \hat{\mathbf{i}})\hat{\mathbf{i}} - 1\right| \tag{5.17}$$

After some algebraic manipulations, Eq. (5.12) gives the value of B_0 required to obtain the collinear AEPs at $\boldsymbol{\rho}_{\odot_0} = \rho_{\odot_0}(\hat{\boldsymbol{\rho}}_{\odot_0} \cdot \hat{\mathbf{i}})\hat{\mathbf{i}}$. The result is

$$B_0 = \frac{\mu \rho_{\odot_0}^\eta}{(1-\mu)\hat{\boldsymbol{\rho}}_{\odot_0} \cdot \hat{\mathbf{i}}} \left(1 + \frac{\rho_{\odot_0}\hat{\boldsymbol{\rho}}_{\odot_0} \cdot \hat{\mathbf{i}} - 1}{\|\rho_{\odot_0}\hat{\boldsymbol{\rho}}_{\odot_0} \cdot \hat{\mathbf{i}} - 1\|^3}\right) + \rho_{\odot_0}^{\eta-2} - \frac{\rho_{\odot_0}^{\eta+1}}{1-\mu} \tag{5.18}$$

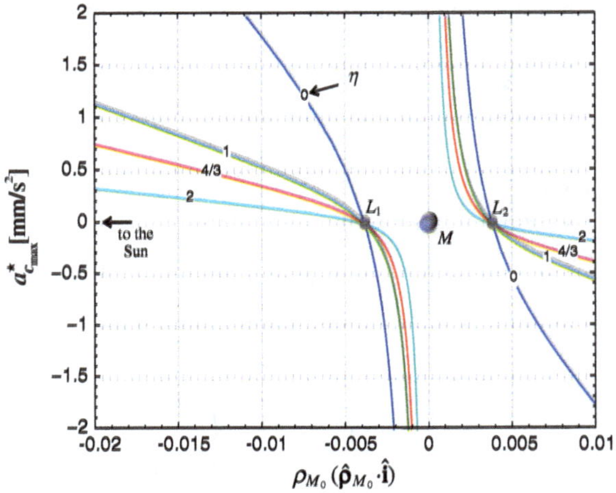

Fig. 5.4 Position of L_1-type and L_2-type AEPs in the Sun-Mercury system as a function of the maximum required characteristic acceleration

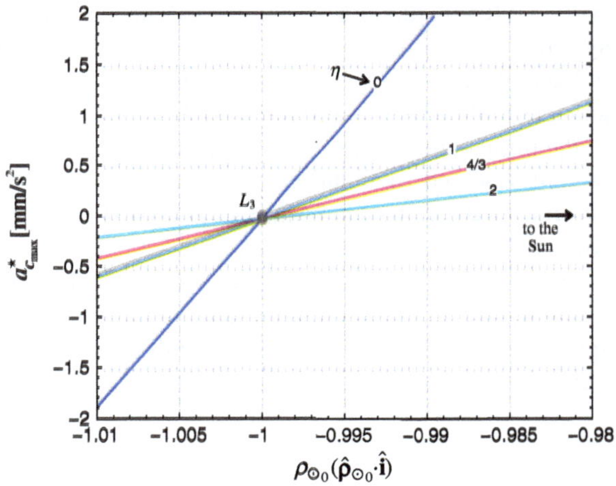

Fig. 5.5 Position of L_3-type AEPs in the Sun-Mercury system as a function of the maximum required characteristic acceleration

The latter relationship can be substituted into Eq. (5.13) to get the maximum characteristic acceleration required to generate the collinear AEP, as shown in Figs. 5.4 and 5.5.

In general, increasing (decreasing) the maximum characteristic acceleration moves the equilibrium points toward (away from) the Sun. Note that for a given value of the maximum characteristic acceleration, the displacement of an AEP with respect to the classical equilibrium points (corresponding to $a_c^\star \equiv 0$) increases with the value of η.

5.3.2 Triangular Points

The triangular points are located along the circle of unitary (dimensionless) radius centered at Mercury (i.e. $\rho_{M_0} \equiv 1$) and their positions are univocally identified by the distance $\rho_{\odot_0}$ from the Sun.

The value of B_0 required to maintain a triangular AEP at $\rho_{\odot_0}$ from the Sun is obtained from Eq. (5.12) as

$$B_0 = \rho_{\odot_0}^{\eta+1}\left(\frac{1}{\rho_{\odot_0}^3} - 1\right) \tag{5.19}$$

while the maximum characteristic acceleration is given by Eq. (5.13), see Fig. 5.6. Similar to what happens for the collinear points, an increasing characteristic acceleration moves the equilibrium points toward the Sun, and the corresponding displacement tends to increase with the value of η.

5.3.3 Linear Stability and Control of AEPs

The dynamical behavior of a spacecraft in the neighborhood of an AEP may be suitably described by linearizing the equation of motion (5.10). To that end, the transformations

$$\mathbf{r} = \mathbf{r}_0 + \delta\mathbf{r} \text{ and } B = B_0 + \delta B \tag{5.20}$$

are introduced into Eq. (5.10), where $\mathbf{r}_0$ and B_0 correspond to the equilibrium conditions defined by Eq. (5.12), $\delta\mathbf{r}$ represents a small displacement with respect to

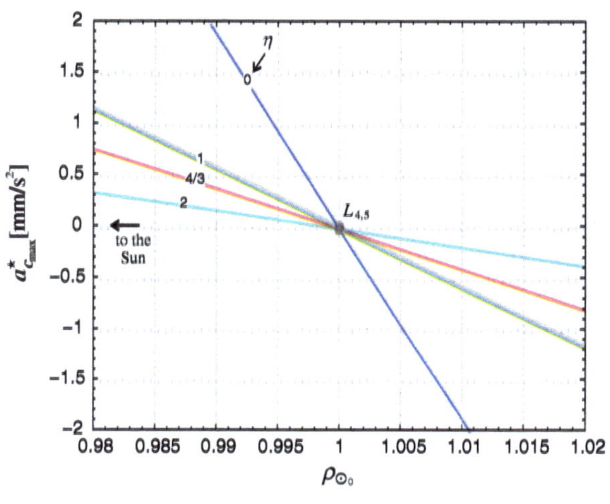

Fig. 5.6 Position of triangular AEPs in Sun-Mercury system as a function of the maximum required characteristic acceleration

the equilibrium position, and δB models the possibility of modifying B to control the system. Such a control strategy has been first proposed by Biggs and McInnes (2010), and it is usually referred to as β-control. From Eq. (5.9) a variation of B induces a corresponding variation $\delta a_c^\star$ of the characteristic acceleration with respect to the nominal equilibrium condition.

Substituting Eq. (5.20) into Eq. (5.10), the variational equation of motion of the system around an AEP is

$$\delta \mathbf{r}'' = g(\mathbf{K}+\mathbf{K}_P)^{\mathrm{T}} \cdot \delta \mathbf{r} - 2\mathbf{E} \cdot \delta \mathbf{r}' + g\mathbf{u}\delta B \tag{5.21}$$

where $\mathbf{K}$, $\mathbf{K}_P$, and $\mathbf{E}$ are second-order tensors defined by the following equations

$$\mathbf{K} \triangleq \nabla \left[-\frac{1-\mu}{\rho_\odot^3} \boldsymbol{\rho}_\odot - \frac{\mu}{\rho_M^3} \boldsymbol{\rho}_M + B \frac{1-\mu}{\rho_\odot^{\eta+1}} \boldsymbol{\rho}_\odot - \hat{\mathbf{k}} \times \left(\hat{\mathbf{k}} \times \mathbf{r} \right) \right] \Bigg|_{\mathbf{r}_0,\, B_0} \tag{5.22}$$

$$\mathbf{K}_P \triangleq \nabla \left[-e \cos \nu \left(\mathbf{r} \cdot \hat{\mathbf{k}} \right) \hat{\mathbf{k}} \right] \Big|_{\mathbf{r}_0,\, B_0} \tag{5.23}$$

$$\mathbf{E} \cdot \mathbf{r} \triangleq \hat{\mathbf{k}} \times \mathbf{r} \tag{5.24}$$

and

$$\mathbf{u} \triangleq \frac{1-\mu}{\rho_{\odot_0}^{\eta+1}} \boldsymbol{\rho}_{\odot_0} \tag{5.25}$$

When $\delta B \equiv 0$, Eq. (5.21) may be used to study the dynamical behavior of an uncontrolled spacecraft around an AEP. On the other hand, a control strategy that varies δB can be chosen to maintain the spacecraft in the neighborhood of an AEP. A simple and effective way to actively control the spacecraft dynamics is by means of a Proportional-Integral-Derivative (PID) feedback control law in the form

$$\delta B = -\mathbf{h}_P \cdot \delta \mathbf{r} - \mathbf{h}_D \cdot \delta \mathbf{r}' - \mathbf{h}_I \cdot \boldsymbol{\zeta} \tag{5.26}$$

where $\mathbf{h}_P$, $\mathbf{h}_D$, and $\mathbf{h}_I$ are constant gains vectors and

$$\boldsymbol{\zeta}' = \delta \mathbf{r} \tag{5.27}$$

Such a choice, albeit simple, combines the stabilizing effect of the proportional term with the damping capability of the derivative term, while the error dynamics is driven to zero by the integral action (Aliasi et al. 2013).

Substituting Eq. (5.26) into (5.21) the resulting closed-loop equation is

$$\delta \mathbf{r}'' = g(\mathbf{K}+\mathbf{K}_P - \mathbf{h}_P \mathbf{u})^{\mathrm{T}} \cdot \delta \mathbf{r} - (2\mathbf{E} + g\mathbf{u}\,\mathbf{h}_D) \cdot \delta \mathbf{r}' - g\mathbf{u}\,\mathbf{h}_I \cdot \boldsymbol{\zeta} \tag{5.28}$$

with which the system stability may be investigated by varying the gain vectors. For the purposes of this chapter, only a feedback term along the x-axis will be used in the control strategy, as this is sufficient to guarantee a stable dynamics. In other terms, only the first components of the vectors $\mathbf{h}_P$, $\mathbf{h}_D$, and $\mathbf{h}_I$ will be assumed different from zero. In particular, those three components will be referred to as $\mathbf{h}_P \cdot \hat{\mathbf{i}} = h_P$, $\mathbf{h}_D \cdot \hat{\mathbf{i}} = h_D$, and $\mathbf{h}_I \cdot \hat{\mathbf{i}} = h_I$, respectively.

Equations (5.27) and (5.28) constitute a system of linear, non-autonomous, homogeneous, periodic differential equations. As such, the stability analysis of the system may be addressed with the aid of Floquet theory (Hale 2009).

Stability of the uncontrolled AEPs

As stated previously, the stability of the uncontrolled AEPs is studied by setting $\delta B = 0$ in Eq. (5.21). The analysis can be performed using different combinations of positions and values of η to build up maps of the regions of stable AEPs for different types of primary propulsion systems (Aliasi et al. 2012).

Simulations show that collinear points are always unstable, except for a range of L_2-type points ($\hat{\boldsymbol{\rho}}_{M_0} \cdot \hat{\mathbf{i}} \equiv 1$), which are placed far away from Mercury ($\rho_{M_0} > 1$), see Fig. 5.7. Some stability regions also exist for triangular points, see Fig. 5.8.

These results were to be expected, since a similar behavior exists in the circular problem and for different values of the mass parameter and eccentricity (Aliasi et al. 2011, 2012).

Stability of the controlled AEPs

The effects of a PID control law on the stability of the AEPs may be studied using Eqs. (5.27) and (5.28). In general, the PID control law exhibits good stabilizing capability and performance. This result still holds even when the x-component alone of $\delta\mathbf{r}$ is fed back (proportional control), as is shown in Fig. 5.9 for L_2-type points obtained with $\eta = 0$.

The stabilizing effectiveness of a full PID controller is illustrated i Fig. 5.10, where the L_2-type point at $\rho_{M_0} = 0.003$, with $\eta = 0$, is stabilized by a PID control acting along the x-components of $\delta\mathbf{r}$ and $\delta\mathbf{r}'$.

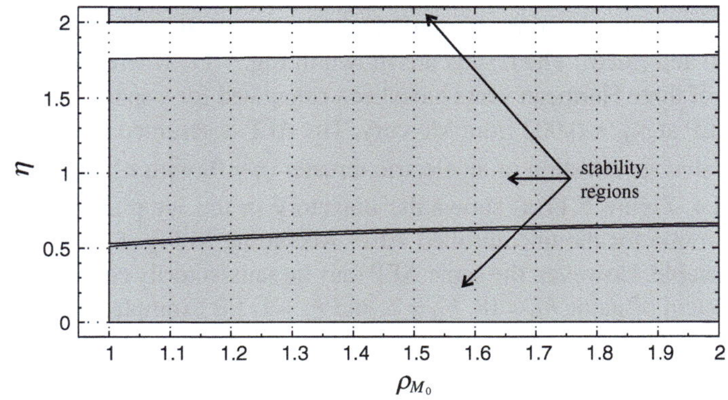

Fig. 5.7 Stability region (in gray color) for L_2-type AEP

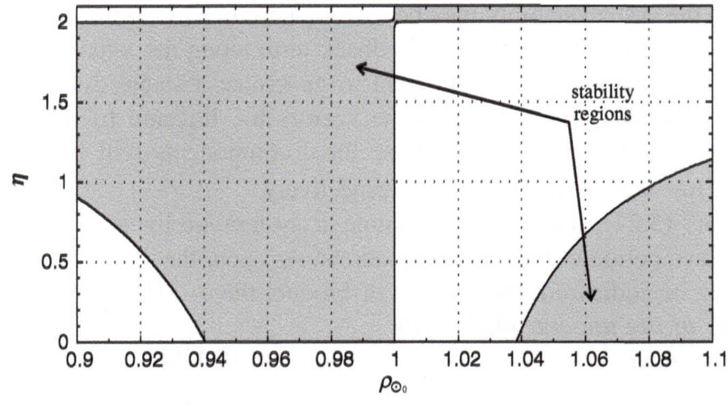

Fig. 5.8 Stability region (in gray color) for triangular AEP

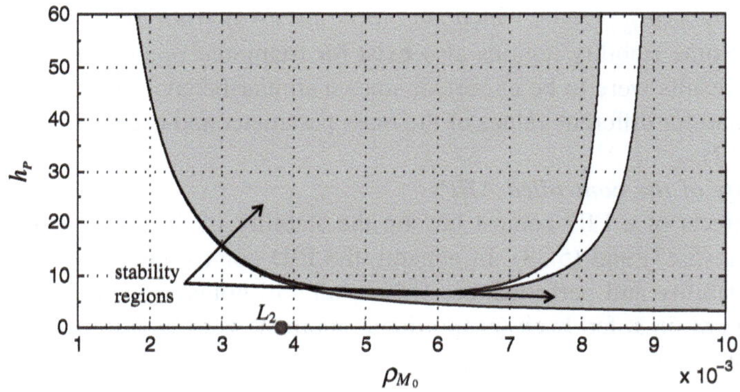

Fig. 5.9 Stability regions (in gray color) for L_2-type AEPs with a proportional feedback control on the x-component of $\delta\mathbf{r}$ when $\eta = 0$

The beneficial effects of a PID control system are also confirmed by a numerical propagation of the equation of motion (5.10), where B is defined by means of Eqs. (5.20) and (5.26). The results are shown in Fig. 5.11, where simulations over a time span of three Hermean years have been calculated for a spacecraft placed at the L_2-type AEP at $\rho_{M_0} = 0.003$ from Mercury. The AEP is assumed to be maintained by the propulsive acceleration of an electric thruster ($\eta = 0$) with $a^{\star}_{c_{\max}} = 0.5939$ mm/s^2, see Fig. 5.4. Figure 5.11(a) shows the trajectory in the x-y plane of the pulsating reference frame for the uncontrolled AEP. According to Fig. 5.7, the uncontrolled AEP is unstable. However, the same AEP may be satisfactorily controlled with a PID control system of gains $h_P = 18$, $h_D = 2$, and $h_I = 1$, for example, see Fig. 5.10. The simulation results, illustrated in Fig. 5.11(b), show that the AEP is now asymptotically stable.

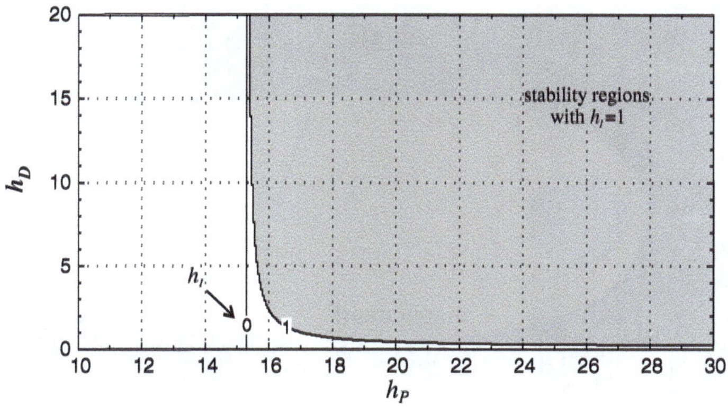

Fig. 5.10 Stability regions (in gray color) for the L_2-type AEPs at $\rho_{M_0} = 0.003$ with a PID feedback control on the x-component of $\delta\mathbf{r}$ and $\delta\mathbf{r}'$ when $\eta = 0$

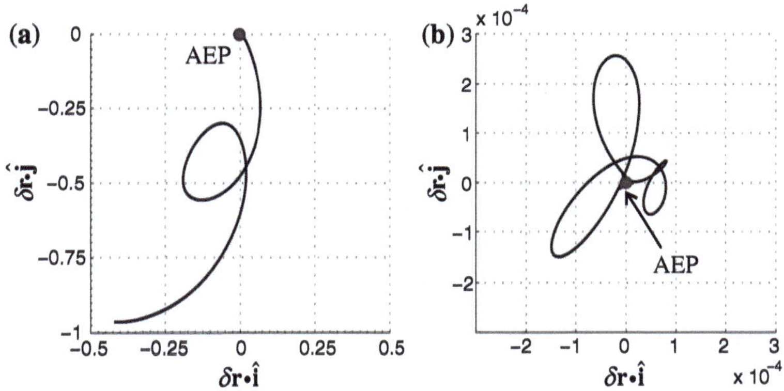

Fig. 5.11 Effects of a PID controller on the AEP's stability

5.4 Observation of Mercury's Magnetotail

A mission devoted to the observation and measurement of the Hermean magnetotail (Ness et al. 1974, 1975) requires the spacecraft to be maintained as much as possible inside the magnetotail and to be moved along a wide set of positions on the night-side of Mercury. These two requirements are indeed necessary to guarantee both a spatial and a temporal mapping of the magnetotail.

A Mercury orbiter mission fulfills only partially the requirements for magnetotail observation, even though it satisfies multidisciplinary aims. For example, the Mercury Magnetospheric Orbiter (MMO) of BepiColombo is going to move around Mercury on a high eccentricity polar orbit. In general, such an orbit will provide a quite long permanence time inside the planet's magnetotail, however the spatial

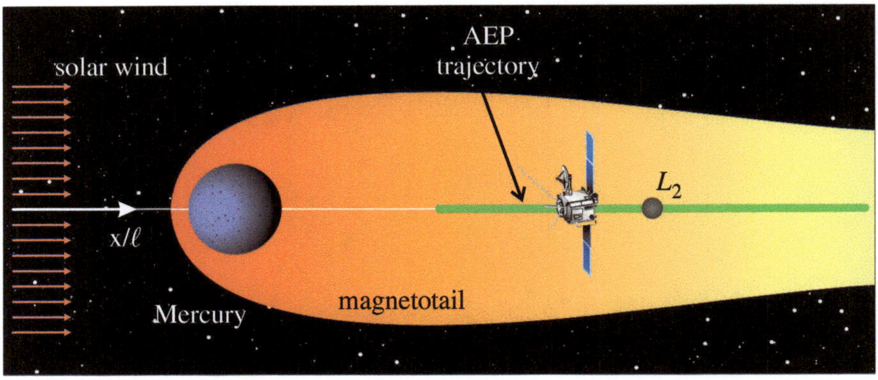

Fig. 5.12 Schematic of the magnetotail mission concept

range covered is limited within six radii of Mercury ($R_M = 2439.7$ km), thus precluding the mapping of a large region of the extended magnetotail (Milan and Slavin 2011).

The two previous requirements can be totally fulfilled by means of a mission exploiting L_2-type AEPs, see Fig. 5.3. The family of L_2-type AEPs is indeed completely immersed in the planet's magnetotail. Therefore, a spacecraft parked at a L_2-type AEP would always be in the Mercury's magnetotail and, if it is moved to another L_2-type point, the magnetotail could be mapped from two different locations. Moreover, if the transfer between the two AEPs is achieved by moving the spacecraft along the L_2-type family (see Fig. 5.12), an even more accurate mapping could be obtained on a region extending between the two L_2-type AEPs.

A mission based on L_2-type AEPs also provides observational opportunities to study the comet-like tail of Mercury, which is mostly composed by neutral sodium atoms stretching more than thousand R_M (Ip 1986; Potter et al. 2002; Schmidt et al. 2010).

5.4.1 Mission Analysis

The fundamental idea behind the magnetotail mission is that of continuously and slowly displacing the AEP along the x-axis. Such an AEP trajectory is used as the reference condition in the feedback controller. With such a strategy the control system is expected to be able to adjust the characteristic acceleration so that the spacecraft may track the AEP displacement. In fact, the AEP is moved between two points that lie along the x-axis on the night-side of Mercury and whose positions (referred to as $\rho^i_{M_0}$ and $\rho^f_{M_0}$) are assessed by the spacecraft's thrust capabilities.

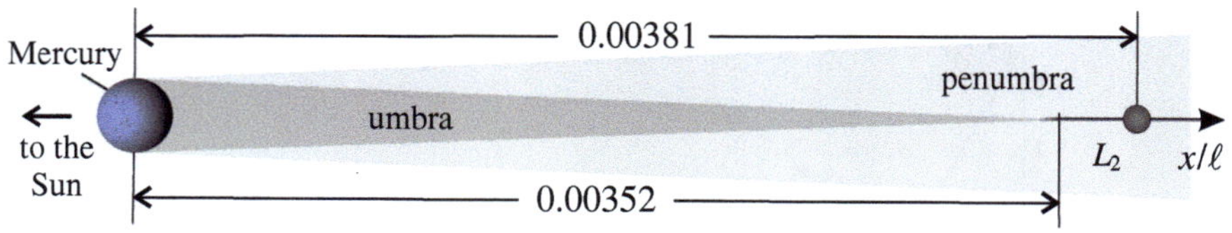

Fig. 5.13 Schematic of the Hermean umbra and penumbra

Propulsion system selection

As stated, the L_2-type AEPs can be generated by means of different continuous thrust propulsion systems. However, the selection of the most suitable thrusters greatly depends on the mission constraints to be met. In fact, the magnetosphere of Mercury shields both the planet and the spacecraft in the magnetotail from the solar wind. As a result, propulsion systems exploiting the solar wind momentum are ineffective. Moreover, since L_2-type points are placed on the night-side of Mercury, the spacecraft will be either in umbra or penumbra, see Fig. 5.13. Such illumination conditions either prevent or sensibly reduce the use of propulsion systems exploiting the solar radiation pressure and that of solar arrays as electric power generators.

A suitable choice for the primary propulsion system of a magnetotail mission is an electric thruster with a power source independent of Sun's radiation, such as, for example, a nuclear generator. This kind of thruster, which is included in the generalized sail model by simply setting $\eta = 0$, is able to provide the control capabilities necessary to stabilize the AEPs and displace them along the L_2-type branch of the family (Figs. 5.9 and 5.10).

Range of reachable AEPs

To perform a preliminary design of a magnetotail mission, the electric thruster is now assumed to provide a thrust level capable of guaranteeing an upper limit on the maximum characteristic acceleration equal to $\pm 2\ \mathrm{mm/s^2}$(the minus sign indicates the possibility of a thrust directed toward the Sun).

In theory, such a propulsion system could maintain L_2-type AEPs positions from $\rho_{M_0} \approx 0.00209$ to $\rho_{M_0} \approx 0.0112$, however a smaller range from $\rho^a_{M_0} \approx 0.00217$ to $\rho^b_{M_0} \approx 0.0102$ is considered in the magnetotail mission design to guarantee a margin of 10 % on the maximum characteristic acceleration, in order to prevent the saturation of the thrust during the control of the AEPs positions, see Fig. 5.14. The chosen thrust level is sufficient to reach a range of distances from Mercury between 40 R_M and 292 R_M, inside the magnetotail, see Fig. 5.15.

Control Law

It is assumed that the AEPs positions are controlled by means of a PID control law acting along the x-axis of the rotating reference frame. As stated previously, such a control strategy is able to guarantee the stability and controllability of the spacecraft motion around the AEPs.

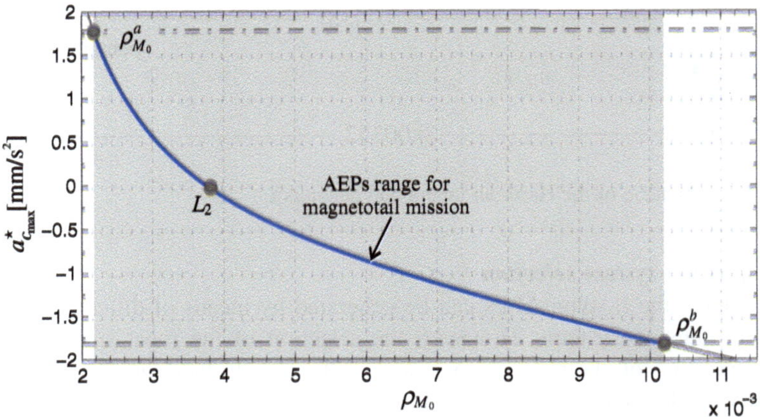

Fig. 5.14 Range of L_2-type reachable AEPs using an electric thruster with $\max|a^\star_{c_{max}}| = 2$ mm/s^2

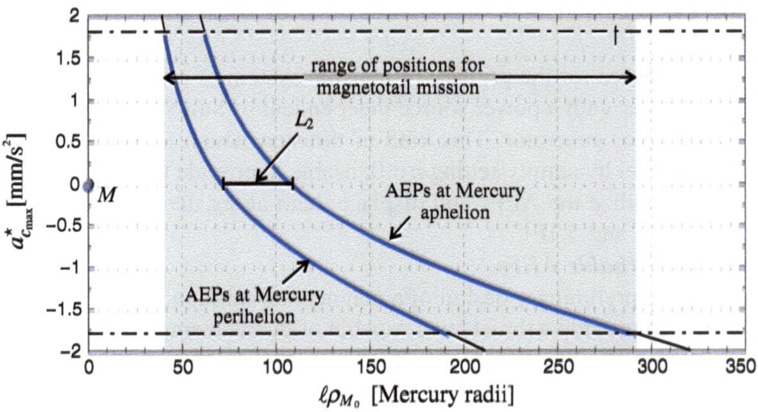

Fig. 5.15 Range of reachable distances from Mercury using an electric thruster with $\max|a^\star_{c_{max}}| = 2$ mm/s^2

To simplify the following discussion, the gains are now chosen as $h_P = h_D$ and $h_I = 1$. Under these assumptions, the stability map for L_2-type AEPs within the considered range of positions is drawn in Fig. 5.16.

The map of Fig. 5.16 can be used to select the value of the proportional gain to obtain stability. Such a map shows that a minimum value of the proportional gain ($h_{P_{min}}$) exists to guarantee a stable dynamics for each AEP position. Using a proportional gain $h_P = 2\,h_{P_{min}}$, corresponding to the blue line in Fig. 5.16, the spacecraft dynamics is stable.

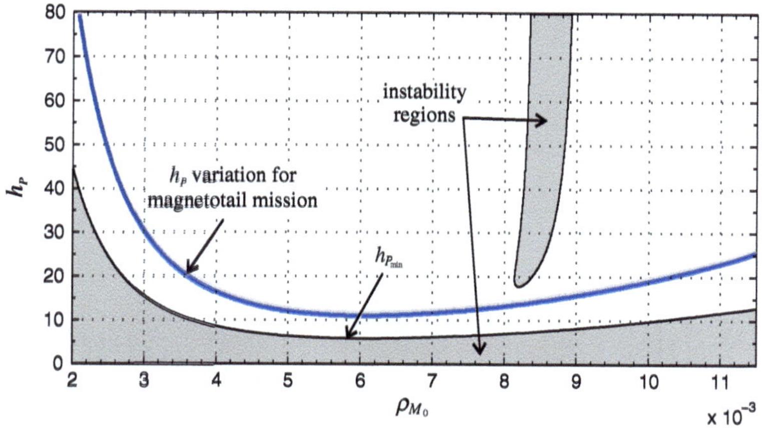

Fig. 5.16 Controllability of L_2-type AEPs with $h_P = h_D$ and $h_I = 1$ when $\eta = 0$

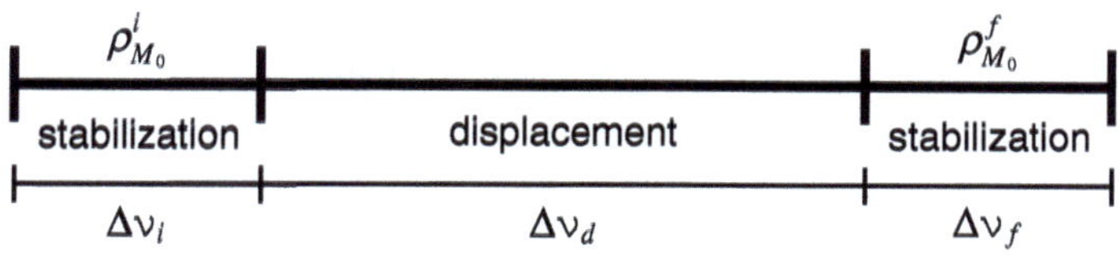

Fig. 5.17 Phases of the magnetotail mission

AEP displacement strategy

The magnetotail mission is assumed to start when the spacecraft is released in the proximity of the initial equilibrium point $\rho^i_{M_0}$ and to finish when the spacecraft moves stably around the final position $\rho^f_{M_0}$. Accordingly, the entire mission can be thought of as being constituted by three phases, see Fig. 5.17.

In the first phase, of length $\Delta\nu_i$, it is necessary to guarantee that the spacecraft dynamics is stable around the initial position. During the second phase, with length $\Delta\nu_d$, the spacecraft is moved along the family of AEPs between $\rho^i_{M_0}$ and $\rho^f_{M_0}$. In this phase a number of measurements and experiments inside the magnetotail may be carried out. Finally, the aim of the third phase is to stabilize the spacecraft around the last equilibrium point at $\rho^f_{M_0}$ in a length $\Delta\nu_f$. After the last phase, an extension of the mission can also be planned, if the spacecraft is still operational.

In a preliminary design, a simple strategy is to displace an AEP with a constant velocity defined by

$$\rho'_{M_0} \triangleq \frac{\rho^f_{M_0} - \rho^i_{M_0}}{\Delta\nu_d} \tag{5.29}$$

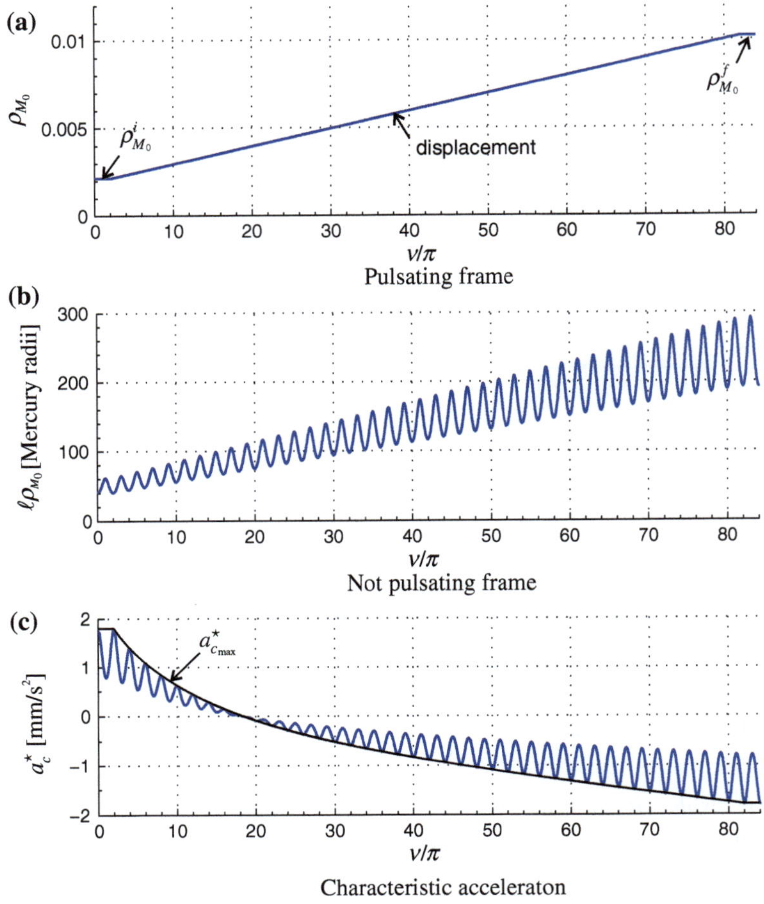

Fig. 5.18 Displacement strategy for the magnetotail mission

Assuming that the entire mission lasts 84π (corresponding to 42 revolutions of Mercury, i.e. about 10 years), using an initial and final phase of length $\Delta\nu_i = \Delta\nu_f = 2\pi$, and a range of positions in accordance with Fig. 5.15, the displacement velocity is $|\rho'_{M_0}| = (\rho^b_{M_0} - \rho^a_{M_0})/(80\pi) \approx 3.195 \times 10^{-5}\,\mathrm{rad}^{-1}$.

This displacement strategy requires that the position of the AEPs is varied in the pulsating frame according to Fig. 5.18(a), starting from $\nu_0 = 0$ (when Mercury is at perihelion) with the initial AEP at $\rho^a_{M_0}$ and the final AEP at $\rho^b_{M_0}$. This variation corresponds to the displacement strategy in the pulsating frame illustrated in Fig. 5.18(b) and to the variation of $a^\star_c$ reported in Fig. 5.18(c). In case of displacing from $\rho^b_{M_0}$ to $\rho^a_{M_0}$, the corresponding results are obtained through a simple left-right mirror image of each figure.

As a final remark, it is worth noting that not only the suggested mission approach allows a wide region of the magnetotail to be mapped, but it also guarantees some recursive passages of the spacecraft at the same distance from Mercury, thus obtaining a temporal mapping of the magnetotail.

5.4.2 *Numerical Simulations*

Numerical simulations of Eq. (5.10) have been performed to verify the spacecraft capability of following the displacement of the AEP at a constant velocity, by means of the previously discussed PID control law.

At the starting time, Mercury is assumed to be at perihelion of its heliocentric orbit, while the spacecraft is assumed to be released in the proximity of the initial AEP with injection errors of 500 km in the position and 0.05 km/s in the velocity. These errors correspond to $||\delta\mathbf{r}||_0 \approx 1.087 \times 10^{-5}$ and $||\delta\mathbf{r}'||_0 \approx 8.478 \times 10^{-4}$. The assumption about the initial position of Mercury is not restrictive because the numerical simulations have shown a negligible dependence of the results on the starting anomaly.

Figure 5.19 shows the simulations of the displacement from $\rho^i_{M_0} = \rho^a_{M_0}$ to $\rho^f_{M_0} = \rho^b_{M_0}$. The trajectory of the spacecraft follows the ideal trajectory with a maximum error in the position of about $15\,R_M$, see Figs. 5.19(a), (b).

The error along the x-axis is in general very small, except for the initial and final phase when the control action is stronger, see Fig. 5.19(c). The motion along the z-axis remains almost unchanged during the mission. Actually, such a motion is nearly insensitive to the control, as to a first order it is uncoupled with respect to the motion in the x-y plane. As a result, a small injection error along the z-component is crucial to avoid large oscillation along the z-axis. A remarkable error is, instead, found on the y-component, especially toward the end of the mission.

In particular, the trajectory exhibits a drift along the negative y-direction. Such a behavior is a consequence of the displacement of the AEPs toward larger distance from Mercury. Indeed, during the displacement, the spacecraft experiences an increasing centrifugal force that is not counterbalanced by the gravitational and propulsive forces. The resulting force adds to the small Coriolis force, thus creating an unbalanced force along the negative y-direction, which moves the spacecraft outward with respect to the x-axis. Despite the control, the drift is not prevented because the control acts directly along the x-direction and only indirectly along the y-direction, as a consequence of the coupling effect in the equations of motion of the x-y plane. The control of the y motion is therefore less effective and a drift exists.

The drifting effect is reversed when the motion happens from $\rho^i_{M_0} = \rho^b_{M_0}$ to $\rho^f_{M_0} = \rho^a_{M_0}$. In this case, reducing the distance from Mercury, the centrifugal force decreases while the gravitational and propulsive forces increase. The resulting force

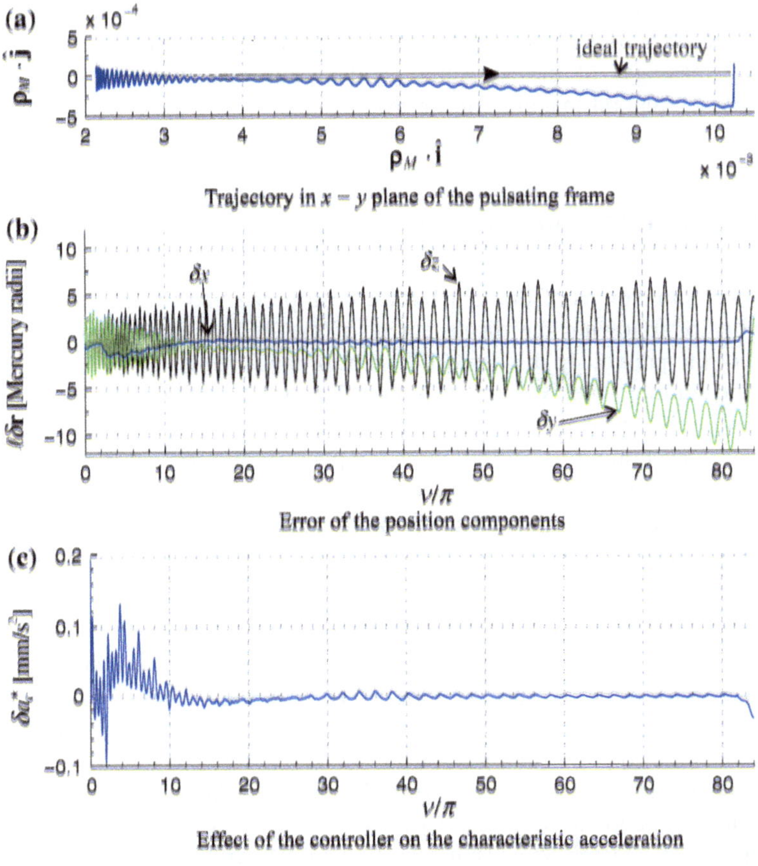

Fig. 5.19 Simulation of the displacement from $\rho^a_{M_0}$ to $\rho^b_{M_0}$

is now reduced by the small Coriolis force, however an unbalanced force in the positive y-direction still exists, which causes a drift inwards. Indeed, in the latter case the trajectory tends toward the final AEP, see Fig. 5.20(a).

Nevertheless the maximum error is now bigger than in the previous case. In fact, Fig. 5.20(b) shows the existence of an error of about 30 R_M at the beginning of the displacement. This is due to the action of the controller and, in fact, a similar error is found in the first phase of the motion also if no AEP displacement is assumed.

In both Figs. 5.19(a) and 5.20(a) 2π long stabilization phase is included at the end of the mission. In fact, as seen in Figs. 5.19(a) and 5.20(c), the trajectory tends toward the final AEP location.

A possible limitation to the previous control strategy is due to the non-negligible costs related to the control actuation, corresponding to a required annual ΔV on the order of some kilometers per second. Nevertheless, the previous simulations confirm the practical importance of the mission concept, which should become a feasible option if supported by a more refined control strategy.

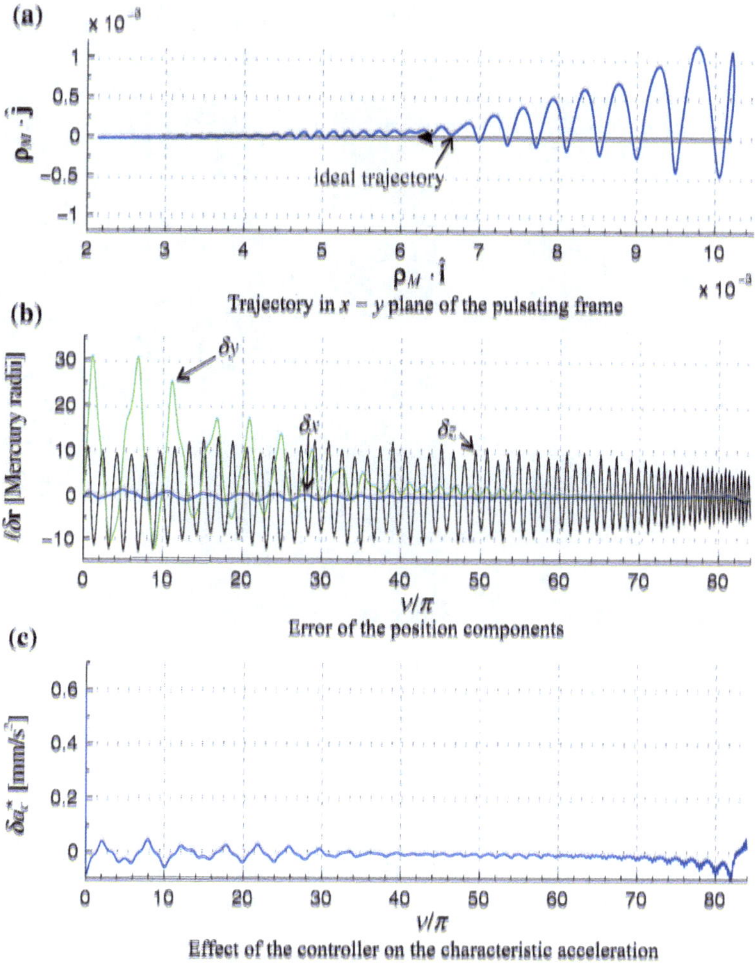

Fig. 5.20 Simulation of the displacement from $\rho^{b}_{M_0}$ to $\rho^{a}_{M_0}$

5.4.3 *Mission Enhancement*

It is now interesting to investigate whether the previous mission concept may be further improved to obtain more advanced scientific results. In this context an important aspect is that the magnetotail originates from the interaction of solar wind with the magnetic field of Mercury and an exhaustive comprehension of the magnetotail structure and dynamics is possible only if a correlation between these two acting causes may be found.

The study of such a correlation requires the use of more than one spacecraft in the magnetotail mission, as is shown in the arrangement of Fig. 5.21, where three spacecraft are used. One of them is immersed in the magnetotail, and accomplishes

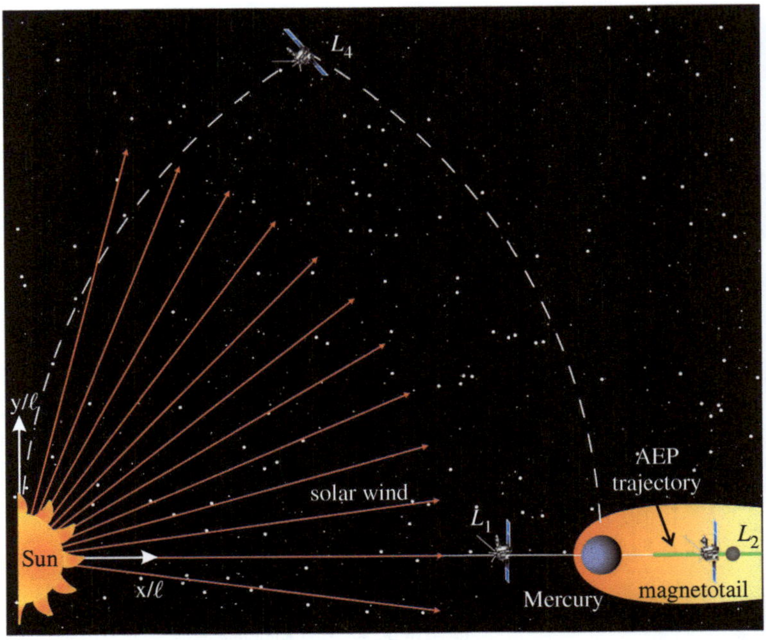

Fig. 5.21 Concept of the enhanced magnetotail mission

the previously described mission of AEPs displacement. The remaining two spacecraft are placed and maintained at suitable AEP positions of the triangular and L_1-type families.

In the example of Fig. 5.21, the classical L_1 and L_4 points are considered as possible locations for the spacecraft. Those locations are remarkable in that they require the use of thrust for control purposes only, but not for the keeping phase.

5.5 Concluding Remarks

Theory and simulations have shown the feasibility of a mission to Mercury's magnetotail that exploits the displacement of AEPs, even though significant position errors with respect to the ideal trajectory may arise.

The choice of a suitable control law, necessary to stabilize the AEPs, is a crucial point to obtain a successful mission. Moreover, rather complex control strategies seem to be necessary to successfully plan a future mission inside the magnetotail of Mercury. This kind of analysis is however beyond the scope of this chapter and is left to future research.

Beside the AEPs displacement scenario, a further mission configuration has been suggested, where three spacecraft are involved. Such a configuration has remarkable features. In fact, it enables the simultaneous observation and measurement of both the magnetotail and the solar wind. In particular, the solar wind could be studied from two different positions, thus providing a temporal and spatial mapping. This is a very relevant feature, as the comprehension of the structure and evolution of the young solar wind plasma is a key point for understanding how solar wind and coronal mass ejections generate from the Sun's corona. Moreover, the closeness of the Sun would guarantee the possibility of observing the heliosphere and testing the principles of general relativity by measuring the aphelion precession of Mercury.

References

Aliasi G, Mengali G, Quarta AA (2011) Artificial equilibrium points for a generalized sail in the circular restricted three-body problem. Celest Mech Dyn Astron 110(4):343–368. doi:10.1007/s10569-011-9366-y

Aliasi G, Mengali G, Quarta AA (2012) Artificial equilibrium points for a generalized sail in the elliptic restricted three-body problem. Celest Mech Dyn Astron 114(1–2):181–200. doi:10.1007/s10569-012-9425-z

Aliasi G, Mengali G, Quarta AA (2013) Artificial Lagrange points for solar sail with electrochromic material panels. J Guid Control Dyn 36(5):1544–1550. doi:10.2514/1.58167

Baoyin H, McInnes CR (2006) Solar sail equilibria in the elliptical restricted three-body problem. J Guid Control Dyn 29(3):538–543. doi:10.2514/1.15596

Battin RH (1987) An Introduction to the mathematics and methods of astrodynamics. Chap 8, pp 371–381. AIAA, Washington. ISBN:1563473429

Baumjohann W, Matsuoka A, Glassmeier KH, Russell CT, Nagai T, Hoshino M, Nakagawa T, Balogh A, Slavin JA, Nakamura R, Magnes W (2006) The magnetosphere of Mercury and its solar wind environment: open issues and scientific questions. Adv Space Res 38(4):604–609. doi:10.1016/j.asr.2005.05.117

Biggs JD, McInnes CR (2010) Passive orbit control for space-based geo-engineering. J Guid Control Dyn 33(3):1017–1020. doi:10.2514/1.46054

Biggs JD, McInnes CR, Waters T (2009) Control of solar sail periodic orbits in the elliptic three-body problem. J Guid Control Dyn 32(1):318–320. doi:10.2514/1.38362

Bourke RD, Sauer CGJ (1972) The effect of solar array degradation on electric propulsion spacecraft performance. In: AIAA 9th electric propulsion conference. Paper AIAA 72–444. Bethesda, MD, USA

Circi C (2004) Mars and Mercury missions using solar sails and solar electric propulsion. J Guid Control Dyn 27(3):496–498. doi:10.2514/1.5425

Hale JK (2009) Ordinary differential equations, Chap 8. Dover Publications, Inc., Mineola, pp 117–121, 280. ISBN: 0–486–47211–6

Hunt ME (1993) High efficiency dynamic radioisotope power system for space exploration—a status report. IEE AES Syst Mag 8(12):18–23. doi:10.1109/62.246037

Ip WH (1986) The sodium exosphere and magnetosphere of Mercury. Geophys Res Lett 13 (5):423–426. doi:10.1029/GL013i005p00423

Jahn RG (1968) Physics of electric propulsion, Chap 1. Dover Publications, Inc., pp 2–11. ISBN: 0-486-45040-6

Janhunen P (2010) The electric solar wind sail status report. In: European Planetary Science Congress. Paper EPSC 2010-297. Rome, Italy

Lyngvi AE, van den Berg ML, Falkner P (2007) Study overview of the interstellar heliopause probe. Technical report, 3 (revision 4). SCI-A/2006/114/IHP. ESA

Macdonald M, Mckay RJ, Vasile M, Bosquillon de Frescheville F, Biggs JD, McInnes CR (2011) Low-thrust enabled highly non-keplerian orbits in support of future Mars exploration. J Guid Control Dyn 35(5):1396–1411. doi:10.2514/1.52602
McInnes CR (1999) Solar sailing: technology, dynamics and mission applications. Springer-Praxis Series in Space Science and Technology. Springer, London. ISBN: 1-852-33102-X
McKay RJ, Macdonald M, Biggs JD, McInnes CR (2011) Survey of highly non-keplerian orbits with low-thrust propulsion. J Guid Control Dyn 34(3):645–666. doi:10.2514/1.52133
Milan S, Slavin J (2011) An assessment of the length and variability of Mercury's magnetotail. Planet Space Sci 59(15):2058–2065. doi:10.1016/j.pss.2011.05.007
Ness NF, Behannon KW, Lepping RP, Whang YC (1975) The magnetic field of Mercury, 1. J Geophys Res 80(19):2708–2716. doi:10.1029/JA080i019p02708
Ness NF, Behannon KW, Lepping RP, Whang YC, Schatten KH (1974) Magnetic field observations near Mercury: preliminary results from Mariner 10. Science 185(4146):151–160. doi:10.1126/science.185.4146.151
Potter AE, Killen RM, Morgan TH (2002) The sodium tail of Mercury. Meteorit Planet Sci 37 (9):1165–1172. doi:10.1111/j.1945-5100.2002.tb00886.x
Quarta AA, Mengali G, Aliasi G (2013) Optimal control laws for heliocentric transfers with a magnetic sail. Acta Astronaut 89:216–225. doi:10.1016/j.actaastro.2013.04.018
Rayman MD, Williams SN (2002) Design of the first interplanetary solar electric propulsion mission. J Spacecr Rockets 39(4):589–595. doi:10.2514/2.3848
Richardson AJ, Warren JW (1971) Solar array degradation due to meteoroid impacts during extended planetary missions. J Spacecr Rockets 8(6):681–683. doi:10.2514/3.59713
Sauer CGJ (1978). Modeling of thruster and solar array characteristics in the JPL low-thrust trajectory analysis. In: 13th international electric propulsion conference. Paper AIAA 78-645. San Diego, CA, USA
Schmidt CA, Wilson JK, Baumgardner J, Mendillo M (2010) Orbital effects on Mercury's escaping sodium exosphere. Icarus 207(1):9–16. doi:10.1016/j.icarus.2009.10.017
Slavin JA, Krimigis SM, Acuña MH, Anderson BJ, Baker DN, Koehn PL, Korth H, Livi S, Mauk BH, Solomon SC, Zurbuchen TH (2007) MESSENGER: exploring Mercury's magnetosphere. Space Sci Rev 131(1–4):133–160. doi:10.1007/s11214-007-9154-x
Szebehely V (1967) Theory of orbits: the restricted problem of three bodies. Chapters 1, 5, 10. Academic Press, Inc., New York, pp 7–25, 250, 587–602. ISBN: 0126806500
Winglee RM, Slough J, Ziemba T, Goodson A (2000) Mini-magnetospheric plasma propulsion: tapping the energy of the solar wind for spacecraft propulsion. J Geophys Res 105 (A9):21067–21077. doi:10.1029/1999JA000334
Zubrin RM, Andrews DG (1991) Magnetic sails and interplanetary travel. J Spacecr Rockets 28 (2):197–203. doi:10.2514/3.26230

Chapter 6
Low-Thrust Earth-Venus Trajectories

Alessandro A. Quarta, Giovanni Mengali and Generoso Aliasi

6.1 Introduction

The remarkable results obtained by the pioneering Deep Space 1 (DS1) mission (Rayman et al. 2000) have demonstrated the practical possibility of using electric thrusters to successfully perform interplanetary robotic missions. The striking advances in ion engines have greatly reduced the costs and risks of using this kind of thrusters as a primary system for spacecraft propulsion. The use of such a technology is also an effective way to render many deep-space missions scientifically more attractive as it enables the use of smaller, less expensive launch vehicles and helps reducing the total propellant mass. It is known that electric propulsion is especially interesting for those missions requiring large changes in orbital energy. However, the use of a low thrust propulsion system, which remains switched on over long time periods, implies a substantial complication of the mission analysis when compared to the more conventional approach involving high thrust propulsion systems. For these reasons, the problem of mission design becomes particularly attractive also from a theoretical point of view.

Starting from the preliminary design phase of a space mission, the analysis of spacecraft trajectory with a low thrust propulsion system is usually tackled by solving a constrained optimal control problem (Russell 2007). Usually the problem aims at finding the spacecraft transfer trajectory that maximizes (or minimizes) a scalar performance index. The latter may be given in terms of propellant mass, flight

A.A. Quarta (✉) · G. Mengali · G. Aliasi
University of Pisa, Pisa, Italy
e-mail: a.quarta@ing.unipi.it

G. Mengali
e-mail: g.mengali@ing.unipi.it

G. Aliasi
e-mail: g.aliasi@dia.unipi.it

V. Badescu and K. Zacny (eds.), *Inner Solar System*,
DOI 10.1007/978-3-319-19569-8_6

time, or a suitable combination of these two quantities (Casalino and Colasurdo 2004; Mengali and Quarta 2008). Significant results are obtained by varying the coefficients that weight the relative importance of flight time and propellant mass: the resulting trade-off study may be chosen to satisfy different mission constraints (Circi 2003, 2004). In particular, assuming a direct transfer (that is, without gravity assist maneuvers) between the starting and the target orbit, a minimum-time trajectory corresponds to the situation in which the required propellant mass is maximized within the set of propellant-optimal trajectories with a given flight time. Accordingly, the minimum-time trajectory represents a valuable information and a necessary starting point for succeeding and more refined mission analyses.

The aim of this Chapter is to discuss the simulation results involving the minimum-time trajectories for an Earth-Venus mission transfer using a spacecraft with an electric propulsion system. The analysis is performed in a parametric way as a function of relevant design parameters, such as the available electric power and the initial in-flight mass. Various models are considered to describe the propulsion system behavior with different levels of approximations and obtain increasingly refined information about the mission performance. Some simplifying assumptions are introduced in order to make the mathematical problem tractable, to reduce the simulation time and guarantee a thorough parametric investigation of mission performance. The main assumptions and the fundamentals of the mathematical model adopted are summarized in the next section.

6.2 Mathematical Model

As stated, in a preliminary mission analysis it is customary to analyze the transfer phase by introducing some simplifying assumptions. In fact, the designer wishes to explore different mission scenarios and collect a number of data in such a way that the problem may be analyzed in a parametric way within a reduced simulation time. Those simplifying assumptions are then gradually removed in a second phase of analysis, after the most promising mission options have been selected, and a more accurate approximation of the spacecraft transfer trajectory may eventually be obtained.

The first important simplification that is used in this study is to assume that, in the heliocentric transfer phase, the Sun is the only source of gravitational attraction, thus neglecting any perturbative effect (including the gravitational effect from other planets) on spacecraft motion. Within this model, the orbits of the two planets involved in the problem (i.e. Earth and Venus) are both Keplerian. The heliocentric orbital parameters of Earth and Venus are summarized in Table 6.1, where the data are taken from JPL ephemerides (Standish 1990, 1998) database.

The second simplification is to assume a point mass spacecraft that is propelled by an electric propulsion system. Finally, the following analysis is performed using an ephemeris-free transfer, i.e. by looking for the globally optimal result independent of the launch date. Note that this is a major simplifying assumption, because the launch date strongly affects transfer time and fuel consumption.

Table 6.1 Reference (Keplerian) orbital elements of Earth+Moon barycenter and Venus at January 1, 2014

Orbital elements	Earth+Moon barycenter	Venus
semimajor axis [au]	1	0.72334
eccentricity	1.66984×10^{-2}	6.77702×10^{-3}
inclination [deg]	1.77591×10^{-3}	3.39461
long. of asc. node [deg]	178.359	76.64356
arg. of pericenter [deg]	284.589	54.91802

6.2.1 Equations of Motion

Under the previous assumptions, the equations of motion of the spacecraft, in a heliocentric inertial reference frame, are conveniently expressed (Walker 1986; Walker et al. 1985; Betts 2000) in terms of Modified Equinoctial Orbital Elements (MEOE) p, f, g, h, k, and L as:

$$\dot{\mathbf{x}} = \eta_P \, (T/m) \mathbb{A} \hat{\mathbf{a}}_T + \mathbf{d} \tag{6.1}$$

where $\mathbf{x} \triangleq [p, f, g, h, k, L, m]^{\mathrm{T}}$ is the state vector, m is the spacecraft mass, $T \geq 0$ is the propulsive thrust modulus, $\hat{\mathbf{a}}_T$ is the thrust unit vector, whose components are expressed in a local-vertical/local-horizontal orbital reference frame $\mathcal{T}_{\mathrm{RTN}}$, and $\eta_P \in [0, 1]$ is the duty cycle. The latter, according to Rayman and Williams (2002), is a constant parameter defined as the fraction of time during deterministic thrust periods in which the primary propulsion system is engaged (i.e. $T \neq 0$). Moreover, matrices $\mathbb{A} \in \mathbb{R}^{7\times 3}$ and $\mathbf{d} \in \mathbb{R}^{7\times 1}$ are defined as:

$$\mathbb{A} \triangleq \begin{bmatrix} 0 & A_{12} & 0 \\ A_{21} & A_{22} & A_{23} \\ A_{31} & A_{32} & A_{33} \\ 0 & 0 & A_{43} \\ 0 & 0 & A_{53} \\ 0 & 0 & A_{63} \\ 0 & 0 & 0 \end{bmatrix}, \quad \mathbf{d} \triangleq \begin{bmatrix} 0 \\ 0 \\ 0 \\ 0 \\ 0 \\ d_6 \\ d_7 \end{bmatrix} \tag{6.2}$$

$$A_{12} = \frac{2p}{1 + f \cos L + g \sin L} \sqrt{\frac{p}{\mu_\odot}} \tag{6.3}$$

$$A_{21} = \sin L \sqrt{\frac{p}{\mu_\odot}} \tag{6.4}$$

$$A_{22} = \frac{(2+f\cos L+g\sin L)\cos L+f}{1+f\cos L+g\sin L}\sqrt{\frac{p}{\mu_\odot}} \tag{6.5}$$

$$A_{23} = -\frac{g(h\sin L-k\cos L)}{1+f\cos L+g\sin L}\sqrt{\frac{p}{\mu_\odot}} \tag{6.6}$$

$$A_{31} = -\cos L\sqrt{\frac{p}{\mu_\odot}} \tag{6.7}$$

$$A_{32} = \frac{(2+f\cos L+g\sin L)\sin L+g}{1+f\cos L+g\sin L}\sqrt{\frac{p}{\mu_\odot}} \tag{6.8}$$

$$A_{33} = \frac{f(h\sin L-k\cos L)}{1+f\cos L+g\sin L}\sqrt{\frac{p}{\mu_\odot}} \tag{6.9}$$

$$A_{43} = \frac{(1+h^2+k^2)\cos L}{2(1+f\cos L+g\sin L)}\sqrt{\frac{p}{\mu_\odot}} \tag{6.10}$$

$$A_{53} = \frac{(1+h^2+k^2)\sin L}{2(1+f\cos L+g\sin L)}\sqrt{\frac{p}{\mu_\odot}} \tag{6.11}$$

$$A_{63} = \frac{h\sin L-k\cos L}{1+f\cos L+g\sin L}\sqrt{\frac{p}{\mu_\odot}} \tag{6.12}$$

$$d_6 = \sqrt{\mu_\odot p}\left(\frac{1+f\cos L+g\sin L}{p}\right)^2 \tag{6.13}$$

$$d_7 = -\eta_P\beta \tag{6.14}$$

where $\mu_\odot = 132712439935.5\ \mathrm{km}^3/\mathrm{s}^2$ is the Sun's gravitational parameter, $\beta \geq 0$ is the propellant mass flow rate, and p is the semilatus rectum of the spacecraft osculating orbit.

The transformations from MEOE to classical orbital elements a, e, i, ω, Ω, and ν are given by Coverstone and Prussing (2003):

$$a = \frac{p}{1-f^2-g^2} \tag{6.15}$$

$$e = \sqrt{f^2+g^2} \tag{6.16}$$

$$i = 2\arctan\sqrt{h^2+k^2} \tag{6.17}$$

$$\sin\omega = gh - fk, \quad \cos\omega = fh + gk \tag{6.18}$$

$$\sin\Omega = k, \ \cos\Omega = h \tag{6.19}$$

$$\nu = L - \Omega - \omega \tag{6.20}$$

where a is the semimajor axis, e is the eccentricity, i the orbital inclination, ω the argument of perihelion, Ω the longitude of the ascending node, and ν is the true anomaly of the spacecraft osculating orbit.

The transformations from MEOE to spacecraft position $\boldsymbol{r}$ and velocity $\boldsymbol{v}$, whose components are expressed in the heliocentric-ecliptic inertial reference frame $\mathcal{T}_\odot$, are:

$$[\mathbf{r}]_{\mathcal{T}_\odot} = \frac{r}{(1+h^2+k^2)}\begin{bmatrix}(1+h^2+k^2)\cos L + 2hk\sin L \\ (1-h^2+k^2)\sin L + 2hk\cos L \\ 2h\sin L - 2k\cos L\end{bmatrix} \tag{6.21}$$

$$[\mathbf{v}]_{\mathcal{T}_\odot} = -\frac{\sqrt{\mu_\odot/p}}{(1+h^2+k^2)}\begin{bmatrix}(\sin L + g)(h^2-k^2+1) - 2hk(\cos L + f) \\ (\cos L + f)(h^2-k^2-1) + 2hk(\sin L + g) \\ -2h(\cos L + f) - 2k(\sin L + g)\end{bmatrix} \tag{6.22}$$

where $r = p/(1 + f\cos L + g\sin L)$ is the Sun-spacecraft distance given as a function of MEOE.

Note that when the propulsive acceleration is zero ($T = 0$ and $\beta = 0$), the spacecraft experiences a Keplerian motion. As a result, Eq. (6.1) states that p, f, g, h, k, and m are all constants of motion. This happens when a coasting phase exists in the interplanetary transfer trajectory.

6.2.2 Propulsive Thrust Model

Two mathematical models are now discussed to approximate the behavior of an electric propulsion system equipped with different types of power generation systems.

Nuclear Electric Propulsion Engine

In this first model, which is simple to implement, the thrust level T provided by the propulsion system and the propellant mass flow β are both assumed to be constant and known parameters. This model approximates the behavior of a Nuclear Electric Propulsion (NEP) engine (Patel et al. 2006; Yam et al. 2004), i.e. an electric propulsion system with a nuclear type power generation system. In this case, the electric power necessary for producing the thrust can be thought of as independent of the distance r of the spacecraft from the Sun. Note, however, that the following mathematical model is only a first order approximation of the real behavior of the propulsion system, since an electric propulsion engine may vary, within prescribed limits, both the thrust level and the propellant mass flow rate.

In this mathematical model the thruster, by assumption, is unable to modulate the thrust level, even if the latter may be set to zero, thus guaranteeing the possibility of obtaining coasting phases during the interplanetary travel. Observing that η_P is constant, the propulsion system performance is now described as a function of two parameters only, i.e. $\eta_P T$ and $\eta_P \beta$. Each parameter, in accordance with the previous assumptions, may take two admissible values only, viz.

$$\eta_P T \in \{0, \tilde{T}\} \ , \ \eta_P \beta \in \{0, \tilde{\beta}\} \tag{6.23}$$

where $\tilde{T}$ and $\tilde{\beta}$ are given values and may be thought of as a sort of nominal thrust and propellant mass flow weighted by the duty cycle value (η_P).

Let m_0 be the spacecraft mass at the initial time $t_0 \triangleq 0$. Bearing in mind that the propulsive thrust is included in Eq. (6.1) in the form of an acceleration and assuming $T \neq 0$, one obtains

$$\eta_P \frac{T}{m} = \frac{\tilde{T}}{m} = \frac{\tilde{a}_{T_0}}{m/m_0} \tag{6.24}$$

where $\tilde{a}_{T_0} \triangleq \tilde{T}/m_0$ is the initial propulsive acceleration weighted with the duty cycle and m/m_0 is the dimensionless spacecraft mass. Recalling Eqs. (6.1) and (6.14), the time variation of the spacecraft mass is

$$\frac{\dot{m}}{m_0} = -\frac{\eta_P \beta}{m_0} = -\frac{\tilde{\beta}}{m_0} \tag{6.25}$$

According to the literature the spacecraft mass variation is usually quantified by introducing the specific impulse of the propulsion system, defined as

$$I_{sp} \triangleq \frac{T}{\dot{m} g_0} \equiv \frac{\tilde{T}}{\tilde{\beta} g_0} = \frac{\tilde{a}_{T_0}}{(\tilde{\beta}/m_0) g_0} \tag{6.26}$$

where $g_0 = 9.81\ \text{m/s}^2$ is the standard gravity acceleration. The equivalent form of Eq. (6.25) is therefore:

$$\frac{\dot{m}}{m_0} = -\frac{\tilde{a}_{T_0}}{g_0 I_{sp}} \tag{6.27}$$

To summarize, the two scaling parameters that govern the problem are the initial propulsive acceleration $\tilde{a}_{T_0}$, and the engine specific impulse I_{sp}.

Having fixed these two design parameters, the spacecraft control variables are the orientation of the propulsive acceleration $\hat{\mathbf{a}}_T$ and the thrust level to be selected within the set of two admissible values (0 or T). The spacecraft transfer trajectory will be found as the solution of an optimization process in which the control variables are chosen in order to minimize the total flight time required to transfer the spacecraft from its initial orbit to a final (prescribed) target orbit.

Solar Electric Propulsion Engine

The mathematical model of a Solar Electric Propulsion (SEP) engine is different from that described in the previous section. In fact, in a SEP spacecraft the thrust level T and the propellant mass flow rate β are closely related to the input power P to the Power Processing Unit (PPU). In particular, an electric thruster has a finite number of operation points (Patterson et al. 2001; Patterson and Benson 2007), each one characterized by a corresponding set of values of T, β, and P.

Following the approach described in detail by Quarta et al. (2011, 2013), for a given propulsion system, whose performance characteristics have been evaluated experimentally, it is possible to build up a throttle table similar to that shown in Table 6.2. The SEP system is assumed to be fully described by a finite number $(n-1)$ of operation points (the generic point will be referred to through the integer number I_d), each one being characterized by a given triplet of values, i.e. the thrust provided by the propulsion system, the amount of electric power required by the PPU, and the propellant mass flow. The last point of the table, (with $I_d = n$, $T_n = 0$ and $\beta_n = 0$) is actually a fictitious operation point, which is added to the actual engine throttle table to model the presence of possible coasting phases in the spacecraft transfer trajectory.

In practice, not all of the points within the throttle table may actually be used by the propulsion system at any time along the spacecraft trajectory. The available points correspond to operation levels in which the electric power necessary to the PPU is less (or equal) to the available input power P_A for the propulsion system. The latter approximately equals the solar array output power P_S less the power allocated to operate the spacecraft systems P_L (including the control system, telecommunication system, payload, etc.), viz.

$$P_A = \begin{cases} P_S - P_L & \text{if } P_S > P_L \\ 0 & \text{if } P_S \leq P_L \end{cases} \tag{6.28}$$

Equation (6.28) implies that when PS is less than PL, the propulsion system cannot be used and the spacecraft follows a Keplerian motion.

Table 6.2 Typical throttle table for an electric thruster

I_d	T	P	β
1	T_1	P_1	β_1
2	T_2	P_2	β_2
3	T_3	P_3	β_3
…	…	…	…
$n-1$	T_{n-1}	P_{n-1}	β_{n-1}
n	0	0	0

The value of P_S is a function of the building characteristics of the solar arrays (such as their specific power and the exposed surface), of the Sun spacecraft distance and of the time, due to the time degradation effect on solar array performance. The model adopted to describe the available power follows that described by Quarta and Mengali (2011), in which the power supplied by the Sun-facing solar arrays is (Rayman and Williams 2002)

$$P_S = P_{S_\oplus} \gamma_t \gamma_r \tag{6.29}$$

where $P_{S_\oplus}$ is the power generated at a distance $r = r_\oplus \triangleq 1$ au from the Sun at the beginning of mission (t_0). In particular, γ_r allows one to take into account the laboratory measurement of solar cell performance made at light intensities and temperatures expected at a Sun-spacecraft distance equal to r (Sauer 1978; Sauer and Atkins 1972; Williams and Coverstone-Carroll 1997). Also, γ_t models the solar array performance degradation with time because of phenomena such as the destructive effects of radiation on the semiconductor material of the solar cells, the decreased optical transmission of concentrator lenses from solar ultraviolet light, and the contamination from spacecraft outgassing (Bourke and Sauer 1972; Rayman et al. 2000; Rayman and Williams 2002).

According to Quarta and Mengali (2011), the following expressions for γ_t and γ_r are assumed for an Earth-Venus transfer:

$$\gamma_t = 0.87 + 0.13 \exp(-3t/T_\oplus) - 0.02(t/T_\oplus) \tag{6.30}$$

$$\gamma_r = \left(\frac{r_\oplus}{r}\right)^2 \left[\frac{1.1063 + 0.1495(r_\oplus/r) - 0.299(r_\oplus/r)^2}{1 - 0.0432(r/r_\oplus)}\right] \tag{6.31}$$

where $T_\oplus \triangleq 1$ year is a reference time used to make γ_t dimensionless. The numerical values for the coefficients in Eqs. (6.30) and (6.31) are of experimental nature and are from Bourke and Sauer (1972), Sauer (1978), Williams and Coverstone-Carroll (1997). Accordingly, the power allocated to operate the spacecraft systems is assumed to take a constant value $P_L \triangleq 0.4$ kW.

As far as the numerical values of the throttle table are concerned, three different electric propulsion systems are considered, whose experimental data are known. The first system coincides with NASA's Evolutionary Xenon Thruster (NEXT), for which a set of 40 operation points is available (Patterson and Benson 2007), see Table 6.3. The second electric thruster is the NASA's Solar electric propulsion Technology Application Readiness (NSTAR), which has been used for DS1 mission and then for Dawn mission. The corresponding throttle table is summarized in Table 6.4, whose data are taken from Brophy (2002). The third propulsion system is a Busek BHT-8000 eight kilowatt Hall thruster, whose characteristics are taken from Szabo et al. (2013) and summarized in Table 6.5. The operation points for the three selected thrusters are graphically shown in Figs. 6.1, 6.2 and 6.3.

To summarize, at a given mission time t, when the spacecraft is at a distance r from the Sun, the solar array power P_S is given by Eq. (6.29), while the available

Table 6.3 Operation levels for NEXT propulsion system (data taken from Patterson et al. 2007)

I_d	T [mN]	P [kW]	β [mg/s]
1	236	7.220	5.76
2	221	6.385	5.76
3	208	5.780	5.76
4	192	4.965	5.76
5	208	6.390	5.12
6	194	5.660	5.12
7	184	5.085	5.12
8	169	4.490	5.12
9	181	5.600	4.46
10	169	4.920	4.46
11	160	4.455	4.46
12	147	3.860	4.46
13	137	3.425	4.46
14	158	4.870	3.92
15	147	4.315	3.92
16	139	3.910	3.92
17	128	3.390	3.92
18	119	3.010	3.92
19	134	4.235	3.16
20	125	3.760	3.16
21	118	3.415	3.16
22	108	2.970	3.16
23	101	2.620	3.16
24	107	3.460	2.60
25	99.9	3.080	2.60
26	94.3	2.765	2.60
27	86.7	2.415	2.60
28	80.6	2.160	2.60
29	80.2	2.585	2.05
30	74.9	2.300	2.05
31	70.7	2.090	2.05
32	65.0	1.825	2.05
33	60.4	1.635	2.05
34	57.8	1.520	2.05
35	55.1	1.415	2.05
36	49.2	1.210	2.05
37	48.1	1.175	2.05
38	37.2	0.865	2.05
39	31.8	0.740	2.05
40	25.5	0.610	1.85
41	0	0	0

Table 6.4 Operation levels for NSTAR propulsion system (data taken from Brophy 2002)

I_d	T [mN]	P [kW]	β [mg/s]
1	92.67	2.567	3.0209
2	87.87	2.416	2.8310
3	83.08	2.272	2.6532
4	78.39	2.137	2.5120
5	73.60	2.006	2.3475
6	68.37	1.842	2.1889
7	63.17	1.712	2.0494
8	57.90	1.579	1.8947
9	52.67	1.458	1.7466
10	47.87	1.345	1.5921
11	42.61	1.222	1.4435
12	37.35	1.111	1.2941
13	32.12	0.994	1.1517
14	27.47	0.825	1.0456
15	24.55	0.729	1.0506
16	20.69	0.577	1.0657
17	0	0	0

Table 6.5 Operation levels for Busek BHT-8000 propulsion system (data taken from Szabo et al. 2013)

I_d	T [mN]	P [kW]	β [mg/s]
1	465	8.060	23.9518
2	507	8.078	27.4320
3	508	8.096	30.0719
4	343	6.158	16.4693
5	365	5.999	18.8199
6	390	6.022	21.7718
7	403	6.016	24.5403
8	395	6.022	26.9873
9	228	4.026	11.3263
10	245	4.063	12.9603
11	265	3.995	15.4185
12	279	3.999	18.0230
13	285	4.040	20.5169
14	143	2.022	9.9163
15	149	2.011	11.7468
16	0	0	0

input power P_A is obtained from Eq. (6.28). The operation level I_d must be chosen under the constraint that the thruster power cannot exceed the available power. As a final remark, for a fixed value of the power necessary to operate the spacecraft systems (P_L), the two design parameters for the SEP engine are the reference electric power $P_{S_\oplus}$ and the spacecraft initial mass m_0.

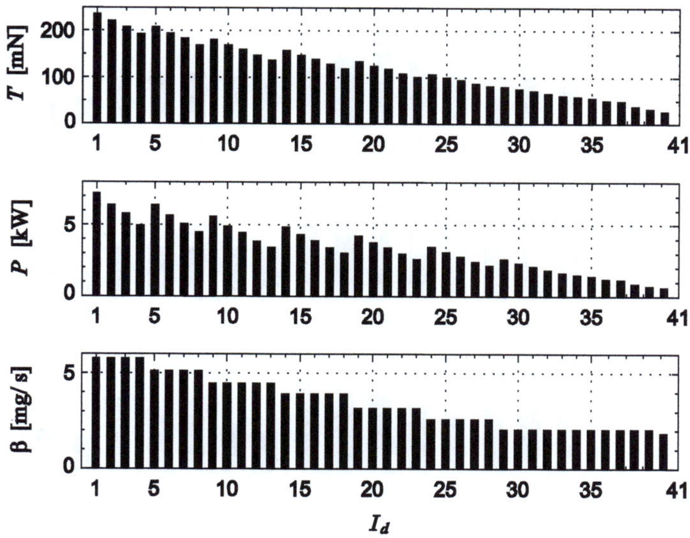

Fig. 6.1 Operation levels for NEXT propulsion system (see also Table 6.3)

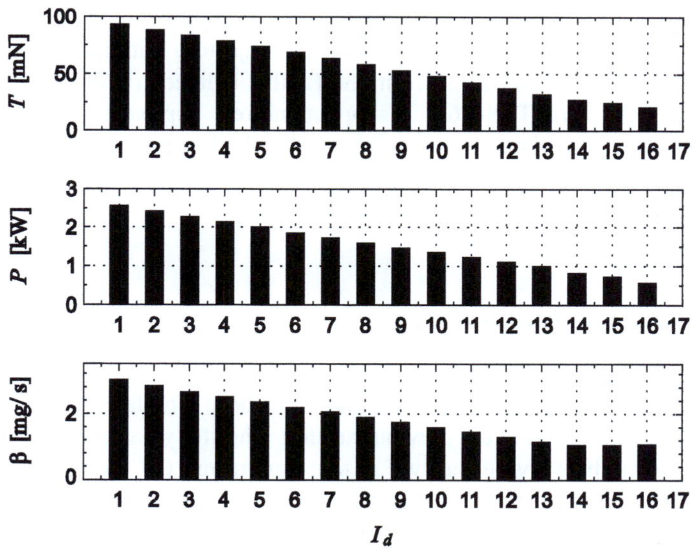

Fig. 6.2 Operation levels for NSTAR propulsion system (see also Table 6.4)

6.2.3 Trajectory Optimization

At a generic point of the transfer trajectory, the control variables, i.e. the thrust level and the orientation of the propulsive acceleration unit vector, are obtained by

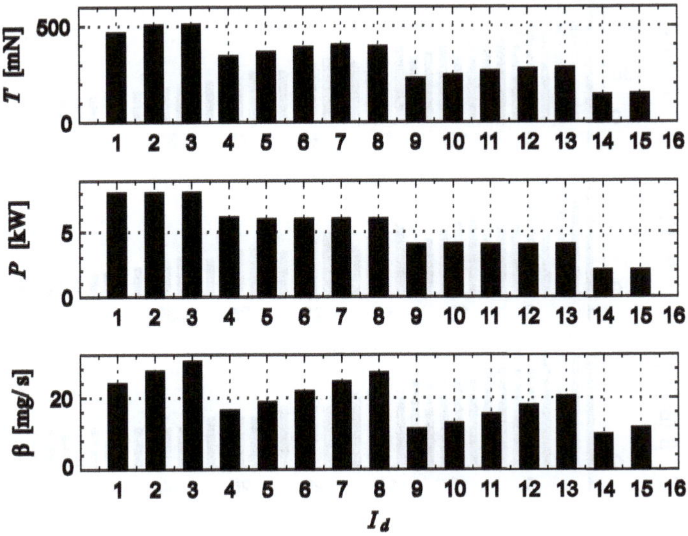

Fig. 6.3 Operation levels for Busek BHT-8000 propulsion system (see also Table 6.5)

solving an optimal control problem. In particular, the optimization problem consists of finding the minimum time trajectory (i.e. the trajectory with the minimum value of t_f) that transfers the spacecraft from the Earth's heliocentric orbit to the Venus' heliocentric orbit, see Table 6.1. Using an indirect approach (Betts 1998), the optimal control time-history is found from the Pontryagin's maximum principle (Chobotov 1996). The optimization algorithm is discussed in detail in Quarta et al. (2013), to which the interested reader is referred.

6.3 Numerical Simulations

A number of transfer trajectories to Venus have been simulated using the mathematical model described in the previous section. The study has been performed in a parametric way, by varying the two design parameters ($\tilde{a}_{T_0}$ and I_{sp} for the NEP engine, $P_{S_\oplus}$ and m_0 for the SEP engine) within a suitable range of variation. The two fundamental performance parameters, i.e. the total flight time t_f and the dimensionless final spacecraft mass $m_f = m_0$, have been calculated and separately collected as a function of the propulsion system used in the simulation.

In all of the simulations, the differential equations (6.1) have been integrated in double precision using a variable order Adams-Bashforth-Moulton solver (Shampine and Gordon 1975; Shampine and Reichelt 1997) with absolute and relative errors of 10^{-12}. The two-point boundary value problem associated to the (indirect) optimization problem, has been solved by means of a hybrid numerical technique that combines the use of genetic algorithms to obtain a rough estimate of the adjoint variables, with gradient-based and direct methods to refine the solution (Mengali and Quarta 2005a, b). A set of canonical units (Bate et al. 1971) has been used in the integration of the differential equations to reduce their numerical sensitivity and the final boundary constraints were set to 100 km for the position error and 0.05 m/s for the velocity error.

6.3.1 Nuclear Electric Propulsion Engine

As previously stated, in a mission performed with a NEP engine, the two design parameters are the initial propulsive acceleration weighted with the duty cycle $\tilde{a}_{T_0}$ and the specific impulse I_{sp}. The parametric analysis has been performed by varying the two parameters within the intervals

$$\tilde{a}_{T_0} \in [0.04,\, 0.4]\ \mathrm{mm/s^2}\ ,\ I_{sp} \in [2000,\, 4000]\ \mathrm{s}$$

For example, the preliminary study of Yam et al. (2004), which analyzes a mission toward the outer planets of the Solar system with a NEP engine providing a thrust of 2.26 N and a specific impulse of 6000 s, estimates an initial spacecraft mass of about 20000 kg. In that case, assuming a duty cycle of $\eta_P = 0.92$, the propulsive acceleration would be slightly greater than 0.1 mm/s^2.

The upper side of Fig. 6.4 clearly shows a hyperbolic behavior of the minimum flight time as a function of $\tilde{a}_{T_0}$ with a modest dependence on the specific impulse. A marked dependence on I_{sp} is instead visible in the bottom of Fig. 6.4, where the final mass is shown as a function of the two design parameters. Recall that m_f is not the maximum spacecraft final mass, but it represents the spacecraft final mass corresponding to the minimum time transfer trajectory. Figure 6.4 shows that the ratio $m_f = m_0$ tends to a constant value when the propulsive acceleration is sufficiently small. This behavior is consistent with the approximate results that can be found using the method of Alfano and Thorne (1994). When the propulsive acceleration is sufficiently small, the variation of m_f/m_0 with $\tilde{a}_{T_0}$ has an oscillatory behavior, and the local minima are reached when the spacecraft carries out an integer number of revolutions around the Sun during the transfer. This intermediate region corresponds to a transition zone from thrust dominance to gravitational-force dominance.

As far as the minimum flight time is concerned, a good approximation of the curve shown in Fig. 6.4 is obtained with the approximate method discussed by Battin (1987). In fact the transfer may be approximated with a circle-to-circle orbit lowering during which the propulsion system is always switched on. Using a circumferential thrust law, which approximates the optimal steering law, the flight time may be estimated as (Battin 1987)

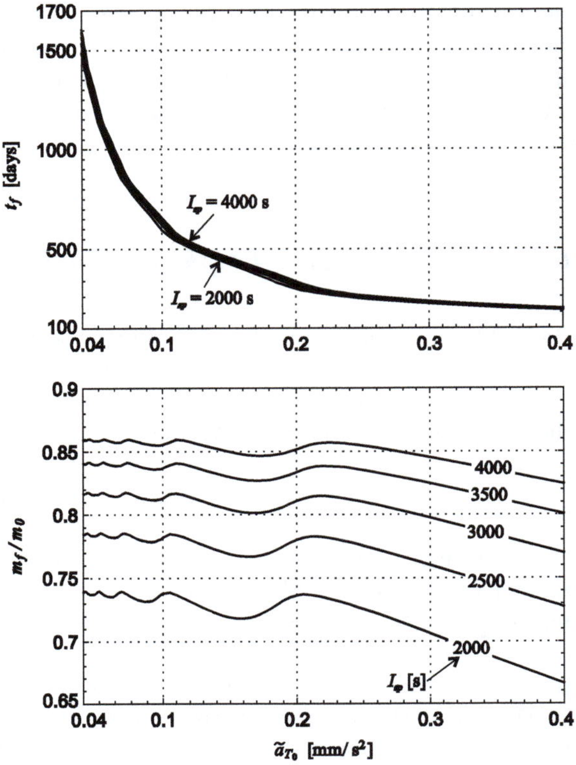

Fig. 6.4 Minimum flight time and final (dimensionless) mass for a mission scenario with a NEP thruster

$$t_f = \frac{\sqrt{\mu_\odot/a_f} - \sqrt{\mu_\odot/a_0}}{\tilde{a}_{T_0}} \tag{6.32}$$

where $a_f = 0.72664$ au and $a_0 = 1$ au are the semimajor axis of Venus's and Earth's heliocentric orbit, respectively, see Table 6.1.

6.3.2 Solar Electric Propulsion Engine

For a SEP system, the three previously described engine models have been used, each one with its own throttle table. For each engine a sensitivity analysis have been performed by varying the two design parameters m_0 and $P_{S_\oplus}$. In particular, a value of $P_{S_\oplus}$ much less than the power required by the typical engine (see column P in Table. 6.3, 6.4 and 6.5) has been assumed to model a (partial) failure of the solar electric power system. Recall that in all of the simulations the duty cycle and the power necessary to operate the spacecraft systems were set equal to $\eta_P = 0.92$ and $P_L = 0.4$ kW, respectively.

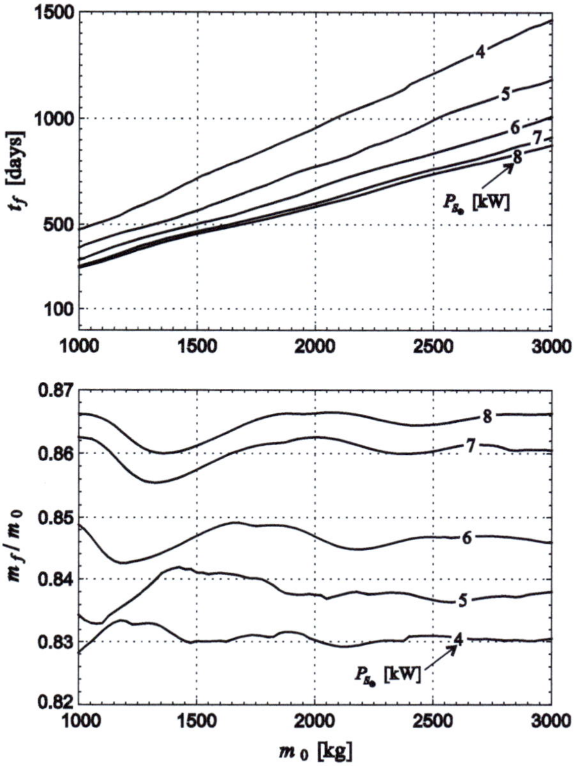

Fig. 6.5 Minimum flight time and final (dimensionless) mass for a mission scenario with a NEXT thruster

For a SEP system of NEXT type, see Table 6.3, the design parameter variation was assumed to be

$$m_0 \in [1000,\ 3000]\ \text{kg},\ \ P_{S_\oplus} \in [4, 8]\ \text{kW}$$

The simulation results are summarized in Fig. 6.5. The figure shows that even a space vehicle with a significant launch mass requires a reasonable propellant mass to complete the mission. For example, if the initial spacecraft mass is $m_0 = 3000$ kg (for comparative purposes the launch mass of NASA's Dawn was about 1210 kg), and assuming that the power generation system provides a nominal power of 8 kW, the required propellant mass is about 13 % of m_0, i.e. 400 kg of propellant. Note that the propellant throughput capability of a NEXT propulsion system (Van Noord 2007) is about 450 kg (qualification-level), which corresponds to 22000 h of operation at maximum thrust (operation point $I_d = 1$, see Table 6.3).

As far as the NSTAR engine is concerned, the design parameters variation was assumed to be

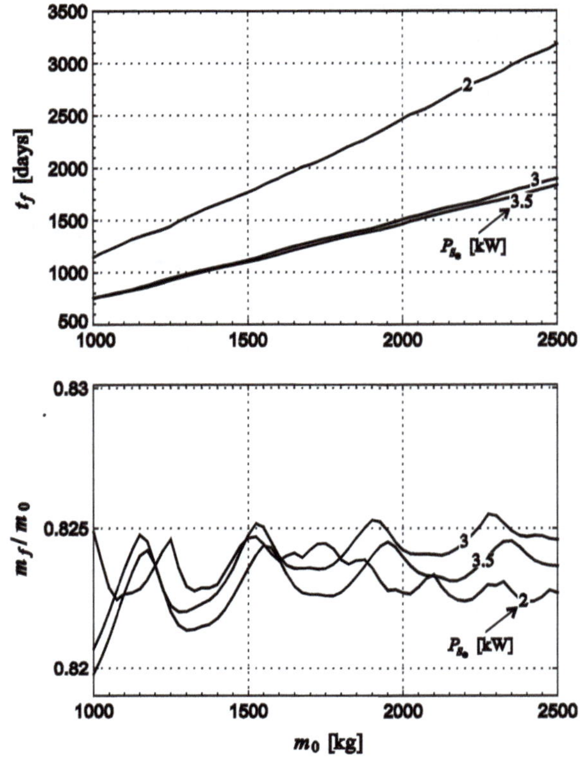

Fig. 6.6 Minimum flight time and final (dimensionless) mass for a mission scenario with a NSTAR thruster

$$m_0 \in [1000, 2500] \text{ kg}, \ P_{S_\oplus} \in [2, 3.5] \text{ kW}$$

and the simulation results are summarized in Fig. 6.6.

In this case the propellant percentage consumption is greater than the previous case, with longer flight times. This is mainly due to the different characteristics of the propulsion system. In particular, the maximum thrust level of NSTAR engine is about 40 % of the corresponding maximum thrust provided by the NEXT engine. This parameter plays a fundamental role in the performance of a minimum-time mission. The percentage consumption of propellant mass for a mission based on a NSTAR engine is also greater than that required by a spacecraft equipped with a NEXT engine. For example, assuming an initial mass $m_0 = 1200$ kg (similar to that used in the Dawn mission) and a reference electric power of 3.5 kW, the minimum flight time is $t_f = 880$ days and the dimensionless final mass is $m_f/m_0 \approx 0.823$, which amounts to a required propellant mass of 212 kg. Such a value is consistent with the actual working capability of a NSTAR engine, see Sengupta et al. (2009).

It is worth noting that for a given value of $P_{S_\oplus}$ the minimum mission time has a nearly linear variation with the initial spacecraft mass. This observation simplifies the construction of the interpolating curves, which provide a simple and quick

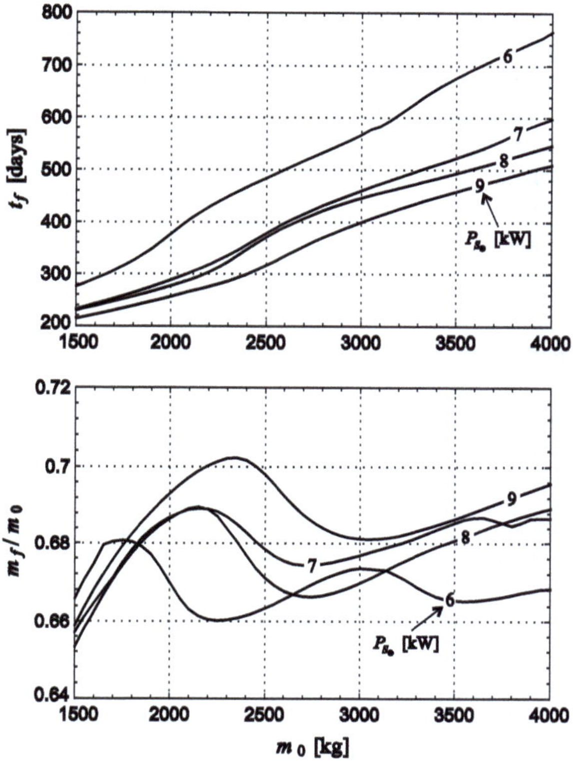

Fig. 6.7 Minimum flight time and final (dimensionless) mass for a scenario with a Busek BHT-8000 thruster

estimation of the minimum flight time. A qualitatively similar behavior is also shown in Fig. 6.7, which summarizes the results obtained for a Busek BHT-8000 engine. In this case the interval of variation of the two design parameters is

$$m_0 \in [1500,\ 4000]\ \text{kg}\ ,P_{S_\oplus} \in [6, 9]\ \text{kW}$$

Not surprisingly, in this case the results are better than those of the previous two engines, because the maximum thrust level provided by the Busek BHT- 8000 engine is greater. At the same time the required propellant mass is greater, due to a lower value of the specific impulse.

6.4 Conclusions

The minimum flight time required to accomplish an Earth-Venus mission has been calculated using different types of electric thrusters. The analysis performed in this Chapter is useful to obtain a first order estimate of the mission requirements as a function of the specific thruster characteristics. The problem has been solved as a function of the design parameters that differ according to the different way with which the electric power supplied to the spacecraft is generated. The results of a parametric analysis involving the mission time and the required propellant mass have been collected into suitable plots. These data represent a first order analysis of the mission, which must then be completed with the constraints related to the planetary ephemerides. A more refined approach should involve tradeoff study in which the performance index to be minimized is a suitable combination of the flight time and the propellant mass consumption.

References

Alfano S, Thorne JD (1994) Circle-to-Circle constant-thrust orbit raising. J Astronaut Sci 42 (1):35–45

Bate RR, Mueller DD, White JE (1971) Fundamentals of Astrodynamics. In: ISBN: 0-486-60061-0. Dover Publications, New York, p 429

Battin RH (1987) An introduction to the mathematics and methods of astrodynamics. In: AIAA Education Series. AIAA, New York, p 418

Betts JT (1998) Survey of numerical methods for trajectory optimization. J Guidance Control Dyn 21(2):193–207. doi:10.2514/2.4231

Betts JT (2000). Very low-thrust trajectory optimization using a direct SQP method. J Comput Appl Math 120(1):27–40. doi:10.1016/S0377–0427(00)00301-0

Bourke RD, Sauer GGJ (1972) The effect of solar array degradation on electric propulsion spacecraft performance. In: AIAA 9th Electric Propulsion Conference. Paper AIAA 72-444. Bethesda, Maryland, USA

Brophy JR (2002) NASA's deep space 1 ion engine. Rev Sci Instrum 73(2):1071–1078. doi:10.1063/1.1432470

Casalino L, Colasurdo, G (2004) Trade-off between payload and trip- time for EP interplanetary trajectories. In: 40th AIAA/ASME/SAE/ASEE Joint Propulsion Conference and Exhibit. Paper AIAA 2004-3805. Fort Lauderdale, Florida

Chobotov VA (1996) Orbital Mechanics. In: Chobotov VA (ed) AIAA Education Series, 2nd edn. American Institute of Aeronautics and Astronautics, New York. Chap. 9, ISBN: 1-563-47179-5

Circi C (2003) Optimal strategy for geostationary orbit acquisition using ion propulsion. J Guidance Control Dyn 26(4):608–614. doi:10.2514/2.5088

Circi C (2004) Mars and mercury missions using solar sails and solar electric propulsion. J Guidance Control Dyn 27(3):496–498. doi:10.2514/1.5425

Coverstone VL, Prussing JE (2003) Technique for escape from geosynchronous transfer orbit using a solar sail. J Guidance Control Dyn 26(4):628–634. doi:10.2514/2.5091

Mengali G, Quarta AA (2005a) Optimal three-dimensional interplanetary rendezvous using nonideal solar sail. J Guidance Control Dyn 28(1):173–177. doi:10.2514/1.8325

Mengali G, Quarta AA (2005b) Time-optimal three-dimensional trajectories for solar photon thruster spacecraft. J Spacecraft Rockets 42(2):379–381. doi:10.2514/1.9417

Mengali G, Quarta AA (2008) Optimal trade studies of interplanetary electric propulsion missions. Acta Astronaut 62(12):657–667. doi: 10.1016/j.actaastro.2008.01.037, ISSN: 0094-5765

Patel P, Scheeres D, Gallimore A (2006) Maximizing payload mass fractions of spacecraft for interplanetary electric propulsion missions. J Spacecraft Rockets 43(4):822–827. doi:10.2514/1.17433

Patterson MJ et al (2001) Development status of a 5/10-kW class ion engine. In: 37th AIAA/ASME/SAE/ASEE Joint Propulsion Conference and Exhibit. Paper AIAA 2001-3489. Salt Lake City, Utah

Patterson MJ, Benson SW (2007) NEXT ion propulsion system development status and performance. In: 43rd AIAA/ASME/SAE/ASEE Joint Propulsion Conference and Exhibit. Paper AIAA 2007-5199. Cincinnati, Ohio

Quarta AA, Mengali G (2011) Minimum-time space missions with solar electric propulsion. Aerosp Sci Technol 15(5):381–392. doi:10.1016/j.ast.2010.09.003

Quarta AA, Izzo D, Vasile M (2013) Time-optimal trajectories to circumsolar space using solar electric propulsion. Adv Space Res 51(3):411–422. doi:10.1016/j.asr.2012.09.012

Rayman MD et al (2000) Results from the deep space 1 technology validation mission. Acta Astronautica 47.2–9, pp. 475–487

Rayman MD, Williams SN (2002) Design of the first interplanetary solar electric propulsion mission. J Spacecraft Rockets 39(4):589–595. doi:10.2514/2.3848

Russell RP (2007) Primer vector theory applied to global low-thrust trade studies. J Guidance Control Dyn 30(2):460–472. doi:10.2514/1.22984

Sauer CG Jr (1978) Modeling of thruster and solar array characteristics in the JPL low-thrust trajectory analysis. In: 13th International Electric Propulsion Conference. Paper AIAA 78-645. San Diego, California

Sauer CG Jr, Atkins KG (1972) Potential advantages of solar electric propulsion for outer planet orbiters. In: AIAA 9th Electric Propulsion Conference. Paper AIAA 72-423. Bethesda, Maryland

Sengupta A et al (2009) Deep space 1 flight spare ion thruster 30,000- hour life test. J Propul Power 25(1):105–117. doi:10.2514/1.36549

Shampine LF, Gordon MK (1975) Computer solution of ordinary differential equations: the initial value problem. In: ISBN: 0-716-70461-7. San Francisco: W. H. Freeman & Co Ltd. Chap. 10

Shampine LF, Reichelt MW (1997) The MATLAB ODE Suite. SIAM J Sci Comput 18(1)1–22. doi:10.1137/S1064827594276424

Standish EM (1990) The observational basis for JPL's DE200, the planetary ephemerides of the astronomical almanac. Astron Astrophys 233(1):252–271

Standish EM (1998) JPL planetary and lunar ephemerides, DE405/LE405. Interoffice Memoradum IOM 312.F-98-048. Jet Propulsion Laboratory

Szabo J et al (2013) Eight Kilowatt Hall thruster system characterization. In: 33rd International Electric Propulsion Conference. Paper IEPC-2013-317. Washington D.C

Van Noord J (2007) Lifetime assessment of the NEXT ion thruster. In: 43rd AIAA/ASME/SAE/ASEE Joint Propulsion Conference and Exhibit. Paper AIAA 2007-5274. Cincinnati, Ohio

Walker MJ (1986) Erratum - a set of modified equinoctial orbit elements. Celest Mech 38 (4):391–392. doi:10.1007/BF01238929

Walker MJH, Owens J, Ireland B (1985) A Set of Modified Equinoctial Orbit Elements. Celestial Mechanics 36:409–419. doi:10.1007/BF01227493

Williams SN, Coverstone-Carroll V (1997) Benefits of solar electric propulsion for the next generation of planetary exploration missions. J Astronaut Sci 45(2):143–159

Yam CH et al (2004) Preliminary design of nuclear electric propulsion missions to the outer planets. In: AIAA/AAS Astrodynamics Specialist Conference and Exhibit. Paper AIAA 2004-5393. Providence, Rhode Island. doi:10.2514/6.2004-5393

Chapter 7
Estimation of the Fuel Consumption for Space Trip to Mercury and Venus

Alexander A. Bolonkin

7.1 Introduction

Mercury is the smallest and closest to the Sun of the eight planets in the Solar System. It has an orbital period of about 88 Earth days. Seen from Earth, it appears to move around its orbit in about 116 days, which is much faster than any other planet. This rapid motion may have led to it being named after the Roman deity Mercury, the fast-flying messenger to the gods. Because it has almost no atmosphere to retain heat, Mercury's surface experiences the greatest temperature variations of all the planets, ranging from 100 K (−173 °C; −280 °F) at night to 700 K (427 °C; 800 °F) during the day at some equatorial regions. The poles are constantly below 180 K (−93 °C; −136 °F). Mercury's axis has the smallest tilt of any of the planets of the Solar System (about 1/30 of a degree), but it has the largest orbital eccentricity. As such, it does not experience seasons in the same way as most other planets such as Earth do. At aphelion, Mercury is about 1.5 times as far from the Sun as it is at perihelion. Mercury's surface is heavily cratered and similar in appearance to the Moon, indicating that it has been geologically inactive for billions of years.

Mercury is gravitationally locked and rotates in a way that is unique in the Solar System. Relative to the fixed stars, it rotates exactly three times for every two revolutions it makes around its orbit. As seen from the Sun, in a frame of reference that rotates with the orbital motion, it appears to rotate only once every two Mercurian years. An observer on Mercury would therefore see only one day every two years.

Because Mercury's orbit lies within Earth's orbit, it can appear in Earth's sky in the morning or the evening, but not in the middle of the night. Also, like Venus and the Moon, it displays a complete range of phases as it moves around its orbit

---

A.A. Bolonkin (✉)
C&R, Brooklyn, NY, USA
e-mail: aBolonkin@juno.com; aBolonkin@gmail.com

V. Badescu and K. Zacny (eds.), *Inner Solar System*,
DOI 10.1007/978-3-319-19569-8_7

relative to the Earth. Although Mercury can appear as a very bright object when viewed from Earth, its proximity to the Sun makes it more difficult to see than Venus. Two spacecraft have visited Mercury: Mariner 10 flew by in the 1970s and Messenger, launched in 2004, remains in orbit.

Venus is the second planet from the Sun, orbiting it every 224.7 Earth days. It has no natural satellite and is named after the Roman goddess of love and beauty. After the Moon, it is the brightest natural object in the night sky, reaching an apparent magnitude of −4.6, bright enough to cast shadows. Because Venus is an inferior planet to Earth, it never appears to venture far from the Sun: its elongation reaches a maximum of 47.8°. Venus reaches its maximum brightness shortly before sunrise or shortly after sunset, for which reason it has been referred to by ancient cultures as the Morning Star or Evening Star.

Venus is a terrestrial planet that is sometimes called Earth's "sister planet" because of its similar size, gravity, and bulk composition (Venus is both the closest planet to Earth and the planet closest in size to Earth). However, it has also been shown to be very different from Earth in other respects. It has the densest atmosphere of the four terrestrial planets, consisting of more than 96 % carbon dioxide. The atmospheric pressure at the planet's surface is 92 times that of Earth's. With a mean surface temperature of 735 K (462 °C; 863 °F), Venus is by far the hottest planet in the Solar System. It has no carbon cycle to lock carbon back into rocks and surface features, nor does it seem to have any organic life to absorb in biomass. Venus is shrouded by an opaque layer of highly reflective clouds of sulfuric acid, preventing its surface from being seen from space in visible light. Venus might have had oceans in the past, but these would have vaporized as the temperature rose due to a runaway greenhouse effect. The water has most probably photo-dissociated and, because of the lack of a planetary magnetic field, free hydrogen has been swept into interplanetary space by the solar wind. Venus's surface is a dry desert-scape, interspersed with slab-like rocks and periodically refreshed by volcanism.

7.2 Methods of Launch and Interplanetary Flight

There are a lot of launch and propulsion systems for the interplanetary space ship and apparatus (Bolonkin 2005, 2006, 2011a, 2013a, 2009a, 2009b, 2009c, 2011b). Currently, the following launch and propulsion systems are known:

1. Rocket launch and flight propulsion systems. These are well developed and widely used at present time. They are chemical and electric. Chemical rocket engines are very expensive, electric rocket engines have a small trust, hence they may be used only as flight propulsion or correction engines as they require a lot of electrical energy.
2. Non-Rocket launch systems and Propulsion Systems. There are many non-rocket launch and flight systems (Bolonkin 2005, 2006, 2011a, 2013a, 2009a, 2009b, 2009c, 2011b, 2012, 2014, 2013b; Wikipedia 2013). For example:

- Cable space accelerators
- Circle launcher and space keeper
- Gas tube hypersonic launchers
- Earth–Moon and Earth-Mars cable transport systems
- Kinetic anti-gravitator (Repulsator)
- Centrifugal space launcher
- Asteroids as propulsion systems of space ships
- Multi-reflex propulsion systems for space and air vehicles and energy transfer for long distance
- Electrostatic Solar wind propulsion
- Electrostatic utilization of asteroids for space flight
- Electrostatic levitation on Earth and artificial gravity for space ships and asteroids
- Guided solar sail and energy generator
- Radioisotope space sail and electro-generator
- Electrostatic solar sail
- Electrostatic AB-Ramjet Space Propulsion
- Beam Space Propulsion
- Magnetic Sail
- High Speed AB-Solar Sail
- Simplest AB-Thermonuclear Space Propulsion and Electric Generator
- Electrostatic Linear Engine and Cable Space AB Launcher
- Magnetic Space Launcher
- Ground catapult launcher

7.3 Theory of Space Flight

Escape velocity. The total specific orbital energy:

$$\frac{v^2}{2} - \frac{GM}{r} \tag{7.1}$$

does not depend on the distance, r, from the center of the central body to the space vehicle in question. Therefore, the object can reach infinite r only if this quantity is nonnegative, which implies:

$$v \geq \sqrt{\frac{2GM}{r}} \tag{7.2}$$

The escape velocity from the Earth's surface is about 11.2 km/s, but that is insufficient to send the body to an infinite distance because of the gravitational pull of the Sun.

Formula for free orbits. Orbits are conic sections so, naturally, the formulas for the distance of a body for a given angle corresponds to the formula for that curve in polar coordinates, which is:

$$\mu \approx GM, \quad p = \frac{h^2}{\mu}, \quad r = \frac{p}{1 + e \cdot \cos\theta}, \tag{7.3}$$

where $\mu = 4 \times 10^{14}$ m^3/s^2 is called the gravitational parameter of the Earth; $G = 6{,}674 \times 10^{-11}$ $m^3/kg\ s^2$ is the gravitational constant; M the mass (kg) of objects 1 (planet, Earth, Sun, etc.); and h is the specific angular momentum of object 2 (debris, satellite, space ship, etc.) with respect to object 1 (Fig. 7.1). The parameter θ is known as the true anomaly, p is the semi-latus rectum, while e is the orbital eccentricity, all obtainable from the various forms of the six independent orbital elements.

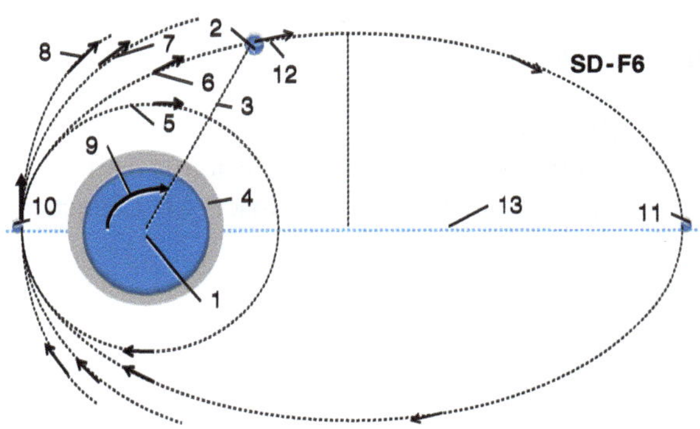

Fig. 7.1 Trajectories of space object (satellite, space ship, space debris, so on) from value of the rocket impulse in perigee (point 10). *Notations*: 1 – planet or star (Earth, Sun); 2 – space object; 3 – radius-vector; 4 – atmosphere of Earth; 5 – circular trajectory; 6 – elliptic trajectory; 7 – parabolic trajectory; 8 – hyperbolic trajectory; 9 - angle θ; 10 – perigee; 11 – apology; 12 – speed of space object; 13 – semi-major axis "*a*"

Circular orbits. All bounded orbits where the gravity of a central body dominates are elliptical in nature. A special case of this is the circular orbit, which is an ellipse of zero eccentricity. The formula for the velocity of a body in a circular orbit at distance r from the center of gravity of mass M is

$$v = \sqrt{\frac{GM}{r}} \tag{7.4}$$

The quantity GM is often termed the standard gravitational parameter, which has a different value for every planet or moon in the Solar System. Sun has $\mu = 1.3276 \times 10^{20}$ m^3/s^2.

Once the circular orbital velocity is known, the escape velocity is easily found by multiplying by the square root of 2:

$$v = \sqrt{2}\sqrt{\frac{GM}{r}} = \sqrt{\frac{2GM}{r}}. \tag{7.5}$$

Elliptical orbits. If $0 < e < 1$, then the denominator of the equation of free orbits varies with the true anomaly θ, but remains positive, never becoming zero. Therefore, the relative position vector remains bounded, having its smallest magnitude at *periapsis (perigee)* r_p which is given by:

$$r_p = \frac{p}{1+e} \tag{7.6}$$

The maximum value r is reached when $\theta = 180^{\circ}$. This point is called the *apoapsis (apology)*, and its radial coordinate, denoted r_a, is:

$$r_a = \frac{p}{1-e} \tag{7.7a}$$

Let $2a$ be the distance measured along the apse line from periapsis P to apoapsis A, as illustrated in the equation below:

$$2a = r_p + r_a \tag{7.7b}$$

Substituting the equations above, we get:

$$r = \frac{a(1-e^2)}{1+e\cos\theta}, \quad a = \frac{p}{1-e^2} \tag{7.8}$$

Orbital period. Under standard assumptions the orbital period (T) of a body traveling along an elliptic orbit can be computed as:

$$T = 2\pi\sqrt{\frac{a^3}{\mu}} \tag{7.9}$$

Velocity. Under standard assumptions the orbital speed (v) of a body traveling along an *elliptic orbit* can be computed from the Vis-viva equation as:

$$v = \sqrt{\mu\left(\frac{2}{r} - \frac{1}{a}\right)} \tag{7.10}$$

where r is the distance between the orbiting bodies. The velocity equation for a hyperbolic trajectory has either $1/(\pm a)$, or it is the same with the convention that in that case a is negative.

Energy. Under standard assumptions, specific orbital energy (ε) of elliptic orbit is negative and the orbital energy conservation equation (the Vis-viva equation) for this orbit can take the form:

$$\frac{v^2}{2}-\frac{\mu}{r}=-\frac{\mu}{2a}=\varepsilon<0 \tag{7.11}$$

Conclusions: For a given semi-major axis the specific orbital energy is independent of the eccentricity.

Parabolic orbits. If the eccentricity equals 1 ($e = 1$), then the orbit equation becomes:

$$r=\frac{h^2}{\mu}\frac{1}{1+\cos\theta},\quad \varepsilon=\frac{v^2}{2}-\frac{\mu}{r}=0,\quad v=\sqrt{\frac{2\mu}{r}} \tag{7.12}$$

Hyperbolic orbits. If $e > 1$, the orbit formula,

$$r=\frac{h^2}{\mu}\frac{1}{1+e\cos\theta} \tag{7.13}$$

describes the geometry of the hyperbolic orbit. The system consists of two symmetric curves. The orbiting body occupies one of them. The other one is its empty mathematical image. Clearly, the denominator of the equation above goes to zero when $\cos\theta = -1/e$. We denote this value of true anomaly $\theta_\infty = \cos^{-1}(-1/e)$ since the radial distance approaches infinity as the true anomaly approaches θ_∞. θ_∞ is known as the *true anomaly of the asymptote*. Observe that θ_∞ lies between 90° and 180°. From the identity $\sin^2\theta + \cos^2\theta = 1$ it follows that:

$$\sin\theta_\infty=\left(e^2-1\right)^{1/2}/e. \tag{7.14}$$

Energy. Under standard assumptions, specific orbital energy (ε) of a hyperbolic trajectory is greater than zero and the orbital energy conservation equation for this kind of trajectory takes form:

$$\varepsilon=\frac{v^2}{2}-\frac{\mu}{r}=\frac{\mu}{-2a} \tag{7.15}$$

Hyperbolic excess velocity. Under standard assumptions the body traveling along hyperbolic trajectory will attain in infinity an orbital velocity called hyperbolic excess velocity (v_∞) that can be computed as:

$$v_{\infty} = \sqrt{\frac{\mu}{-a}} \tag{7.16}$$

The hyperbolic excess velocity is related to the specific orbital energy or characteristic energy by

$$2\varepsilon = C_3 = v_{\infty}^2 \tag{7.17}$$

Here C_3 is sign of *the escape orbit (C3).*

The rocket speed equals:

$$\Delta V = -w \ln \frac{M}{M_0} \quad or \quad \Delta V = w \ln \frac{M_0}{M}, \tag{7.18}$$

where ΔV is increasing the speed of rocket (delta-v), m; w is the velocity of the combustion gas of the rocket fuel (specific impulse), m/s; M_0 is initial rocket mass, kg; M is final rocket mass, kg.

Liquid rocket propellant has the specific $w \approx 3200 \div 3720$ m/s, (for example (top value), N_2H_4/UDMH $\approx$ 3500 m/s, liquid oxygen + kerosene w $\approx$ 3720 m/s, for liquid hydrogen – liquid oxygen $w \approx$ 4630 m/s), for the solid rocket booster $w \approx 2600 \div 3400$ m/s.

Delta-v. In astrodynamics a Δv or *delta-v* (literally "change in velocity") is a measurement of the amount of "effort" that is needed to change from one trajectory to another by making an orbital maneuver. It is a scalar that has units of speed. Delta-v is produced by the use of propellant by reaction engines to produce a thrust that accelerates the vehicle.

Producing delta-v. Delta-v is typically provided by the thrust of a rocket engine, but can be created by other reaction engines. The time-rate of change of delta-v is the magnitude of the acceleration *caused by the engines*, i.e., the thrust per total vehicle mass. The actual acceleration vector would be found by adding thrust per mass on to the gravity vector and the vectors representing any other forces acting on the object. The total delta-v needed is a good starting point for early design decisions since consideration of the added complexities are deferred to later times in the design process.

The rocket equation shows that the required amount of propellant dramatically increases, with increasing delta-v. Therefore, in modern spacecraft propulsion systems, considerable study is put into reducing the total delta-v needed for a given spaceflight, as well as designing spacecraft that are capable of producing a large delta-v. Increasing the Delta-v provided by a propulsion system can be achieved by: staging, increasing specific impulse, improving propellant mass fraction.

Let us assume a multi-stagy rocket having n stages. If we computed the request ΔV, the need ratio $\Delta \overline{M} = M/M_0$ is:

$$\Delta\overline{M} = \frac{M}{M_0} = e^{-\Delta V/w}. \quad \text{If} \quad \frac{\Delta V}{w} < 0.1 \quad \text{than} \quad \Delta\overline{M} \approx 1 - \frac{\Delta V}{w}. \tag{7.20}$$

If i is number of the rocket stage, then:

$$\begin{aligned} \Delta\overline{M} &= \frac{M_1}{M_{0,1}} \cdot \frac{M_2}{M_{0,2}} \cdot \ \ldots \ \cdot \frac{M_n}{M_{0,n}} \\ &= e^{-\Delta V_1/w_1} \cdot e^{-\Delta V_2/w_2} \cdot \ \ldots \ \cdot e^{-\Delta V_n/w_n}. \end{aligned} \tag{7.21}$$

If $w = w_1 = w_2 = \ldots = w_n$, $\Delta V = \sum_{i=1}^{i=n} \Delta V_i$ then $\Delta\overline{M} \approx e^{-\Delta V/w}$.

Delta-v budgets. *Delta-v budget* is the total of all changes in speed made by the propulsion system during the mission. Delta-v is a particularly useful measurement since it is independent of the mass of the space vehicle. Tables of the delta-v required to move between different space venues are useful in the conceptual planning of space missions. In the absence of an atmosphere, delta-v is typically the same for changes in orbit in either direction; in particular, gaining or losing speed means an equal effort. An atmosphere can be used to slow a spacecraft by aerodynamic braking.

A typical delta-v budget might enumerate various classes of maneuvers, delta-v per maneuver, and number of maneuvers required over the life of the mission. It simply sums the total delta-v, much like a typical financial budget. Because delta-v needed to achieve the mission usually varies with the relative position of the gravitating bodies, launch windows are often calculated from pork-chop plots that show delta-v plotted against the launch time.

When designing a trajectory, delta-v budget is used as a good indicator of how much propellant will be required. Propellant usage is an exponential function of delta-v in accordance with the rocket equation; it will also depend on the exhaust velocity.

It is not possible to determine delta-v requirements from conservation of energy by considering only the total energy of the vehicle in the initial and final orbits since energy is carried away in the exhaust. For example, most spacecrafts are launched into an orbit at an inclination similar to that at the launch site, to take advantage of the Earth's rotational surface speed. If it is necessary for mission-based reasons, to put the spacecraft in an orbit of a different inclination, a substantial delta-v is required, though the specific kinetics and potential energies in the final orbit and the initial orbit are equal.

When rocket thrust is applied in short bursts, the other sources of acceleration may be negligible and the magnitude of the velocity change of one burst may simply be estimated by delta-v. The total delta-v to be applied can then be found by adding each of the delta-v's needed at the discrete burns, even though between bursts the magnitude and direction of the velocity changes due to gravity, e.g. in an elliptic orbit.

For examples of calculating delta-v, see Hohmann transfer orbit, gravitational slingshot, and Interplanetary Transport Network. It is also notable that large thrust can reduce gravity drag.

Delta-v is also required to keep satellites in orbit and is expended in propulsive orbital station-keeping maneuvers. Since the propellant load on most satellites cannot be replenished, the amount of propellant initially loaded on a satellite may well determine its useful lifetime.

Launch window. In the context of a spaceflight, *launch window* is a time period during which a particular vehicle (rocket, Space Shuttle, etc.) must be launched in order to reach its intended target. If the rocket is not launched within the "window", it has to wait for the next one.

To go to another planet using the simple low-energy Hohmann transfer orbit, if eccentricity of orbits is not a factor, launch windows are periodic according to the synodic period.

7.4 Computation the Space Flught

7.4.1 Impulse Method

If we have the speed V_1 in apology r_1 and want to reach the perigee in r_2 (Fig. 7.2, from point 3 to point 7) by optimal trajectory, we can get the speed V_2 in point 3:

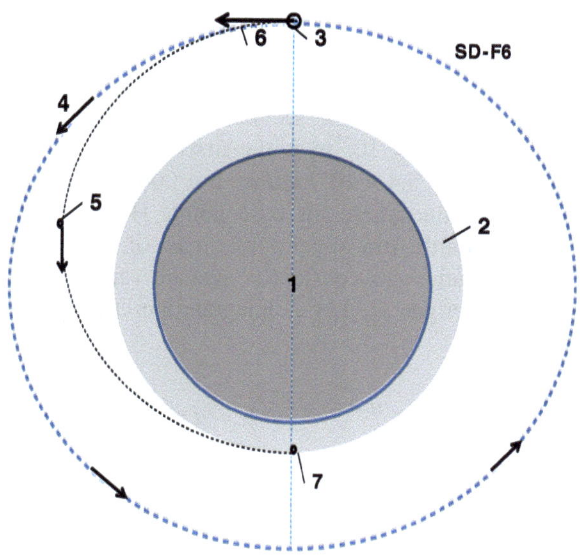

Fig. 7.2 Transfer space object from the circular high orbit to the low circular orbit or to Earth atmosphere. *Notations:* 1 – planet or star (Earth, Sun); 2 – atmosphere of Earth; 3 – space object; 3 – radius-vector; 4 – speed of initial orbit; 5 – speed of the transfer trajectory; 6 – rocket brake impulse

$$V_1^2 - \frac{2\mu}{r_1} = V_2^2 - \frac{2\mu}{r_2}, \quad V_1 = V_2\frac{r_2}{r_1}, \quad V_2 = \left(\frac{2\mu r_1}{r_2(r_1 - r_2)}\right)^{1/2}, \tag{7.22}$$

Hence the requested rocket speed is

$$\Delta V = V_2 - V_1 , \tag{7.23}$$

Unfortunately, this method requires a lot of fuel and significantly decreases the useful load.

7.4.2 Air Brake Parachute Method

In this method we calculate the lifetime of the space object (satellite, space ship, space debris, etc.) located in the top planet atmosphere and braking by the atmosphere drag. For circulate orbit from Energy Law we have:

$$mgdH = DdL \tag{7.24}$$

where m is mass of the space object, kg; g is the planet gravity acceleration, m/s^2; H is altitude, m; D is atmosphere drag at space speed in planet atmosphere, N/m^2; L is the path (way) of the space object, m.

The magnitude of Eq. (7.23) approximately equal:

$$D = C_D\rho(H)V^2S, \quad \rho = a_1e^{-(H-H_0)/b}, \quad q = \frac{m}{S}, \\ -\frac{dH}{e^{-(H-H_0)/b}} = \frac{C_Da_1V^2}{gq}dL \tag{7.25}$$

where ρ is the atmosphere density, N/m^2; V is object speed, m/s; $C_D \approx 1$ is drag coefficient; S is cross section area of object (perpendicular on flight), m^2; H_0 = 100 km = 10^5 m is the first base point of atmosphere (the 2-nd point is 300 km); $a_1 = 5.4 \times 10^{-7}$ kg/m^3, b = 20640 m are the approximation coefficients of the Earth atmosphere for base points H_1 = 100 km and H_2 = 300 km of the Standard Earth Atmosphere; m is mass of space object, kg; q is mass load, kg/m^2.

We assume g = constant = average value on altitude. The speed V equals the speed in perigee V_p. Let us integrate the last equation in Eq. (7.24) from H to H_0. We have:

$$L \approx \frac{bq}{C_Da_1V_p^2}\left(e^{(H-H_0)/b} - 1\right), \quad t = \frac{L}{V}, \quad t \approx \frac{bq}{C_Da_1V_p^3}\left(e^{(H-H_0)/b} - 1\right). \tag{7.26}$$

Here t is lifetime (flight time) of the space object, s; H is perigee altitude, m. If the orbit is ellipse, the lifetime of space object is

$$
\begin{aligned}
& t \approx \frac{bq}{C_D a_1 V_p^3}\left(e^{(H-H_0)/b}-1\right)\cdot\left(\frac{H_a}{H_p}\right) \\
& or \quad t \approx \frac{bq}{C_D V_p^3}\left(\frac{H_a}{H_p}\right)\cdot\left(\frac{1}{\rho}-\frac{1}{\rho_0}\right)
\end{aligned}
\tag{7.27}
$$

where ρ is the air density at the perigee altitude, kg/m^3; $\rho_0 = a_1 = 5.4 \times 10^{-7}$ kg/m^3 is the Earth air density at $H = 100$ km $= 10^5$ m; H_a, H_p is apology and perigee of the initial orbit respectively.

The lifetime from Eqs. (7.25), (7.26) may be recalculated from seconds in days or years:

$$
1 \text{ day } = 8.64\times 10^4\text{s}, \ 1 \text{ year } = 3.145\times 10^7\text{s} \tag{7.28}
$$

The second Eq. (7.26) is used if you have a standard top atmosphere or know the density of the top atmosphere (a_1 and b for given planet). In the range of perigee altitude $H = 200 \div 400$ km we can use the more simple equations:

$$
\begin{aligned}
& t \approx \frac{bq}{C_D a_1 V_p^3}\left(e^{(H-H_0)/b}\right)\cdot\left(\frac{H_a}{H_p}\right) \\
& or \quad t \approx \frac{bq}{C_D V_p^3}\left(\frac{H_a}{H_p}\right)\cdot\left(\frac{1}{\rho}\right)
\end{aligned}
\tag{7.29}
$$

Example. The first soviet satellite had the following data: $m = 83.7$ kg; $d = 0.56$ m; $q = 340$ kg/m^2; $H_p = 288$ km, $H_a = 947$ km. Let us substitute this data in Eq. (7.29), we receive the flight time 95 days. In reality, one had 92 days. We did not consider an air drag of the 4 three-meter antennas.

If we want to delete a satellite as the space debris, we can connect it to a small parachute having diameter $d = 5.6$ m. The flight time decreases by 100 times and in one day, the satellite (debris) comes in the lower Earth atmosphere.

7.4.3 Braking by Solar Radiation

Over the altitude of 400 km, the Earth's atmospheric pressure is very small. At $H = 400 \div 500$ km it approximately equals the solar pressure $p = 4 \times 10^{-6}$ N/m^2. For changing the high orbit we can use only the solar radiation. To increase the efficiency of space usage, the object must have a big, light film surface with different reflectivity (color) on each side (for example: black – white or black – mirror). This reflector is turned one side to the Sun. The solar pressure is different on the two sides of the object and we can increase or decrease the perigee or apology of the space object. We can decrease the perigee up H = 400 km and further to use the air drag.

The author researched and offers the following equations for computation the changing orbit for right control:

$$t \approx \frac{q\mu}{2c_2\delta pV} \cdot \left(\frac{1}{r_2} - \frac{1}{r_1}\right) \quad or \quad t \approx \frac{q\mu}{2c_2c_1\delta p}\ln\left(\frac{r_1}{r_2}\right) \tag{7.30}$$

where $c_2 \approx 1/6$ is coefficient activity (part of used orbit); $\delta = 0 \div 1$ is coefficient reflectivity; V is average orbit speed, m/s; r_1 is initial radius of orbit, m; r_2 is final radius of orbit, m; $c_1 = rV$ is orbit parameter (r is average radius).

Example: Let us to estimate the time needed for decreasing the orbit perigee r_1 = 700 km to r_2 = 400 km by the reflector having q = 1 kg/m^2, δ = 0.5, $p = 4 \times 10^{-6}$ N/m^2. The average orbit speed is 7.15 km/s. Equation (7.30) gives $t \approx 16$ years.

Trajectories. The trajectories to Mercury and Venus are shown in Fig. 7.3.

Results. A general trajectory contains the following parts:

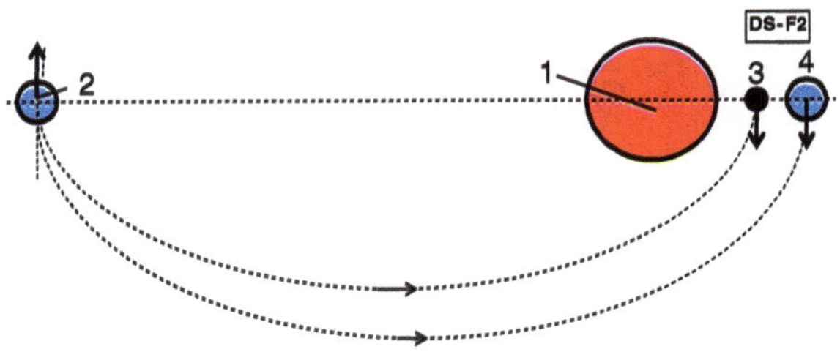

Fig. 7.3 Trajectories to Mercury and Venus. Notations: 1 is Sun, 2 is Earth, 3 is Mercury, 4 is Venus

1. Escape trajectory from the Earth to interplanetary space $\Delta V = 11.2 + (1 \div 1{,}3)$ km/s. Start has two parts: fly to Earth orbit and start from Earth orbit interplanetary orbit. $\Delta V = (1 \div 1{,}3)$ km/s is the loss due to the gravitation and atmosphere drag.
2. Correct orbit during interplanetary flight.
3. Entry to the planetary orbit. Further may be:
4. (a) landing on planet's surface, research of the planet's surface.
   (b) Start and return to the Earth.
   Or
5. Flight to the orbit of the planet's satellite. Further may be:
   (a) landing to the surface of the planet's satellite.
   (b) Start and return to the Earth.

Example of Estimation of space flight to Venus. Let us to show how we estimate ΔV for flight to Venus. Data used for calculation (Table 7.1):

$$r_1 = 15 \cdot 10^{10}\text{m}, \quad r_2 = 10.8 \cdot 10^{10}\text{m}, \quad \mu = 1.3276 \cdot 10^{20}\,,$$
$$V_{or,\,Earth} = 29.78\ \text{km/s}\,, \quad V_{or,\,Venus} = 35.02\ \text{km/s}$$

1. Start from Earth to LEO (Low Earth Orbit). Speed V_2 in end of the transfer orbit and speed V_1 in beginning transfer orbit are:

$$V_2 = \left[\frac{2\mu r_1}{r_2(r_1 + r_2)}\right]^{1/2} = 37.7\ \text{km/s}\,, \quad V_1 = V_2\left(\frac{r_2}{r_1}\right) = 27.14\ \text{km/s}$$

Requested impulses are:

$$\Delta V_1 = V_1 - V_{or,Earth} = 27.14 - 29.78 = -2.64\,\text{km/s}$$
$$\Delta V_2 = V_2 - V_{or,Venus} = 37.7 - 35.02 = -2.7\,\text{km/s}$$
$$\Delta V_3 = V_e - V_o = 10.36 - 7.25 = 3.11\,\text{km/s}$$

Sign minus means that the impulse is oppose the flight direction. If the flight is in atmosphere the gas drag may be used for braking.

Time of flight in one direction is (semi-transfer orbit):

$$a = 0.5 \cdot (r_1 + r_2)\,, \quad T = \frac{\pi}{\sqrt{\mu}} a^{3/2} = \frac{3,14}{\sqrt{1.32 \cdot 10^{20}}} (12.9 \cdot 10^{10})^{3/2}$$
$$= 126.4 \cdot 10^5\ \text{s} = \frac{126.4 \cdot 10^5}{8.64 \cdot 10^4} = 146\ \text{days}.$$

Results of computation for Mercury and Venus are presented in Table 7.2.

The summary impulse for flight from Earth to circular Venus low orbit is

$$\Delta V = \Delta V_0 + \Delta V_1 + \Delta V_2 + \Delta V_3 = 11.2 + 2.64 + 2.7 + 3.11 = 19.65\ \text{km/s}$$

If space apparatus lands on Venus's surface, we must add the brake impulse $V_{0,\ \text{Venus}} = -7.25$ km/s. We can get this impulse by braking in Venus atmosphere.

Let us estimate the requested mass of rocket. The Soviet space apparatus "Venus 5–8" had mass ≈ 1180 kg and used the four-stage launch rocket "Molnia". All stages used same fuel, $w \approx 3.55$ km/s. Substitute these values in Eqs. (7.18)–(7.21) we have

$$\overline{M} = \frac{M}{M_0} = e^{-\Delta V/w} = e^{-19.65/3,55} = 3.945 \cdot 10^{-3},\ M_0 = \frac{G}{\overline{M}} = \frac{1180}{3.945 \cdot 10^{-3}} \approx 300\ \text{tons}$$

In reality "Molnia" has M_0 = 305 tons.

Table 7.1 Data for Sun, Mercury, Venus and Earth

Space body	Distance from Sun, 10^{10} m	Mass, 10^{24} kg	Gravitation constant μ m^3/s^2	Circular speed, V_0 km/s	Escape speed, V_e km/s	Orbital speed, V_{or} km/s	Gravity Constant. g, m/s^2	Planet radius R_p, 10^6 m
Sun	–	$1.99 \cdot 10^6$	$1.3276 \cdot 10^{20}$	437	617.7	–	274	696
Mercury	5.79	0.33	$2.2 \cdot 10^{13}$	3	4.25	47.87	3.72	2.43
Venus	10.8	4.87	$3.2 \cdot 10^{14}$	7.25	10.36	35.02	8.69	6.05
Earth	14.96	5.976	$4 \cdot 10^{14}$	7.9	11.18	29.78	9.78	6.378

Table 7.2 Delta-*V* of the interplanetary flight for Mercury, Venus and Earth

Space body	Escape velocity from Earth, ΔV_o km/s	Impulse to transfer trajectory, ΔV_1 km/s	Impulse to planet orbit trajectory, ΔV_2 km/s	Impulse to circular planet orbit, ΔV_3 km/s	Flight time T, days	Circular speed, V_0 km/s	Escape speed, V_e km/s
Mercury	11.2 + 1.1	7.7	9.4	1.25	105	3	4.25
Venus	11.2 + 1.1	2.64	2.7	3.11	146	7.25	10.36
Earth	11.2 + 1.1	–	–	3.28	–	7.9	11.18

Table 7.3 Timeline by NASA Goddard Space Flight Center

Apparatus	Launch system	Elements and result	Mass of apparatus, kg
Venera 4–8	Molnia-M*	Probe, Lander	1106–1184
Venera 9–16	Proton - M	Probe, Lander	4976–5250
Marine-10	Atlas SLV-30D	Fly by	503
Pioner-Venus	Atlas Centaur	Orbiter	517

* Molnia has 4 stages, mass 305 tons, engine RD-107, Fuel: Kerosene – oxygen. 229 launches, Last launch was in 2010

Brief Information about some apparatus launched to Venus and current Launch rocket Systems. Table 7.3 shows a list of some attempted and successful spacecrafts that have left Earth to explore Venus more closely.

Rocket Launchers are suitable for Venus Launch now (2014):

- Proton-M (Russia): mass 705 tons;
- Delta IV, heavy Delta (USA): mass 249.5–733.4 tons;
- Ariane 5 (Europe): mass 777 tons.

Acknowledgments The author wishes to thank Richard Cathcart (USA) for correcting the English and for offering useful advices and suggestions.

References

Bolonkin AA (2005) Non rocket space launch and flight. Elsevier, p 488. http://vixra.org/abs/1504.0011 v4, https://archive.org/details/Non-rocketSpaceLaunchAndFlightv.3. ISBN-13: 978-0-08044-731-5, ISBN-10:0-080-44731-7

Bolonkin AA (2006) New concepts, ideas, innovations in aerospace, technology and the human sciences, p 510. http://viXra.org/abs/1309.0193, http://www.archive.org/details/NewConceptsIfeasAndInnovationsInAerospaceTechnologyAndHumanSciences. ISBN-13:978-1-60021-787-6

Bolonkin AA (2009a) Review of new ideas, innovations of non-rocket propulsion systems for space launch and flight (Part 1). http://intellectualarchive.com #1368, http://www.archive.org/details/ReviewOfNewIdeasInnovationOfNonrocketPropulsionSystemsForSpace

Bolonkin AA (2009b) Review of new ideas, innovations of non-rocket propulsion systems for space launch and flight (Part 2). http://www.archive.org/details/Review2OfNewIdeasInnovationsOfNonrocketPropulsionSystemsForSpace
Bolonkin AA (2009c) Review of new ideas, innovations of non-rocket propulsion systems for space launch and flight (Part 3). http://intellectualarchive.com #1369, http://www.archive.org/details/Review3OfNewIdeasInnovationsOfNonRocketPropulsionSystemsForSpace
Bolonkin AA (2011a) Femtotechnologies and revolutionary projects. Lambert, USA, p 538. http://viXra.org/abs/1309.0191, http://www.archive.org/details/FemtotechnologiesAndRevolutionaryProjects
Bolonkin AA (2011b) Air Catapult Transportation. NY, USA, IJEE, 2011. http://intellectualarchive.com #1375, http://www.archive.org/details/AirCatapultTransport, http://viXra.org/abs/1310.0065
Bolonkin AA (2012) Re-entry space apparatus to Earth. http://viXra.org/abs/1212.0003, http://archive.org/details/ReentryOfSpaceCraftToEarthAtmosphere, http://www.scribd.com/doc/115174092/REENTRY-OF-SPACE-CRAFT-TO-EARTH-ATMOSPHERE
Bolonkin AA (2013a) Innovations and new technologies. Lulu, 2014, p 465. https://archive.org/details/Book5InnovationsAndNewTechnologiesv2102014/, http://archive.org/details/InnovationsAndNewTechnologies, http://intellectualarchive.com/ #1115, http://viXra.org/abs/1307.0169. ISBN:978-1-312-62280-7
Bolonkin AA (2013b) Provisional patent application: method and installation for cleaning the outer space from space debris. http://vixra.org/abs/1403.0669, http://www.IntellectualArchive.com #1388, http://intellectualarchive.com #1242
Bolonkin AA (2014) New methods of removing space debris. http://viXra.org, Ref. 7936378, http://intellectualarchive.com
Wikipedia (2013). Space launch

Chapter 8
Drilling and Sample Transfer Mechanisms for Potential Missions to Venus

Yoseph Bar-Cohen, Xiaoqi Bao, Mircea Badescu, Stewart Sherrit, Hyeong Jae Lee, Kris Zacny, Nishant Kumar and Erik Mumm

8.1 Introduction

Increasingly, NASA is launching missions with in situ exploration objectives that are aimed at planets in the solar system with extreme ambient conditions including very high temperatures. Venus (Fig. 8.1) and Mercury (Fig. 8.2) are the hottest planets in the Solar system with surface temperatures of hundreds of degrees Centigrade (Bar-Cohen 2014).

Venus is the nearest neighbor in Solar System located next to Earth from the Sun side and it has an ambient temperature of about 460 °C. Similar to Earth, its surface is subjected to geological processes and climate that are driven by feedback

Y. Bar-Cohen (✉) · X. Bao · M. Badescu · S. Sherrit · H.J. Lee
Jet Propulsion Laboratory (JPL)/California Institute of Technology, Pasadena, USA
e-mail: yosi@jpl.nasa.gov

X. Bao
e-mail: xiaoqi.bao@jpl.nasa.gov

M. Badescu
e-mail: mircea.badescu@jpl.nasa.gov

S. Sherrit
e-mail: stewart.sherrit@jpl.nasa.gov

H.J. Lee
e-mail: Hyeong.Jae.Lee@jpl.nasa.gov

K. Zacny · N. Kumar · E. Mumm
Honeybee Robotics, Pasadena, CA, USA
e-mail: zacny@honeybeerobotics.com

N. Kumar
e-mail: kumar@honeybeerobotics.com

E. Mumm
e-mail: mumm@honeybeerobotics.com

V. Badescu and K. Zacny (eds.), *Inner Solar System*,
DOI 10.1007/978-3-319-19569-8_8

Fig. 8.1 A synthetic image of the volcano Maat Mons on the Venus surface. This image was created from Magellan orbital radar data. Credit: Wikimedia commons that is freely licensed media file repository. http://en.wikipedia.org/wiki/File:Venus_-_3D_Perspective_View_of_Maat_Mons.jpg

between the surface, the atmosphere, and its subsurface interior. Over the last half century, many missions have been launched to Venus and it is an objective for future exploration in order to help understand the origin of Earth, and possibly other bodies in the solar system and beyond. One of the questions to be answered is – has Venus ever been like Earth and will Earth someday become like Venus? In addition, efforts are sought to determine the factors that make a rocky body evolve to a warm wet world like Earth, or to a sulfurous, desiccated, extremely hot planet like Venus. The Soviet Union (Venera and Vega) and U.S. past missions to Venus did not answer all of these questions and there have been several recent *in-situ* exploration mission proposals including the Surface and Atmosphere Geochemical Explorer (SAGE) and Venus *In-Situ* Explorer (VISE) to address these questions. A synthetic image of the volcano Maat Mons on the Venus surface that was created from Magellan orbital radar data is shown in Fig. 8.1. This figure shows an example an image taken from the surface of Venus.

Mercury is the other hot planet in the Solar System and it is only slightly larger than Earth's Moon. It has very little atmosphere to stop impacts from meteorites and it is covered with many craters. Its dayside is super-heated by the Sun to temperatures as high as 427 °C, but due to the thin atmosphere, at night the temperature drops significantly to as low as −173 °C, where ice may even exist in its craters. A topographic image of the Goethe basin which is located in Mercury's northern region is shown in Fig. 8.2 and it illustrates an example of Mercury's topography.

The state-of-the-art and the limitations of operating at high temperatures and pressures dictated the scope of the in situ exploration missions to Venus. Technologies have been developed to slow the destructive effect of heating on the landers. Some of these technologies include novel thermal insulation and phase-change heat sink materials temporarily maintaining the inside of the lander at the phase-transition temperature. Over the short length of the Russian's and the US past missions (<2 h), various tasks were executed including making measurements, drilling, and taking photos.

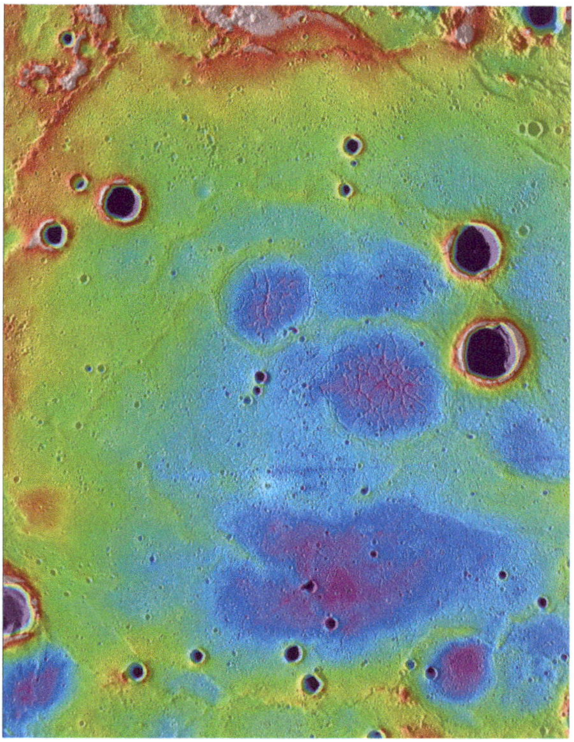

Fig. 8.2 Topographic colorized image from the Mercury Laser Altimeter (MLA) showing the Goethe basin which is located in Mercury's northern region. Credit: NASA/Johns Hopkins University Applied Physics Laboratory/Carnegie Institution of Washington. http://photojournal.jpl.nasa.gov/catalog/PIA15536

Improvements in technology allow for planning missions with longer durations. For example, the proposed SAGE mission is considering a lander survival for 2–3 h. The surface will be analyzed remotely by a suite of instruments that include: cameras, spectrometers, and meteorology package. Also, instruments are proposed for determining the mineralogy and surface texture that will be done after trenching the surface and exposing the sub-surface.

8.2 Drilling on Venus and the Related Challenges

At very high temperatures, drilling or penetrating rocks or subsurface is a significant challenge. On Earth, this is done when drilling very deep oil and gas reservoirs, as well as geothermal or enhanced geothermal wells, and in space when exploring hot planets. In geothermal drilling, for example, the temperature of formations can reach 300 °C or more. Due to the relatively small market, equipment manufactures do not have an incentive to develop geothermal-specific technologies or products. Many of the techniques or tools that are in use were adopted from the oil and gas industry and do not work well past 300 °C (GTP 2010).

In planetary explorations, drilling is required to acquire subsurface samples for in situ analysis or return to Earth (Bar-Cohen and Zacny 2009; Bar-Cohen et al 2011; Bar-Cohen et al. 2014; Gershman and Wallace 1999; Zacny et al. 2008). Venus is one of the planets that, in the latest NASA Decadal Survey study, have been specifically identified as potential targets for a Frontier-class mission. Venus has a very harsh environment and hence its exploration has been postponed due to the unavailability of technologies able to function in the extremely high temperature. The average surface temperature on Venus is about 460 °C and its atmosphere consists mostly of supercritical carbon dioxide due to its pressure of approximately 90 bars. Several spacecraft in the Soviet Venera and Vega series and the NASA Pioneer Venus 2 (Pioneer Venus Multiprobe) made it to the surface and survived long enough to perform surface measurements. While Venera 13 and 14 had small drills, the other missions used only passive instruments such as cameras, seismometers or spectrometers.

Future exploration of Venus will require accomplishing significantly more challenging objectives than past missions and will need to include a means of collecting deeper samples and delivering them to onboard instruments. Unfortunately, under the harsh conditions of Venus, the existing actuation technology cannot maintain functionality and this is one of the major obstacles to the capabilities to sample and to conduct robotic manipulation, as well as other tasks that require moving parts.

When developing devices for operation at high temperature, there are many issues that need to be considered. These are issues that may affect the chemical and physical nature and interactions of the various components and they include materials compatibility, chemical reactions, alloy make-up, annealing and diffusion characteristics. One of the key concerns is thermal expansion mismatch that can be catastrophic to components that need to fit precisely inside a structure or mechanism.

Actuators such as brush, brushless, and stepper motors require magnetic materials. The temperature limitation of electromagnetic motors is dictated by the transition temperature (i.e., Curie temperature) where the material switches from ferro- to paramagnetic causing the loss of actuation capability. Magnetic materials are available commercially but they can be operated only up to approximately 200–300 °C. Some motors, which have been designed for extraction of smoke and deadly toxic fumes during fire emergencies, are available for operation at the range of several hundred degrees Centigrade. However, they are usually large and have lifetimes of a couple of hours. In recent years, electromagnetic actuators that can operate at much higher temperatures were reported by Honeybee Robotics (Bar-Cohen and Zacny 2009) and NASA JPL (Troy 2011). These actuators could be used to actuate drills, robotic arms or enable mobility. The Honeybee Robotics 200 W Switch Reluctance motor (Kumar 2014) was developed for high temperature applications – it is relatively small (2 inch × 2 inch) and has been tested for over 20 h at 460 °C without any degradation in its performance (Bar-Cohen et al. 2014).

Advances in developing electro-mechanical materials, such as piezoelectric and electrostrictive materials, at Penn State University have enabled potential actuation capabilities that can be used to support missions to hot planets (Sherrit et al. 2014). Specifically, development in piezoelectric materials has led to actuators that can maintain operation at temperatures higher than 500 °C (Sherrit et al. 2004). High temperature piezoelectric materials have the advantage that they can be attached to actuated structures, do not require windings of electric wires and electrical or mechanical commutation. An ultrasonic/sonic driller/corer (USDC) mechanism that is a high temperature piezoelectric actuated drill was developed by the authors from JPL. The piezoelectric materials that were used in their investigation include lithium niobate ($LiNbO_3$), which has a Curie temperature of 1150 °C, and bismuth titanate materials with various levels of tungsten doping, which were developed at Penn State University.

The following sections describe the drills that have been developed for potential use on Venus.

8.2.1 Rotary Drill Powered by High Temperature Actuators

In general, drill systems require actuators to rotate the bit, drive a hammering mechanism (if present) and deploy the drill to the surface of the rock or regolith. For Venus high temperature and high pressure applications, conventional actuators would not work. For this reason, over the past decade, Honeybee Robotics developed a suite of high temperature motors. They include Switched Reluctance Motors, Brushless DC Motors, and Hybrid Stepper Motors as shown in Fig. 8.3 (Kumar 2014).

During the development process, material selection was crucial. All the materials were selected to survive and to maintain their critical properties at temperatures above 460 °C. In addition, during the design stage, Coefficients of Thermal Expansion were investigated to make sure the motor will not cease rotating once the motor temperature reaches 460 °C.

The Switched Reluctance Motor (SRM) has no rotor windings and permanent magnets; this simple construction makes it ideal for high fault tolerance applications. If one or more of the motor's phases fail, the motor will still operate, though at a lower torque. The motor also requires no feedback sensor (encoder or resolver) for operation, making it much simpler than DC Brushless motor. A prototype Switched Reluctance Motor developed by Honeybee Robotics is 5 cm in diameter and 5 cm in length (Fig. 8.3A). The motor has been tested non-continuously for over 20 h in a Venus chamber (i.e. at 460 °C and mostly CO_2 gas environment). The motor was fully functional throughout the test duration with no signs of degradation of motor torque or rpm. This motor prototype is comparable to the COTS Maxon RE-25 motor as shown in Table 8.1.

The Brushless DC (BLDC) motor requires feedback for commutating. For this reason, during the motor development effort, a resolver was also designed and manufactured. The prototype BLDC motor is 5 cm in diameter × 5 cm long, while

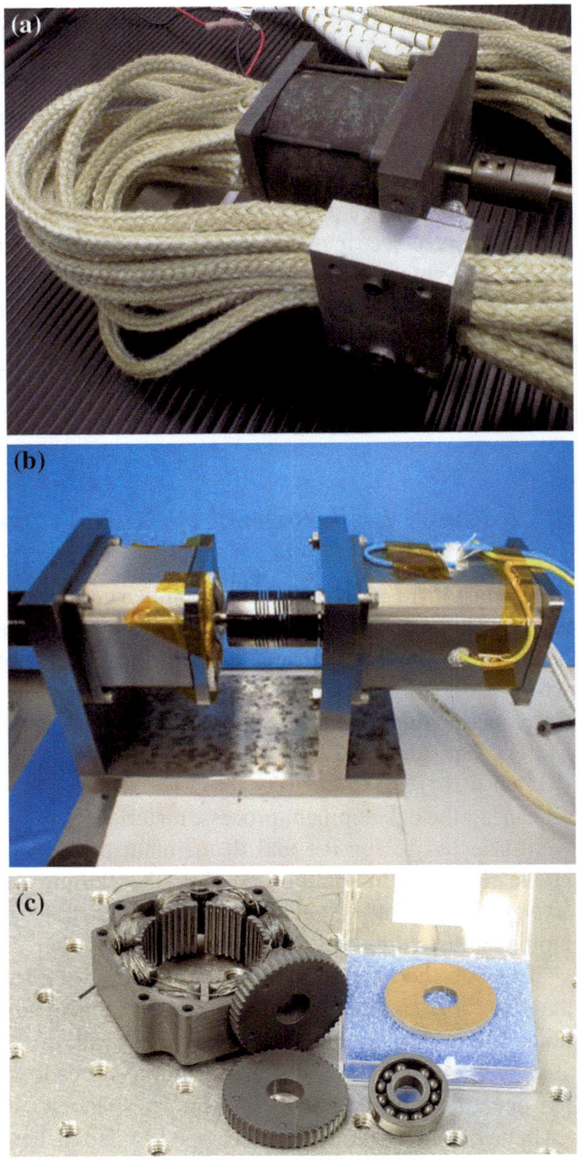

Fig. 8.3 Three high temperature motors developed by Honeybee Robotics. A: Switch reluctance motor; B: Brushless DC motor (left) and Resolver (right); C: Stepper motor

the prototype resolver is 5 cm in diameter × 7.5 cm (Fig. 8.3B). The resolver is coupled to the motor via a shaft coupling. The motor and resolver were successfully tested at 460 °C and 82 atm (8 MPa) in CO_2 atmosphere inside NASA Jet Propulsion Laboratory's Venus Chamber.

Table 8.1 Specifications of the Maxon RE-25 and the Switch Reluctance Motor

Characteristics	Units	Maxon RE-25	Switch reluctance motor
		Range at 25 °C	Range at 460 °C
Applied voltage	V	4.5–48	40
Maximum speed	rpm	5500	5000
No-Load speed	rpm	4790–5500	1000–5000
No-Load current	mA	7–80	1000–4000
Stall torque	mNm	119–144	100–150

A disadvantage that BLDC and SR motors have is they require sophisticated control electronics, while the BLDC also requires a shaft feedback sensor. The Hybrid Stepper Motor, on the other hand, is an open-loop device and thus does not require feedback sensors or complex control electronics. This in turn minimizes wire count required for an actuator. This is especially important when the motor is used to actuate robotic arms as it reduces the number of conductors to be routed across robotic arm joints. The stepper motor has a small step angle of 1.5° (Fig. 8.3C). Since drilling is a high-speed and high torque application, the stepper motor would not be ideal for the drill. For these applications a BLDC or SR motor is a better choice.

A Venus prototype drill was developed with two SR motors to demonstrate drilling at 460 °C temperature. The first actuator rotated an auger with a drill bit at its end while the second actuator rotated the ball screw and in turn provided linear motion (Z-axis) of the bit and auger into and out of a rock as well as the feed force, also called Weight-on–Bit (WOB). Both motors were geared down at the output shaft to accommodate relatively low auger speed and Z-axis rate. The drill volume was 18 cm × 11 cm × 48 cm, while its drill stroke (i.e. maximum drilling depth) was 25 cm. The drill included a 1.27 cm (0.5 inch) diameter carbide bit. All drill components such as gears, bearings and bushings were chosen based on the requirement to survive temperatures of 460 °C.

In order to validate the drill operation at Venus surface temperature, a small test chamber was developed that could maintain 460 °C. The chamber was purged with CO_2 gas during the duration of the rest to roughly simulate Venus surface conditions (Fig. 8.4). The drilling system was tested for more than 15 h and during this time the drill successfully completed three tests by drilling into a solid block of chalk up to ~20 cm depth. The average penetration rate was 1 cm per minute and the average power was 50 W at a bit rotational speed of 400 rpm. Therefore, the energy required to drill to 20 cm was approximately 17 Whr.

Figure 8.5 shows Power and Time as a function of Depth for one of the tests. Note that the rate of penetration was slowing down with depth (the gradient of the Time curve was increasing). This penetration rate decrease was also accompanied by gradual increase in drilling power due to accumulation of cuttings on the auger flutes. Every so often the power peaked and immediately dropped – this is an indication of the auger choking up with cuttings. When the drill initially penetrated

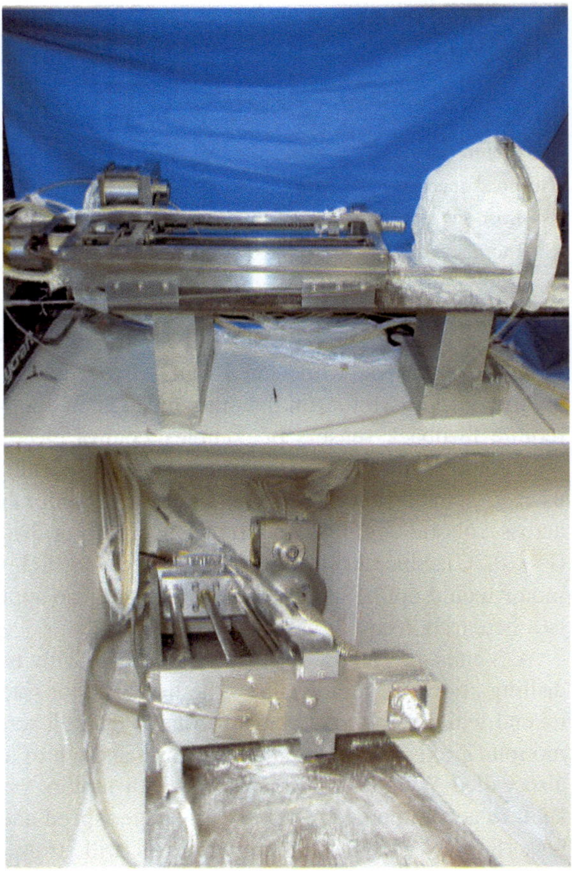

Fig. 8.4 Top: High temperature drill with 2 SR motors before the drill test. Bottom: The drill was placed inside an oven at 460 °C

the rock, the power was ~20 W. This was the actual drilling power to penetrate the target, while the remaining power increase was due to conveyance of cuttings out of the borehole and parasitic losses from drill auger rubbing against the side wall.

An integral part of an actuator is the gearbox. The purpose of the gearbox is to convert high speed and low torque rotation of the motor shaft into lower speed and higher torque rotation of the output shaft. Depending on the load the actuator is required to drive, the gearbox can be custom made to fit the application. Honeybee Robotics has developed a High Temperature gearbox for its Hybrid Stepper Motor (Fig. 8.6). The gearbox is a 2 stage planetary drive with 36:1 reduction ratio. The system was designed taking into account the coefficient of thermal expansion (CTE) of all its gears, housing and bearings. It was successfully tested at 460 °C for over 20 h and provided an output torque of 80 Nm.

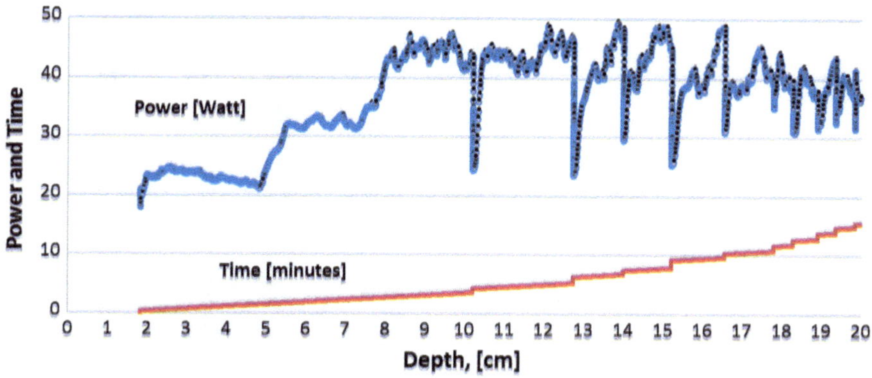

Fig. 8.5 Typical drilling data with a high temperature drill

Fig. 8.6 High temperature gearbox developed by Honeybee Robotics

8.2.2 Piezoelectric Actuated High Temperature Drills

Since 1999, the authors from JPL have been developing piezoelectric actuated drills while addressing the need for drilling at low gravity conditions using low preload (Bar-Cohen et al. 1999; Bao et al. 2003). The applicability of such drills to the exploration of high temperatures planets is the result of the fact that piezoelectric materials can have Curie temperatures that are much higher than the ambient temperatures on Venus. The developed drills were designed to produce both core and powdered cuttings, operate as a sounder to emit elastic waves and serve as a platform for sensors (Bar-Cohen and Zacny 2009). The requirement for low axial load allows for the operation of this drill from lightweight robotic platforms such as rovers. One of the earliest developed drills has been the ultrasonic/sonic driller/corer (USDC) using an intermediate free-mass between the actuator horn and the bit converting high frequency vibrations to low frequency hammering (Fig. 8.7).

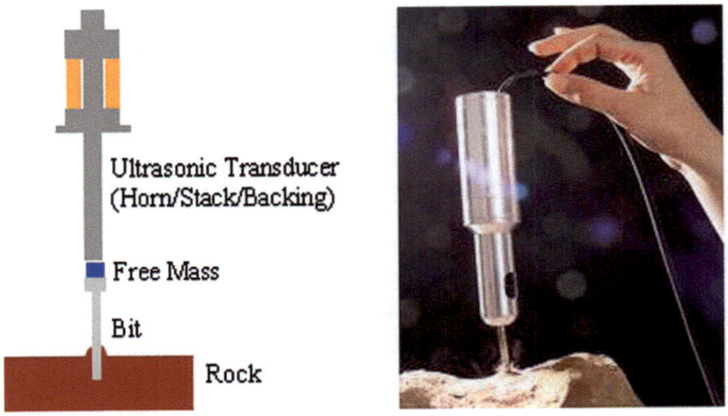

Fig. 8.7 Schematic cross-section view (left) of the USDC and a photo showing its ability to core with minimum axial force (right)

The USDC consists of three key components: actuator, free-mass, and bit (Fig. 8.7) (Bao et al. 2003). The actuator operates as an ultrasonic vibration mechanism that transfers energy to the free-mass and in turn impacts the bit, producing stress impulses onto the drilled medium that fracture rocks when their ultimate strength is exceeded. The actuator consists of a piezoelectric stack, backing, and horn. A stress bolt between the backing and the horn is used to maintain the piezoelectric stack in compression. The backing layer and stress bolt are free to expand and are used to enhance power delivery to the horn. The horn is used to amplify the induced displacement. In the nominal configuration, the actuator is driven at a resonance frequency of about 20 kHz. Using software or hardware control, the drive electronics maintains tuning of the actuator to ensure maximum electric current input. This control allows adjusting the resonance frequency to address the effect of the drilled medium in reducing the Q (the Q is mechanical energy stored divided by the energy loss in each cycle) of the resonator and shifting the resonance frequency as the actuator heats slightly. In addition, there is a need to give attention to the impacts that cause time variations in the current signal. Its effect is minimized using various control algorithms including hill climbing, extremum seeking, and others (Aldrich et al. 2008). Unlike typical ultrasonic drills where the bit is acoustically coupled to the horn, in the USDC the actuator drives a free-mass that converts ultrasonic impacts to hammering at sonic frequencies. This enables imparting higher energy blows to the bit.

Following the development of the USDC, numerous novel designs were conceived and disclosed in NASA New Technology Reports and patents (e.g., A Badescu et al. 2006a, b; Bao et al. 2004; Bar-Cohen et al. 2003a, b, 2005, 2008, 2010; Bar-Cohen and Sherrit 2003a, b; Dolgin et al. 2001; Sherrit et al. 2001, 2002, 2003, 2005, 2006, 2008, 2010a, b). These include the Ultrasonic/sonic Rock Abrasion Tool (URAT), Ultrasonic/Sonic Gopher and the Auto-Gopher for deep drilling (Bar-Cohen et al. 2012), the Lab-on-a-drill, and many others (Bar-Cohen and Zacny 2009). Further, it was demonstrated to drill ice and various rocks including granite, diorite, basalt, and limestone. In this configuration, where the bit vibrates longitudinally and does not rotate, sensors (e.g., thermocouple and fiber optics) were integrated into the bit to examine the borehole during drilling. Another benefit that non-rotating bits enabled has been the ability to produce non-round shape cores. While developing the analytical capability to predict and optimize its performance, efforts were made to enhance its ability to drill at power levels of tens of Watts thereby reaching higher drilling rates.

8.2.2.1 HT Piezo-Ceramic Actuators

Generally, there are many piezoelectric materials with high Curie temperature, which dictates the highest temperature that such materials can heated and still maintain functionality as actuators. The single crystal $LiNbO_3$ has the highest known Curie temperature and it can reach the level of more than 1,100 °C. Increasingly, at a lower temperature levels, new piezoelectric materials with a high Curie temperature are being developed (Sherrit et al. 2014). To determine the aging characteristics of $LiNbO_3$ at 500 °C, an isothermal test was done for 1000 h (Bar-Cohen et al. 2011, 2012, 2014; Sherrit et al. 2014) and, to the level of the measurement error, no appreciable change in properties was observed demonstrating its high stability.

Since the electrodes need to sustain the high temperature environment too, three different types were subjected to 500 °C, including sputtered platinum film; gold-palladium thick film; and sputtered gold film. Evaluating the results of the electrodes exposure to 500 °C showed that the sputtered gold electrodes showed severe degradation while the Pd-Au and Pt electrodes operated quite effectively. Based on these results and the fabrication ease, the electrodes made of sputtered platinum films were chosen for the actuators.

For the purpose of making piezoelectric stacks for driving high temperature percussive drills, a series of bismuth titanate discs with various composition levels of titanate and dopants (W, Fe, Ca, Sr, and Mn) were made at Penn State University (PSU) (Bar-Cohen et al. 2014). The discs were tested to measure the electrical characteristics at room and high temperatures of up to 500 °C. The test results have shown significantly better electromechanical conversion capability compared to the previously known materials. Following these tests, efforts were made to optimize the piezoelectric responses using dopant additive of tungsten into the powder mixture. Effective mixtures were identified and HT piezoelectric rings were made using hot isostatic pressing to ensure the production of robust low porosity ceramic compositions. Specifically, bismuth titanate rings were made with diameters of 25.4 and 38.1 mm (1.0 and 1.5 inch) and, in parallel, $LiNbO_3$ rings with 25.4 and 50.8 mm (1.0 and 2.0 inch) diameter were used.

8.2.2.2 Materials for Fabricating Piezoelectric Actuated Drills

There are many materials that can be used to produce devices for operation at high temperatures. The making of drills requires the materials to sustain impacts challenging the durability of the devices (Fergus and Hoffmann 2014; Smialek and Jacobson 2014). Several materials were used to produce piezoelectric actuated drills that were found effective at room temperatures, including CPM-3 V with hardness (HRC 59) that is a tool steel. This alloy is made by the Crucible Particle Metallurgy process, designed to provide maximum resistance to breakage and chipping in a highly wear-resistant steel. However, the bits that were made of Crucible hardened CPM-3 V and exposed to 500 °C have degraded during drilling. As a replacement for the drill bit material, Tungsten Carbide was used and this grade of carbide is considered one of the hardest commercially available materials having hardness of HRA 92. To make sure that the actuator can be operated at 500 °C, the horn was fabricated of Titanium; the stress-bolt and backing were made of stainless steel; while the Belleville washers were made of Inconel. The use of Belleville washers between the stress bolt and backing allowed for increasing the thermal stability and thermal expansion matching, while maintaining the pre-stress of the piezoelectric stack as the actuator is heated.

For testing the performance of the high temperature drills limestone basalt and bricks were used. Based on a study by Schultz (1993), basalt rock that had unconsolidated compressive strength of 262 MPa at room temperature was measured at 450 °C to have strength of 210 MPa. Generally, large variability was found between the different rocks that were drilled as well as along the depth while drilling. To drill samples with more uniform properties at high temperatures, clay bricks were used to perform the drilling tests.

8.2.2.3 Modeling and Analysis of Actuation Transducers

The piezoelectric transducer that drives the USDC is the key to its operation and to maximize its performance it is essential to optimize its design. For this purpose, finite element (FE), equivalent circuit models and impedance spectrum measurements were used to predict, analyze, and characterize the performances of various configurations of the transducers. ANSYS finite element package, which is capable of dealing with piezoelectric materials, was used to evaluate and optimize various transducer designs. The FE modeling was first performed using the available materials data at room temperature and then for verification the corresponding equivalent circuit was calculated with lumped components (Bao et al. 2003). The verification was performed by impedance spectrum measurement at room temperature. Then, the impedance data of the transducer were measured at high temperature to obtain the ratios of changes of the components of the circuit. For various designs of transducers having similar configurations, the corresponding equivalent circuits calculated by FE were modified using the changing ratios determined with the first fabricated transducer to estimate the performances at high temperature. Details of the analysis and the experimental corroboration are given in (Bar-Cohen et al. 2014). The results were used to come up with effective design of the drill.

8.2.2.4 Testbed Setup

In order to test the developed samplers at high temperatures, a testbed chamber was needed. The chamber has to provide a controlled elevated temperature environment and enable placement of the drill and samples inside the chamber and perform tests while controlling the preload. The JPL authors tested their piezoelectric actuated drill at temperatures as high as 500 °C and tracked the drilling rate as a function of various test parameters. Specifically, the testbed was designed to allow placing the drill inside the chamber with the bit pushed against a test rock while controlling the preload of the drill from outside the chamber. The testbed consisted of a commercial horizontal tube shape furnace (made by Carbolite, UK) with customized insulation side caps having a center hole (see the sketch in Fig. 8.8). This hole was used in the initial studies for inserting the bit through the side cap while the drill was kept outside the chamber. Later, the drill was inserted completely inside the chamber while the cable to its drive electronics was inserted through the hole. To control the load on bit and feed rate, a pushrod was used (Fig. 8.9) that was controlled by a pneumatic cylinder. This fixture allows sliding the drill while driving it and securing the drilled rocks along the bit path. The drill was pushed from outside the chamber using a controlled preload system.

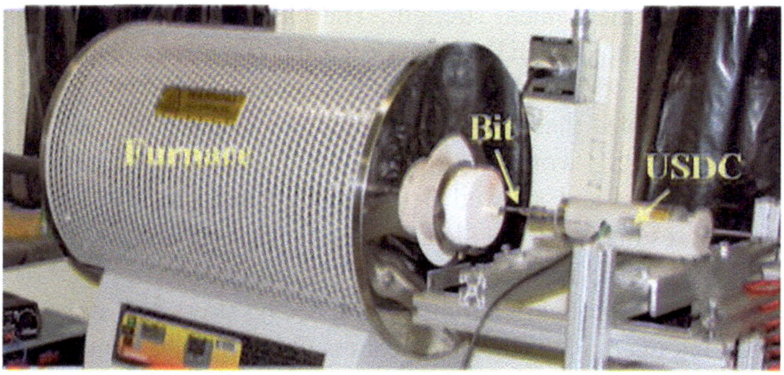

Fig. 8.8 Photographic view of the HT testbed. For the initial testing, rock samples were placed inside the chamber and were drilled at various temperatures while the USDC was placed outside and the preload was controlled pneumatically

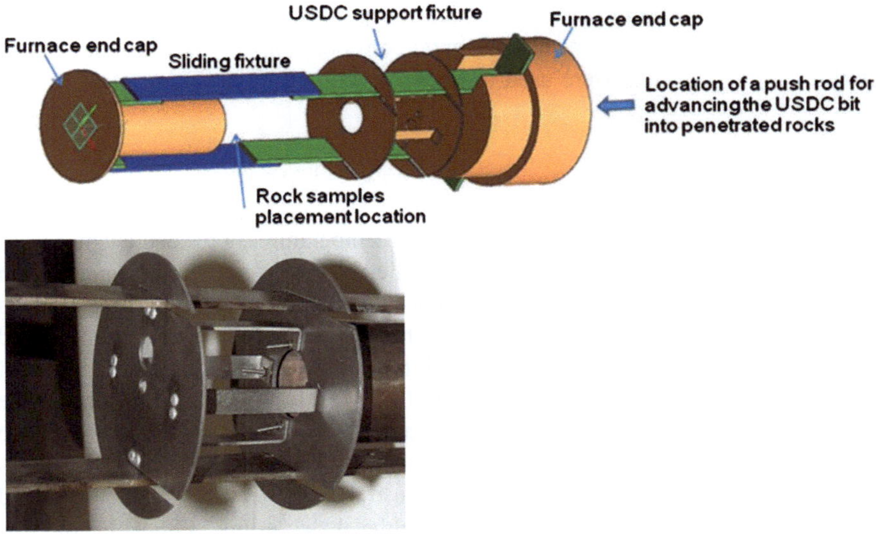

Fig. 8.9 The support fixture for testing the USDC inside the HT chamber. The photograph shows a close-up of the USDC mount

8.2.2.5 HT Piezoelectric Actuated Drills

Basic Designs of the HT Piezo-Actuated Drills

Components of a USDC that can operate at high temperatures were produced to support the testing for the developed piezoelectric stacks. The key components of the USDC that have to be considered for high temperature operation are the actuator, free-mass and bit. The use of titanium allows for producing a low mass device that has high mechanical strength. The piezoelectric transducers that were used are bismuth titanate (BiT) polycrystalline ceramics made by Ferroperm (PZ46), tungsten (W) doped BiT ceramics (PSU), and $LiNbO_3$, single crystals. Preliminary data showed that the samplers that were driven by both BiT ceramics from Ferroperm and PSU worked at 500 °C, but the BiT made by PSU were found to be relatively robust and showed better performance in terms of drilling rate. Based on the test results it was decided to make larger diameter samplers using PSU BiT and $LiNbO_3$. $LiNbO_3$ single crystals and BiT ceramic discs (PSU) were mounted as actuators and held in compression by the stress bolt to prevent fracture of the discs during operation. The analytical results were used to guide the design of transducers with 2.5 cm (1 inch), 38.1 mm (1.5 inch) and a 48.3 mm (2 inch) diameter.

2.5 cm (1 inch) Diameter Samplers

Samplers with 2.5 cm (1 inch) actuators were tested drilling brick at both room temperature and 500 °C. A photograph of the $LiNbO_3$ actuator after being exposed to the 500 °C environment is shown in Fig. 8.10. Several cracks were found on the crystal after test and it is believed to be due to overstressing the crystals while tightening the stress bolt. Note that the single crystals are relatively fragile compared to polycrystalline ceramics and, therefore, care must be taken during fabrication and operation process. However, the test results were encouraging since the resonance characteristics of the $LiNbO_3$ transducer indicated good response in spite of the fracture of the individual discs. To overcome the cracking issue, the discs were manufactured with beveled edges to minimize stresses on the corners and softer electrodes were used to avoid stress concentrators. In addition, a more stringent assembly procedure was initiated. Another issue that was identified is the electric shorting to the stress bolt from the electrodes through the mica film electrical insulation. To prevent applying transverse stresses on the mica insulation and possibly damaging the film and causing a short, great efforts were made to assure the alignment of the electrodes. Also, a thicker layer of mica was used to increase its insulation capability.

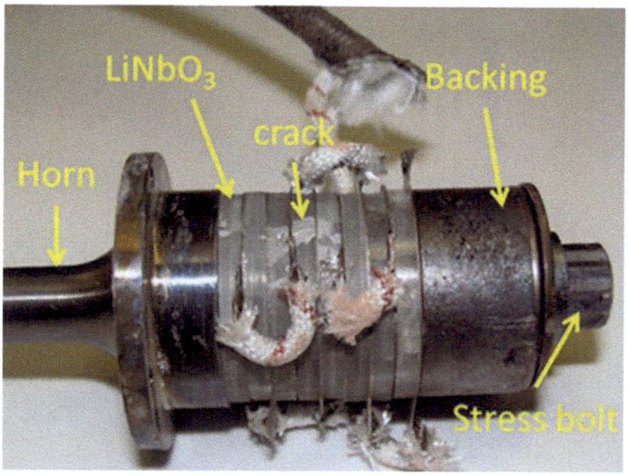

Fig. 8.10 Photographic view of the first HT USDC actuator breadboard. The piezoelectric stack was made of $LiNbO_3$ and was tested at 500 °C

3.81 cm (1.5 inch) Diameter Bismuth Titanate Samplers

A 3.81 cm (1.5 inch) diameter bismuth titanate driven USDC-based sampler was made and tested drilling a 26 mm thick brick at 460 °C. The weight-on-bit was 20 N and the preload for free-mass was 17 N. The sampler was driven in a duty cycle of 50 % (i.e., half the time on and half off). The bit was manually rotated at a low speed of about 4–5 rpm and it was much easier to rotate when the sampler was active than inactive. The torque needed for the rotation was estimated to be less than 0.02 Nm (0.2 inch-lbs). The test results are presented in Figs. 8.11 and 8.12. A starting hole of about 1 mm deep was drilled initially at room temperature and the drilling was continued at high temperature. The driving frequency was 20.9–21.0 kHz, and the averaged power was 26 W. The power factor was low (approximately ~0.094) and the sampler drilled through the sample in 21 active minutes and reached the other end of the brick sample. The average drilling rate was found to be 1.2 mm/min.

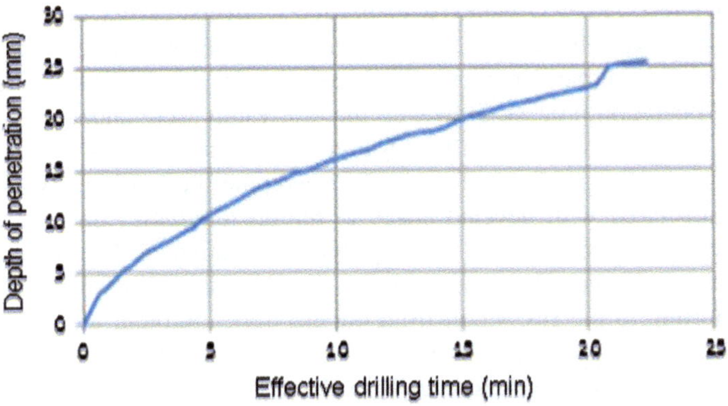

Fig. 8.11 Penetration vs. time at 460 °C

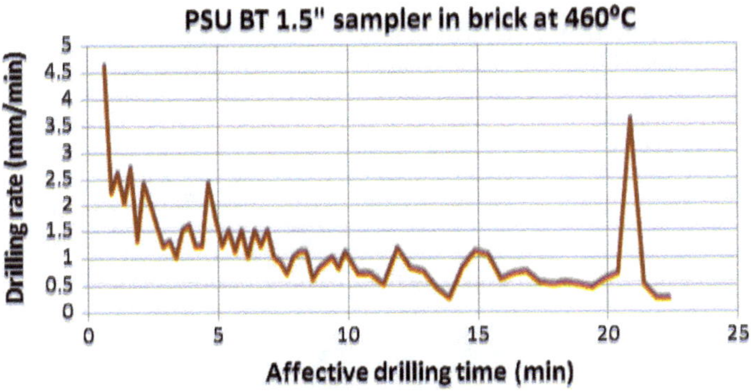

Fig. 8.12 Drilling rate vs. time at 460 °C

4.83 cm (2 inch) Diameter $LiNbO_3$ Samplers

A 5.08 cm (2 inch) diameter $LiNbO_3$ actuator was assembled, where the stack was compressed circumferentially by wrapping fiberglass around it. Also, two de-poled PZT rings were used to provide further protection of the $LiNbO_3$ stack from cracking and a self-rotating bit assembly was added. Tests that were conducted at room temperature with a rate of 7–8 mm/min in a brick sample were recorded. Then, the drill was tested at high temperature (460 °C) and the result is given in Fig. 8.13. The driving frequency, voltage and weight-on-bit were adjusted during the test in order to search for the best operating parameters. In the last 5 min of the test, the driving frequency was 19.9 kHz, the average power was 78 W, the weight-on-bit was 18.7 N, the preload for free-mass was 40 N, and the average drilling rate was found to be 0.43 mm/min. After the test, cracks were seen on the stack accompanied with a large degraded drilling performance at room temperature, indicating that the high temperature drilling operation without cracks for $LiNbO_3$

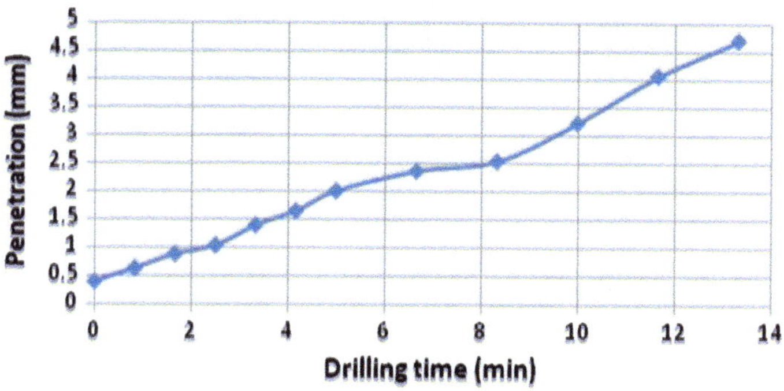

Fig. 8.13 The penetration depth as a function of drilling time at 460 °C for the drill with 4.83 cm (2 inch) diameter $LiNbO_3$ actuator

single crystals remains to be solved. One approach to avoid cracking is to use 1–3 $LiNbO_3$/epoxy composites. This configuration allows for improved mechanical flexibility and more robustness in thermal cycling, while maintaining piezoelectric properties. This approach will be considered for future actuator designs.

8.2.3 Thermal Drilling

Penetrating the subsurface by heating is another method of penetration and there are two principal methods of thermal drilling: thermal-spalling and thermal-melting. Thermal–spalling occurs at temperatures of about 400–600°C and it uses the resulting thermal expansion and mismatch. On the other hand, thermal-melting uses vaporization at temperatures at the range of 1,100–2,200 °C (Maurer 1980).

8.2.3.1 Thermal Spalling

Thermal spalling is a natural process of rock breakage and it is also known as "exfoliation", where the resulting stresses cause flaking of rock fragments due to the fracture and degradation of the rock. Such thermal stresses can be generated when heating a heterogeneous rock and they are caused by mismatch in the thermal expansion of the constituents and the grains within its structure (Just 1963). The fracturing process results from the thermal gradients that are produced in the rock and they dictate the effectiveness of the fracture but this process is limited in capability since not all rocks are sufficiently heterogeneous to sustain spalling. Since the thermal spalling method needs a temperature differential to work, one will need to reach temperatures significantly higher than Venus ambient at the hole bottom to make this work.

Fig. 8.14 Example of the effect of thermal spalling - granite dome exfoliation form due to the surface layers cyclic temperature increase and decrease. Credit: This image is from Wikimedia commons, a freely licensed media file repository. http://en.wikipedia.org/wiki/File:GeologicalExfoliationOfGraniteRock.jpg

Generally, in the desert rock surfaces are heated and expand during the day and contract during the night due to cooling (Fig. 8.14).

The expansion and contraction creates small cracks that grow with time until entire layers of the rock break off. The process is accelerated in the presence of water when it reaches below freezing temperatures due to the increase in the expansion coefficient of water ice. Heating a rock non-uniformly may also cause development of internal stresses and cracks. Laser or plasma can be used as forms of heating (Xu et al. 2003) and the degraded rock material can be removed by such techniques as water jet erosion.

8.2.3.2 Melting and Vaporization

A rock that is subjected to very high temperatures can melt and the process is followed by vaporization, where using a high intensity laser beam allows forming controlled-shape holes (Ready 1997). The laser does not necessarily have to be close to the material and on the NASA's Curiosity rover a laser is used to vaporize rocks from a distance of up to 7 m. This laser is part of a Laser Induced Breakdown Spectroscopy (LIBS) instrument that is called the Chemistry and Camera (ChemCam) analyzer. Once the laser beam strikes a rock it creates plasma and the resulting emission spectrum is captured by a remote micro-imager (RMI) to determine the composition of the rock. An example of holes produced by the laser in a Mars rock at a distance of 3.5 m (11.5 feet) from the Curiosity rover is shown in Fig. 8.15.

Melting rocks requires very high power but, in contrast, melting ice requires lower power. At the Earth's poles, this method is used to drill ice allowing penetration rates as high as hundred meters per hour. Using ice melting drills has been

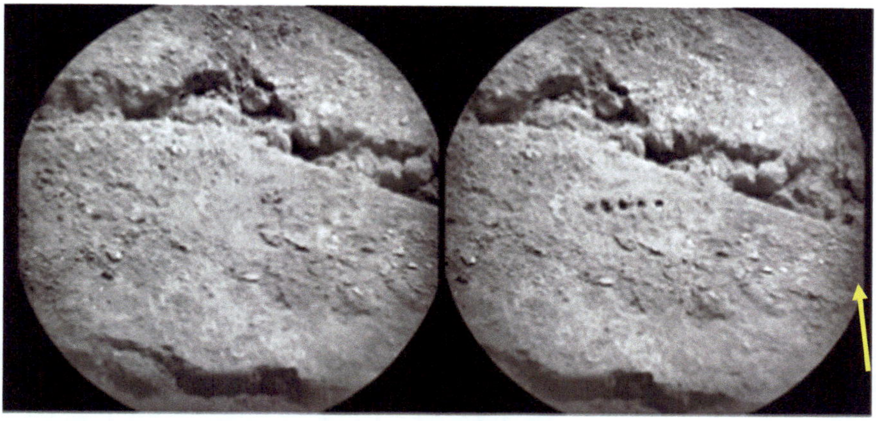

Fig. 8.15 Laser holes that were produced by the chemistry and camera instrument from the NASA's curiosity Mars rover. The photo on the left shows the rock before it was interrogated by the laser while on the right 5 holes are shown that were generated. Image credit: NASA/JPL-Caltech/LANL/CNES/IRAP http://photojournal.jpl.nasa.gov/catalog/PIA16075

proposed for NASA exploration missions to Europa (one of the icy moons of Jupiter), and to the water-ice cap at the poles of Mars (Smith et al. 2006; Rapp 2007).

As an alternative to direct heating one can use microwaves to melt objects (Maurer 1968) and researchers at Tel Aviv University, Israel (Jerby 2002) have investigated the use of a coaxial near-field radiator driven by a conventional microwave source (Fig. 8.16). The center electrode of a coaxial radiator was used as the bit and drilled by softening and melting the penetrated medium. The drill bit serves as an antenna that focuses microwave energy onto a small spot below the surface of the drilled material.

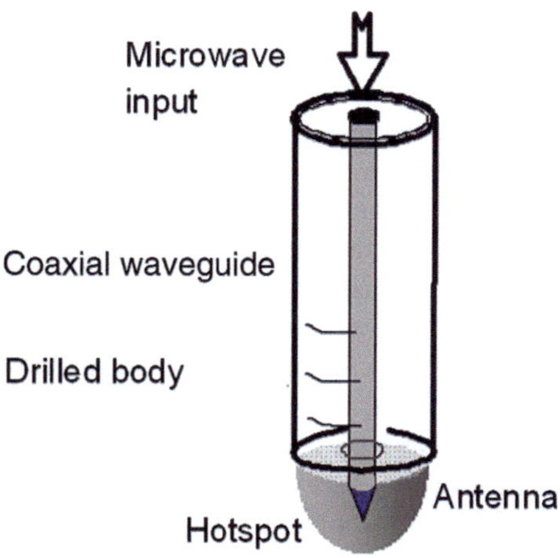

Fig. 8.16 Schematic illustration of a microwave-drill. Courtesy of Eli Jerby, Tel Aviv University, Israel

Fig. 8.17 A 12-mm-diameter, 10-cm-depth hole in a concrete slab produced by a microwave drill. Courtesy of Eli Jerby, Tel Aviv University, Israel

An example of a 12-mm diameter and 10-cm deep hole in concrete is shown in Fig. 8.17. The benefits of using this method include the elimination of the need to rotate the drill bit and it does not produce dust or noise. The latter advantage is important for such applications as construction in urban areas or for some military applications. However, this method is not ideal for *in-situ* planetary exploration due to its high power requirement that can be in the range of hundreds of Watts or more and the thermal and physical/chemical alteration or destruction of the drilled material and the material surrounding the hole. This implies that the method will only be of use for studies of elemental composition, with no structural info.

8.3 Sample Handling

The extreme conditions on Venus (460 °C and 90 atm) prevent the use of any existing science instruments outside of the lander (Bar-Cohen 2014). Due to these conditions, the current lander configurations that are proposed for Venus consist of a pressure vessel where the internal science instruments are kept at close to Earth-like temperature and pressure. This configuration has limited the duration and the possible science investigations that are feasible for the mission under consideration to about 3 h.

Sampling and sample handling is a significant challenge due the extremely high temperature that is involved. A pneumatic sample transfer system was conceived by several of the authors (Badescu et al. 2013) and it is illustrated in Fig. 8.18. The system allows sample powder acquisition, encapsulation, rapid transfer inside the spacecraft and analysis for planets and bodies with extreme environments including Venus, Mercury and others. Development of the conceived breadboard will significantly enhance the science capability of such missions by enabling direct analysis of sampled surface materials using instruments that require close contact and cannot be subjected to the ambient extreme conditions. A high temperature drilling mechanism will produce the required powder from the planetary surface and a pneumatic system (using the pressure difference between the ambient and external reservoirs or the spacecraft internal low pressure) will transfer the powder via capsules to a carousel inside the lander where they will be tested by science instruments onboard.

The samples will be inserted into a series of 6 small capsules and transferred into the lander for analysis. While the Soviet Union used capsules to deliver samples, their approach was limited in terms of repeatability and required significant power. Although Soviet landers determined the presence of major elements in the crust at several locations, the measurements had quite large error bars, and provided only a limited suite of elements. Current technology can provide a vastly better

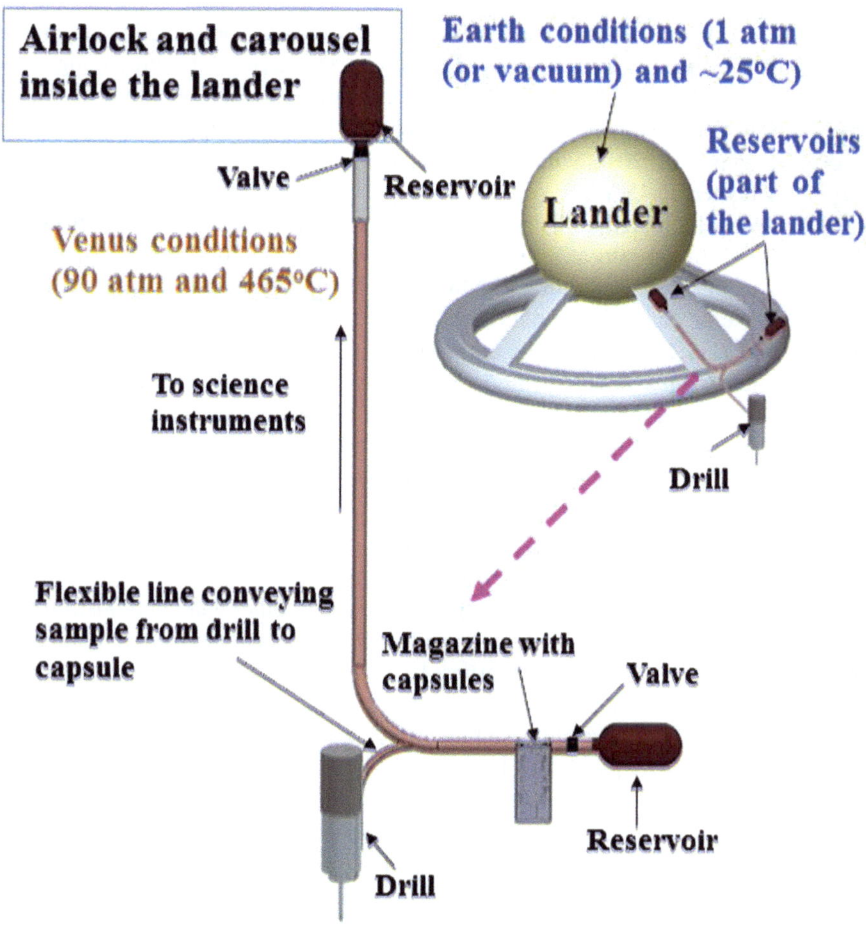

Fig. 8.18 Schematic view of the proposed pneumatically driven system

understanding of the geochemistry of the crust, and yield insights on the depth and origin of volcanism, the differentiation of the crust and core, and geologic history of the planet.

The conceived breadboard will enable studies of the crystallography, chemistry, mineralogy, and petrology of the planetary surface by future exploration missions to bodies with extreme environment.

8.4 Summary and Conclusions

This chapter covered methods of drilling in a high temperature environment. The described methods include the use of high temperature piezoelectric materials and electromagnetic actuators as well as thermal techniques.

Two switch reluctance motors (SRM), which were developed for high temperature applications (Honeybee Robotics), have been used for actuating a conventional drill. The motors were employed to activate a rotating bit with an auger and for advancing the bit into the rock. Tests were conducted at temperatures as high as 460 °C that have shown that this conventional drilling approach is feasible. A drilling depth of 20 cm has been reached in approximately 20 min with a maximum power of 45 W.

Piezoelectric actuated drills with various diameters have been described that were designed to operate at temperatures as high as 500 °C. The use of various piezoelectric materials including $LiNbO_3$ and Bismuth Titanate (made by Ferroperm and Penn State University) was tested by fabricating Ultrasonic/Sonic Driller/Corer (USDC) based drills and testing them at high temperatures. The drill was designed with a novel method of operating the USDC as a rotary-hammer where the rotation is induced by the vibration of the piezoelectric actuator. The drill was actuated by piezoelectric materials that have higher Curie temperature than the temperature on the surface of Venus. The drill that was driven by Bismuth Titanate (PSU) showed the best results where a brick sample was drilled to a depth of 25 mm in 21 active minutes (accounting for the use of duty cycling). A key benefit of the developed drill is its ability to create fine powdered cuttings that are ideal for X-Ray Diffraction and X-Ray Fluoroscopy analysis and many other analytical instruments. The high pressure on Venus was not considered since it is not expected to affect the operation of the USDC. Many significant challenging issues had to be overcome including cracking when using a $LiNbO_3$ crystal as an actuator for the USDC-based sampler. The cracking was addressed by applying radial compression via fiberglass wrapping of the piezoelectric stack.

Beside the mechanically actuated drills there are thermal drills including thermal-spalling and thermal-melting. The thermal–spalling drills operate at about 400–600 °C causing fracture of rocks due to grain expansion and thermal mismatch; whereas thermal-melting uses vaporization at temperatures in the range of 1,100–2,200 °C. Only rocks that contain minerals with different coefficient of thermal expansion are suitable for thermal-spalling. On the other hand, thermal melting requires temperatures that are high enough to melt the rock. In both cases, some means of rock removal out of the hole is required.

In addition to the drills that were described, the chapter also covered a mechanism that was conceived for the transfer of acquired powder samples in a potential Venus lander. The mechanism uses pneumatic actuation and involves the use of capsules. The inclusion of such a mechanism in a future mission will allow the use of science instruments to analyze the acquired powders, which otherwise will not be feasible due to the temperature limitation of existing analytical instruments.

Acknowledgments Some of the research reported in this manuscript was conducted at the Jet Propulsion Laboratory (JPL), California Institute of Technology, funded by the NASA Planetary Instrument Definition and Development Program (PIDDP), under a contract with the National Aeronautics and Space Administration (NASA).

References

Aldrich J, Bar-Cohen Y, Sherrit S, Badescu M, Bao X, Scott J (2008) Percussive augmenter of rotary drills (PARoD) for operating as a rotary-hammer drill. NTR Docket No. 46550

Badescu M, Sherrit S, Olorunsola A, Aldrich J, Bao X, Bar-Cohen Y, Chang Z, Doran PT, Fritsen CH, Kenig F, McKay CP, Murray A, Du S, Peterson T, Song T (2006a) Ultrasonic/sonic gopher for subsurface ice and brine sampling: analysis and fabrication challenges, and testing results. In: Proceedings of the SPIE smart structures and materials symposium, Paper #6171-07, San Diego, CA

Badescu M, Sherrit S, Bar-Cohen Y, Bao X, Kassab S (2006b) Ultrasonic/sonic rotary-hammer drill (USRoHD). US Patent 7,740,088, 22 June 2010. NASA New Technology Report (NTR) No. 44765

Badescu M, Avellar L, Sherrit S, Bar-Cohen Y, Zimmerman W (2013) Pneumatic sample acquisition and transfer. NTR Docket No. 49167, Submitted on May 14, 2013

Bao X, Bar-Cohen Y, Chang Z, Dolgin BP, Sherrit S, Pal DS, Du S, Peterson T (2003) Modeling and computer simulation of ultrasonic/sonic driller/corer (USDC). IEEE Trans Ultrason Ferroelectr Freq Contr 50(9):1147–1160

Bao X, Bar-Cohen Y, Chang Z, Sherrit S, Stark R (2004) Ultrasonic/sonic impacting penetrator (USIP). NASA NTR No. 41666

Bar-Cohen Y, Sherrit S, Dolgin B, Peterson T, Pal D, Kroh J (1999) Smart-ultrasonic/sonic driller/corer. US Patent No. 6,863,136, 8 March 2005. NASA NTR No. 20856

Bar-Cohen Y, Sherrit S (2003a) Self-mountable and extractable ultrasonic/sonic anchor (U/S-Anchor). NASA NTR No. 40827

Bar-Cohen Y, Sherrit S (2003b) Thermocouple-on-the-bit a real time sensor of the hardness of drilled objects. NASA NTR No. 40132

Bar-Cohen Y, Sherrit S, Herz JL (2003a) Ultrasonic/sonic jackhammer (USJ). NASA New Technology Report (NTR), Docket No. 40771

Bar-Cohen Y, Sherrit S, Bao X, Chang Z (2003b) Enabling actuation capabilities for operation at venus ambient temperature. Spontaneous R&TD final report, Pasadena, CA, 4 Dec 2003

Bar-Cohen Y, Sherrit S, Dolgin B, Bao X, Askin S (2005) Ultrasonic/sonic mechanism of deep drilling (USMOD). US Patent No. 6,968,910

Bar-Cohen Y, Badescu M, Sherrit S (2008) Rapid rotary-percussive auto-gopher for deep subsurface penetration and sampling. NASA NTR No. 45949

Bar-Cohen Y, Zacny K (eds) (2009) Drilling in extreme environments—penetration and sampling on Earth and other planets. Wiley-VCH, Hoboken, pp 1–827. ISBN-10: 3527408525, ISBN-13: 9783527408528

Bar-Cohen Y, Badescu M, Sherrit S (2010) Acquisition and retaining granular samples via rotating coring bit. NASA NTR No. 47606

Bar-Cohen Y, Badescu M, Bao X, Sherrit S, Shrout T, Zhang S, Jones B (2011) High temperature piezoelectric actuated sampler for operation on venus. Final report for the PIDDP task, NASA WBS No. 811073.02.06.55.13, Period of performance from January 2007 – January 2011, Report No. D-68419 (27 Jan 2011)

Bar-Cohen Y, Badescu M, Sherrit S, Zacny K, Paulsen G, Beegle L, Xiaoqi Bao (2012) Deep drilling and sampling via the wireline auto-gopher driven by piezoelectric percussive actuator and EM rotary motor. In: Proceedings of the SPIE smart structures and materials/NDE symposium, San Diego, CA, 12–15 March 2012

Bar-Cohen Y (ed) (2014) High temperature materials and mechanisms. CRC Press, Taylor & Francis Group, Boca Raton, Florida, pp 1–551. ISBN 13: 9781466566453, ISBN 10: 1466566450

Bar-Cohen Y, Bao X, Badescu M, Sherrit S, Zacny K, Kumar N, Shrout TR, Zhang S (2014) High-temperature drilling mechanisms, Chapter 14. In: Bar-Cohen Y (ed) High temperature materials and mechanisms. CRC Press, Taylor & Francis Group, Boca Raton, Florida pp 427–466. ISBN 13: 9781466566453, ISBN 10: 1466566450

Dolgin B, Sherrit S, Bar-Cohen Y, Rainen R, Askins S, Sigel D, Bickler D, Carson J,. Dawson J, Bao X, Chang Z, Peterson T (2001) Ultrasonic rock abrasion tool (URAT). NASA NTR No. 30403

Fergus JW, Hoffmann WP (2014) Refractory metals, ceramics, and composites for high-temperature structural and functional applications, Chapter 3. In: Bar-Cohen Y (ed) High temperature materials and mechanisms. CRC Press, Taylor & Francis Group, Boca Raton, Florida, pp 39–68. ISBN 13: 9781466566453, ISBN 10: 1466566450

Geothermal History Reports (GTP) (2010) A history of geothermal energy research and development in the United States, 1976–2006, Volume 2: Drilling

Gershman R, Wallace RA (1999) Technology needs of future planetary missions, acta astronautica, 45(4–9):329–335

Jerby E, Dikhtyar V, Aktushev O, Grosglick U (2002) The microwave drill. Science 298 (5593):587–589

Just GD (1963) The Jet piercing process. Quarry Managers' J, Institute of Quarrying Transactions, pp 219–226 (June 1963)

Kumar N (2014) High-temperature motors, Chapter 9. In: Bar-Cohen Y (ed) High Temperature Materials and Mechanisms. CRC Press, Taylor & Francis Group, Boca Raton, Florida, pp 281–296. ISBN 13: 9781466566453, ISBN 10: 1466566450

Maurer WC (1968) Novel drilling techniques. Pergamon Press, New York pp 1–114. ISBN-10: 0080036155, ISBN-13: 978-0080036151

Rapp D (2007) Human missions to Mars: enabling technologies for exploring the red planet. Springer/Praxis Publishing, ISBN: 3-540-72938-9, Appendix C, "Water on Mars"

Ready JF (1997) Industrial applications of lasers. Academic Press, New York, 2nd edition. ISBN-10: 0125839618, ISBN-13: 978-0125839617

Schultz RA (1993) Brittle strength of basaltic rock masses with applications to Venus. J Geophys Res 98(E6):10,883. doi:10.1029/93JE00691

Sherrit S, Askins SA, Gradziel M, Dolgin BP, Bar-Cohen Y, Bao X, Cheng Z (2001) Novel ultrasonic horns for power ultrasonics. NASA Tech Briefs, vol. 27, No. 4, 2003, pp 54–55, NASA NTR No. 30489

Sherrit S, Bar-Cohen Y, Dolgin B, Bao X, Chang Z (2002) Ultrasonic crusher for crushing, milling, and powdering. NASA NTR No. 30682

Sherrit S, Bar-Cohen Y, Bao X, Chang Z, Blake D, Bryson C (2003) Ultrasonic/sonic rock powdering sampler and delivery tool. NASA NTR No. 40564

Sherrit S, Bao X, Bar-Cohen Y, Chang Z (2004) Resonance analysis of high temperature piezoelectric materials for actuation and sensing. In: SPIE smart structures and materials symposium, Paper #5388-34, San Diego, CA, March pp 15–18

Sherrit S, Badescu M, Bar-Cohen Y, Chang Z, Bao X (2005) Portable rapid and quiet drill (PRAQD), Patent disclosure submitted on Feb 2006. US Patent No. 7,824,247, 4 Nov 2010. NASA NTR No. 42131

Sherrit S, Bar-Cohen Y, Badescu M, Bao X, Chang Z, Jones C, Aldrich J (2006) Compact non-pneumatic powder sampler (NPPS). NASA NTR No. 43614

Sherrit S, Badescu M, Bar-Cohen Y (2008) Miniature low-mass drill actuated by flextensional piezo-stack (DAFPiS). NASA NTR No. 45857

Sherrit S, Bao X, Badescu M, Bar-Cohen Y (2010a) Single piezo-actuator rotary-hammering (SPaRH) Drill. NASA NTR No. 47216, (2009). In: Sherrit S, Bao X, Badescu M, Bar-Cohen Y, Allen P (ed) Monolithic flexure pre-stressed ultrasonic horns. A Provisional Patent Application 61/362,164 was filed on July 8, 2010. Patent application #12/877,390 was filed on 8 Sept 2010. NASA NTR No. 47610

Sherrit S, Bao X, Badescu M, Bar-Cohen Y, Ostlund P, Allen P, Geiyer D (2010b) Planar rotary piezoelectric motor using ultrasonic horns. A Provisional Patent was filed on July 7, 2011, NASA NTR No. 47813

Sherrit S, Lee HJ, Zhang S, Shrout TR (2014) High-temperature electromechanical actuators, Chapter 10. In: Bar-Cohen Y (ed) High temperature materials and mechanisms. CRC Press, Taylor & Francis Group, Boca Raton, Florida, pp 297–330. ISBN 13: 9781466566453, ISBN 10: 1466566450

Smialek JL, Jacobson NS (2014) Oxidation of high-temperature aerospace materials, Chapter 5. In: Bar-Cohen Y (ed) High Temperature Materials and Mechanisms. CRC Press, Taylor & Francis Group, Boca Raton, Florida, pp 95–162. ISBN 13: 9781466566453, ISBN 10: 1466566450

Smith M, Cardell G, Kowalczyk R, Hecht MH (2006) The chronos thermal drill and sample handling technology. In: The 4th international conference on Mars polar science and exploration, Davos, Switzerland, Abstract 8095, 2–6 October 2006

Troy R (2011) Private communication at JPL. Pasadena

Xu Z, Reed CB, Konercki G, Gahan BC, Parker RA, Batarseh S, Graves RM, Figueroa H, Skinner H (2003) Specific energy for pulsed laser rock drilling. J Laser Appl 15(1):25–30. doi:10.2351/1.1536641

Zacny K, Bar-Cohen Y, Brennan M, Briggs G, Cooper G, Davis K, Dolgin B, Glaser D, Glass B, Gorevan S, Guerrero J, McKay C, Paulsen G, Stanley S, Stoker C (2008) Drilling systems for extraterrestrial subsurface exploration. Astrobiol J 8(3). doi:10.1089/ast.2007.0179

Chapter 9
Pneumatic Drilling and Excavation in Support of Venus Science and Exploration

Kris Zacny, Justin Spring, Gale Paulsen, Stephen Ford, Philip Chu and Steve Kondos

9.1 Introduction

9.1.1 Introduction to Venus Exploration

Venus is considered to be Earth's sister planet hence we can learn a lot about Earth by investigating Venus tectonics, volcanism, and atmosphere. As opposed to Mars which lost most of its atmosphere but retained a lot of water, Venus has extremely dense and hot, carbon dioxide atmosphere (95 % CO_2, >90 atm pressure, and ~480 °C temperature) and lost most of its water. One day Earth could end up looking just like Venus or Mars. Mars has been mapped by multitudes of spacecraft and we learn more about that planet each year. In comparison, understanding of Venus is relatively poor. The science objectives for Venus exploration are expressed in various reports by the Venus Exploration Analysis Group (Vexag 2014).

The most successful missions to date were the USSR Venera and Vega programs. For example, Venera 13 launched 1981 was a soft-lander that survived over 127 min

K. Zacny (✉) · J. Spring · G. Paulsen · S. Ford · P. Chu · S. Kondos
Honeybee Robotics, Pasadena, CA, USA
e-mail: zacny@honeybeerobotics.com

J. Spring
e-mail: spring@honeybeerobotics.com

G. Paulsen
e-mail: paulsen@honeybeerobotics.com

S. Ford
e-mail: ford@honeybeerobotics.com

P. Chu
e-mail: chu@honeybeerobotics.com

S. Kondos
e-mail: bruinbro@sbcglobal.net

V. Badescu and K. Zacny (eds.), *Inner Solar System*,
DOI 10.1007/978-3-319-19569-8_9

Fig. 9.1 Venera 13 mockup at the Cosmos Pavillion in Moscow. The lander module in the foreground would sit inside the brown sphere atop the Venera spacecraft (background). Courtesy of Sky & Telescope. https://heasarc.gsfc.nasa.gov/Images/misc_missions/venera13.jpg

on the Venusian surface (Fig. 9.1). The spacecraft was equipped with a 26 kg drill for acquisition of samples for x-ray fluorescence spectrometer (XRF) analysis.

The most recent Planetary Decadal Survey (NRC 2011) recommended Venus In Situ Explorer (VISE) as one of the five candidates for the New Frontiers (NF) type mission (NF missions with a budget of $1B, address high priority and technically complex science goals that are beyond the capabilities of a smaller NASA Discovery mission that cost $500 M). The VISE mission was reaffirmed by the Decadal Survey because many questions about Venus cannot be addressed by an orbiter. The VISE mission would focus on detailed characterization of the surface, deep atmosphere, and their interaction.

In support of the latest Decadal Survey, two NASA Flagship class mission architecture studies were also developed (the Flagship missions address high priority investigations and cost several $Bs). These are the Venus Mobile Explorer (Glaze 2009) and the Venus Intrepid Tessera Lander (Gilmore and Glaze 2010).

All three studies/mission concepts considered XRD/XRF and Raman/LIBS as potential mineralogy/chemistry instruments. The XRD/XRF requires a sampling capability to bring a sample inside the lander pressure vessel while Raman/LIBS instrument are remote instruments but could take advantage of a trench or even a drilled hole for analysis of subsurface rocks as capacity to investigate stratigraphy is vital to planetary exploration. The ability to investigate a borehole wall with LIBS has already been demonstrated by the ChemCam instrument on the Mars Science

Fig. 9.2 A vertical array of pits in the side of the hole in a rock "John Klein" resulted from using the ChemCam (LIBS) to assess composition at those points. Courtesy NASA

Laboratory Curiosity rover (Fig. 9.2). The XRF instrument was successfully used by the Venera 13 and 14 and Vega 2, while Clegg et al. (2012) demonstrated the use of LIBS/Raman at Venus pressures (Clegg et al. 2012).

9.1.2 Past Venus Surface Missions

Table 9.1 shows a list of past Venus surface missions (Siddiqi 2002; Wilson 1987). The first successful surface mission was Venera 7 in 1970 which survived for over 23 min on the Venusian surface. The last successful surface mission was Vega 2 in 1984 which survived over 56 min on the Venusian surface. Only six missions included a sampling device. These were Venera 11, 12, 13 and 14, as well as Vega 1 and 2. However, of those six, only three successfully captured and delivered samples for XRF analysis: Venera 13, 14, and Vega 2.

The Venera 11 drill managed to collect soil for analysis, but sample analysis was unsuccessful because the sample was not correctly placed in an instrument container for analysis. It seems the same problem occurred on Venera 12 mission. With respect to the Vega 1, the reason for the failed attempt to capture a sample was due to high turbulence during the descent phase which triggered premature deployment of the drill.

Venera 13 lander successfully landed on the Venus surface on 1st of March 1982 after a 1 h descent, and transmitted data for the record 127 min. The lander took 360° panoramic pictures, which were also the first color pictures of the Venusian surface (Fig. 9.3). The images revealed orange-brown rocks and loose soil. Russian geologists concluded that the spacecraft probably landed on the most common type of Venusian terrain, plains with wrinkle ridges. The plains on the images are of the oldest volcanic plains which were formed about 750 million years ago, fractured and buckled by tectonic pressures in the crust.

Table 9.1 Venus Landers

Surface landed missions	Launch year	Surface time* (min)	Surface sample acquisition capabilities
Venera 7	1970	23	No
Venera 8	1972	50	No
Venera 9	1975	53	No
Venera 10	1975	65	No
Venera 11	1978	95	Yes. Failed to deposit sample
Venera 12	1978	110	Yes. Failed to deposit sample
Pioneer Venus 2	1978	60	No
Venera 13	1981	127	Yes
Venera 14	1981	57	Yes
Vega 1 Lander	1984	56	Yes. Activated during descent by error
Vega 2 Lander	1984	57	Yes

*Data was transmitted to flyby spacecraft which relayed data back to Earth. Hence Surface Time is linked to the time window for the relay flyby spacecraft while in range. The actual Survival Time of the lander was probably longer

Fig. 9.3 Venera-13 Terrain. The rotary penetrometer can be seen extended out to measure physical and electrical soil properties. Courtesy USSR Academy of Sciences

Investigation of soil particles blown onto the spacecraft helped to estimate the wind speed to be approximately 1–2 km/h. The temperature and pressure at the landing site were 465 °C and 89.5 atmospheres, respectively. The lander deployed a rock drill and successfully captured samples for XRF analysis, a first for this planet. Sample analysis showed the rock was similar to terrestrial leucitic basalt with high potassium content.

Venera 14 arrived at Venus four days after its twin, on 5 March 1982 and transmitted data for 57 min. Temperature and pressure at the landing site were considerably higher than at the Venera 13 site: 470 °C and 93.5 atmospheres, respectively. Higher pressure implied the spacecraft landed at a lower altitude than Venera 13. The Venera-14 landing site (Fig. 9.4) displayed a flat expanse of rock with no soil. Radar imaging also revealed a younger volcanic plain with lobate flows of lava, probably formed a few million years ago. The rock, which was first considered to be solidified lava, was actually layered and crunchy (as determined by impact deceleration analysis). One theory suggests that it is a pumice-like material, probably formed out of volcanic ash or dust generated by meteorite strikes. When

Fig. 9.4 Venera 14 Terrain. The rotary penetrometer can be seen extended out to measure physical and electrical soil properties. Courtesy USSR Academy of Sciences

large meteorites enter Venus' dense atmosphere, they generate pulverizing shock waves and tend to burn up before impacting.

The sample acquisition system on Venera 14 was also successful. The drill penetrated to 3 cm depth and delivered a sample to a chamber with a temperature of only 30 °C for the XRF instrument. The material was determined to be similar to basaltic rocks on Earth formed at mid-ocean ridges by underwater volcanoes.

Vega 2 touched down without problems on 15 June 1985 and transmitted from the surface for 57 min. The drill system also managed to collect samples for analysis – the same way as in Venera 13 and 14. The material was identified as an anorthosite-troctolite rock - rarely found on Earth, but present in the lunar highlands (Surkov et al. 1986, 1987).

These missions attest to the successful development of a range of technologies for surviving high temperature, pressure, and a very corrosive atmosphere. The technologies included heat-resistant materials, actuators, electronics, lubricants (based on molybdenum disulphide and microscopic metal flakes), and even thermopile batteries (though these were never deployed). However, since the last successful mission occurred over 30 years ago, most of the technologies and know-how is no longer available and in turn little heritage exists. There are a number of publications describing various instruments, but the descriptions are relatively high level and/or focused on concepts of operations. There are no hardware details regarding the exact type of material used, design drawings, or manufacturing processes. Probably the most insightful statement was that the sampling drills on Venera and Vega spacecrafts were designed to work only at 500 °C, and this approach significantly solved design problems related to two or more materials having dissimilar coefficient of thermal explanations. New technologies, materials and components need to be developed from scratch. Some materials properties have been evaluated at high temperature for other high temperature applications and these could serve as a good starting point for material selection for Venus drilling system (see Appendix).

9.1.3 The 3 cm Drill on Venera 13 and 14, and Vega 2

A critical payload element of the Venera 11, 12, 13 and 14 as well as Vega 1 and 2 landers was a drill referred to as the GZU soil sampling drill as shown in Fig. 9.5

and Table 9.2 (Barmin and Shevchenko 1983). The 26 kg drill system was capable of drilling up to 3 cm into igneous rocks. This was the only mechanism with moving parts and hence required development of a high temperature electric motor, lubricants and new alloys. As mentioned earlier, a very clever engineering approach helped to solve a major hurdle associated with the different coefficient of thermal expansions of two materials in close proximity (e.g. gears and shaft). These machine parts were in fact designed to function properly only after thermal expansion to 500 °C.

The drill used a series of pre-coded steps – there was no control feedback. The drilling operation started as the hollow telescoping drill head was lowered to the surface. A common 90 W electric motor advanced the drill into the formation and also rotated the drill bit at 50 rpm. Hence the drill was forced into the subsurface at 0.3 mm per revolution. Although the drill operated for 126 s, the sample transfer stage started at 120 s time mark (i.e. before the drill stopped rotating) with a firing of four pyrotechnic charges in sequence. At the 120 s mark, the first pyro charge broke a series of seals that allowed the high pressure atmosphere of Venus to flow into an assembly of tubes.

Soil was carried in stages, into a soil transfer tube and onto a sample container by the atmospheric gas. The second pyro closed the soil feeding system to the outside atmosphere at the 128 s mark. At the 188 s mark, the third pyro broke the seal between the vacuum tank and the soil feeding system resulting in a pressure drop from 9 MPa to 5 kPa which was required for the XRF measurement. At the 200 s time mark, the fourth pyro moved the sample tray through an airlock and into the XRF chamber. During the forward motion, the tray hit a hard stop and the sample was sprayed into XRF examination cups.

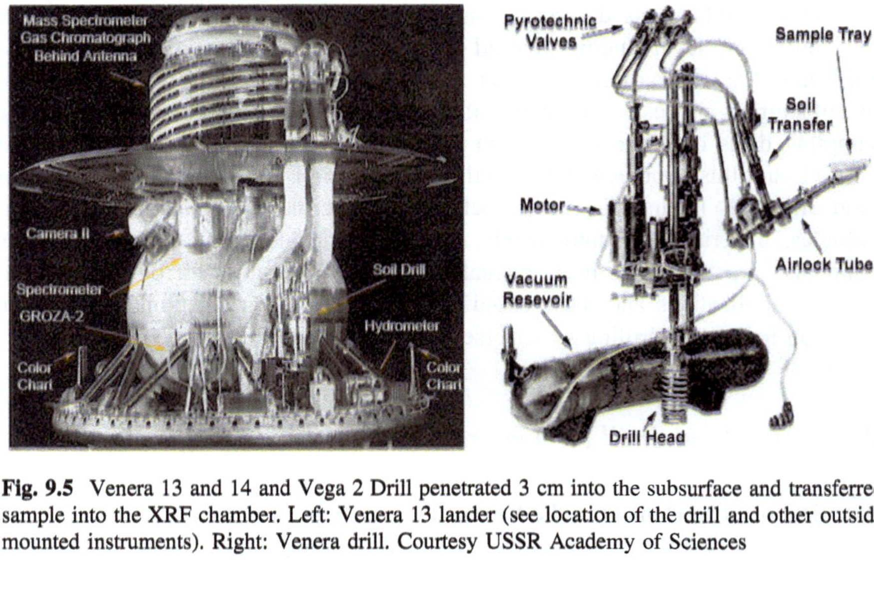

Fig. 9.5 Venera 13 and 14 and Vega 2 Drill penetrated 3 cm into the subsurface and transferred sample into the XRF chamber. Left: Venera 13 lander (see location of the drill and other outside mounted instruments). Right: Venera drill. Courtesy USSR Academy of Sciences

Table 9.2 Venera and Vega drill specifications

Specification	Value	Units
Mass	26.2	kg
Max consumable power	90	W
Power supply voltage	27	V
Time to carry all operations	200	sec
Rotational speed	50	rpm
Penetration per revolution (assumes 30 mm in 120 s)	0.3	mm/rev
Number of gas generators	4	
Depth of drilling	30	mm
Diameter of drill	16	mm
Volume of sample fed into the zone of analysis	1-6	cc
Total working path of instrument	400 ± 10	mm
Method of drilling	Rotary	
Regime of operation	Automatic	

During the Venera 13 and 14 missions, drill telemetry revealed the drill penetrated to the target depth of 3 cm. In addition, analysis of the drill telemetry at the Venera 13 and 14 landing sites indicated that the physical strength of the drilled formations corresponded to that of weathered porous basalt or compacted ashy volcanic tuff-type material. However, since the drill was mounted close to the lander body, it was out of the camera field of view. Hence, it is not possible to determine whether the drill in fact penetrated Venus rock or regolith.

9.2 Cutter Materials and Rocks Used During Tests

9.2.1 Considerations for Cutter Material

Polycrystalline Diamond Compacts (PDCs), Diamond Impregnated cutters (DI), and Thermally Stable Polycrystalline Diamonds (TSP) have been used for decades in rock cutting and drilling. However, diamond is an unstable allotrope of graphite. As a result diamonds graphitize and also oxidize (if oxidizing atmosphere is present) at high enough temperature (Wilks and Wilks 1994). Since Venus does have an oxidizing atmosphere, this poses significant risk. Venus also has high ambient temperature, which means the temperature at the cutter-rock interface will be even higher (higher temperature occurs at the cutter-rock interface due to friction). For cutting rocks at these high temperatures other hard materials such as tungsten carbide are better suited.

9.2.2 Rocks Strength at High Temperature

Five different types of rocks were used for testing Venus drilling and Venus trenching breadboards. These included (from weakest to strongest): Firebrick, Bishop Tuff, Indiana Limestone, Berea Sandstone, and Saddleback Basalt. The actual strength of the rocks at ambient temperature was obtained from Unconfined Compressive Strength tests performed by the suppliers of the rocks and is as follows: Firebrick: 1 MPa, Tuff: 15 MPa, Limestone: 45 MPa, Sandstone: 80 MPa, Basalt: 105 MPa.

Only two of those rocks (Tuff and Basalt), are potential Venus analog rocks. The other three rocks (Firebrick, Limestone and Sandstone), were selected for their intermediate strengths, since they fill the strength gap between very soft (Tuff) and very hard (Basalt).

In order to determine whether a rock changes its strength at high temperature, a Schmidt hammer was used to obtain rock's strength at 20 °C and then at 500 °C. In each case, five measurements were taken and averaged out. The results summarized in Table 9.3 show that in general the strength of Indiana Limestone and Berea Sandstone drop at 500 °C, while the strength of Saddleback Basalt is the same at both temperatures. It was difficult to determine strength of the Brick and the Tuff since these rocks are very weak. It is believed that the Firebrick properties won't change since the brick is used at high temperature applications. Tuff strength is more likely to decrease based on findings reported by Wang and Wang (2010).

An explanation for the strength drop in Indiana Limestone might be explained by the fact that Calcium Carbonate (the main mineral component of the limestone) breaks down at high temperature to Calcium Oxide and Carbon Dioxide, following the formula: $CaCO_3 \rightarrow CaO + CO_2$. Calcium Oxide is very soft and is a 3 on the Mohs Scale (note, copper and gold are also 3).

Berea Sandstone strength comes from the bonding agent; clay minerals that 'glue' quartz grains together. At high temperature, clay loses adsorbed water and can no longer provide bonding strength. It should be noted that strength of dry sandstone and limestone (with 0 wt%) is higher (Mellor 1971), hence bulk water loss cannot explain a strength drop at 500 °C for these two rocks. The strength of basalt remained the same since its constituent minerals were not affected by the 500 °C temperature.

It is therefore expected that Venus rocks will either maintain similar or have lower strength to the 'analog' rocks on Earth. This is in agreement with observations by Ying et al. (2009).

Table 9.3 Normalized strength of rocks at room and Venus temperatures

	BNZ-23 firebrick	Bishop tuff	Indiana limestone	Berea sandstone	Saddleback basalt
20 °C	1	1	1	1	1
500 °C	1	0.8	0.66	0.79	1

9.3 High Temperature Drilling

Figure 9.6 shows a high level concept and a starting point for the Venus drill development project. The concept also includes a pneumatic based (suction-type) sample delivery system. The sampling steps are as follows: Atmospheric air (1) is drawn in by a spinning impeller connected to a high speed spinning drill (2). The air flows down the double wall drill pipe and enters the center of the drill, causing a low pressure area to develop (3), which creates a Venturi effect suction at the tip of the bit (4). The Venturi effect causes atmospheric air to be drawn down the hole and into holes within the bit (4). That air carries soil and rock cuttings up the drill pipe and into an instrument such as an XRF/XRD (X-Ray Fluorescence/X-Ray Diffraction), for example.

The goal of the research work reported in this section was to determine whether this particular concept would work in Venus conditions, to investigate other promising methods for sample delivery, and to determine required spacecraft resources such as time, energy, and power that would be required for developing an actual Venus mission concept.

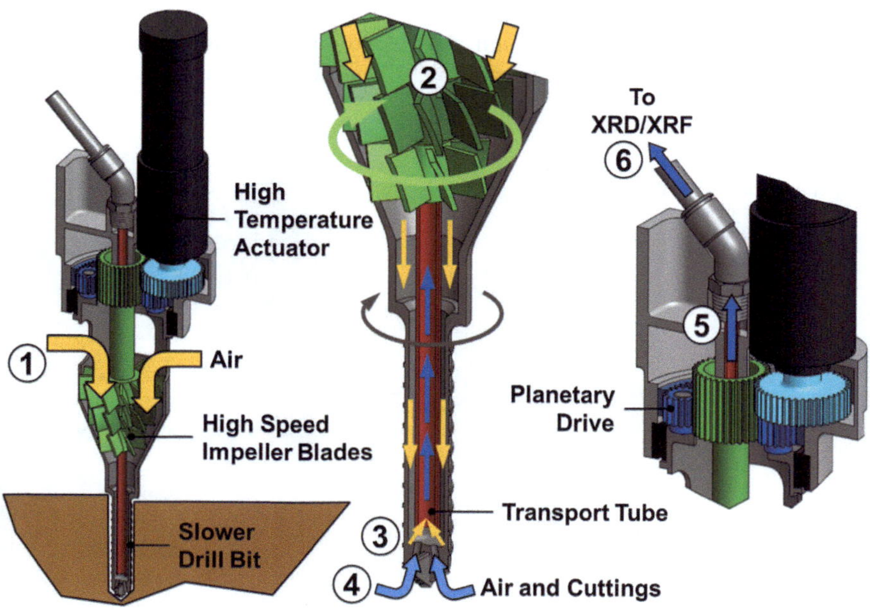

Fig. 9.6 A notional concept for a Venus drill and a pneumatic sample delivery system

9.3.1 Drill Test Setup

Figure 9.7 shows the test setup for 20 and 500 °C drilling tests. The setup included a rotary-percussive drill developed for previous drilling projects (Zacny et al. 2011). The SASSI drill allows for the investigation of both rotary and rotary-percussive drilling approaches since the rotary and percussive subsystems are driven by two independent actuators. However, the drill was not designed for high temperature operation and in turn had to be located outside of an oven. To thermally isolate the drill from a drill bit, a ceramic adapter was inserted in-line between the bit and the bit extension, through a small hole on top of the oven.

All drilling tests used Relton 0.5 in. Groo-V masonry drill bits, which in previous ambient tests performed very well in hard and soft rock formations. All rock samples were cut to size to remove possible weathered surface layers and also to allow ease of clamping inside the oven.

For 20 °C tests, the oven was turned off and left at room temperature. For 500 °C tests, the oven was turned on, the door was closed and the rock samples were left in the oven at ~500 °C for at least 5 h prior to drilling. This time duration was sufficient for the entire rock sample to reach a homogeneous temperature distribution.

The drill was programmed for a 50 mm depth of penetration. The rotational speed was set at 300 rpm for all the tests. Software limits were imposed on penetration rate not to exceed 2 mm/s (to prevent auger choking) and on Weight On Bit not to exceed 50 N. The WOB limit, was based on the assumption that a Venus mission drill itself will provide needed WOB. As such it was assumed that a future

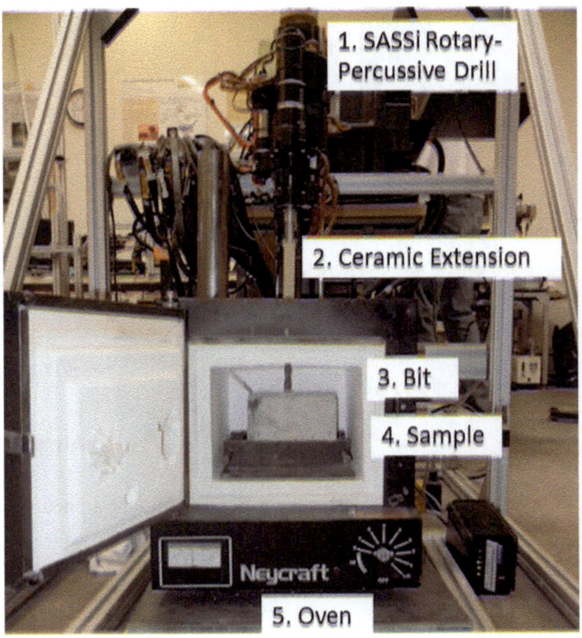

Fig. 9.7 Set up for high temperature drilling tests

Table 9.4 Test matrix for the drill tests

Test material	Approximate strength at 20 °C MPa	Rotary 500 °C	20 °C	Rotary-percussive 500 °C	20 °C	Total
BNZ-23 insulating firebrick	1	1	1	0	0	2
Bishop tuff	15	1	1	0	0	2
Indiana limestone	45	1	2	1	1	5
Berea sandstone	80	0	0	11	1	2
Saddleback basalt	105	0	0	1	1	2

Venus drill would weight approximately 100 N on Venus and a factor of safety of 2 was used to prevent the drill from bouncing up during actual drilling operations. The percussive system was designed to provide 0.6 J/blow of percussive energy and up to 1500 blows per minute. The tests, which included percussion, have been clearly labeled. A total of 13 tests were performed in 5 different materials. The list of test rocks and the test matrix is shown in Table 9.4.

9.3.2 Drill Test Results

Table 9.5 shows a detailed summary of 13 tests, while Fig. 9.8 shows Rate of Penetration (ROP) in mm/min as a function of rock strength in MPa and test temperature (20 and 500 °C). The data in Fig. 9.8 accounts for the decrease in rock strength at 500 °C. The test data shows a drop in penetration rate with increase in rock strength, as expected. However, it also showed that penetration rate in rocks at Venus temperature is on average 4x lower. This was also observed in high temperature Ultrasonic drilling tests performed by Bar Cohen et al. (2014), except that the drop in the penetration rate in that case ranged from 2 to 100 and was mainly attributed to degradation of piezo material.

The drop in penetration rate in the current tests, cannot be attributed to the thermally-induced changes in the drill, since the drill was outside of the chamber and it was also thermally isolated using a ceramic bit coupler. As documented earlier in section 0, rocks either retain their strength at high temperature or have lower strength at high temperature. Hence a drop in penetration rate cannot be explained by changes in rock properties either (in fact, penetration rate should have been greater at Venus temperature). Visual examinations of the carbide drill bit did not show any significant wear. In addition, two tests were performed at 20 °C before and after the 500 °C test to determine if the 500 °C test caused any significant changes to the bit itself. The two 20 °C tests yielded similar results, meaning the 500 °C tests did not degrade the bit. The drill bit could also be easily extracted from the hole during 500 °C tests indicating that the bit binding due to thermal expansion of steel auger was not the cause of the drop in penetration rate.

Table 9.5 Drill test results

Rock	Rock strength*	Ternp	R or RP	ROP	Anger power	Perc. power	Anger net power (excl. NLC and losses ~20 %)	Anger net torque (excl. NLC and losses ~20 %)	Anger net friction	WOB	Energy (incl. losses)
					(hid. losses)		(excl. NLC and losses ~20 %)				
	[MPa]	[C]		[mm min]	[Watt]	[Watt]	[Watt]	[mNm]		[N]	[Whr/cm]
Firebrick	1	20	R	240	5.1		1.04	33	0.5	10	0.004
	1	500	R	240	4.3		0.35	11	0.2	11	0.003
Bishop tuff	15	20	R	120	11.4		6.88	219	1	33	0.016
	12	500	R	156	13.1		8.24	262	1.5	28	0.014
Indiana limestone	45	20	R	12	15.5		9.84	313	1.1	47	0.215
	30	500	R	0.36	11.1		6.32	201	0.6	52	5.139
	30	500	RP	6	10.4	45	5.76	183	0.5	53	1.539
	45	20	R	10.8	24		16.48	525	1.7	49	0.370
	45	20	RP	31	10.3	45	5.92	189	0.6	46	0.297
Berea sandstone	80	20	RP	12	16.6	45	8.96	285	0.9	49	0.856
	64	500	RP	4.8	20.3	45	12.8	408	1.3	51	2.267
Saddleback basalt	105	20	RP	3.6	15.6	45	8.96	285	0.7	61	2.806
	105	500	RP	0.84	12.9	45	6.8	217	0.7	52	11.488

*Rock strength has been adjusted based on Table 9.3

R = Rotary; RP = Rotary-Percussive; NLC = No Load Current; WOB = Weight on Bit; ROP = Rate of Penetration

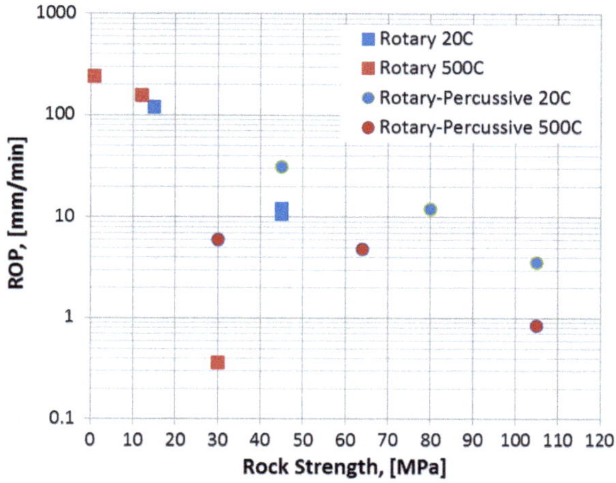

Fig. 9.8 Rate of Penetration (log scale) vs. Rock Strength at 20 and 500 °C. Decrease in rock strength at 500 °C included

Material changes in the steel bit auger (e.g. Young's Modulus, E – see Table 2) could explain drop in performance using Rotary-Percussion, but not in Rotary drilling (Young's Modulus affects the efficiency of percussive wave). In addition, Young's Modulus drops by approximately 20 %, and in turn this is insufficient to explain drop in propagation of compressive wave that would had been required to explain the large drop in penetration rate (Table 9.6).

Decrease in penetration rate could potentially be explained by higher friction between the bit and the rock at 500 °C. High bit to rock friction at high temperature, has been observed before by Zacny and Cooper (2007). They measured almost a threefold increase in sliding friction coefficient as temperature was increased from approx. 200 to over 400 °C. In all likelihood, high temperature plays a similar role to a very high vacuum, in that once the oxide or thin surface film layer is removed by abrasive action, it cannot be readily replaced. Under high vacuum, the dominant driver is lack of molecules to form new surface films. At high temperature, it is the activation energy in the form of heat that causes lowering of the thermodynamic stability of the surface films and in turn their desorption and dissociation.

One way to determine drilling efficiency is to investigate drilling Apparent Coefficient of Friction or ACoF which is measured from the auger torque and WOB. The ACoF has two components: Sliding and Rock Breaking. The sliding component generates heat while the Rock Breaking component is almost solely responsible for drill penetration. Hence, large ACoF should imply large ROP (since a larger cut is excavated per revolution), unless the Sliding component is extremely high while the Rock Breaking component is very low. Figure 9.9 shows a ratio of ACoF to ROP as a function of rock strength. If the ratio is higher for a given ROP that means that Sliding component is much higher. To better view the results, the data is plotted against the UCS of rock, which allows comparison of values in the same rock type. UCS values at 500 °C have been decreased as before to account for the decrease in the rock strength at high temperature (see 0). It needs to be noted that drilling parameters such as RPM and WOB, which would have otherwise

Table 9.6 Carbide Blade trenching test matrix

	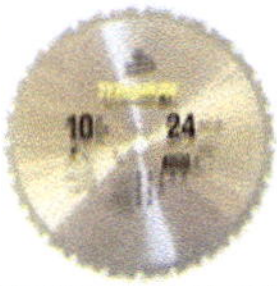			
Blade	Titanium 24 Tooth (Rip)	King Carbide 40 Tooth (General Purpose)	DeWalt 40 Tooth (General Purpose)	King Carbide 60 Tooth (Crosscut)
Width	1.78 mm	2.18 mm	1.77 mm	2.22 mm
Mass	655 g	804 g	640 g	836 g

Blade	Trade Duty 60 Tooth (Crosscut)	Freud 80 Tooth (Crosscut)	Bosch 80 Tooth (Ultra-Fine Crosscut)	
Width	1.77 mm	2.54 mm	1.83 mm	
Mass	657 g	911 g	678 g	

affected ROP, were kept constant in all but Firebrick and Tuff experiments. The data for Tuff was not included since the penetration limit set in the drilling software was reached in both cases at 20 and 500 °C.

Figure 9.9 clearly shows that the ratio of ACoF/ROP is much higher at 500 °C than at 20 °C, implying that the Sliding component at 500 °C is much higher. This explains lower ROP even though the drilling Power and Torque were similar at both temperatures.

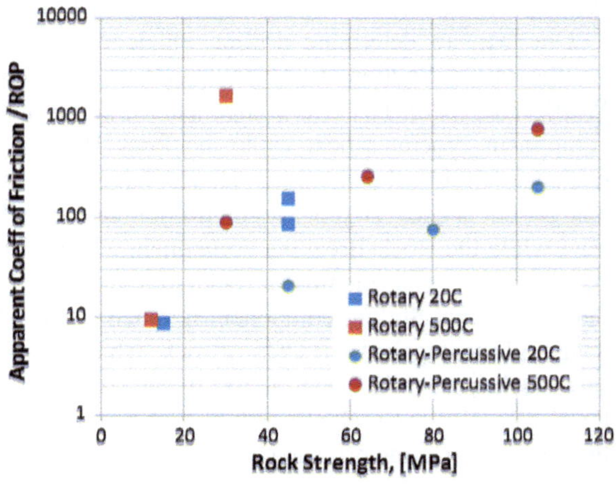

Fig. 9.9 Log (Coefficient of Friction /ROP) vs. UCS of Rock. Decrease in rock strength at 500 °C included

Figure 9.10 shows Total Drilling Power in Watts vs. Rock Strength in MPa. The power gradually increased during rotary-only mode for harder rocks until it reached approx. 20 W for 40 MPa rocks. Above that strength, the Percussive system had to be engaged to maintain penetration rate. The drilling power for harder rocks was lower than 70 W and did not change much with respect to rock strength.

Figure 9.11 shows WOB vs. Rock Strength. The WOB was software limited to approx. 50 N. In general, the WOB was lower than 50 N for rocks weaker than approximately 30 MPa.

Figure 9.12 shows Specific Energy in Whr/cm as a function of rock strength in MPa. This graph is useful in determining battery size for an actual mission. For example, if a rock was very hard (100 MPa), it would take 10 Whr to penetrate to

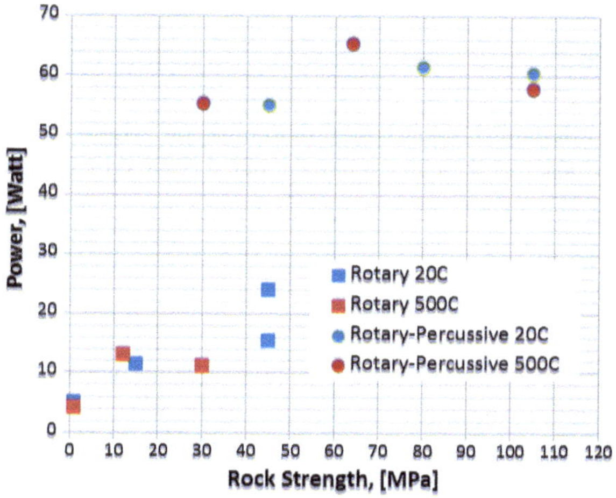

Fig. 9.10 Total drilling power (incl. losses) vs. UCS of Rock. Decrease in rock strength at 500 °C included

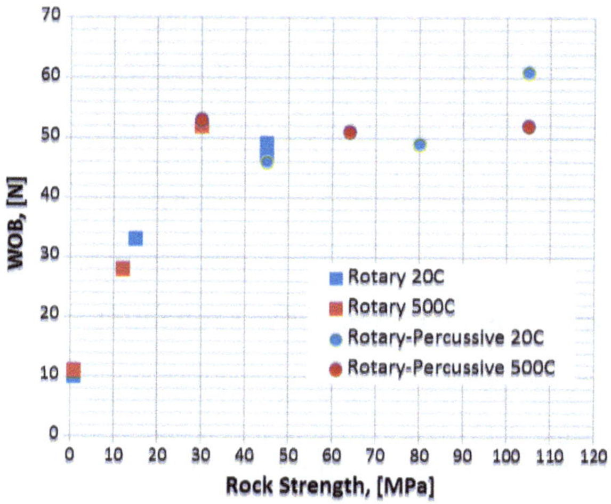

Fig. 9.11 WOB vs. UCS of Rock. Decrease in rock strength at 500 °C included

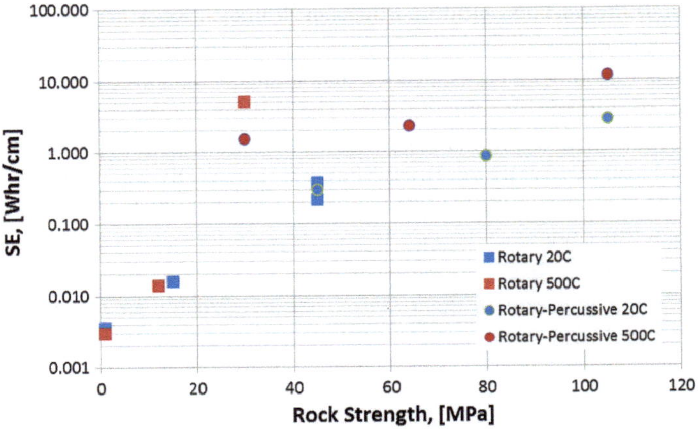

Fig. 9.12 Log (Specific Energy) vs. UCS of Rock. Decrease in rock strength at 500 °C included

1 cm depth. Hence, for a 5 cm hole, the required battery capacity would need to be at least 50 Whr (an iPhone 5 has a 5.45 Whr battery).

9.3.3 Conclusion

In conclusion, unless rocks on Venus are weak (less than 30 MPa), a Venus drill needs rotary-percussive action to be effective. A Venus Rotary-Percussive drill would need approx. 50 N Weight on Bit and 70 W of Power to penetrate ~100 MPa basalt rock to a depth of 5 cm (nominal) in 50 min with 50 Whr of electrical energy.

To reduce the drilling time, the drilling rotational speed can be increase but this would directly increase the drilling power. Alternatively, Weight on Bit could be increased, which could happen if the weight of the drill system is used for applying WOB (Venus gravity is 90 % of Earth's, hence a 20 kg drill would apply 180 N of WOB). Another way of decreasing penetration time, would be to decrease the drill bit diameter. In the current tests, the bit diameter was 0.5 in. (12.7 mm) and if this is reduced to 10 mm, the drilling time could be reduced by as much as 60 % (since drilled volume drops as diameter squared). The exact drilling time, however, would need to be determined from drilling tests.

9.4 High Temperature Trenching

The second of the two surface penetration options researched was trenching. Figure 9.13 shows a notional concept of a Venus trencher that incorporates a suction type sample delivery system.

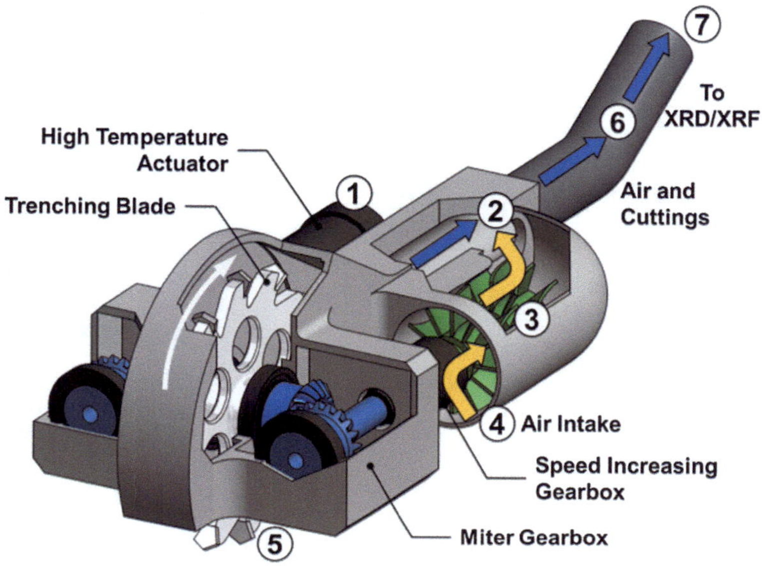

Fig. 9.13 A notional concept for a Venus trencher and a pneumatic sample delivery system

The concept of the operation would be as follows: The high temperature motor rotates the saw which cuts the trench. The motor also drives an impeller (3), which draws air from the outside (4). The air flows through a small gap at the bottom of the Blade Housing (5). As the air is drawn in, it creates suction at (2) and lofts the cut rock or soil into the delivery vent (6), and the XRD/XRF instrument port (7).

The goal of the work described in this section was to determine whether this particular concept would work at Venus conditions, to investigate alternative methods of sample transfer, and to determine required spacecraft resources such as trenching time, energy, and power that would be required for developing of an actual Venus mission.

9.4.1 Trencher Test Setup

The general test setup for the trencher testing can be seen in Fig. 9.14. An off-the-shelf miter saw was securely mounted inside a clear protective case. To control WOB, a known weight was attached to the end of a string, which was connected to the saw handle via an arrangement of pulleys. This also allowed the operator to hold the saw above the rock to be cut and release it once the blade started to rotate, thus simulating actual trencher operation on Venus.

A vacuum bag was installed at the back of the saw in order to capture cuttings. Later, to mimic the effects of a continuous pneumatic transfer system, a vacuum

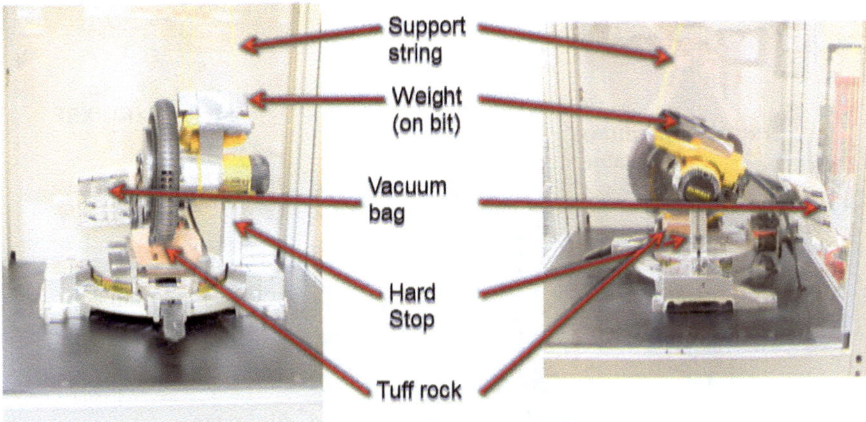

Fig. 9.14 Venus trencher test setup

cleaner was connected to this location to collect cuttings. Two important pieces of equipment not seen in Fig. 9.14 are a Variac, which allowed adjustment of the AC voltage into the AC powered saw, and an AC current sensor. These two instruments were used to determine electrical power usage. For measurement of peak power and ROP, a camera was set up to film each test, including the readout of the current sensor. A stroboscope was used to verify the RPM of the saw blades.

9.4.2 Trencher Tests Results

9.4.2.1 Carbide Blade Testing

Initially, a number of commercial 10 in. (25.4 cm) diameter carbide tipped steel blades were procured and tested by sawing into a number of rocks without using any coolant. The selection of carbide rather than other cutting materials was done, because the carbide cutters performed well in the high temperature drill tests (see 0). Each of the seven blades were tested at 5000 and 2355 RPM, and cuttings were collected using the vacuum bag method. The Weight On Blade (WOB) was set at 20 N. Initial tests were done on blocks of Bishop Tuff cut from the same piece of rock used for the drill tests. The maximum depth possible with these diameter blades was 31 mm.

Figure 9.15 shows a Cut Rate in mm/s for all seven carbide blades. The penetration rate for the King Carbide 40 tooth blade (KC40) was two times higher than for any other blades, while power (not shown) was similar for all blades.

One of the reasons why the KC40 blade performed better, might be due to its higher mass and in turn angular momentum. Other reasons might include different

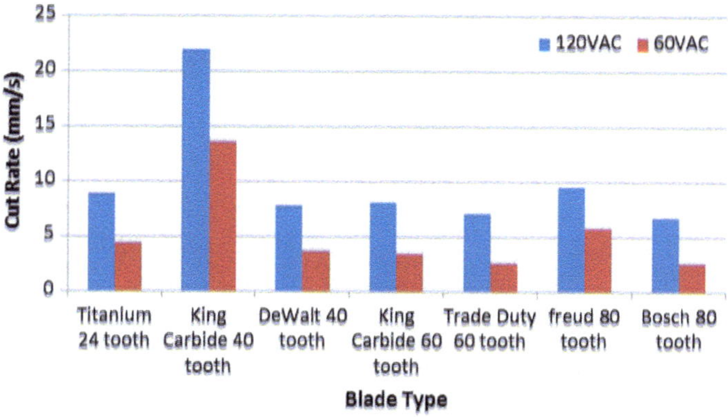

Fig. 9.15 Cut rate for Carbide Blade types

Fig. 9.16 An example of sawed Berea Sandstone rock

carbide grades or tooth design, although based on visual assessment all teeth looked very similar.

The KC 40 blade was next tested in other rocks at 5000 rpm and 20 N WOB to determine its performance across the range of rock strengths (Fig. 9.16). During the tests, a vacuum cleaner was attached to the back of the saw to collect cuttings. A percent of Mass of Cuttings was estimated for each test based off of the cuttings collected, assumed rock density and the measured volume of the hole left by the saw blade.

Table 9.7 summarizes test results. It can be seen that the Cut Rate drastically dropped in harder rocks.

In particular, the Cut Rate was 17.3 mm/s in 15 MPa Tuff, 1.1 mm/s in 45 MPa Limestone, and 0.08 mm/s in Sandstone. In addition, in Sandstone, at one stage the

Table 9.7 King Carbide 40 test results

Rock type	Tuff	Indiana limestone	Berea sandstone
Compressive strength (MPa)	15 MPa	45 MPa	80 MPa
Depth of cut (mm)	32 mm	33 mm	8 mm
Cut rate (mm/s)	17 mm/s	1 mm/s	0.1 mm/s
Peak power (W)	1391 W	1066 W	1040 W
Mass of cuttings (g)	10 g	24 g	4 g
%Mass of cuttings	70 %	70 %	97 %

blade did not make any further progress and after two minutes of no penetration, the test was aborted. Increasing the WOB from 20 to 40 N did not improved penetration rate but instead, the rock began to glow red hot and sparks were seen coming off.

9.4.2.2 Masonry Blade Testing

Since the best carbide blade did not penetrate 80 MPa rock, other blade types were selected for testing. Diamond impregnated masonry blades were initially avoided, because they would be susceptible to diamond graphitization at Venus temperatures. However, the decision was made to test these after all, and if they indeed performed well, to re-test them at 500 °C. In addition, a 7 in. cutoff wheel with Aluminum Oxide abrasive was procured. Table 9.8 shows the three blades. A vacuum cleaner was used again for cuttings collection.

Table 9.9 shows results from the three saw tests in Indiana Limestone and Berea Sandstone rocks.

The two diamond blades did not do as well as the carbide tipped KC40 blade at 20 N WOB in Limestone. The Cut Rate, with the best performing diamond wheel,

Table 9.8 Masonry Blade test parameters

Blade type	7″ DeWalt Diamond	7″ Vector Seg. Diamond	7″ Flexovit Cutoff
Blade mass	320 g	310 g	160 g
Width of teeth	1.6 mm	1.6 mm	3.4 mm
WOB.	20 N, 40 N	20 N, 40 N	40 N
Voltage	120 VAC	120 VAC	120 VAC
No load speed	5000 rpm	5000 rpm	5000 rpm

Table 9.9 Diamond and aluminum oxide wheels test results

Blade Type	7″ DeWalt Diamond	7″ Vector Seg. Diamonc	7″ Flexovit Cutoff	
20 N WOB, Indiana Limestone				
Depth of cut	5 mm (DNF)	10 mm (DNF)	32 mm	
Width of cut	3.55 mm	2.75 mm	4 mm	
Cut rate	0.06 mm/s	0.16 mm/s	1 mm/s	
Peak power	930 W	895 W	1188 W	
%Mass of cuttings collected	50 %	37 %	38 %	
40 N WOB, Indiana Limestone				40 N, Berea SS
Depth of cut	7 mm	31 mm	32 mm	34 mm
Width of cut	3.55 mm	2.75 mm	4 mm	3.65 mm
Cut rate	0.1 mm/s	0.4 mm/s	2.9 mm/s	1.4 mm/s
Peak power	933 W	991 W	1464 W	1706 W
%Mass of cuttings collected	94 %	34 %	22 %	56 %

was still 7x lower that the Cut Rate with the KC40 (even though the KC40 blade was wider). The cutoff wheel penetrated as fast as the KC40 blade. Hence the diamond wheels overall did not perform well and in any case would need to be tested at 500 °C.

As opposed to the diamond masonry blades, the cut off wheel managed to penetrate through the sandstone at a relatively high Cut Rate of 1.4 mm/s. Hence, the Aluminum Oxide cutoff wheel seemed (from the penetration stand point) to be the best choice for further testing.

The major concern for the cutoff wheel was contamination. These wheels are designed to break as they cut, leaving behind a large fraction of abrasive and bond material. To quantify the level of contamination, the mass of the cutoff wheel was measured before and after every test, and reported in The two diamond blades did not do as well as the carbide tipped KC40 blade at 20 N WOB in Limestone. The Cut Rate with the best performing diamond wheel, was still 7x lower that the Cut Rate with the KC40 (even though the KC40 blade was wider). The cutoff wheel penetrated as fast as the KC40 blade. Hence the diamond wheels overall did not perform well and in any case would need to be tested at 500 °C.

As opposed to the diamond masonry blades, the cut off wheel managed to penetrate through the sandstone at a relatively high Cut Rate of 1.4 mm/s. Hence,

Table 9.10 Measured loss of cutoff wheel

Rock	WOB [N]	Mass of wheel lost [g]	Mass of rock cut [g]	Ratio wheel/rock
Berea sandstone	40	1.7	26	0.06
Indiana limestone	40	0.5	29	0.02
Indiana limestone	40	0.9	51	0.02

the Aluminum Oxide cutoff wheel, seemed (from the penetration stand point) to be the best choice for further testing.

Table 9.10 one can see that contamination from the wheel material is in the range of 2–6 %.

Although it was logistically very difficult to test a saw at 500 °C because of the oven size, several cutoff wheels were placed inside the oven by themselves to determine if high temperature causes any significant damage to the wheels. It was found that all of them disintegrated due to the thermal damage to the bond holding abrasive material together.

9.4.3 Conclusions

The test results showed that carbide and diamond wheels cannot penetrate very hard rocks during dry sawing, while the Cut Rate with an aluminum oxide cutoff wheel in all rocks is competitive with carbide tipped blades. However, the bonding material in the aluminum abrasive in the cutoff wheels disintegrates at very high temperature. In addition, sample capture was relatively difficult when using a saw. Hence, a trencher system might not be the best approach for the Venus mission unless more fundamental work related to cutter wheel development and sample capture is undertaken.

9.5 Pneumatic Sample Transfer

Creating an excavation and capturing a sample is only part of the sample investigation mission. A more complicated task, in fact, is moving a sample to an instrument within the low pressure, low temperature environment of the lander body. This section focuses on pneumatic methods of sample delivery. As opposed to the approach used on Venera 13, 14 and Vega 2 where the sample was sucked into evacuated tubes, this section investigates creating suction through an active means such as turbine.

Three options were considered for sample transfer as detailed in Table 9.11. The first option integrates a Venturi suction system directly into the drill bit (see also Fig. 9.6).

The positive pressure for the system would be created by a fan within the drill (ideally powered by the same motor), and the sample would flow through the tube

Table 9.11 Basic trade for pneumatic sampling methods

Venturi in drill bit	Venturi behind drill bit	Vacuum behind cyclone
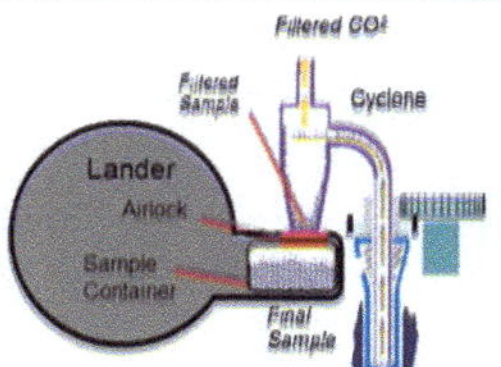	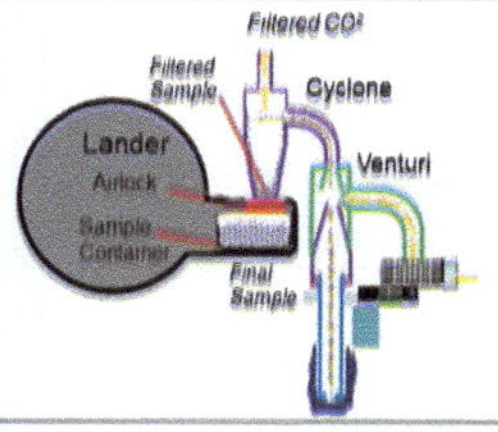	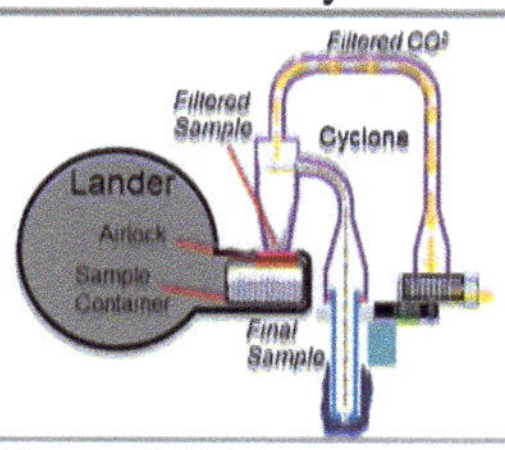
Complex drill bit design	Simple drill bit design	Simple drill bit design
Compact volume	Semi-compact volume	Large volume
Requires making pressure above atmosphere (maybe difficult)	Requires making pressure above atmosphere (maybe difficult)	Requires making pressure below atmosphere
Very short path of gas flow	Intermediate path for gas flow	Longer path for gas flow
Suction created with same actuator as drill	Suction created with same actuator as drill	Suction may need to be made with separate actuator to reduce complexity
Fan may get damaged by cuttings	Fan is away from cuttings	Fan may be damaged by fine cuttings

to a cyclone separator on the lander. The second option would use Venturi suction system directly behind the drill bit – the blower could still be powered by the same drill motor. The third option would use suction at the back end of the cyclone separator on the lander (either by using a suction fan or a Venturi system).

From a system level standpoint, it would appear as though the best option might be the Vacuum Behind the Cyclone option since it creates negative pressure instead of positive. All three options were further explored and breadboarded to determine if there are any other factors that should be considered.

9.5.1 Venturi in Drill Bit

A prototype of the Venturi in Drill Bit was made out of printed plastic as shown in Fig. 9.17. The goal of this breadboard was to determine effectiveness and the efficiency of sample transfer.

The system was comprised of two parts: the bit system and a hole filled with limestone powder. Initial testing was unsuccessful as no sample was collected. Subsequently, the breadboard was modified by changing the shape of the cutter and replacing the sample tube. These changes helped with sampling effectiveness. Additional tests were conducted at different flow rates and with the bit inside the hole at different depths. These parameters were found to affect sampling efficiency. These promising results suggest that for this sampling approach to work, more breadboarding and testing is needed since the Venturi design is difficult to analyze.

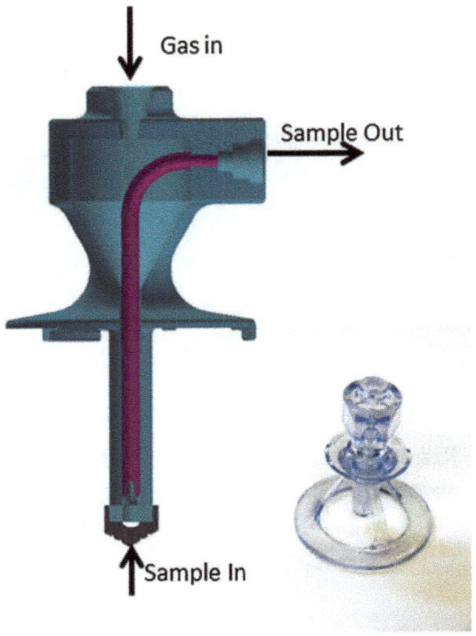

Fig. 9.17 Integrated Venturi

9.5.2 Venturi Behind Drill Bit

For the Venturi Behind Bit experiments, an off the shelf (COTS) Venturi was used with a cyclone separator and a drill bit. During the tests pressure and flow into the Venturi were measured in order to get an estimate of energy efficiency. A known sample was placed in the borehole mockup and after a 30 s run, sample captured by a cyclone was weighed to determine sampling efficiency. Test powders included drill cuttings of Indiana Limestone, Saddleback Basalt, and Bishop Tuff.

Test results are plotted in Fig. 9.18. Clearly visible is a cutoff point in cuttings collection at approx. 100 W; above that point collection efficiency reached its peak. The mechanical power needed to create suction is a product of the pressure and flow rate and does not include any system loses. It is reasonable to assume that an actuator and an impeller would need more than that power to create the necessary particle movement. It is important to note that these experiments were done with a single set up, whereas many factors such as hose diameter and length, not investigated here, affect efficiency of cuttings collection and the power required.

9.5.3 Suction Behind Cyclone Separator

For the Suction Behind Cyclone tests, a shop vacuum was connected to the exhaust port of the cyclone separator. The shop vacuum was controlled via Variac in order to adjust the voltage and in turn the air flow. A current sensor was also used to take measurements of electrical power. As in the previous tests, a known mass of sample was placed inside the borehole mockup, and the amount collected was measured

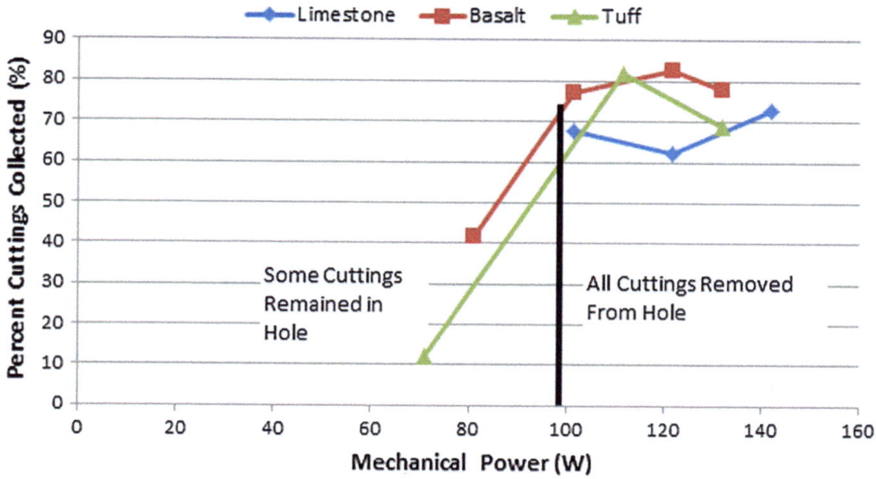

Fig. 9.18 Results from Venturi behind bit tests

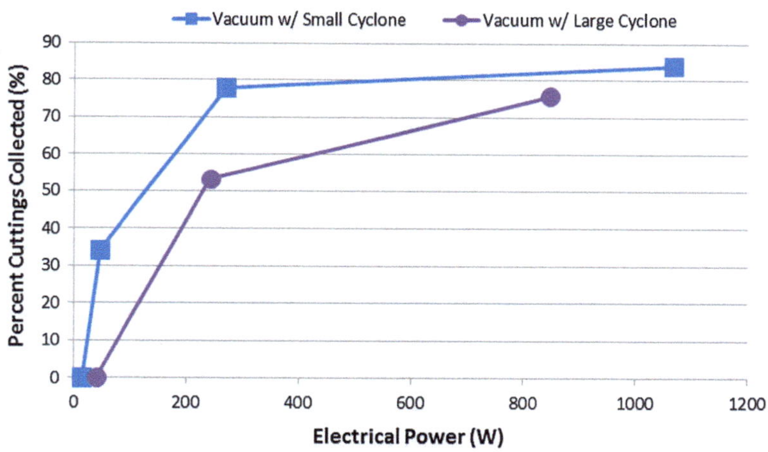

Fig. 9.19 Results from vacuum behind Cyclone tests

after the test. In these tests only Limestone cuttings were used. In addition, two different sizes of cyclone separator were tested to determine the effect of the cyclone design on sampling efficiency.

Figure 9.19 shows test results. It can be seen that to capture most of the samples the required power needs to be over 200 W, twice that of the Venturi Behind Drill Bit. In addition, it was observed that a small cyclone performed better than the larger cyclone, implying further optimization is possible by redesigning the cyclone system.

9.5.4 Conclusions

The overall results of this series of tests showed that pneumatic sample transfer might be a viable means of collecting a sample. The tests also suggest that the second or the third option could be the best choice.

9.6 Fluid Flow Analysis: 1 atm Versus 92 atm

This section focuses on a high level investigation of gas flow at ambient and Venus atmospheric conditions at ground level. Data from Table 9.12 has been used in all calculations.

9.6.1 Particle Transfer in a Tube

For a particle to remain in suspension, force due to drag equals the force due to gravity and the gas velocity is called choking velocity, v:

$$v = (\mathrm{mg})/(6\pi\,\mu\,\mathrm{R}), \tag{9.1}$$

where, m = particle mass, g = gravitational acceleration, μ = fluid viscosity, and R = particle radius.

When considering the Venus environment, only two parameters change in the above equation: gravity (lower on Venus) and fluid viscosity (higher on Venus). Hence choking velocity on Venus would be 50 % of that on Earth.

To validate this principle, a series of tests were performed to determine the minimum air flow needed for the particles to exit a 20 cm long, 7 mm diameter tube. Two particle sizes of Saddleback Basalt cuttings were used: 303 micron and 428 micron. It was found that an actual velocity needs to be two times greater than theoretical for particles to move up the vertical tube. The measured collection velocity for the 303 micron particle was ~20 m/s while the theoretically derived velocity was ~10 m/s.

Table 9.12 Gas parameters (Petroppoulos and Telonis 1988)

Parameter	Units	Earth	Venus	Ratio [V/E]
Atmospheric pressure	[kPa]	105	9322	88.8
Atmospheric temperature	[C]	22.0	500.0	22.7
Atmospheric density	[kg/m^3]	1.2	65.6	53.6
Atmospheric viscosity (GVC, 2014)	[kg/(s*m)]	0.0000179	0.0000312	1.7
Gravity	[m/s^2]	9.8	8.9	0.9

9.6.2 Cyclone Separator

In designing a cyclone separator, the first step is to start with well-defined models and then validate cyclone performance and make additional changes based on actual vs. required efficiencies. In this case, a MATLab program was developed to model performance parameters of a given cyclone separator geometry. In this program, the Muschelknautz modeling method was used which outputs pressure drop across the cyclone separator, and the efficiency of cuttings collection as a function of particle size (Hoffmann and Stein 2008). For Venus considerations, these equations factor in gravity, viscosity, and density of the atmosphere. As mentioned earlier, the difference in gravity is small, and the viscosity on Venus is roughly 2x greater. The density of the Venus atmosphere, however, is roughly 50x greater than that of Earth. The MATLab script was used to develop an initial cyclone prototype for high level tests.

Figure 9.20 shows the test results from collection efficiency experiments. It can be seen that theoretical efficiency overestimates actual collection efficiency if particles are small and velocity is relatively low. Otherwise, theoretical efficiency is very close to actual.

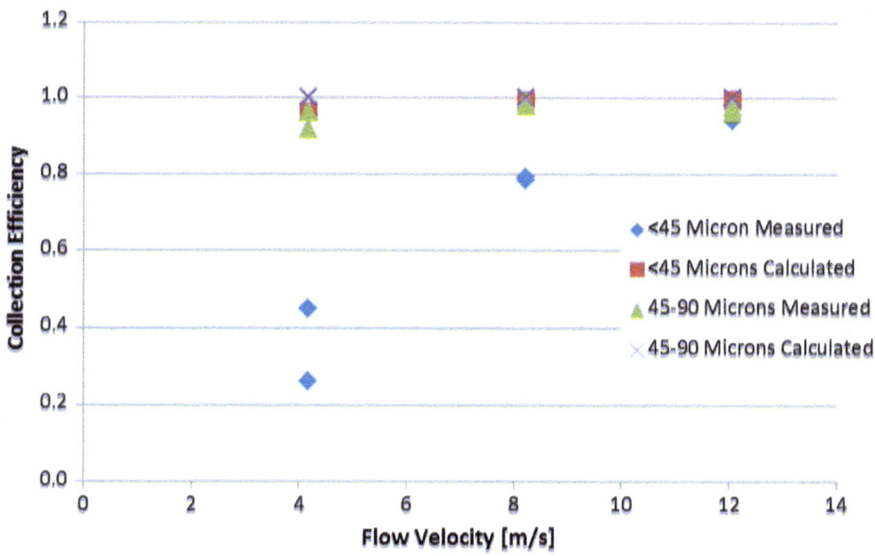

Fig. 9.20 Results from Cyclone Separator Collection Efficiency tests

9.6.3 Impeller Power

Table 9.13 lists a number of equations that need to be considered to determine the ratio between impeller power on Venus and on Earth. In both cases (Earth and Venus) it has been assumed that particle sizes are the same and the main driver with respect to fluid flow is Choking velocity (minimum velocity required to keep particles suspended in gas flow in vertical direction). It has also been assumed that impeller diameters are the same. In these calculations a tube length of 1 m with internal diameter of 5 mm was assumed, and the minimum particle size of 100 micron was used.

The results indicate that the Choking velocity and also impeller velocity are 50 % lower for Venus (because gas density is higher on Venus). The head loss on Venus however, is ~5x higher. The head loss includes a friction factor which is a function of many parameters that are different between Earth ambient and Venus conditions.

Figure 9.21 shows the power ratio as a function of largest particle size lofted at a set choking velocity. The impeller power at Venus conditions increases as particle diameter increases. Based on the ambient tests (0) the measured choking velocity was 2x theoretical. Hence, the power ratio will be greater than what the figure indicates.

9.6.4 Conclusions

Modern fluid flow software allows for complex design of pneumatic devices in different environmental situations. Without using this kind of software however, some simplifications can be made in order to estimate the effects of environmental

Table 9.13 Estimating impeller power

Parameter	Units or equation	Ratio [Venus/Earth]
Choking velocity [m/s]	$v = \frac{mg}{6\pi\mu R}$	0.52
Headloss [kPa]	$\Delta P_{hl} = 0.5 * \rho_{air} * f_{pipe} * \frac{L_{pipe}}{D_{pipe}} * v^2$	~5
	$f_{pipe} = F(\mathrm{Re}) = G\left(\frac{1}{\rho_{air}}, \frac{1}{v}, \frac{1}{D_{pipe}}, \mu_{air}\right)$	Assume the same
Impeller HeadLoss	$\Delta P_{fan} \propto \rho_{air} * n^2 * d^2$ $\Delta P_{fan} = Pressure\ bump\ across\ fan$ $\rho_{air} = Air\ Density$ $n = Rotational\ speed$ $d = Impeller\ Diameter$	
Impeller speed	$\frac{n_V}{n_E} = \sqrt{\frac{P_V}{P_E} * \frac{\rho_E}{\rho_V}}$ Assumed impeller diameter is the same.	0.3
Impeller power [W]	$Power \propto \rho_{air} * n^3 * d^5$	~1.5

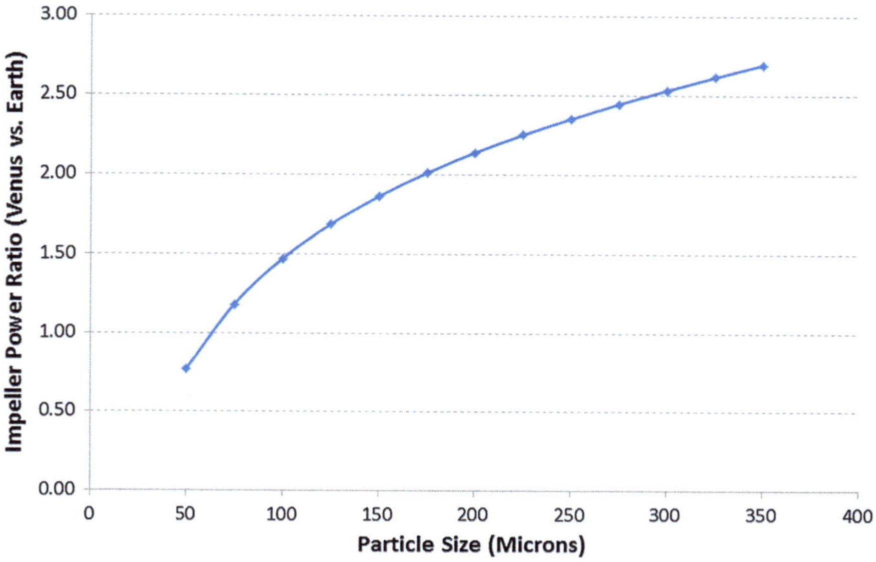

Fig. 9.21 Impeller power ratio (Venus vs. Earth) vs. Particle Size

conditions. At Venus atmospheric conditions, the choking velocity is 50 % lower. The exact value depends on particle diameter. The head loss at Venus conditions is higher and so is impeller power. The exact power values depend on many factors and should be derived experimentally.

9.7 System-Level Trade Study

To aid in the selection process a system level trade study was performed for different aspects of sample acquisition. A sample acquisition and delivery system has three functional subsystems: a deployment subsystem, a sample acquisition subsystem and a sample delivery subsystem. Deployment could be achieved using a simple gravity driven Z-stage or spring loaded arm and an actuated arm (minimum a 3 Degree of Freedom or DOF arm) (Section 0). Sample acquisition includes a drill and a trencher (Section 0). Sample transfer includes a pneumatic sample transfer system and a robotic arm (Section 0).

9.7.1 Selection Criteria

Tables 9.14, 9.15, 9.16, 9.17 and 9.18 outline the selection criteria which were used in this study, along with a description, weight number, and justification.

Table 9.14 Operation criteria

Criteria	Description	Weight	Justification
End-to-end operation time	From start to finish, how long does a single operation take (deployment, cutting, delivery)?	10	All mission objectives must finish quickly before the heat/environment force termination
Required level of automation	How much must the system rely on software and sensor feedback? The lower the required level of automation, the better.	10	All electronics, controllers, actuators, sensors must function properly under Venus conditions, so the fewer that are required (and required to work together), the better.
Placement flexibility	How much manipulation is needed to position the sampler? Less is better	10	The sampler that does not need special deployment system is better
Ground loop independence	To what degree mission control command the operation? Better is no 'human in the loop'	10	Time is critical under Venus conditions
Sensor dependency	Does the operation depend on feedback from sensors? The less dependent, the better	9	Sensors can be damaged in flight, on landing, during operation, etc.
Risk to spacecraft	Would the operation risk damage to the lander, its sensors or actuators?	7	Mission failure risk by damaging external sensors, actuators

Table 9.15 Venus conditions criteria

Criteria	Description	Weight	Justification
High temperature tolerance, mechanism	Can actuators and electro-mechanical systems withstand temperatures exceeding 500 °C? The higher the tolerance to temperature, the better	10	An ambient temperature of 500 °C drives many mechanism design choices, including materials, electronics, actuators, thermal expansion, etc.
High temperature tolerance, cutter	Can the cutter withstand temperatures of 500 °C and more? The higher the tolerance to temperature, the better	10	The cutter must not melt, change composition, expand and seize, etc.
Pressure tolerance	How damaging will the high pressure be to the operation? The higher the tolerance to pressure, the better	10	Venus's 90 atm pressure will drive our design choices
Corrosion resistance	How damaging atmosphere of Venus be to the operation? The higher the resistance to corrosion, the better	10	Venus's atmosphere, comprised mainly of super-critical CO_2, is known to be highly corrosive and is the main concern for spacecraft failure
Robust to surface conditions	How indifferent is the system to the slope, roughness of rock? The less that the surface conditions matter, the better	10	The terrain will be largely unknown, the lander will be stationary, and arm might not be deployed, so the system must be able to sample from a wide range of slopes

Table 9.16 Sample criteria

Criteria	Description	Weight	Justification
Contamination of sample	Could any part of the sample collection process contaminate the sample collected?	10	If a sample is contaminated by any part of the s/c, it could lead to false data
Sample alteration	Could the system fundamentally change the sample and thus give false information about the surface composition?	9	Sample alteration could occur from excessive mechanical processes, additional heat or pressure, etc.
Sample bias	Does the process remove fines, lighter or heavier particles? The lower the chance of creating a bias, the better	8	Sample should be representative of the location
Ability to cut hard rocks	Does the cutter demonstrate the ability to cut through hard rocks? The harder the rock the cutter can penetrate the better	5	The device must be able to cut typical Venus simulants, which could be hard basalts
Sampling redundancy	Can some kind of sample be acquired even if critical parts fail? Higher redundancy better	5	In case of critical part failure, any sample is better than no sample
Multiple samples	Is there potential to acquire more than one sample? If more samples could be acquired the better	2	More data is a plus, but by doing so there is a risk of cross contamination
Cross contamination	Could particles from one sample contaminate the second sample? Lower cross-contamination is better	1	Two rocks might look similar even if they are not

Table 9.17 Mechanism criteria

Criteria	Description	Weight	Justification
Physical robustness	Structural integrity, resistance to all types of external forces, factor of safety. The more robust the system, the better	10	Any component or structural failure will compromise mission success
Mechanism redundancy	Can the system easily be outfitted with a backup? The more available the redundancy, the better	5	It is always good to have redundancy
Mass	How massive is the deployment/cutting/delivery system? The lower the mass, the better	4	Minimizing mass is desirable on flight missions
Number of actuators	Minimum number of actuators required for operation? The fewer actuators, better	4	More actuators mean more points of failure, and higher cost
Energy	How much energy is required to complete the operation? The lower the energy, the better	3	The energy drill uses is relatively low
Peak power	The lower the peak power, the better	3	The higher the power thicker cables, more loses
Volume	How large is the deployment/cutter/delivery system? The lower the volume, the better	2	TBD
Consumables	Does this option require many one-time-use components, such as pyros, gas, or parts that destruct? The fewer consumables, the better	1	Consumables reduce the reusability of any part of the system

Table 9.18 Designer-side criteria

Criteria	Description	Weight	Justification
System complexity	How many parts, how many moving parts, how intricate is the required design? The lower the system complexity, the better	7	This is a high-risk mission. The fewer things that can go wrong, the better
Flight heritage	Has this technology been used in flight operations before? The more times something has been used in flight, the better	5	Venus conditions are extreme and difficult to replicate in earth-based tests, so it would help the designers to know what has and hasn't worked in similar flight situations before
Lowest TRL	Which option has more precedence and thus will require less research and testing on our part? The higher the overall TRL level, the better	5	The less time spent on research and testing, the more time spent in fine-tuned development
Cost	Monetary cost of components, raw material, man hours, etc. The lower the cost, the better	2	Cost is always an issue

In each of the three sections below, the options being compared are graded on a scale from 0 to 3. "0" means that the criterion is not applicable. A score of 3 vs. 1 means that the first option is significantly superior. These scores are multiplied by the weight at the end for a final score.

9.7.2 Sampler Deployment: Spring Actuated Arm Versus Motor Actuated Robotic Arm

Tables 9.19, 9.20, 9.21, 9.22 and 9.23 use the initial criteria set to compare deploying the penetrator (Section 0) with a simple spring-loaded device (such as an arm or lander deployment) versus a 3 DOF robotic arm. This section is meant to be comparing the options independently of a choice for penetrator (drill, trencher) or sample delivery method.

The final tally puts the Spring at 392 and the Arm at 255. The simple deployment is the clear winner in this case under the criteria previously identified. Venera 13 and 14 and Vega 2 used simple, one time, one location deployments for their drills. However, if a mission were to put more emphasis on multiple samples the arm would be required. In addition an arm could be used for sample drop off and in turn, a pneumatic system would not be required.

Table 9.19 Operation: spring vs. arm deployment

Criteria	Weight	Spring	Arm	Justification
End-to-end operation time	10	3	1	Spring firing is fast, arm must manipulate over to rock
Required level of automation	10	3	1	Arm requires control software
Ground loop independence	10	3	1	Arm might require human feedback/decisions
Placement flexibility	10	1	3	Arm can manipulate and adjust position/orientation. Spring loaded mechanism has no control where it lands
Sensor dependency	9	3	2	May be able to deploy an arm blind, can definitely deploy spring loaded system blind
Risk to spacecraft	7	1	3	Reaction from tightly coiled spring launch could be significant
Weighted score	134	99		

Table 9.20 Venus conditions: spring vs. arm deployment

Criteria	Weight	Spring	Arm	Justification
High temperature tolerance, mechanism	10	3	1	Spring loaded mechanism will hold up better than the arm at 500 °C temperatures
High temperature tolerance, cutter	10	0	0	Not applicable
Pressure tolerance	10	3	1	Arm seals could succumb to pressure
Corrosion resistance	10	3	1	Corrosion will happen to both, but its damage will be more severe on the arm
Robust to surface conditions	10	1	3	Arm can manipulate and adjust position/orientation
Weighted score	100	60		

9.7.3 Sample Acquisition: Drill vs. Trencher

Tables 9.24, 9.25, 9.26, 9.27 and 9.28 use the criteria from the section above to compare a drill and a trencher.

The final tally puts Drill at 415, and Trencher at 416. The main advantages of the trencher are fast sampling time, overall simplicity, and robustness to surface conditions. However because cutoff wheels disintegrate at high temperatures while carbide wheels cannot cut harder rocks, it was decided that the drill would be a better immediate choice.

Table 9.21 Sample: spring vs. arm deployment

Criteria	Weight	Spring	Arm	Justification
Contamination of sample	10	0	0	Not applicable
Sample alteration	9	0	0	Not applicable
Sample bias	8	0	0	Not applicable
Ability to cut hard rocks	5	2	2	Arm is limited by reaction torque it can apply, spring loaded mechanism is limited in its blind deployment
Sampling redundancy	5	1	3	If the spring fails, no deployment and no sample can be gathered. If arm remains even partially functional, it may be able to manipulate to rock
Multiple samples	2	1	3	Once the spring is deployed, it cannot deploy again. Arm can manipulate back and forth
Cross contamination	1	0	0	Not applicable
Weighted score	17	31		

Table 9.22 Mechanism: spring vs. arm deployment

Criteria	Weight	Spring	Arm	Justification
Physical robustness	10	3	1	Spring loaded system has 1 moving part
Mechanism redundancy	5	3	1	Much easier to implement back up springs/mechanisms than include a backup arm
Mass	4	3	1	Robotic arms are heavy
Number of actuators	4	3	1	1 pyro to release spring, at least one for every joint in arm
Energy	3	3	1	Spring relies on stored mechanical energy, arm requires electrical energy
Peak power	3	3	1	Arm requires more power
Volume	2	3	3	Similar – depending on design
Consumables	1	1	3	Being spring loaded is consumable (one-time use)
Weighted score	94	38		

9.7.4 Sample Delivery: Pneumatic Versus Robotic Arm

Tables 9.29, 9.30, 9.31, 9.32 and 9.33 below compare the two options for sample delivery of a sample collected from an excavator. The Robotic arm assumes a robotic arm for deployment.

The final score for the Pneumatic system is 378 while the final score for the Arm system is 254. It should be noted that this comparison does not take into account

Table 9.23 Designer-side: spring vs. arm deployment

Criteria	Weight	Spring	Arm	Justification
System complexity	7	3	1	Spring loaded is strictly mechanical, arm is an intricate electromechanical device
Flight heritage	5	3	1	Venera 14 had spring deployed arms to sample soil
Lowest TRL	5	1	3	Honeybee robotics has conducted extensive research with robot arms with tool end effectors
Cost	2	3	1	Robot arms are very expensive compared to spring and release mechanisms
Weighted score	47	29		

Table 9.24 Operation: drill vs. trencher

Criteria	Weight	Drill	Trencher	Justification
End-to-end operation time	10	1	3	Circular saw testing has demonstrated cutting trenches in seconds, while a drill takes on the order of minutes
Required level of automation	10	1	3	A drill may require additional feedback, percussion control, z stage control, etc., while a trencher can just drop and go
Ground loop independence	10	2	3	A drill may require some parameter changes and adjustment from ground control
Placement flexibility	10	2	3	A circular saw will cut well at many angles, while a drill bit must advance along its z-axis fairly precisely
Sensor dependency	9	2	3	A drill relies a bit more on sensor feedback, and since it is more sensitive to placement, visual feedback may be necessary
Risk to spacecraft	7	3	2	Neither is too risky, but the behavior of the circular saw to kick particles backwards could pose a threat
Weighted score	99	161		

deployment method, such that a pneumatic system could still be used with robotic arm deployment.

9.7.5 Conclusions

Table 9.34 shows the final tallies for each of the six possible combinations in this study. This number is six and not eight because a robotic arm sample delivery system cannot be considered for a spring-loaded deployment. As far as deployment

Table 9.25 Venus conditions: drill vs. trencher

Criteria	Weight	Drill	Trencher	Justification
High temperature tolerance, mechanism	10	1	2	The drill is likely to have more actuators, electronics
High temperature tolerance, cutter	10	3	1	Off the shelf saw blades which can penetrate well have a high sensitivity to heat, but testing has shown that drill bits can withstand the heat
Pressure tolerance	10	3	3	Both options will not be significantly affected by HP
Corrosion resistance	10	1	1	Super-critical CO_2 is so highly corrosive that neither option has an advantage over the other
Robust to surface conditions	10	2	3	A drill's performance may vary depending on surface conditions but a trencher is more surface tolerant
Weighted score	100	100		

Table 9.26 Sample: drill vs. trencher

Criteria	Weight	Drill	Trencher	Justification
Contamination of sample	10	2	1	The aggressive cutting of the circular saw leads to bit wear (extreme in the case of cutoff wheels)
Sample alteration	9	3	1	The circular saw's high-speed, high-inertia, high-friction, aggressive cuts are likely to generate more heat
Sample bias	8	2	1	The way a circular saw digs and kicks back material is likely to lead to particle size bias
Ability to cut hard rocks	5	2	3	The right circular saws can cut through hard rock quickly, while a drill may be slower due to lower rpm
Sampling redundancy	5	3	1	A drill can still be used to suck a sample
Multiple samples	2	3	2	Each tool has the potential for multiple samples
Cross cont	1	1	2	Trencher can be cleaned by spinning at high rpm
Weighted score	95	53		

and delivery is concerned, a spring loaded option with pneumatic delivery is best. This is not surprising, and in fact this is what has flown on Venera 13/14 and Vega 2. However, it is important to note that depending on the requirements of the

Table 9.27 Mechanism: drill vs. trencher

Criteria	Weight	Drill	Trencher	Justification
Physical robustness	10	3	2	Drill bits can be made compact and rugged, while a circular saw blade is thin and can snap under external forces
Mechanism redundancy	5	1	2	It is unlikely that a mission would have redundant samplers, but since the trencher is less complex, it would be easier to add redundancy (such as an additional drive actuator)
Mass	4	2	3	Drill percussive system could add extra mass
Number of actuators	4	2	2	A drill could be made with more actuators (percussion, z stage) but in there simplest form, both require one core actuator
Energy	3	2	2	Drill is slow but low power, trencher is fast but high power
Peak power	3	3	1	Trencher requires high power
Volume	2	2	2	Initial concepts indicate comparable size
Consumables	1	3	2	Circular saw wears over time faster
Weighted score	73	65		

Table 9.28 Designer-side: drill vs. trencher

Criteria	Weight	Drill	Trencher	Justification
System complexity	7	2	3	A drill can be made relatively simply, but may require z stage actuation, percussion, etc. a circular saw just needs rotation
Flight heritage	5	3	1	Venera 13–14 have already drilled on Venus
Lowest TRL	5	3	1	Honeybee Robotics has extensive experience with drills, not as much with trenchers
Cost	2	2	3	Because of the complexity of drill vs. a trencher, it will require more components and thus will cost more
Weighted Score	48	37		

actual mission, this could change in favor of a different combination. It is recommended that the Drill would be optimum solution. However, when comparing sample delivery and deployment, a 3-DOF arm could satisfy both operations.

Table 9.29 Operation: pneumatic vs. arm delivery

Criteria	Weight	Pneumatic	Arm	Justification
End-to-end operation time	10	3	1	An arm must physically transport the sample back, while suction occurs nearly instantaneously
Required level of automation	10	3	1	A pneumatic system would require the opening of valves, whereas the movement of a robotic arm is more complex
Ground loop independence	10	3	1	Little is required from ground control on pneumatic delivery; arm may require commands to deliver properly
Placement flexibility	10	0	0	Not applicable
Sensor dependency	9	3	1	Pneumatic system can operate with no feedback, but the arm may require extensive visual and joint feedback
Risk to spacecraft	7	3	1	The pneumatic delivery is largely passive, but the arm has the potential to strike the lander
Weighted score	138	46		

Table 9.30 Venus conditions: pneumatic vs. arm delivery

Criteria	Weight	Pneumatic	Arm	Justification
High temperature tolerance, mechanism	10	2	1	Arm has more actuators. Pneumatic transfer requires HT pyros
High temperature tolerance, cutter	10	0	0	Not applicable
Pressure tolerance	10	1	3	Pneumatic system parameters are very dependent on ambient pressure
Corrosion resistance	10	3	1	Corroding joints become inoperable;
Robust to surface conditions	10	0	0	Not applicable
Weighted score	60	50		

9.8 Prototype Venus Sampler System

Table 9.35 lists various combinations that are possible with respect to sample acquisition and delivery. If just one sample is needed (Option 1), the architecture could include a Rotary-Percussive Drill with a Vertical deployment stage and

Table 9.31 Sample: pneumatic vs. arm delivery

Criteria	Weight	Pneumatic	Arm	Justification
Contamination of sample	10	1	3	Abrasive particles can shed tubing material system
Sample alteration	9	2	3	Chemical reactions could occur inside tubing
Sample bias	8	1	3	Pneumatic parameters can bias particle size
Ability to cut hard rocks	5	0	0	Not applicable
Sampling redundancy	5	2	1	As long as can create suction, you can gather surface dust
Multiple samples	2	1	2	Arm can deliver multiple samples
Cross contamination	1	1	3	Previous sample can coat the inside of the tubing
Weighted score	49	93		

Table 9.32 Mechanism: pneumatic vs. arm delivery

Criteria	Weight	Pneumatic	Arm	Justification
Physical robustness	10	3	1	Many joints, actuators, components, etc. can break in an arm
Mechanism redundancy	5	3	1	Too big, expensive, and complex to have multiple arms. Easy to hook up multiple tube systems
Mass	4	3	1	Robotic arms are heavy
Number of actuators	4	3	1	Arm requires an actuator for each joint
Energy	3	2	1	The actuators of a robotic arm will consume a great deal of energy
Peak power	3	2	1	Arm requires electrical power
Volume	2	2	2	Robotic arms are long, but could be folded up to occupy a similar volume to more static pneumatic system
Consumables	1	1	3	Pneumatic systems may require vacuum tank for suction
Weighted score	86	36		

Table 9.33 Designer-side: pneumatic vs. arm delivery

Criteria	Weight	Pneumatic	Arm	Justification
System complexity	7	2	1	Robotic arms are more complex
Flight heritage	5	3	1	Venera 13-14 used vacuum tank and tubes
Lowest TRL	5	3	3	Both systems have low TRL but pneumatic system only requires HT pyros
Cost	2	3	1	Robot arms are very expensive
Weighted score	50	29		

Table 9.34 Trade study cumulative scores

	Spring loaded arm deployment		Robotic arm deployment	
	Pneumatic Delivery	Robotic arm delivery	Pneumatic delivery	Robotic arm delivery
Drill	1185	Not an option	1048	924
Trencher	1186	Not an option	1049	925

Table 9.35 Venus sample acquisition system (VSAS) decision matrix

Penetration		Deployment				
		Sample transfer				
Drill	Trencher	Vertical slide	Spring loaded arm	3 DOF Arm	Pneumatic – Blower	Pneumatic – Suction
1, 2		2		2		1

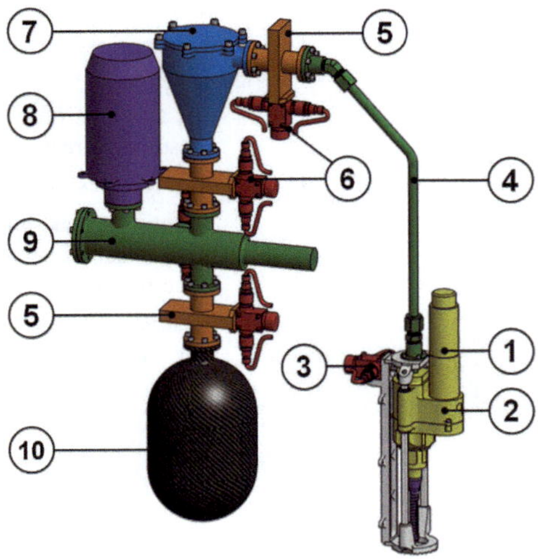

Fig. 9.22 A Venus drill concept, including sample delivery for the "single sample" architecture. The main are: 1. High Temperature Actuator; 2. Percussive Drill Head Assembly; 3. Pyrotechnic Initiator for Releasing Drill Head; 4. Sample Transfer Tubing; 5. Gate Valve; 6. Pyrotechnic Initiator for Actuating Gate Valve; 7. Inlet Funnel; 8. Generic Science Instrument; 9. Sample Analysis Chamber; 10. Low Pressure (LP) Reservoir

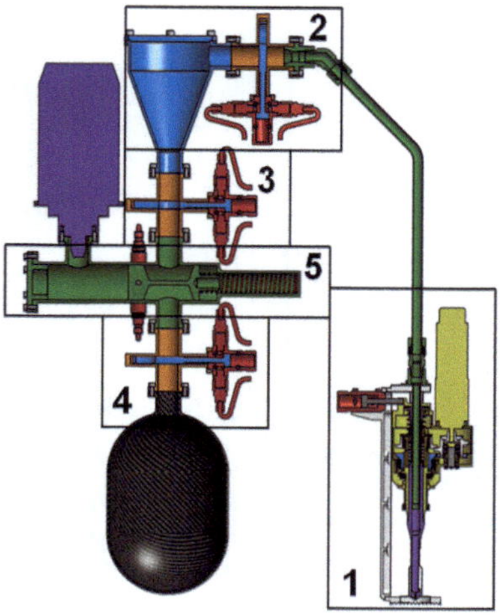

Fig. 9.23 High level sequence of operations of a Venus drill concept. 1. Drill acquires sample. 2. Gate valve opens at cyclone inlet. Initial low pressure (LP) within cyclone is enough to create enough suction to pull in sample; 3. Gate valve closes at cyclone outlet, trapping sample in the Sample Analysis Chamber; 4. Gate valve opens at LP reservoir using pyrotechnic initiator, creating LP atmosphere; 5. Pyrotechnic Initiator releases Sample Transfer "Shuttle". Spring force moves

pneumatic suction based sample delivery (i.e. Venera 13/14 and Vega 2 like system with a percussive rather than rotary drill). If multiple samples are needed (Option 2) the architecture could include a Rotary-Percussive Drill with a 3 DOF Robotic Arm for Drill deployment and sample delivery.

A high level concept for a Venus drill and sample delivery system for a "single sample" architecture is shown in Fig. 9.22. Figure 9.23 shows high level operational sequence while Fig. 9.24 details the concept of operation. The concept is very similar to the one used on Venera 13 and 14 as well as Vega 2 mission, reinforcing the fact that the approach used back in 1980s was probably the optimal for the Venus sampling application.

The major improvement in the current design is the use of a percussive drill system (as opposed to purely rotary) in order to have ability to penetrate harder formations. However, if the drill motor and Weigh on Bit is large enough, a pure rotary drilling approach might be sufficient to penetrate hard rock in a relatively short period of time. This trade (percussive vs. rotary) would need to be made based on spacecraft available resources.

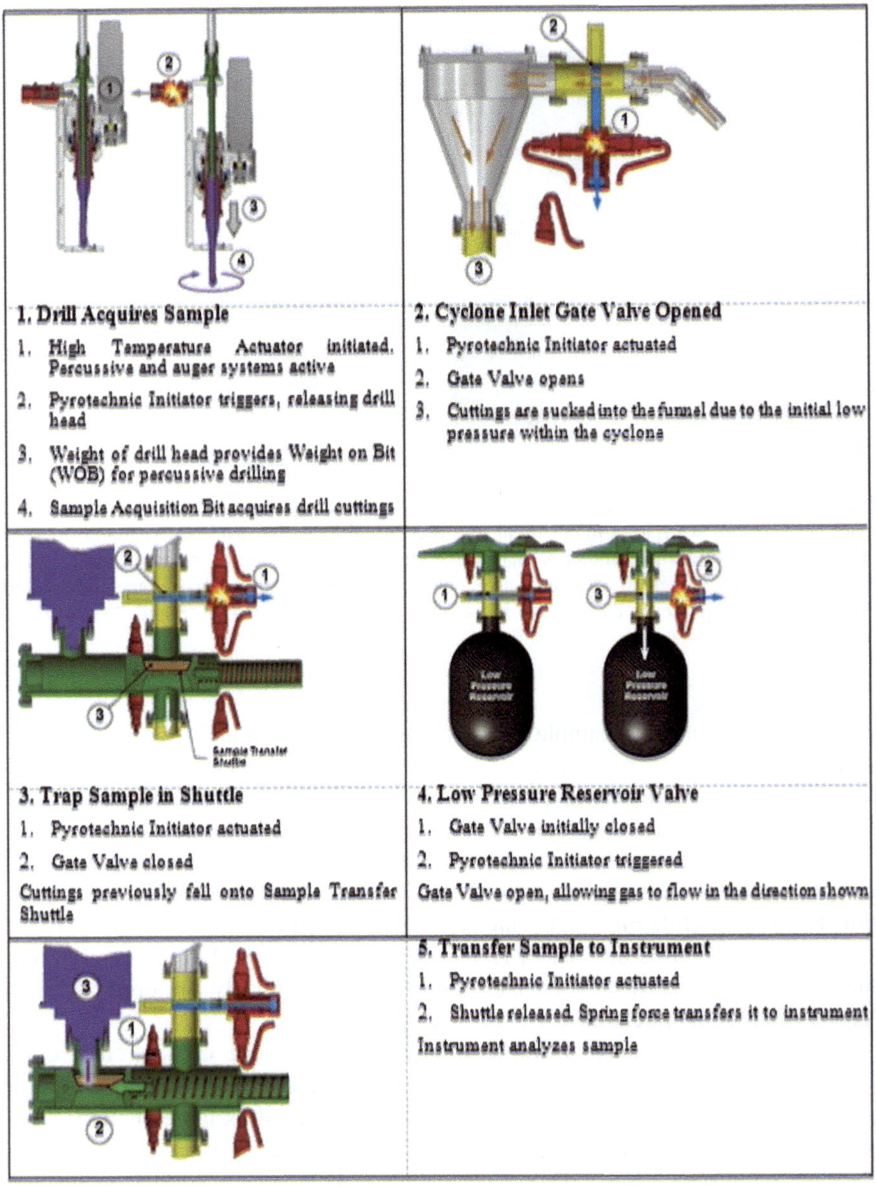

Fig. 9.24 Sample acquisition and delivery steps for the Venus drill concept

9.9 Conclusions

This chapter presented numerous studies and experimentally derived data in support of Venus sampling acquisition and sample delivery systems. The main findings are grouped into Sampler Deployment, Sample Acquisition, and Sample Transfer categories.

Sampler Deployment - A 3 DOF Robotic Arm has numerous advantages but also several disadvantages. If the drill or the trencher is deployed from a robotic arm, the system could be used multiple times. The same robotic arm could also be used to place the sampler above the instrument inlet port for sample delivery. Since the sampler can be retracted out of the rock, a hole or a trench will be accessible to stand off instrument investigation such as LIBS/Raman. The main disadvantages of the actuated arm are that it requires actuators and therefore electronics, it also needs kinematic software for arm positioning and possible human in the loop for decision making.

More passive deployment systems include a body mounted vertical deployment stage as in Venera 13 and 14 and Vega 2 drills or a spring loaded arm similar to one used to deploy the soil probe on Venera 13 and 14. These passive systems can be deployed only once, they have no control over what part of the landed area they are deployed to and require a pneumatic sample delivery system or separate arm system.

In this case, there is no clear winner. Depending on requirements and spacecraft resources either of the systems could be used.

A Trencher might not be suitable for Venus operations because progress in hard rocks is slow with carbide cutters and very aggressive cutoff wheels that can penetrate hard rock disintegrate at high temperature. For the trencher to be a viable option a new excavation wheel would need to be developed that incorporates abrasive material (alumina, silica) in a matrix that can withstand high temperature. Diamond abrasive is not suitable since Venus temperature is close to diamond graphitization temperature.

A Drill is suitable for this Venus application. The drill prototype was found to penetrate hard materials at 500 °C, though at reduced penetration rate compared to 20 °C tests. A percussive system had to be engaged since pure rotary system was unable to penetrate hard rocks. The carbide bit withstood high temperature and did not show signs of wear after several tests.

Based on the above conclusions, an optimum Venus sample acquisition system is a Rotary-Percussive drill.

If the sampler is deployed using a 3 DOF Robotic Arm, the same arm could be used to deliver a sample. A blower based pneumatic sample delivery might be feasible but needs more power than the same blower at ambient conditions. Both systems, a 3 DOF arm and the Blower, could be used multiple times and in turn multiple samples could be delivered for analysis. A suction based pneumatic sample transfer has been successfully used three times: Venera 13 and 14 and Vega 2. This approach does not require motors and in turn electrical energy. High temperature pyrotechnic devices needed to puncture membranes leading to the transfer tube need development, though, as the historical engineering data is no longer available. Also, the pyro system works only once. Depending on mission requirements and spacecraft resources, all three options are viable.

Acknowledgments The research reported in this chapter was conducted at Honeybee Robotics under a contract with the National Aeronautics and Space Administration (NASA). We would like to thank for the advice and support from Jeff Hall, Satish Khanna, Kristo Kriechbaum, Dara Sabahi, and Lori Shiraishi of NASA Jet Propulsion Laboratory.

Appendix: Material Properties at High Temperature

Table 9.36 shows properties of selected structural materials while Table 9.37 shows percent decrease of these properties at 500 °C. Table 9.38 shows a list of material parameters considered in this study. It should also be noted that all properties taken from the Metallic Materials Properties Development and Standardization (MMPDS-05). Material properties are for tubing unless not available, then

Table 9.36 Properties of selected material at room temperature

Material	Specification	Density	Room temp (20 °C)							
			Ftu	Fly	E	G	Fey	Fsu	Fbru	Fbry
			ksi	ksi	msi	msi	ksi	ksi	ksi	ksi
CP Titanium	AMS 4901. AMS-T-9046 CP-I	0.163	80	70	15.5	6.5	70	42	120	101
Ti-8Al-lMo-lV	AMS-T-9047	0.158	130	120	17.5	6.7				
T i-6 Al-2 Sn -4Zr-2Mo	AMS 4975	0.164	130	120	16.5	6.2				
TI-6AL-4 V	*MIL-T-9047*	*0.160*	*160*	*ISO*	*16.9*	*6.2*				
TI-6AL-4 V	AMS 493 5	0.160	130	120	16.9	6.5	128	83	214	ISO
TI-6AL-4 V	AMS 4934	0.160	151	138	16.9	6.5	147	92	237	208
Ti-6Al-6 V-2Sn	*MIL-T-81556&AMS-T-81556, Comp. AB-3 annealed*	*0.164*	*142*	*129*	*16*	*6.2*	*137*	*93*	*218*	*196*
A-286	AMS 5731, AMS 5732 STA	0.287	130	85	29.1	11.1	85	85	195	127
Inconel 600	ASTMB166 CW	0.304	110	85	30	11				
Inconel 600	ASTMB166 HW	0.304	90	40	30	11				
Inconel 625	AMS 5666	0.305	120	60	29.8	11.8	60	79	192	88
Inconel 706	AMS 5701	0.292	170	140	30.4	11	146	106	263	188
Inconel 718	AMS 5662, AMS 5663	0.297	185	150	29.4	11.4	156	111	309	216
Inconel X-750	AMS 5667	0.298	165	105	30.6	11.8	105	102	247	157
René 41	AMS 5712, AMS 5713	0.298	170	130	31.6	12.1	133	110		
Waspalov	AMS 5704	0.298	175	120	30.6					
Waspalov	AMS 5706, AMS 5707	0.298	160	110	30.6					
AISI4130	AMS-T-6736	0.283	180	165	29	11	173	108	277	257
AISI 4135	AMS-I-6735	0.283	200	165	29	11	181	120	308	274
AISI4340	AMS 6414	0.283	260	217	29	11	235	156	347	312
0.42C 3COM	AMS 6257, AMS 6419	0.283	280	230	29	11	247	16S	430	360
5Cr-Mo-V	*AMS 6487, AMS 6488*	*0.281*	*280*	*240*	*30*	*11*	*260*	*168*	*465*	*363*
15-5PH	AMS 5659	0.283	190	170	28.5	11.2				

Table 9.37 Changes in material property at 500 °C as a % of value at room temperature

Material	Venus temp (500 °C)							
	Ftu % of RT	Fty % of RT	E % of RT	G % of RT	Fcy % of RT	Fsu % of RT	Fbru % of RT	Fbry % of RT
	%	%	%	%	%	%	%	%
CP Titanium	32.0	20.0			34.0	32.0	32.0	32.0
Ti-8Al-lMo-lV	60.0	54.0	75.0					
Ti-6Al-2Sn-4Zr-2Mo	68.0	58.0						
T1-6AL-4 V	*62.0*	*52.0*	*73.0*		*47.0*	*61.0*	*66.0*	*63.0*
TI-6AL-4 V	58.0	50.0	65.0		50.0	58.0	63.0	56.0
TI-6AL-4 V	62.0	52.0	73.0		47.0	61.0	66.0	63.0
Ti-6Al-6 V-2Sn	*58.0*	*50.0*			*56.0*	*58.0*		
A-286	85.0	86.0	82.0	82.0			82.0	85,0
Inconel 600	79.0	75.0	84.0		78.0	90.0	91.0	
Inconel 600	79.0	75.0	84.0		78.0	90.0	91.0	
Inconel 625	84.0	68.0	82.0					
Inconel 706	87.0	84.0	76.0					
Inconel 718	90.0	90.0	88.0	87.0				
Inconel X-750	88.0	88.0	86.0	84.0	83.0	73.0	76.0	93.0
René 41	86.0	86.0	88.0	–	84.0	84.0	88.0	90.0
Waspalov	86.0	86.0	89.0	–	–	–	–	–
Waspaloy	86.0	86.0	89.0	–	–	–	–	–
AISI4130	66.0	53.0	75.0		48.0	64.0	62.0	66.0
AISI4135	66.0	53.0	75.0		48.0	64.0	62.0	66.0
AISI4340	66.0	53.0	75.0		48.0	64.0	62.0	66.0
0.42C 300 M	66.0	53.0	75.0		48.0	64.0	62.0	66.0
5Cr-Mo-V	*72.0*	*70.0*	*81.0*		*72.0*	*80.0*	*72.0*	*69.0*
15-5PH	77.0	69.0	81.0					

Table 9.38 Legend for Tables 9.36 and 9.37

Heading	Description	Units
Density	Weight per unit volume	lbf/in^3
Ftu	Tensile ultimate stress	ksi
Fty	Tensile yield stress	ksi
E	Modulus of elasticity	10^3 ksi
G	Modulus of rigidity	10^3 ksi
Fcy	Compressive yield stress	ksi
Fsu	Shear ultimate stress	ksi
Fbru	Bearing ultimate stress	ksi
Fbry	Bearing yield stress	ksi
%	Percent of room temperature properties	%

properties for rod and/or bar are shown. “A” basis properties selected unless not available, then “S” basis properties are shown. Worst case grain or E/d properties are shown. There are other materials that have possible utility for high temperature operations that are not shown due to lack of high temperature data.

It can be seen that in all cases, material properties decrease at 500 °C and this needs to be considered when designing structural elements such as a drill or a trencher. *Preferred materials based on specific ultimate tensile strength and specific compression yield strength are in italics.* Preferred materials based on specific tensile stiffness are underlined.

References

Bar-Cohen Y, Zacny K (2009) Drilling in extreme environments penetration and sampling on earth and other planets. Wiley, New York

Bar-Cohen et al (2014) High-temperature drilling mechanisms, Chap. 14. In: Bar-Cohen Y (ed) High temperature materials and mechanisms. CRC Press, Taylor & Francis Group, Boca Raton

Barmin I, Shevchenko AA (1983), Soil scooping mechanism for the Venera 13 and 14 Unmanned Interplanetary Spacecraft. Translated from Kosmicheskie Issledovaniya (Cosmic Research), vol 21, no 2, pp 171–175

Clegg et al (2012) Raman and laser-induced breakdown spectroscopy (LIBS) remote geochemical analysis under venus atmospheric pressure, #2105, LPSC 2012, Houston, TX

Gas Viscosity Calculator (2014) http://www.lmnoeng.com/Flow/GasViscosity.php

Glaze L (2009) Venus Mobile Explorer. http://ia700600.us.archive.org/22/items/VenusMobileExplorerMissionConceptStudy/02_Venus_Mobile_Explorer.pdf

Gilmore M, Glaze L (2010) Venus Intrepid Tessera Lander. http://ia700505.us.archive.org/30/items/VenusIntrepidTesseraLanderConceptStudy/03_Venus_Intrepid_Tessera_Lander.pdf

Hoffmann A, Stein L (2008) Gas cyclones and swirl tubes, principles, design and operation. Springer, Berlin

Mellor M (1971) Strength and deformability of rocks at low temperature. CRREL RR 294

NRC (2011) Planetary decadal survey 2013-2022: the future of planetary science. https://solarsystem.nasa.gov/2013decadal/ Earth, Moon, and Planets (ISSN 0167-9295), vol 43, Oct 1988, pp 45–60

Petroppoulos B, Telonis P (1988) The lower atmosphere of Venus after VEGA 1 and 2 missions

Siddiqi A (2002) Deep space chronicle: a chronology of deep space and planetary probes, 1958–2000. Monographs in Aerospace History, No 24

Surkov et al (1986) The water vapor content profile in the Venusian atmosphere according to the results of experiments from Vega 1 and 2. J Geophys Res 91(B13):E219–E221

Surkov et al (1987) Uranium, thorium, and potassium in the Venusian rocks at the landing sites of Vega 1 and 2. J Geophys Res 92(B4):E537–E540

Vexag (2014) Venus exploration analysis group goals, objectives, investigations and priorities. http://www.lpi.usra.edu/vexag/reports/GOI-140625.pdf

Wang H, Wang C (2010) Complete stress-strain test of basalt tuff under high-temperature and high confining pressure conditions and its tectonic significance. AGU Fall Meeting Abstracts

Wilks J, Wilks E (1994) Properties and applications of diamond. Paperback

Wilson A (1987) Solar system log. Jane's Publishing Co., Ltd

Ying L, Xian Z, Mao B, Lu AH (2009) Experimental study on the mechanical properties of rocks at high temperature. Sci China Ser E: Technol Sci 52(3):641–646

Zacny K, Cooper GA (2007) Friction of drill bits under Martian pressure. J Geophys Res 112: E03003

Zacny K, Wilson J, Chu P, Craft J (2011) Prototype rotary percussive drill for the mars sample return mission, paper #1125. In: IEEE Aerospace conference, 5–12 March 2011, Big Sky, Montana

Chapter 10
Power System Options for Venus Exploration Missions: Past, Present and Future

Simon D. Fraser

10.1 Introduction

Venus, the second planet from the Sun, has always fascinated mankind. One of the reasons for this is that Venus can be seen particularly well shortly before sunrise or shortly after sunset, as it is the second-brightest natural object in the night sky after the Moon. Venus is therefore commonly referred to as *Morning/Evening Star*. Venus is also considered as a sister planet of Earth, being the planet closest in distance and size to Earth.

The dawn of space exploration has enabled mankind to switch from observation to active exploration also with Venus. Venus has been targeted by a multitude of primarily Soviet and American missions since the early 1960s; in addition, the first European and the first Japanese mission have been recently launched. Many of the early missions failed during launch, in Earth orbit, en route to Venus, or after reaching Venus. A considerable number of missions, however, was highly successful and contributed significantly to our current understanding of the planet and its highly interesting and challenging atmosphere.

Despite being very close to Earth in terms of distance and size, the ambient conditions present on the Venusian surface are very challenging, even for robotic exploration. The surface pressure is in the order of 92 bars, and the average temperature is in the range of 464 °C; the atmosphere is primarily composed of Carbon Dioxide (96.5 %), Nitrogen (3.5 %) and some minor traces of Sulfur Dioxide (150 ppm), Argon (70 ppm), Water (20 ppm), Carbon Monoxide (17 ppm), Helium (12 ppm) and Neon (7 ppm) (NASA Venus Fact Sheet 2014). These ambient conditions are prohibitive for many space-proven technologies and systems. Mission profiles and power systems for Venus surface exploration applications are thus

S.D. Fraser (✉)
Mechanical Engineering Consultant, Graz, Austria
e-mail: simon.fraser@gmx.at

V. Badescu and K. Zacny (eds.), *Inner Solar System*,
DOI 10.1007/978-3-319-19569-8_10

very much different to what has been developed and implemented with Lunar or Mars surface exploration applications; this is also reflected in the operational lifetime of previous descent and landing systems.

This chapter summarizes the history of Venus exploration starting in the 1960s and including missions currently discussed for launch dates within the coming years. Selected exploration missions and their technological approaches are described in more detail to provide the reader an overview of how the challenges of Venus exploration have been addressed in the past, and how the power systems of these missions have been designed.

10.2 The Timeline of Venus Exploration

Within this chapter, Venus exploration missions are discussed in four different time periods:

- The 1960s: flyby and probe missions
- The 1970s: orbiters and landers
- The 1980s & 1990s: orbiters, landers and balloons
- 2000+: present and future missions

At first, a tabulated overview of all Venus exploration missions launched within the different time periods is given. Missions performing a Venus fly-by while heading for a different destination are also included in this overview.

Each of the Venus exploration missions is briefly presented with mission name, launch date and mission result. This overview includes successful as well as failed fly-by missions, probes, orbiters, atmospheric systems and landers. In this, the term *lander* is exclusively used for soft-landing mission elements designed to operate on the Venusian surface over a certain period of time upon soft-landing.

The timeline presented in the following is modified from the National Space Science Data Center (NSSDC) Venus exploration chronology (NSSDC Chronology 2011).

10.3 The 1960s: Flyby and Probe Missions

The first man-made object intended for Venus exploration was Sputnik 7, launched on February 4, 1961. This first, and unfortunately failed, mission was followed by a total of 20 missions designed and launched for Venus exploration purposes until the end of the decade.

All of the Venus exploration missions launched in the 1960s are summarized in Table 10.1.

Table 10.1 Venus exploration timeline of the 1960s; adapted from the National Space Science Data Center *Chronology of Venus Exploration* (NSSDC Chronology 2011)

Launch date	Mission name	Outcome
February 4, 1961	Sputnik 7	Failed to escape Earth orbit
February 12, 1961	Venera 1	Failure of mid-course correction motors, passed Venus within 100,000 km
July 22, 1962	Mariner 1	Launch failure
August 25, 1962	Sputnik 19	Failed to escape Earth orbit
August 27, 1962	Mariner 2	Passed Venus at a closest distance of 34,773 km
September 1, 1962	Sputnik 20	Failed to escape Earth orbit
September 12, 1962	Sputnik 21	Failed to escape Earth orbit
November 11, 1963	Cosmos 21	Failed to escape Earth orbit
February 19, 1964	Venera 1964A	Launch failure
March 1, 1964	Venera 1964B	Launch failure
March 27, 1964	Cosmos 27	Failed to escape Earth orbit
April 2, 1964	Zond 1	Contact lost en route to Venus
November 12, 1965	Venera 2	Contact lost en route to Venus
November 16, 1965	Venera 3	Contact lost en route to Venus
November 23, 1965	Cosmos 96	Failed to escape Earth orbit
November 23, 1965	Venera 1965A	Launch failure
June 12, 1967	Venera 4	Successfully entered the Venusian atmosphere
June 14, 1967	Mariner 5	Passed Venus at a closest distance of 4,094 km
June 17, 1967	Cosmos 167	Failed to escape Earth orbit
January 5, 1969	Venera 5	Successfully transmitted atmospheric data
January 10, 1969	Venera 6	Successfully transmitted atmospheric data

10.3.1 1960s Mission Example: Venera 4

Mission Profile
Venera 4 was launched on June 12, 1967, and successfully arrived at Venus on October 18, 1967 (NSSDC Chronology 2011).

The Venera 4 mission consisted of an interplanetary and a descending stage, which were separated before entering the Venusian atmosphere. The descending stage was decelerated by aerodynamic drag and deployed a parachute system at an altitude of 26 km to transmit data during the descent (Vakhnin 1968).

Scientific Equipment
The Venera 4 mission was designed to address three main scientific goals (Vakhnin 1968):

- Cosmophysical investigations along the trajectory of the flight
- Cosmophysical probing of the near-planet region
- Measurements of the atmosphere's physical and chemical characteristics

In order to achieve the scientific objectives of the mission, descending and interplanetary stage carried a multitude of instruments. The descending stage carried, among others, thermometers, a barometer, a radio altimeter, an atmospheric density gauge as well as gas analyzers (NSSDC/COSPAR ID: 1967-058A).

Venera 1-3 failed to achieve their mission objective, but Venera 4 successfully reached Venus, deployed the descending stage and transmitted atmospheric data to Earth until being crushed by the atmospheric pressure, and also before reaching the surface of the planet (Poncy et al. 2010).

Power System
The interplanetary stage was equipped with two solar panels. Solar irradiance actually increases en route to Venus, as an average of 2613.9 W/m^2 is available in Venus orbit versus an average of 1367.6 W/m^2 available in Earth orbit (NASA Venus Fact Sheet 2014). This makes solar panels a good choice, provided that the solar cells and/or the solar panels are chosen or designed to cope with the high irradiance available in Venus orbit, and the elevated operating temperatures it causes.

The capsule of the descending stage was equipped with a rechargeable battery having a capacity sufficient for powering the measurement and transmitter systems for 100 min (NASA Venera 4 2014).

10.4 The 1970s: Orbiters and Landers

The final mission targeting Venus in the 1960s, Venera 6, launched in 1969, was followed by Venera 7, launched in 1970; ten more missions followed until the end of the decade. The majority of Venus exploration missions was again launched by the Soviet Union's space program, as was the case in the previous decade. All of the Venus exploration missions planned and launched in the 1970s are summarized in Table 10.2.

10.4.1 1970s Mission Example: Venera 9

Mission Profile
Venera 9 was launched on June 8, 1975, and entered Venus orbit on October 20, 1975 (NSSDC Chronology 2011). The orbiter was the first spacecraft to orbit Venus, and acted as a communications relay for the lander. In addition, it also carried a scientific payload to investigate the Venusian atmosphere.

The lander successfully performed a soft-landing maneuver and transmitted data for 53 min. It was the first lander to return images from the surface of another planet (NSSDC/COSPAR ID: 1975-050A).

Table 10.2 Venus exploration timeline of the 1970s; adapted from the National Space Science Data Center *Chronology of Venus Exploration* (NSSDC Chronology 2011)

Launch date	Mission name	Outcome
August 17, 1970	Venera 7	Returned signal for 35 min during descent and 23 min after landing on Venus
August 22, 1970	Cosmos 359	Failed to escape Earth orbit
March 27, 1972	Venera 8	The lander transmitted data for 50 min after landing
March 31, 1972	Cosmos 482	Failed to escape Earth orbit
November 4, 1973	Mariner 10	Passed Venus at a closest distance of 5,768 km
June 8, 1975	Venera 9	The lander transmitted data for 53 min after landing
June 14, 1975	Venera 10	The lander transmitted data for 65 min after landing
May 20, 1978	Pioneer Venus 1	Successfully operated in orbit around Venus
August 8, 1978	Pioneer Venus 2	One of four probes successfully transmitted signals after impact for over an hour
September 9, 1978	Venera 11	The lander transmitted data for 95 min after landing
September 14, 1978	Venera 12	The lander transmitted data for 110 min after landing

Scientific Equipment
The orbiter mission carried a UV photometer, a photo-polarimeter, an infrared spectrometer and an infrared radiometer; in addition, the orbiter also carried a magnetometer and charged particle traps (NSSDC/COSPAR ID: 1975-050A 2013).

The lander carried temperature and pressure sensors, an accelerometer, a visible/IR photometer, backscatter and multi-angle nephelometers, a mass spectrometer, telephotometers, an anemometer, a gamma ray spectrometer and a gamma ray densitometer (NSSDC/COSPAR ID: 1975-050D 2013).

Power System
Two solar panels, each having a size of $1.25 \times 2.1\ m^2$, were installed to provide electrical power en route to Venus as well as in orbit around Venus (Venera 75 2014). Batteries were used to provide electrical energy to the lander systems during the short operational period of 53 min (Venera 75 2014).

10.4.2 1970s Mission Example: Pioneer Venus 1&2

Mission Profile
The Pioneer Venus program actually consisted of two submissions, which were launched independently: Pioneer Venus 1, deploying an orbiter, and Pioneer Venus 2, deploying a multiprobe (NASA Pioneer Venus 2005).

Pioneer Venus 1 was launched on May 30, 1978, and inserted into an elliptical orbit around Venus on December 4, 1978. The orbiter ran out of fuel and was destroyed upon entering the atmosphere in August 1992.

Pioneer Venus 2 was launched on August 8, 1978, and released a large probe on November 16, 1978, and three small probes on November 20, 1978 (NASA Pioneer Venus 2005).

The large probe had a total mass of 302 kg (109 kg for the deceleration module and 193 kg for the pressure vessel) and aeroshell/pressure vessel diameters of 142/78 cm. The small probes had a total mass of 94 kg (33 kg for the deceleration module and 61 kg for the pressure vessel) and aeroshell/pressure vessel diameters of 76/47 cm (Bienstock 2003).

All four probes entered the Venusian atmosphere on December 9, 1978, followed by the bus.

An image of a Pioneer Venus small probe is shown in Fig. 10.1.

Scientific Equipment

The Pioneer Venus 1 orbiter carried a total of 17 instruments, including a cloud photopolarimeter, a surface radar mapper, an infrared radiometer, a mass spectrometer and many other instruments (NASA Pioneer Venus 2005).

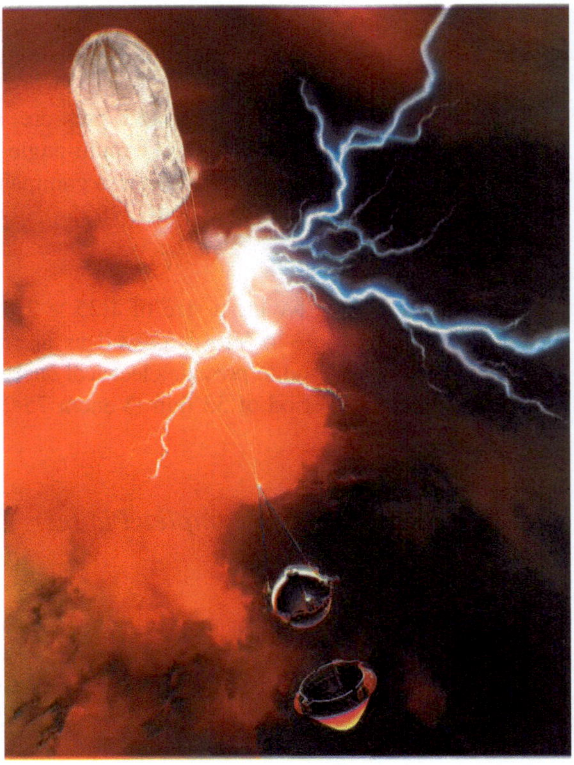

Fig. 10.1 Pioneer Venus small probe (Image credit: NSSDC/COSPAR ID: 1978-078G)

The Pioneer Venus 2 large probe carried a total of seven instruments in a sealed pressure vessel. The range of instruments included a mass spectrometer, a gas chromatograph, a solar flux radiometer, an infrared radiometer, a cloud particle size spectrometer, a nephelometer as well as temperature, pressure, and acceleration sensors (NASA Pioneer Venus 2005).

The three Pioneer Venus 2 small probes had spherical pressure vessels surrounded by an aeroshell. Unlike the large probe, they had no parachutes and the aeroshells did not separate from the probe. Each small probe carried a nephelometer and temperature, pressure, and acceleration sensors as well as a net flux radiometer experiment (NASA Pioneer Venus 2005).

The Pioneer Venus bus, previously carrying the multiprobe to Venus, also carried two experiments: a neutral mass spectrometer and an ion mass spectrometer. The bus made measurements to about 110 km altitude before disintegrating (NASA Pioneer Venus 2005).

Power System

A solar array with a power output of 312 W was installed around the circumference of the cylindrical Pioneer Venus 1 orbiter (NASA Pioneer Venus 2005).

The Pioneer Venus 2 large probe was powered by a 19 cell silver–zinc (AgZn) battery with 40 Ah; the scientific instruments had a power consumption of 106 W (Bienstock 2003).

The Pioneer Venus 2 small probes were also powered by silver–zinc batteries. A 20-cell battery with 11 Ah capacity was installed with each small probe. The scientific payload of the small probes had a power consumption of 10 W (Bienstock 2003).

10.5 The 1980s & 1990s: Orbiters, Landers and Balloons

The first Venus exploration mission launched in the 1980s was again a Soviet mission, Venera 13. A total of nine missions heading for Venus were launched in the 1980s, followed by a single mission in the 1990s. All of the Venus exploration missions launched in the 1980s and 1990s are summarized in Table 10.3.

10.5.1 1980s & 1990s Mission Example: Vega 1&2

Mission Profile

The two identical Vega 1 and Vega 2 spacecraft were launched on December 15, 1984 (Vega 1) and December 21, 1984 (Vega 2), respectively. The mission profile combined a Venus swingby, releasing Venus entry probes in the vicinity of Venus, and a Comet Halley flyby in 1986. Approximately half of the VEGA spacecraft was devoted to the Halley module, and the other half to the Venus lander package (NSSDC/COSPAR ID: 1984-125A 2013).

The Venus descent module separated two days before arrival at Venus and entered the Venusian atmosphere. The lander was identical to the design applied with Venera 9-14, and was again intended to study the Venusian atmosphere and surface (NSSDC/COSPAR ID: 1984-125A 2013).

Table 10.3 Venus exploration timeline of the 1980s and 1990s; adapted from the National Space Science Data Center *Chronology of Venus Exploration* (NSSDC Chronology 2011)

Launch date	Mission name	Outcome
October 30, 1981	Venera 13	The lander transmitted data for 127 min after touchdown
November 4, 1981	Venera 14	The lander transmitted data for 57 min after touchdown
June 2, 1983	Venera 15	Successfully operated in a nearly polar orbit around Venus
June 7, 1983	Venera 16	Successfully operated in a nearly polar orbit around Venus
December 15, 1984	Vega 1	Successfully delivered a lander and a balloon aerostat before intercepting Comet Halley
December 21, 1984	Vega 2	Identical to Vega 1
May 4, 1989	Magellan	Successfully operated in a nearly polar orbit around Venus
October 18, 1989	Galileo	Passed Venus at a distance of 16,106 km
October 15, 1997	Cassini	Performed two fly-bys of Venus

The descent module also piggybacked a balloon aerostat designed to collect data while floating in the Venusian atmosphere. Each balloon had a nominal diameter of 3.4 m and a total mass of 21 kg (Blamont and The Venus Balloon Science Team 1987). The balloon itself weighed 12.5 kg, the helium required to fill the balloon with a slight overpressure of 30 mbars weighed 2.0 kg, and some 6.5 kg were installed with the gondola (Kremnev et al. 1986).

Scientific Equipment

In addition to temperature and pressure sensors, the descent probe also carried a UV spectrometer, an instrument dedicated to the measurement of the concentration of H_2O, a gas-phase chromatograph, an X-ray spectrometer and a mass spectrograph. After landing, the surface was to be analyzed by gamma spectroscopy and an X-ray spectrometer (NSSDC/COSPAR ID: 1984-125A 2013).

The balloon aerostats carried pressure and illumination sensors, temperature sensors, an anemometer as well as a nephelometer (NSSDC/COSPAR ID: 1984-125F 2013).

The balloons successfully floated at an altitude of approximately 50 km within the Venusian cloud systems and provided data from a highly interesting region of the Venusian atmosphere (Blamont and The Venus Balloon Science Team 1987; Seiff and The VEGA balloon science team 1987; Zasova et al. 2007). Data from the balloon instruments were transmitted directly to Earth throughout the more than 46 h of operational lifetime that both balloons achieved.

Power System

The power systems applied with the Vega mission elements essentially resembled the approach already implemented in many previous missions and consisted of two solar panels providing electrical energy en route to Venus and on to comet Halley, respectively, and battery systems installed with the landers and balloons. The VEGA landers resembled the Venera 9-14 lander, and thus did not survive significantly longer than the Venera landers, each operating for nearly one hour (Dyson et al. 2009).

The batteries installed in the balloons are assumed to be the lifetime limiting factor of these unique atmospheric exploration systems. Each of the balloons transmitted data for approximately 46 h (Blamont and The Venus Balloon Science Team 1987).

10.6 2000+: Present and Future Missions

The launch rate of space missions heading for Venus has decreased since the late 1980s, with a single mission in the 1990s (Cassini, which was not primarily intended for Venus exploration, but made two Venus fly-bys). Up to date, the only successful mission solely designed for Venus exploration since the late 1980s is Venus Express, launched in 2005.

All of the Venus exploration missions launched since the year 2000 and planned in the near future are summarized in Table 10.4.

Table 10.4 Venus exploration timeline 2000+ ; adapted from the National Space Science Data Center *Chronology of Venus Exploration* (NSSDC Chronology 2011)

Launch date	Mission name	Outcome
August 03, 2004	Mercury Surface, Space Environment, Geochemistry and Ranging (MESSENGER)	Performed two fly-bys of Venus
November 09, 2005	Venus Express	European mission successfully operated in a polar orbit around Venus
May 20, 2010	Akatsuki	Failed to enter orbit around Venus
Previously proposed for launch in 2013	Venus In Situ Explorer	Discussed, but currently removed as candidate from the New Frontiers program (NASA New Frontiers 2014)
Scheduled for launch in July 2016	BepiColombo	Joint mission of ESA and JAXA; Venus fly-by en route to Mercury
Discussed for a launch around 2020	Venera-D	Previously scheduled for launch in 2016 (Zasova et al. 2009)

10.6.1 2000s Mission Example: Venus Express

Mission Profile
Venus Express was launched on November 9, 2005, and entered a Venusian orbit on April 11, 2006. The nominal mission started on June 4, 2006, and, as of April 2014, is still operational (ESA Venus Express Objectives 2014).

Venus Express is a European Space Agency mission to study atmosphere and plasma environment of Venus from orbit. In this, a comprehensive study of the Venusian atmosphere is being made, and plasma environment as well as interaction between upper atmosphere and solar wind is being investigated. Aspects of the surface and surface-atmosphere interaction are also considered.

An image of the Venus Express spacecraft is shown in Fig. 10.2.

Scientific Equipment
The scientific payload was designed to enable a comprehensive analysis of the Venusian atmosphere. This will include studying the chemistry and the complex dynamics of the atmosphere, analysing interactions between the atmosphere and the surface as well as interactions between the atmosphere and the interplanetary environment (Baines et al. 2006; ESA Venus Express Objectives 2014).

The following instruments are installed (location of installation indicated in Fig. 10.2.):

- Analyser of Space Plasma and Energetic Atoms (ASPERA)
- Venus Express Magnetometer (MAG)
- Planetary Fourier Spectrometer (PFS)
- Ultraviolet and Infrared Atmospheric Spectrometer (SPICAV/SOIR)

Fig. 10.2 Venus Express (Image credit: ESA)

- Venus Radio Science Experiment (VeRa)
- Visible and Infrared Thermal Imaging Spectrometer (VIRTIS)
- Venus Monitoring Camera (VMC)

Power System
Power is provided by two solar panels with a total surface of 5.7 m^2 equipped with triple-junction GaAs solar cells. GaAs cells are not only comparably tolerant to high operating temperatures (considering that Venus is much closer to the Sun than the Earth), but are applied to only about half of the solar panel surface. Aluminium strips were installed between the solar cells to support heat rejection.

The solar panels have a nominal power output of 800 W in Earth orbit, and 1100 W in Venus orbit (ESA Venus Express Spacecraft 2014). Three 24 Ah lithium-ion batteries were installed to provide energy storage capacity for peak power applications and when the spacecraft systems have to be operated in the shadow of Venus.

10.7 Conclusions and Outlook

Up to date, Venus has been targeted by a considerable number of flyby missions, orbiters, descent probes, landers, and even two balloon probes.

Proven power system options for Venus exploration can essentially be broken down into two sub-categories: firstly, power systems for the time spent en route to Venus and/or in orbit around Venus; secondly, descent systems designed to enter the atmosphere and to (soft-)land on the surface of Venus.

Power systems for transfer to - and orbit around - Venus benefit from the high solar irradiance level. An average of 2613.9 W/m^2 is available in Venus orbit versus 1367.6 W/m^2 available in Earth orbit (NASA Venus fact sheet 2014). This makes solar panels equipped with solar cells capable of working at elevated temperatures and/or with additional design features for heat rejection a good choice. By including rechargeable battery modules for peak power applications and/or short-time operation in the shadow of Venus, this kind of power system essentially resembles the proven and cost-effective approach also established in the commercial terrestrial satellite market. This approach will also be the option of choice with future missions and applications.

Power systems for landers, however, are far more challenging. Up to date, Venus surface landers had an average operational lifetime in the order of a single hour. The longest-lived lander was the Soviet Venera 13, and it only survived two hours on the surface (Dyson et al. 2009). The limiting factor, however, was not the power system or the installed battery capacity, but the fact that the landers simply were not designed to survive the corrosive, high-temperature and high-pressure surface conditions for much longer.

Venus is traditionally not associated with the same level of interest in terms of scientific exploration - or even commercial exploitation - as Mars. The far more hostile environmental conditions on Venus simply prevent any kind of human exploration or even settlement, as envisioned for Mars. Despite being limited to robotic systems in the foreseeable future, Venus is nevertheless a very relevant and worthwhile target for future exploration.

A significant extension of the lander lifetime will only be possible by applying advanced thermal insulation and/or active cooling to keep thermo-sensitive equipment and systems within an acceptable operating temperature range as long as possible. Radioisotope generators were thus considered as power source for operating a Stirling-cycle cooling system designed to outbalance the heat leak in from the environment, for instance (Dyson et al. 2009). Studies for designing a lander architecture capable of surviving at least one Venus day (this equals roughly 243 terrestrial days) by applying Stirling-cycle cooling are available and prove that, in theory, landing systems can be operated with extended operational lifetime (Dyson et al. 2009). Such a scenario is far more challenging in terms of power system design than the short-lived descent and landing elements applied up to date.

The balloon aerostats may be followed by more advanced airborne systems designed to directly analyse the atmosphere in relevant heights. Aircraft systems are, for instance, discussed for such an application (Landis et al. 2002). A solar-powered aircraft system could be operated autonomously over longer periods of time, cover larger distances and be actively directed to investigate the atmosphere in relevant regions or heights.

Sample return missions, as previously demonstrated in returning samples from the Moon and currently planned for Mars, will also play a role in mid-term plans for Venus exploration. Landing site selection for in situ analysis and sample return missions have already been discussed for many years (Basilevsky 2007). The limiting factor in realisation, however, may not be technical implementation, but budget restrictions.

In summary, the future of Venus exploration will most likely see a straight continuation in power system design for transfer and orbiting systems. Many challenges will, however, have to be faced in terms of developing lightweight and robust power systems for mission elements operating in the atmosphere and on the surface of the Earth's demanding sister planet.

References

Basilevsky AT, Ivanov MA, Head JW, Aittola M, Raitala J (2007) Landing on Venus: past and future. Planet Space Sci 55:2097–2112

Baines KH, Atreya S, Carlson RW, Crisp D, Drossart P, Formisano V, Limaye SS, Markiewicz WJ, Piccioni G (2006) To the depths of Venus: exploring the deep atmosphere and surface of our sister world with Venus Express. Planet Space Sci 54:1263–1278

Bienstock BJ (2003) Pioneer Venus and Galileo entry probe heritage. In: Proceedings of the international workshop planetary probe atmospheric entry and descent trajectory analysis and science, Lisbon, Portugal, 6–9 Oct 2003

Blamont JE, The Venus Balloon Science Team (1987) The VEGA Venus balloon experiment. Adv Space Res 7:295–298

Dyson RW, Schmitz PG, Penswick LB, Bruder GA (2009) Long-lived Venus lander conceptual design: how to keep it cool. In: 7th international energy conversion engineering conference, Denver, Colorado AIAA 2009-4631, 2–5 Aug 2009

ESA Venus Express Objectives (2014) http://www.esa.int/Our_Activities/Space_Science/Venus_Express/Venus_Express_objectives. Accessed 25 May 2014

ESA Venus Express Spacecraft (2014) http://www.esa.int/Our_Activities/Space_Science/Venus_Express/The_spacecraft. Accessed 25 May 2014

Kremnev RS, Selivanov AS, Linkin VM, Blamont J, Bakitko RV, Lipatov AN, Tarnoruder IY, Puchkov VI, Kustodiev VD, Shurupov AA, Ignatova SP, Kerzhanovich VV, Khlyustova LI, Frank GA, Turkova AI, Karyagin VP, Kostin AV, Mashkov VI, Pichkadze KM, Terterashvili AV, Malique C, Ragent B, Preston R (1986) The Vega balloons - a tool for studying atmosphere dynamics on Venus. Sov Astron Lett 12:7–9

Landis GA, Colozza A, LaMarre CM (2002) Atmospheric flight on Venus. In: 40th aerospace sciences meeting and exhibit, Reno, Nevada, AIAA–2002–0819, NASA/TM—2002-211467, 14–17 Jan 2002

NASA New Frontiers (2014) New Frontier Program – missions. http://discoverynewfrontiers.nasa.gov/missions/index.cfml. Accessed May 25 2014

NASA Pioneer Venus (2005) Pioneer Venus project information. http://nssdc.gsfc.nasa.gov/planetary/pioneer_venus.html. Accessed May 25 2014

NASA Venera 4 (2014). http://solarsystem.nasa.gov/missions/profile.cfm?Sort=Nation&MCode=Venera_04&Nation=USSR&Display=ReadMore. Accessed May 25 2014

NASA Venus Fact Sheet (2014) http://nssdc.gsfc.nasa.gov/planetary/factsheet/venusfact.html. Accessed May 25 2014

NSSDC Chronology (2011) Chronology of Venus Exploration http://nssdc.gsfc.nasa.gov/planetary/chronology_venus.html. Accessed May 25 2014

NSSDC/COSPAR ID: 1975-050A (2013) Venera 9. http://nssdc.gsfc.nasa.gov/nmc/spacecraftDisplay.do?id=1975-050A. Accessed May 25 2014

NSSDC/COSPAR ID: 1975-050D (2013) Venera 9 Descent Craft. http://nssdc.gsfc.nasa.gov/nmc/masterCatalog.do?sc=1975-050D. Accessed May 25 2014

NSSDC/COSPAR ID: 1984-125A (2013) Vega 1. http://nssdc.gsfc.nasa.gov/nmc/spacecraftDisplay.do?id=1984-125A. Accessed May 25 2014

NSSDC/COSPAR ID: 1984-125F (2013) Vega 1 Balloon. http://nssdc.gsfc.nasa.gov/nmc/spacecraftDisplay.do?id=1984-125F. Accessed May 25 2014

NSSDC/COSPAR ID: 1978-078G (2013) Pioneer Venus Small Probe (Day). http://nssdc.gsfc.nasa.gov/image/spacecraft/pv_probe.jpg. Accessed May 25 2014

Poncy J, Lebleu D, Arfi P, Schipper AM (2010) Entry descent and landing systems for future missions. Acta Astron 67:173–179

Seiff A, The VEGA balloon science team (1987) Further information on structure of the atmosphere of Venus derived from the VEGA balloon and lander mission. Adv Space Res 7:323–328

Vakhnin VM (1968) Review of the Venera 4 flight and its scientific program. J Atmos Sci 25:533–534

Venera 75 (2014) http://www.russianspaceweb.com/venera75.html. Accessed May 25 2014

Zasova LV, Ignatiev N, Khatuntsev I, Linkin V (2007) Structure of the Venus atmosphere. Planet Space Sci 55:1712–1728

Zasova LV, Korablev OI, Zelenyi LM, Zakharov AV, Polischuk GM, Pichkhadze KM, Vorontsov VA, Elkin KS, Voron VV, Marov MYa, Basilevsky AT, Cheremukhina ZP, Gavrik AL, Gerasimov MV, Gotlib VM, Grigoriev A, Gromov V, Ignatiev NI, Khavroshkin OB, Klimov SI, Lipatov AN, Linkin VM, Liash A, Mitrofanov IG, Moshkin BE, Rodin AV, Sanko NF, Skalsky AA, Vaisberg OL, Zastenker GN (2009) Mission Venera D (2016) - scientific goals and payload. http://www.lpi.usra.edu/vexag/oct2009/presentations/zasovaVeneraD.pdf. Accessed May 25 2014

Chapter 11
Production of Energy for Venus by Electron Wind Generator

Alexander A. Bolonkin

11.1 Introduction

Wind power is the conversion of wind energy into a useful form of energy, such as using wind turbines, to make electrical power, windmills for mechanical power, wind pumps for water pumping or drainage, or sails to propel ships. Large wind farms consist of hundreds of individual wind turbines which are connected to the electric power transmission network. Wind power, as a viable alternative to fossil fuels, is plentiful, renewable, widely distributed, clean, produces no greenhouse gas emissions during operation and uses little land. The effects on the environment are generally less problematic than those from other power sources.

Economy of conventional utilization of wind energy. Current wind power plants have low ongoing costs, but moderate capital cost. The estimated average cost per unit incorporates the cost of construction of the turbine and transmission facilities, borrowed funds, returns to investors (including cost of risk), estimated annual production, and other components averaged over the projected useful life of the equipment, which may be in excess of twenty years. Unfortunately, current ground wind energy systems have deficiencies which limit their commercial applications on Earth. Wind energy is unevenly distributed and has relatively low energy density (on Earth). Huge turbines cannot be placed on the ground; many small turbines must be used instead. However, while small turbines are relatively inefficient, very big turbines placed at ground are also inefficient due to the relatively low wind energy density and their high cost. The current cost of wind energy is higher than energy of thermal power stations. Wind power is a function of the cube of wind velocity. At surface level, wind has low speed, and it is non-steady. If wind velocity decreases in half, the wind power decreases by a factor of 8. The productivity of

A.A. Bolonkin (✉)
C&R, Brooklyn, NY, USA
e-mail: aBolonkin@juno.com; aBolonkin@gmail.com

V. Badescu and K. Zacny (eds.), *Inner Solar System*,
DOI 10.1007/978-3-319-19569-8_11

wind-powered systems depends heavily on the prevailing weather. Wind turbines produce noise and visually detract from the landscape.

This chapter proposes a solution to the main technological challenge of this system when used on Venus surface; the transfer of energy to the ground. While there are many research programs and proposals for wind driven power generation systems, all of them are ground or tower based. The system proposed in this chapter is located at high altitude, where strong, permanent, and steady streams are located. The reader can find the new ideas in the electronic utilization of the wind energy in [1–7], innovations in airborne wind energy in [8–13], a general information about wind energy and data for estimations and computations in [14–20].

11.2 Venus Atmosphere

11.2.1 Composition

The atmosphere of Venus is composed mainly of carbon dioxide (96.5 %), along with a small amount of nitrogen (3.5 %) and other trace elements. The amount of nitrogen in the atmosphere is relatively small compared to the amount of carbon dioxide, but because the atmosphere is so much thicker than Earth's, its total nitrogen content is roughly four times higher than Earth's, even though on Earth nitrogen makes up about 78 % of the atmosphere. The atmosphere contains a range of interesting compounds in small quantities, including some based on hydrogen, such as hydrogen chloride (HCl) and hydrogen fluoride (HF). There are carbon monoxide, water vapor and molecular oxygen as well. Hydrogen is in relatively short supply in the Venusian atmosphere. A large amount of the planet's hydrogen is theorized to have been lost to space, with the remainder being mostly bound up in sulfuric acid (H_2SO_4) and hydrogen sulfide (H_2S). The loss of significant amounts of hydrogen is proved by a very high D/H ratio measured in the Venusian atmosphere. The ratio is about 0.025, which is much higher than the terrestrial value of 1.6×10^{-4}. In addition, in the upper atmosphere of Venus D/H ratio is 1.5 higher than in the bulk atmosphere.

The atmosphere is divided into a number of sections depending on altitude. The densest part of the atmosphere, the troposphere, begins at the surface and extends upwards to 65 km. At the furnace-like surface the winds are slow, but at the top of the troposphere the temperature and pressure reaches Earth-like levels and clouds pick up speed to 100 m/s.

11.2.2 Temperature and Pressure

The atmospheric pressure at the surface of Venus is about 92 times that of the Earth, similar to the pressure found 910 meters below the surface of the ocean. The atmosphere has a mass of 4.8×10^{20} kg, about 92 times the mass of the Earth's total atmosphere. The density of the air at the surface is 67 kg/m^3, which is 6.5 % that of liquid water on Earth. The pressure found on Venus's surface is high enough that the carbon dioxide is technically no longer a gas, but a supercritical fluid. This supercritical carbon dioxide forms a kind of sea that covers the entire surface of Venus. This sea of supercritical carbon dioxide transfers heat very efficiently, buffering the temperature changes between night and day (each lasts 56 terrestrial days).

The large amount of CO_2 in the atmosphere, together with water vapor and sulfur dioxide, create a strong greenhouse effect, trapping solar energy and raising the surface temperature to around 740 K (467 °C), hotter than any other planet in the solar system, even that of Mercury, despite being located farther out from the Sun and receiving only 25 % of the solar energy (per unit area) that torrid Mercury does. The average temperature on the surface is above the melting points of lead 600 K (327 °C), tin 505 K (232 °C), and zinc 693 K (420 °C). The thick troposphere also makes the difference in temperature between the day and night, even though the slow retrograde rotation of the planet causes a single solar day to last 116.5 days on Earth. The surface of Venus is in the darkness for 58.3 days before the sun rises again from behind the clouds.

The temperature and atmospheric pressure are shown in Table 11.1 and Fig. 11.1.

Table 11.1 The temperature and atmospheric pressure on Venus

Height (km)	0	5	10	15	20	25	30	35	40
Temperature (°C)	462	424	385	348	306	264	222	180	143
Pressure (xEarth)	92.1	66.65	47.39	33.04	22.52	14.93	9.851	5.917	3.501
Height (km)	45	50	55	60	65	70	80	90	100
Temperature (°C)	110	75	27	−10	−30	−43	−76	−104	−112
Pressure (xEarth)	1.97	1.06	0.531	0.235	0.0976	0.0369	$4.76 \cdot 10^{-3}$	$3.73 \cdot 10^{-4}$	$2.6 \cdot 10^{-5}$

The troposphere on Venus contains 99 % of the atmosphere by mass. Ninety percent of the atmospheric mass of Venus is within 28 km of its crustal surface; by comparison, 90 % of the Earth's atmospheric mass lies just 10 km above our planet's crustal surface. At the height of 50 km, the atmospheric pressure is approximately equal to that at the surface of the Earth. On the side of Venus where there is night, clouds can still be found at 80 km above the surface.

The altitude of the troposphere most similar to Earth, is near the tropopause—the boundary between troposphere and mesosphere. It is located slightly above 50 km. According to measurements by the Magellan and Venus Express probes, the altitude from 52.5 to 54 km has a temperature between 293 K (20 °C) and 310 K

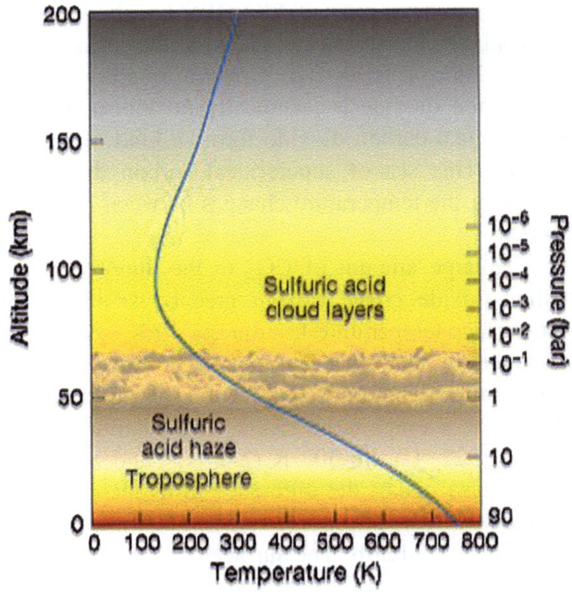

Fig. 11.1 The temperature and pressure of Venus' atmosphere

(37 °C), and the altitude at 49.5 km above the surface is where the pressure becomes the same as that on the Earth at sea-level. As manned ships sent to Venus would be able to compensate for differences in temperature to a certain extent, anywhere from about 50 to 54 km or so above the surface would be the easiest altitude in which to base an exploration or colony. The temperature there, would be in the crucial "liquid water" range of 273 K (0 °C) to 323 K (50 °C) and the atmospheric gas pressure would be the same as habitable regions of Earth. As CO_2 is heavier than Earth's air, the colony's air (nitrogen and oxygen), could keep the structure floating at that altitude like a dirigible.

11.2.3 Circulation of Venus' Atmosphere

The atmosphere is in a state of vigorous circulation and super-rotation. The whole atmosphere circles the planet in just four Earth days, much faster than the planet's sidereal day of 243 days. Winds supporting super-rotation, blow as fast as 100 m/s (~220 mph or 360 km/h). On the other hand, the wind speed becomes increasingly slower as the elevation from the surface decreases, with the breeze barely reaching the speed of 10 km/h on the surface. Anticyclonic structures called polar vortices are located near the poles. Each vortex is double-eyed and shows a characteristic S-shaped pattern of clouds.

The circulation in Venus's troposphere follows the so-called cyclostrophic approximation. Its wind-speeds are roughly determined by the balance of the pressure gradient and centrifugal forces in almost purely zonal flow. In contrast, the circulation in the Earth's atmosphere is governed by the geostrophic balance. Venus's wind-speeds can be directly measured only in the upper troposphere (tropopause), between 60–70 km altitude, which corresponds to the upper cloud deck. The cloud motion is usually observed in the ultraviolet part of the spectrum, where the contrast between clouds is the highest. The linear wind speeds at this level are about 100 ± 10 m/s at lower than 50° latitude. They are retrograde in the sense that they blow in the direction of the retrograde rotation of the planet. Winds quickly decrease towards the higher latitudes, eventually reaching zero at the two poles. Such strong, cloud-top winds cause a phenomenon known as the super-rotation of the atmosphere. In other words, these high-speed winds circle the whole planet zonally, faster than the planet itself rotates axially. The super-rotation on Venus is differential, which means that the equatorial troposphere super-rotates more slowly than the troposphere at the mid-latitudes. The winds also have a strong vertical gradient. They decline deep in the troposphere with the rate of 3 m/s per km. Winds near the surface of Venus are much slower than those on Earth. They actually move at only a few kilometers per hour (generally less than 2 m/s and with an average of 0.3 to 1.0 m/s), but due to the high density of the atmosphere at the surface, this is still enough to transport dust and small stones across the surface, much like a slow-moving current of water.

All winds on Venus are ultimately driven by convection (Fig. 11.2). Hot air rises in the equatorial zone, where solar heating is concentrated, and flows to the poles. Such an almost-planet wide overturning of the troposphere is called Hadley circulation, just as it is in the Earth. However, the meridional air motions are much slower than zonal winds. The pole ward limit of the planet wide Hadley cell on Venus is near ±60° latitudes. Here, air starts to descend and return to the equator below the clouds.

This interpretation is supported by the distribution of the carbon monoxide, which is also concentrated in the vicinity of ±60° latitudes. Pole ward of the Hadley cell, a different pattern of circulation is observed. In the latitude range 60°–70°, cold polar collars exist. They are characterized by temperatures about 30–40 K lower than in the upper troposphere at nearby latitudes.

The lower temperature is probably caused by the upwelling of the air in them, and by the resulting adiabatic cooling. Such an interpretation is supported by the denser and higher clouds in the collars. The clouds lie at 70–72 km altitude in the collars—about 5 km higher than at the poles and low latitudes. A connection may exist between the cold collars and high speed, mid-latitude jets in which winds blow as fast as 140 m/s. Such jets are a natural consequence of the Hadley–type circulation and should exist on Venus between 55–60° latitude. Expected change in wind speed on Venus via altitude is shown in Fig. 11.3.

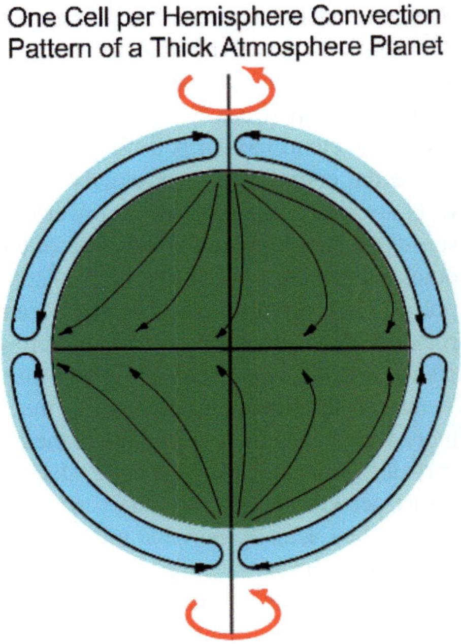

Fig. 11.2 Meridional (north-south) component of the atmospheric circulation in the atmosphere of Venus. Note that the meridional circulation is much lower than the zonal circulation, which transports heat between the day and night sides of the planet

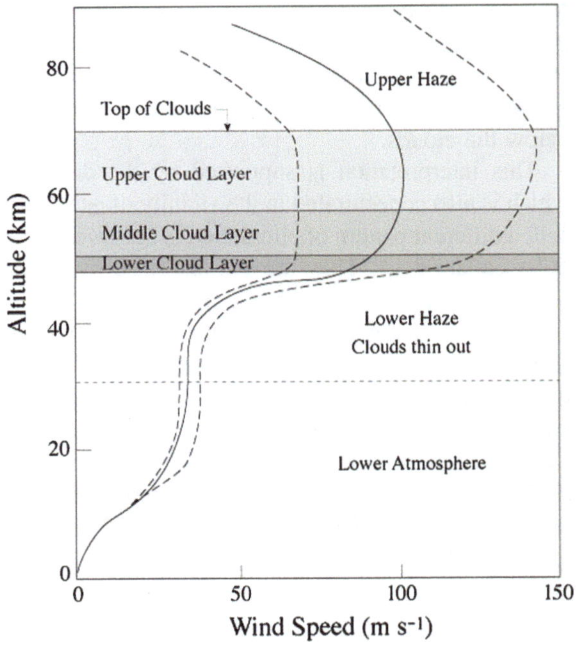

Fig. 11.3 Expected change in wind speed on Venus via altitude

Odd structures, known as polar vortices, lie within the cold polar collars. They are giant, hurricane-like storms, four times larger than their terrestrial analogs. Each vortex has two "eyes"—the centers of rotation, which are connected by distinct S-shaped cloud structures. Such double eyed structures are also called polar dipoles. Vortexes rotate with the period of about 3 days in the direction of general super-rotation of the atmosphere. The linear wind speeds are 35–50 m/s near their outer edges and zero at the poles. Temperatures at the cloud-tops in the polar vortexes are much higher than in the nearby polar collars, reaching 250 K (−23 °C).

The conventional meteorological interpretation of the polar vortexes is that they are anticyclones, with down-welling in the centre and upwelling in the cold polar collars. This type of circulation resembles the winter polar anticyclonic vortexes on Earth, especially the one found over Antarctica.

The observations in the various infrared atmospheric windows indicate, that the anticyclonic circulation observed near the poles may penetrate as deep as to 50 km altitude, i.e. to the base of the clouds. The polar upper troposphere and mesosphere are extremely dynamic; large bright clouds may appear and disappear over a period of a few hours. One such event was observed by Venus Express between 9 and 13 January 2007, when the South Pole region became brighter by 30 %. This event was probably caused by an injection of sulfur dioxide into the mesosphere, which then condensed forming a bright haze. The two eyes in the vortexes have yet to be explained.

The first vortex on Venus was discovered at the North Pole by the Pioneer Venus mission in 1978. A discovery of the second large 'double-eyed' vortex at the South Pole of Venus, was made in the summer of 2006 by Venus Express.

11.3 Description of Innovation

One, simplest version of the offered electron wind generator (ABEG), is presented in Fig. 11.4. Installation contains: electron injectors 2 established in column 6 and electron collector (net) 4 having the conductive leaves 5 (metallic foil). They have a large surface, which helps to collect the electrons from big area. Network connects with the electron injectors through a useful load 7.

Work of ABEG. The ABEG generator works the following way: injector injects the electrons into atmosphere, the wind catch them and moves to collector (network) 4. Network 4 has negative charge, electron injector has positive charge. The electric field breaks the electrons (negative ions) and decreases the wind speed. But the electric ion speed is less than wind speed and electrons when they reach the collector settle into collector and increase its negative charge. Those additional charges (electrons), return through the electric load 7 and make the useful work.

The injectors may be up on a mast (Fig. 11.5a) or located also on planet's surface (Fig. 11.5b). The efficiency of these will be different. The surface collector is a conductivity film 11 (Fig. 11.3) (for example, metal foil), isolated from planet.

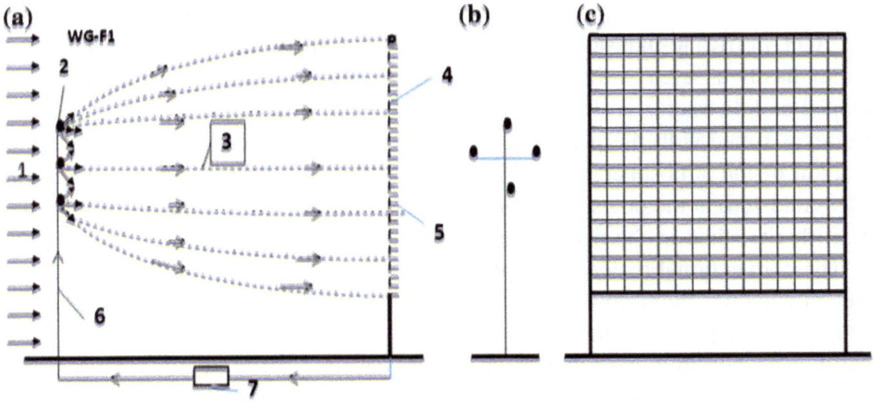

Fig. 11.4 One version of Electron Wind Electric Generator (ABEG). ***a*** – side view of the installation; ***b*** – front view of the electron injector column; ***c*** – front view of the collect net. *Notations:* 1 is wind; 2 is electron injector; 3 is trajectories of electrons; 4 is net collecting the electrons; 5 is conductive leaves (metallic foil); 6 is column (post) for supporting of the electron injectors; 7 is the outer electric load

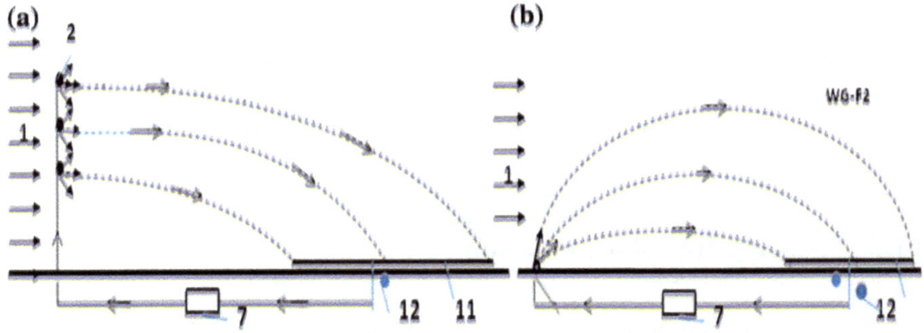

Fig. 11.5 The horizontal conductivity film as collector of electrons. ***a*** – injectors in column; ***b*** - injectors at Venus surface. *Notations:* 11 - conductivity film (for example, metallic foil); 12 (optional) positive isolated charge (for example, electrets)

For increasing the efficiency of collector we can (optionally) place under collector the isolated positive charge 12 (or positive electrets) (Fig. 11.5). If we want to use wind energy at high altitudes, a special parachute can be used. Two versions of these designs are shown in Fig. 11.6. In the first version, the electron injector is supported by wing 13 (Fig. 11.6a), in the second version (Fig. 11.6b), the electron injector is supported by a unique parachute 15 which creates also the lift force. Special parachute is a net, containing the conductive leaves.

Advantages of the proposed electron wind systems (ABEG), in comparison with the conventional air wind systems.

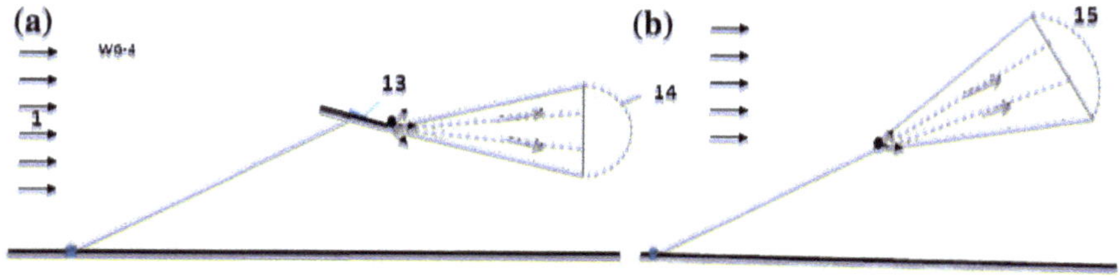

Fig. 11.6 Airborne (flight) high altitude Electron wind generator. ***a*** - wing support; ***b*** – wind parachute support. *Notations* 13 is wing; 14 is parachute; 15 is parachute having lift force.

The suggested new principle, electron wind generator (ABEG), has the following advantages in comparison with conventional wind systems:

- Offered installations are very simple.
- Offered system is very cheap (by hundreds of times). No tower, propeller, magnetic-electric generator, gear box.
- Offered system can cover a large area and has enormous power.
- Offered installations are suitable for planets having hot atmosphere.
- Offered installations produce high voltage direct electricity. That is advantage if energy is being transferred long distances.
- Offered system is very suitable for airborne wind installation because it is very light and produces high voltage electricity.
- Offered system may be used as brake and can supply power to the electric system of aircraft.

11.4 Estimations and Computation of Wind Power

11.4.1 Power of a Wind Energy N

$$N = 0.5\eta\rho AV^3[\mathrm{W}] \tag{11.1}$$

The coefficient of efficiency, η, equals about $0.2 \div 0.25$ for ABEG; $0.15 \div 0.35$ for low speed propeller rotors (ratio of blade tip speed to wind speed equals $\lambda \approx 1$); $\eta = 0.45 \div 0.5$ for high speed propeller rotors ($\lambda = 5 - 7$). The Darrieus rotor has $\eta = 0.35 - 0.4$. The gyroplane rotor has $0.1 \div 0.15$. The air balloon and the drag (parachute) rotor has $\eta = 0.15 - 0.2$. The Makani rotor has $0.15 \div 0.25$. The theoretical maximum equals $\eta \approx 0.6$. Theoretical maximum of the electron generator is 0.25. A - front (forward) area of the electron corrector, rotor, air balloon or parachute [m^2]. ρ - density of planet atmosphere. At Venus surface $\rho \approx 65$ kg/m^3. V is average annually wind speed, m/s. The salient point here is that the strength of wind power depends upon the wind speed (by third order). If the wind speed increases by two times, the power increases by 8 times. If the wind speed increases 3 times, the wind power increases 27 times.

The energy, E, produced in one Earth year is (1 year ≈ 30.2 × 10^6 sec):

$$E = 3600 \times 24 \times 350 N \approx 30 \times 10^6 N, \text{ [J]} \quad (11.2)$$

The density of atmosphere and wind speed on Venus is significantly higher than on Earth. The specific wind power on Venus is considerably more than on Earth. If for the typical Earth wind speed $V = 6$ m/s we have $N = 0.13$ kW/m^2 at altitude 10 m and $N = 3.1$ kW/m^2 at altitude 4 km ($V = 25$ m/s), then on Venus we have $N = 73$ kW/m^2 at altitude 12 km (for $V = 20$ m/s, $\rho = 36.6$ kg/m^3), $N = 217.6$ kW/m^2 at altitude 50 km (for $V = 80$ m/s, $\rho = 1.7$ kg/m^3) and $N = 250$ kW/m^2 at altitude 55 km (for $V = 100$ m/s, $\rho = 1$ kg/m^3). This means that on Venus we will have 2000 times more wind energy from 1 m^2 then we have now at Earth's surface.

11.5 Theory of Jet Electric Generator

11.5.1 Ion and Electron Speed

Ion mobility. The ion speed onto the gas (air) jet may be computed by equation:

$$j_s = qn_-b_-E + qD_-(dn_-/dx), \quad (11.3)$$

where j_s is density of electric current about jet, A/m^2; $q = 1.6 \times 10^{-19}$ C is charge of single electron, C; n_- is density of injected negative charges in 1 m^3; b_- is charge mobility of negative charges, m^2/sV; E is electric intensity, V/m; D_- is diffusion coefficient of charges; dn_-/dx is gradient of charges. For our estimation we put $dn_-/dx = 0$. In this case:

$$j_s = qn_-b_-E, \; Q = qn, \; v = bE, \; j_s = Qv, \quad (11.4)$$

where Q is density of the negative charge in 1 m^3; v is speed of the negative charges about jet, m/s. The air negative charge mobility for normal pressure and temperature $T = 20$ °C is:

$$\text{in dry air } b_- = 1.9 \times 10^{-4}\,\text{m}^2/\text{sV}, \text{ in humid air } b_- = 2.1 \times 10^{-4}\,\text{m}^2/\text{sV}. \quad (11.5)$$

In Table 11.2 is given the ions mobility of different gases for pressure 700 mm Hg and for $T = 18$ °C.

In range of pressure from 13 to 6 × 10^6 Pa, the mobility follows the law bp = const, where p is air pressure. When air density decreases, the charge mobility

Table 11.2 Ions mobility of different gases (pressure 700 mm Hg and T = 18 °C)

Gas	Ion mobility 10^{-4} m²/sV, b_+, b_-		Gas	Ion mobility 10^{-4} m²/sV, b_+, b_-		Gas	Ion mobility 10^{-4} m²/sV, b_+, b_-	
Hydrogen	5.91	8.26	Nitrogen	1.27	1.82	Chloride	0.65	0.51
Oxygen	1.29	1.81	CO_2	1.10	1.14			

Source [8] p.357

increases. The mobility strength depends upon the purity of gas. The ion gas mobility may be recalculated in other gas pressure p and temperature T by equation:

$$b = b_0 \frac{T p_0}{T_0 p}, \tag{11.6}$$

where lower index "$_o$" means the initial (known) point. At the Earth's surface H = 0 km, T_0 = 288 K, p = 1 atm; at altitude H = 10 km, T_0 = 223 K, p = 0.261 atm; For normal air density, the electric intensity must be less than 3 MV (E < 3 MV/m) and depend on pressure. These values change in case of Venus applications.

Electron mobility. The ratio $E/p \approx$ constant. Conductivity σ of gas depends on density of charged particles n and their mobility b, for example:

$$\sigma = neb, \quad \lambda = 1/n\sigma, \tag{11.7}$$

where b is mobility of the electron, λ is a free path of electron. Electron mobility depends from ratio E/n. This ratio is given in Table 11.3. The electrons may connect to the neutral molecules and produce the negative ions (for example, affinity of electron to O_2 equals 0.3–0.87 eV, to H_2O equals 0.9 eV [7] p.424). This way, the computation of the mobility of a gas containing electrons and ions is a complex problem. Usually, calculations are made for all electrons converted to ions.

The maximal electric intensity in air at the Earth surface is E_m = 3 MV/m. If atmospheric pressure changes, then E_m also changes by law E_m/p = constant. For example, if $E = 10^5$ V/m, than v = 20 m/s in Earth conditions. Different values apply in case of Venus.

Table 11.3 Electron mobility b_e in gas vs E/n

Gas	$E/n \times 10^{-17}$ 0.03 V·cm²	$E/n \times 10^{-17}$ 1 V·cm²	$E/n \times 10^{-17}$ 100 V·cm²	Gas	$E/n \times 10^{-17}$ 0.03 V·cm²	$E/n \times 10^{-17}$ 1 V·cm²	$E/n \times 10^{-17}$ 100 V·cm²
N_2	13600	670	370	He	8700	930	1030
O_2	32000	1150	590	Ne	16000	1400	960
CO_2	670	780	480	Ar	14800	410	270
H_2	5700	700	470	Xe	1980	–	240

From: Physics Encyclopedia; http://www.femto.com.ua/articles/part_2/2926.html

11.5.2 Electron Injectors

There are some methods for getting the electron emissions: hot cathode emission, cold field electron emission (edge cold emission, edge cathode). The photo emission, radiation emission, radioisotope emission, and so on, usually produce the positive and negative ions together. We consider only the hot emission and the cold field electron emission (edge cathodes), which produces only electrons.

Hot electron emission. Current i of diode from potential (voltage) U is

$$i = CU^{3/2} \tag{11.8}$$

Here C is constant which depends on form and size cathode. For plate diode

$$C = \frac{4}{9}\varepsilon_0 \frac{S}{d^2}\sqrt{\frac{2e}{m_e}} \approx 2.33 \cdot 10^{-6}\frac{S}{d^2}, \tag{11.9}$$

where $\varepsilon_o = 8.85 \cdot 10^{-12}$ F/m; S is area of cathode (equals area of anode), cm^2; d is distance between cathode and anode, cm; e/m_e is the ratio of the electron charge to electron mass, C/kg. Results obtained by using Eq. (11.9) are shown in Fig. 11.7.

The maximal *hot cathode* emission calculated by equation:

$$j_s = BT^2 exp(-A/kT) \tag{11.10}$$

where B is coefficient, A/cm^2K^2; T is cathode temperature, K; $k = 1.38 \times 10^{-23}$ [J/K] is Bolzmann constant; $A = e\varphi$ is thermoelectron exit work, J; φ is the exit work (output energy of electron) in eV, $e = 1.6 \cdot 10^{-19}$. Both values A, B depend from material of cathode and its cover. The "A" changes from 1.3 to 5 eV, the "B" changes from 0.5 to 120 A/cm^2K^2. Boron thermo-cathode produces electric current

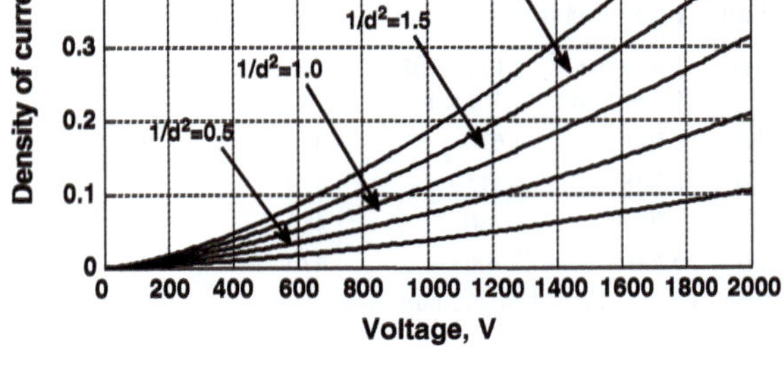

Fig. 11.7 Electric current via voltage the plain cathodes for different ratio of the distance

Table 11.4 Exit work (eV) from cathode is covered by the optimal layer(s)

Cr – Cs	Ti – Cs	Ni – Cs	Mo – Cs	W – Ba	Pt - Cs	W - O – K	Steel- Cs	Mo_2C-Cs	WSi_2-Cs
1.71	1.32	1.37	1.54	1.75	1.38	1.76	1.52	1.45	1.47

Source [8]

up 200 A/cm^2. For temperature 1400–1500 K the cathode can produce current up 1000 A/cm^2. The life of cathode can reach some years.

Exit energy from metal are (eV):

$$\text{W } 4.5, \text{ Mo } 4.3, \text{ Fe } 4.3, \text{ Na } 2.2 \text{ eV}, \tag{11.11}$$

From cathode covered by optimal layer(s) the exit work is in Table 11.4.

Results of computation of the maximal electric current (in vacuum) via cathode temperature for the different exit work of electrons f are presented in Fig. 11.8.

Method of producing electrons and positive ions is well developed in the ionic thrusters for space apparatus.

The field electron emission. (The edge cold emission). The cold field electron emission uses the edge cathodes. It is known that the electric intensity E_e in the edge (needle) is

$$E_e = U/a. \tag{11.12}$$

Here a is radius of the edge. If voltage between the edge and nears net (anode) is $U = 1000$ V, the radius of edge $a = 10^{-5}$ m, electric intensity at edge is the $E_e = 10^8$ V/m. This is enough for the electron emission. The density of electric current may reach up to 10^4 A/cm^2. For getting the required current we make the needed number of edges. The density of electric current is calculated by:

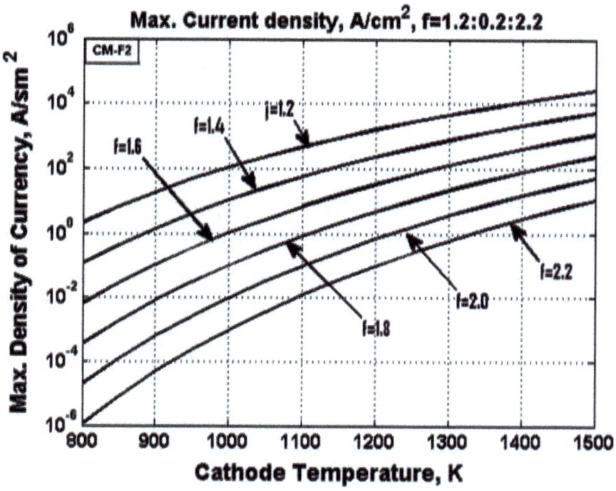

Fig. 11.8 The maximal electric current via cathode temperature for the different exit work of electrons f

Table 11.5 Density of current. Source: http://www.femto.com.ua/articles/part_1/0034.html

φ = 2,0 eV		φ = 4,5 eV		φ = 6,3 eV	
E × 10^{-7}	lg j	E × 10^{-7}	lg j	E × 10^{-7}	lg j
1,0	2,98	2,0	−3,33	2,0	−12,9
1,2	4,45	3,0	1,57	4,0	−0,88
1,4	5,49	4,0	4,06	6,0	3,25
1,6	6,27	5,0	5,59	8,0	5,34
1,8	6,89	6,0	6,62	10,0	6,66
2,0	7,40	7,0	7,36	12,0	7,52
2,2	7,82	8,0	7,94	14,0	8,16
2,4	8,16	9,0	8,39	16,0	8,65
2,6	8,45	10.0	8,76	18,0	9,04
		12,0	9,32	20,0	9,36

$$j \approx 1.4 \cdot 10^{-6} \frac{E^2}{\varphi} 10^{(4.39\varphi^{-1/2} - 2.82 \cdot 10^7 \varphi^{3/2}/E)} \tag{11.13}$$

where j is density of electric current, A/cm^2; E is electric intensity near edge, V/cm; φ is exit work (output energy of electron, field electron emission), eV. The density of current is computed by Eq. (11.11) and results are shown in Table 11.5.

Example: Assume we have needle with edge $S_1 = 10^{-4}$ cm^2, $\varphi = 2$ eV and net $S_2 = 10 \times 10 = 10^2$ cm^2 located at a distance $L = 10$ cm. The local voltage between the needle and net is $U = 10^2$ volts. Then electric intensity at edge of needle, current density and the electric current are:

$$E = \frac{S_2 U}{S_1 L} = \frac{10^2 10^2}{10^{-4} 10^1} = 10^7 \text{ V/cm}, \quad j = 10^3 \text{ A/cm}^2,$$
$$i = jS_1 = 10^3 10^{-4} = 0.1 \text{ A} \tag{11.14}$$

Here j is taken from Table 11.5 or computed by Eq. (11.11). If we need the electric current 10 A, we must locate 100 needles in the entrance area 1 × 1 m of generator. Results obtained by using Eq. (11.11) are presented in Fig. 11.9.

Loss of energy and matter to ionization. Let us estimate the energy and matter requested for ionization and discharge in the offered ABEG generator. Assume we have ABEG generator having the power $P = 10{,}000$ kW and a work voltage $V = 1$ MV. In this case the electric current is $i = P/V = 10$ A = 10 C/s.

Assume we use the nitrogen N_2 for ionization (a very bad gas for it). It has exit work about 5 eV and relative molecular weight 14. One molecule (ion) of N_2 weights $m_N = 14{\cdot}1.67{\cdot}10^{-27} = 2.34{\cdot}10^{-26}$ kg. The 1 Ampere has $n_A = 1/e = 1/1.6{\cdot}10^{-19} = 6.25{\cdot}10^{18}$ ions/s. Consumption of the ion mass is: $M = m_N i\ n_A = 2.34{\cdot}10^{-26}{\cdot}10{\cdot}\ 6.25{\cdot}10^{18} = 1.46{\cdot}10^{-6}$ kg/s $= 1.46{\cdot}10^{-6} \cdot 3.6{\cdot}10^{-3} = 5.26{\cdot}10^{-3}$ kg/hour ≈ 5 gram/hour. If electron exit work equals $\varphi = 4.5$ eV the power spent

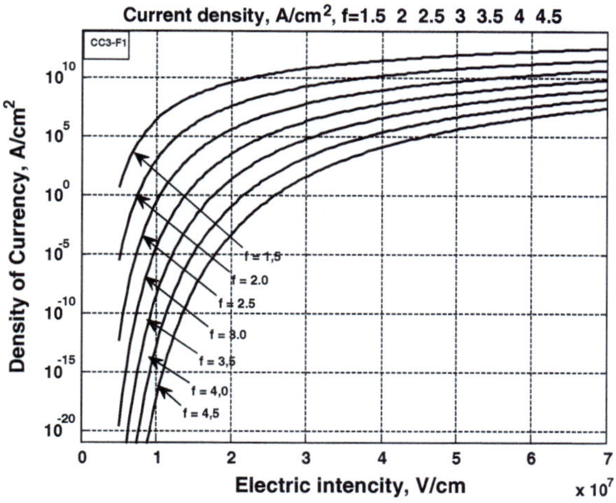

Fig. 11.9 Density of electric current the noodle injector via the electric intensity for different the field electron emissions *f*

extraction of one electron is: $E_1 = \varphi e = 4.5{\cdot}1.6{\cdot}10^{-19} = 7.2{\cdot}10^{-19}$ J. The total power for the electron extraction is $E = i{\cdot}n_A{\cdot}E_1 = 10{\cdot}6.25{\cdot}10^{18}\ {\cdot}7.2{\cdot}10^{-19} = 45$ W. The received values: mass *M* and power *E*, are very small in comparison with conventional consumption of fuel (tons in hour), and generator power (thousands of kW).

11.6 Conclusions

Relatively, no progress has been made in wind energy technology in the last years. While the energy from wind is free, its production is more expensive than its production in conventional electric power stations. Conventional wind energy devices have approached their maximum energy extraction potential, relative to their installation cost. Current wind installations cannot significantly decrease a cost of kWh and provide the stability of energy production. They cannot continue to increase the power of single energy units.

In case of Venus applications, the energy industry needs revolutionary ideas that improve performance parameters (installation cost and power per unit), and that significantly decrease (by 5–10 times) the cost of energy production. The electron wind installations, delineated in this chapter, can move the wind energy industry to revolutionary potential.

As with any new idea, the suggested concept is in need of research and development. The theoretical problems do not require fundamental breakthroughs. Before implementation on Venus, it is necessary to design small, cheap installations to study and get an experience in the design electron wind generator.

References

1. Bolonkin AA (2013) Electronic wind generator. Electr Power Eng Front 2(3):64–71. http://www.academicpub.org/epef/Issue.aspx?Volume=2&Number=3&Abstr=false, http://www.academicpub.org/epef/Issue.aspx?Volume=2&Number=3&Abstr=false, http://viXra.org/abs/1306.0046, http://www.IntellectualArchive.com, http://intellectualarchive.com #1387, https://archive.org/details/ArticleElectronWindGenerator6613AsterShmuelWithPicture
2. Bolonkin AA (2013) http://viXra.org/abs/1306.0046, www. Intellectual Archive.com
3. Bolonkin AA (2014) Jet electric generator. Glob Sci J 23 June 2014. https://archive.org/details/ArticleJetElectricGenerator, http://gsjournal.net/Science-Journals-Papers/Author/1481/Alexander,%20Bolonkin
4. Bolonkin AA (2013) Electron hydro electric generator. Int J Adv Eng Appl ISSN: 2321–7723, Special Issue I. http://fragrancejournals.com/?page_id=18 (Online), http://viXra.org/abs/1306.0196
5. Bolonkin AA (2013) Electron super speed hydro propulsion. Int J Adv Eng Appl 1:15–19. http://viXra.org/abs/1306.0195, http://archive.org/details/ElectronSuperSpeedHydroPropulsion
6. Bolonkin AA (2012) Electron air hypersonic propulsion. Int J Adv Eng Appl 1:42–47. http://viXra.org/abs/1306.0003
7. Bolonkin AA (2014) Electric hypersonic space aircraft. Glob Sci J. http://viXra.org/abs/1407.0011
8. Bolonkin AA (2004) Utilization of wind energy at high altitude, AIAA-2004-5756, AIAA-2004-5705. In: International energy conversion engineering conference at providence, RI, USA, 16–19 Aug 2004
9. Bolonkin AA (2001) Method of utilization a flow energy and power installation for it, USA patent application 09/946,497 of 09/06/2001
10. Bolonkin AA (2013) Energy transfers from airborne wind turbine: review and comparison of airborne turbines. Int J Adv Eng Appl 1:44–64. http://viXra.org/abs/1304.0159, http://archive.org/details/EnergyTransfersFromAirborneWindTurbine
11. Bolonkin AA (2006) New concepts, ideas, innovations in aerospace, technology and the human sciences, NOVA, 2006, pp 510. http://viXra.org/abs/1309.0193, http://www.archive.org/details/NewConceptsIfeasAndInnovationsInAerospaceTechnologyAndHumanSciences
12. Bolonkin AA (2008) New technologies and revolutionary projects, Lulu, pp 324. http://www.archive.org/details/NewTechnologiesAndRevolutioaryProjects
13. Bolonkin AA, Cathcart RB (2007) Macro-Projects: environments and technologies, NOVA, pp 536 http://www.archive.org/details/Macro-projectsEnvironmentsAndTechnologies
14. Koshkin NI, Shirkebich MG (1982) Directory of elementary physics. Nauka, Moscow (in Russian)
15. Kikoin IK (1976) Table of physics values. Atomisdat, Moscow (in Russian)
16. Kalashnikov SG (1985) Electricity, Moscow, Nauka (in Russian)
17. Gipe P (1998) Wind power. Chelsea Green Publishing Co., Vermont
18. Cost of renewable energy (2014) http://www.irena.org/DocumentDownloads/Publications/RE_Technologies_Cost_Analysis-WIND_POWER.pdf
19. Thresher RW et al (1999) Wind technology development: large and small turbines, NRFL
20. Wikipedia (2014) Electric generator. http://wikipedia.org

Chapter 12
Photovoltaic Power Resources on Mercury and Venus

T.E. Girish and S. Aranya

12.1 Introduction

Proximity to the Sun and long night periods are distinct disadvantages for photovoltaic power generation on Mercury and Venus. However solar panels are providing uninterrupted power for the Messenger space craft (Mercury flyby) and Venus Express space craft (polar orbiter) for the past ten years (Solomon et al. 2001; Dakermanji et al. 2006). In this article we have discussed the prospects of utilizing photovoltaic power resources in these inner planets suggesting possible ways to overcome the challenges.

12.2 Problems in Photovoltaic Power Generation on Mercury and Venus

Long night periods (28.5 days in Mercury and 121.5 days in Venus), high surface day time temperature (near 450 °C) and hazardous radiation or physical environments poses severe challenges to photovoltaic power generation on Mercury and Venus (Landis 2003).

T.E. Girish (✉) · S. Aranya
University College, Trivandrum, India
e-mail: tegirish5@yahoo.co.in

S. Aranya
e-mail: aranya_s@yahoo.co.in

V. Badescu and K. Zacny (eds.), *Inner Solar System*,
DOI 10.1007/978-3-319-19569-8_12

12.2.1 Temperature Related Performance Loss of Solar Cells on Mercury and Venus

Adopting the values of temperature coefficient of solar cells (Girish and Aranya 2012) reported in the literature (0.45 %/K for Si, 0.21 %/K for GaAs, 0.18 %/K for multijunction Group III-V solar cells and 0.45 %/K for CIGS) we have calculated the temperature of complete performance loss (100 % decrease in efficiency) for different solar cells as follows:

- Si (247 °C),
- GaAs (501 °C),
- Multijunction Group III-V (581 °C)
- CIGS (247 °C).

The maximum thermal degradation for different types of solar cells expressed as percentage decrease in efficiency and calculated for 450 °C (applicable for both Mercury and Venus) are shown in Table 12.1.

Table 12.1 Maximum thermal and extreme radiation related performance loss of Solar cells in Mercury and Venus

Types of solar cells	Performance loss (% decrease in efficiency) at 450 °C	Performance loss due to extreme SPE in Mercury	Performance loss due to extreme SPE in Venus
Si	100 %	100 %	45 %
GaAs	73.5 %	52 %	15 %
Gr III-V	63 %	52 %	15 %
CIGS	100 %	18 %	5 %

These calculations suggest that some performance is still present for GaAs and multijunction Group III-V solar cells in the high surface noon temperature on Venus and Mercury near equator (Surampudi et al. 2013; Ercol et al. 2010).

12.2.2 Radiation Related Performance Loss of Solar Cells on the Inner Planets

The space environment near Mercury and Venus are more hazardous compared to earth. The intensity of solar radiation and flux of energetic particles show a $1/r^2$ decrease in the inner heliosphere from sun to earth (Aran et al. 2003). However solar wind density shows a $1/r^3$ dependence (Baker et al. 2013). At the mean distance from the Sun the intensity of solar radiation on Mercury is about seven times (6.67) and on Venus is two times (1.92) that near Earth.

In Table 12.2 we have given the fluence of major solar proton events (SPE's) observed near earth during the years 1956–2005 AD (Kim and Cucinotta 2006).

Table 12.2 Estimated fluences of major solar proton events on Mercury and Venus during the years 1956–2005 AD

Date of observation	Fluence on Earth ($x10^9$ protons/cm^2)	Fluence on Venus ($x10^9$ protons/cm^2)	Fluence on Mercury ($x10^9$ protons/cm^2)
12th Nov 1960	9	18	63
2nd Aug 1972	5	10	35
19th Oct 1989	4.23	8.5	30
14th July 2000	3.74	7.5	26
26th Oct 2003	3.25	6.5	23
4th Nov 2001	2.92	6	20.5
10th July 1959	2.3	4.6	16
8th Nov 2000	2.27	4.5	16
23rd March 1991	1.74	3.5	12
12th Aug 1989	1.51	3	10.6
29th Sept 1989	1.35	2.7	9.5
16th Jan 2005	1.04	2	7.3
23rd Feb 1956	1	2	7

Proton events with a fluence of at least 10^9 protons/cm^2 is only included here. The estimated fluence on Venus and Mercury for these proton events are also given in Table 12.2. Here we assumed $1/r^2$ dependence for proton fluence in the inner heliosphere so that earth fluence is multiplied by a factor of two to estimate Venus values and by a factor of seven to estimate Mercury values.

Assuming the values of radiation related decrease in efficiency for different solar cells published in our earlier paper for extreme proton irradiation observed near earth, with energy greater than 30 MeV and fluence 18×10^9 protons/cm^2(Girish and Aranya 2012), we have estimated the corresponding radiation (Odenwald and Green 2007) related performance loss on Mercury and Venus. The calculated values are also given in Table 12.1.

12.2.3 Long Term Degradation of Solar Cells in the Space Environments of Inner Planets

Limited observations of long term degradation of solar cells near Venus and Mercury exist in literature. However we have some useful data (Doody 2010) from the Magellan Mission of NASA to Venus during 1990's. During five years operation from 1990 to 1994 the Magellan Si solar panels suffered an EOL (End of Life) degradation of 67 % in efficiency. This suggests an annual degradation of 13.4 % per year for Si solar cells near Venus.

12.3 Prospects of Utilizing Photovoltaic Power Resources on Mercury and Venus

As explained in previous sections PV power generation has severe challenges to overcome on Mercury and Venus. But we can find two promising possibilities for utilizing photovoltaic power resources on these inner planets using current technology which will be discussed below.

12.3.1 Photovoltaic Power Generation in the Polar Regions of Mercury

Water ice is detected on the polar regions of Mercury by earth based microwave observations and recent Messenger space craft observations (Lawrence et al. 2013). The maximum estimated temperature on the polar regions of Mercury is 400 K or 127 °C (Vasavada et al. 1999). Group III-V multijunction solar cells suffer only a thermal degradation of about 23 % for this temperature ranges when unshielded. There are several parts of Mercury's Polar Regions which are 80–94 % sunlit (Chabot et al. 2012). The maximum intensity of solar radiation here is expected to be around 1 Sun comparable to that of Lunar noon time solar irradiance. The average radiation related performance loss in Mercury can be estimated to be within 20 % per year for triple junction solar cells (near Earth data for these solar cells is found to 2–3 % per year, (Girish and Aranya 2012). The net degradation for these solar cells is unlikely to exceed 40 % in the polar space environment of Mercury. If this is the case we can expect to produce a photovoltaic power of about 2 kW using triple junction Gr III-V solar cell array of 10 m^2 in area and 30 % BOL (Beginning of Life) efficiency on the sunlit polar regions of Mercury. The small obliquity of the Mercury's poles can be utilized to artificially divert the sunlight from the bright side to dark side for night time photovoltaic power generation.

12.3.2 Photovoltaic Power Generation on the Top of Venus Atmosphere

The solar irradiance varies between 1 Sun to 2 Sun between 50 to 70 km altitude in the Venus atmosphere (Landis et al. 2002). The atmospheric pressure is only 0.04–1 bar for these heights so that solar radiation suffers minimum absorption and scattering losses. If balloons or space capsules descended to these heights in Venus, photovoltaic power generation is possible where the temperature is between −43 °C to 77 °C only. Temperature related degradation for triple junction solar cells with cover glass is negligible in these atmospheric regions of Venus. There may be still a radiation related degradation which is however small for Group III-V and GaAs solar cells as seen from Table 12.1.

12.4 Discussion

We have suggested some possibilities of utilizing the photovoltaic power resources in Mercury and Venus in this article. In Venus photovoltaic power generation is not advisable near the surface due to very low intensity of solar irradiance (100 W/m^2) as measured by earlier Venera lander missions and very high temperature (460–470 °C) caused by greenhouse effects. Photovoltaic power generation is possible near the top of Venusian atmosphere (50–70 km) where the air temperature is not high. The solar panels with proper cover glass protection can be accommodated in stratospheric balloons and may be deployed or descended from a suitable Venus orbiter. Triple junction GaAs solar cells of the type used in Venus express space craft is preferable for this application. There are no reported significant long term degradation of Venus Express solar panels (launched in orbit during the year 2005). Due to very long night periods in Venus (121.5 days near equator) the pilot mission can be restricted during the Venusian daytime with a preferable duration of about 80–90 days. The photovoltaic power generated in Venus atmosphere may be beamed to its surface in the form of microwaves to power rover type special vehicles. But this technology is still in the development stage (Landis and Mellott 2007).

The Mercury is known to have hazardous space environment. The average solar EUV flux measured near Mercury is ten times the same measured near Earth. A lander mission to Mercury near its polar regions which is illuminated during the significant part of Mercury's orbital period is preferable for photovoltaic power generation. The reported maximum day time temperature in these regions is estimated to be around 100 °C suggesting a solar irradiance of nearly 1 Sun similar to our lunar day time environment. Messenger mission recently found that 60 sq km of the south polar region of Mercury is illuminated at least 80–94 % during the local daytime (Chabot et al. 2012). Using triple junction GaAs solar cells photovoltaic power generation is feasible near the poles of Mercury during periods of relatively calm space weather like sunspot minima where 10 MeV proton event occurrences is relatively less. Mercury's weak magnetosphere is capable of preventing the entry of energetic particles from space of energy up to 2 MeV (Laurenza et al. 2011).

The probability of occurrence of 10 MeV solar proton events during sunspot minimum can be as low as 0–1 per year as evident from observed earth orbiting satellite data for the past four sunspot cycles (www.swpc.noaa.gov/ftpdir/indices/SPEtxt) (see Table 12.3). The next sunspot minimum (beginning of cycle 25) is predicted to be during 2019–2020 during which a polar mission to Mercury can be preferable planned.

Table 12.3 Periods of minimum occurrence of 10 MeV solar proton events (SPE) during the sunspot cycles 19–24

Sunspot Cycle	Years	Number of 10 MeV SPE observed per year
21	1976–1977	1
22	1987	1
23	1996	0
24	2007–2009	0

The surface temperature of Venus measured by Venera Lander probes are in the range 455 °C–470 °C (Garvin et al. 1984). The adhesives which fix the cover glass to solar cells (which is required to shield from corrosive sulphuric acid atmosphere of Venus) cannot withstand this high temperature. Specular heat shielding is necessary for all instruments in future surface Landers (Dayson et al. 2009). The Venus Express and messenger spacecraft power system are thermally shielded from intense solar irradiance by attaching reflector mirrors to the solar panels. There are several ongoing projects to develop High intensity High Temperature (HIHT) solar cells for the forthcoming solar orbiter mission which will cover very close distances to the Sun (Landis 2008; Zimmermann et al. 2013; Landis and Schmitz 2008). Development of radiation resistant highly efficient multijunction solar cells (Guina 2009) especially using nitride based materials (Akter 2014) are promising for space applications.

12.5 Conclusions

- Photovoltaic power generation near the Venusian surface is not preferable for technological applications. However photovoltaic power generation using triple junction Group III-V solar cells mounted in special balloons near the top of Venus atmosphere (50–70 km) is possible for a day time Venus mission lasting 80–90 days.
- A pilot lander mission to Mercury's polar regions which are significantly illuminated (80 %) during the day time can be planned where photovoltaic power generation with the triple junction solar cells is feasible during periods of calm space weather conditions such as sunspot minima.

References

Akter N (2014) Design and simulation of Indium Gallium nitride multijunction tandem solar cells. IJRET 3:315–321

Aran A, Sanahuja B, Lario D (2003) An engineering model for solar energetic particles in interplanetary space. ESA/ESTEC, December 2003

Baker DN, Poh G, Odstrcil D, Arge CN, Benna M, Johnson CL, Korth H, Gershman DJ, Ho GC, McClintock WE, Cassidy TA, Merkel A, Raines JM, Schriver D, Slavin JA, Solomon SC, Travnicek PM, Winslow RM, Zurbuchen TH (2013) Solar wind forcing at Mercury: WSA-ENLIL model results. J Geophys Res Space Phys 118:45–57

Chabot NL, Ernst CM, Denevi BW, Harmon JK, Murchie, SL, Blewett DT, Solomon SC, Zhong ED (2012) Areas of permanent shadow in Mercury's south polar region ascertained by MESSENGER orbital imaging. Geophys Res Lett 39(9)

Dakermanji G, Jenkins J, Ercol CJ (2006) The MESSENGER spacecraft solar array design and early mission performance. In: IEEE 4th world conference on Photovoltaic energy conversion record of the 2006, vol 2, pp 1919–1922

Dayson RW, Schmitz PG, Penswick LB (2009) Long- lived Venus Lander conceptual design: how to keep it cool. In: 7th International Energy conversion Engineering Conference, Denver, Colorado, 2–5 Aug 2009

Doody D (2010) Deep spacecraft an overview of interplanetary flight, E-Book. Springer Science and Buisness Media, pp 146–147

Ercol CJ, Dakermanji G, Laughery SC (2010) The MESSENGER spacecraft power system: thermal performance through the first mercury flyby. In: 46th AIAA/ASME/SAE/ASEE Joint Propulsion conference and Exhibit, AIAA 2010-6848, AIAA control ID 764024, Nashville, Tennessee, 25–28 July 2010

Garvin JB, Head JW, Zuber MT, Helfenstein P (1984) Venus: the nature of the surface from Venera panoramas. J Geophys Res 89(B5):3381–3399

Girish TE, Aranya S (2012) Photovoltaic power generation on the moon: problems and prospects. In: Badescu V (ed) Moon - prospective energy and material resources. Springer, Heidelberg, pp 367–376

Guina M (2009) High-efficiency III-V solar cells: oppurtunities and challenges. Finnish Academy, Helsinki. www.arc.tut.fi

Kim M, Cucinotta F (2006) Example Solar Proton Data, NASA JSA. http://emmrem.unh.edu/presentation/SPE_Examples_LDR

Landis GA (2003) Colonization of venus. In: Conference on Human Space Exploraion, Space Technology & Applications International Forum, Albuquerque, New Mexico, 2–6 Feb 2003

Landis GA (2008) Set the controls for the heart of the Sun: power & mobility for high temperature environments. In: 6th International Planetary Probe workshop, Atlanta, Georgia, 21–22 June 2008

Landis GA, Mellott KC (2007) Venus surface power and cooling systems. Acta Astronaut 61:995–1001

Landis GA Schmitz PC (2008) Solar power system design for the solar probe mission. In: 6th International Engineering conversion (IECEC), Cleveland, Ohio, 28–30 July 2008. AIAA 2008-5712

Landis GA, Colozza A, LaMarre CM (2002) Atmospheric flight on Venus. In: 40th Aerospace Sciences Meeting and Exhibit sponsored by the American Institute of Aeronautics and Astronautics. Reno, Nevada, 14–17 Jan 2002. NASA/TM-2002-211467, AIAA-2002-0819

Laurenza M, Storini M, Vaini R, Desorgher L (2011) Cutoff rigidities for Mercury-orbiting spacecraft. In: 32nd International Cosmic Ray Conference, Beijing 2011. doi:10.7529/ICRC2011/V11/0230

Lawrence DJ, Feldman WC, Goldsten JO, Maurice S, Peplowski PN, Anderson BJ, Bazell D, McNutt Jr RL, Nittler LR Prettyman TH, Rodgers DJ, Solomon SC, Weider SZ (2013) Evidence for water ice near mercury's north pole from MESSENGER Neutron specrometer measurements. Science 339(6117):292–296. doi:10.1126/science.1229953

Odenwald SF, Green JL (2007) Forecasting the impact of an 1859- caliber superstorm on geosynchronous Earth- orbiting satellites: transponder resources. Space Weather 5:SO6002. doi:10.1029/2006SW000262

Solomon SC, McNutt RL Jr, Gold RE, Acuna MH, Baker DN, Boynton WV, Chapman CR, Cheng AF, Gloeckler G, Head JW III, Krimigis SM, McClintock WE, Murchie SL, Peale SJ, Phillips RJ, Robinson MS, Slavin JA, Smith DE, Strom RG, Trombka JI, Zuber MT (2001)

The MESSENGER mission to Mercury : scientific objectives and implementation. Planet Space Sci 49:1445–1465

Surampudi R, Khanna S, Grandidier J, Bugga K (2013) Power system technologies for future Venus Missions- presentation to Venus technology forum, NASA Space Academy at Glenn/space.academy.grc.nasa.gov. 19 Nov 2013

Vasavada AR, Paige DA, Wood SE (1999) Near-surface temperatures on Mercury and the Moon and the stability of polar ice deposits. Icarus 141:179–193. Article ID icar.1999.6175

Zimmermann CG, Nomayr C, Kolb M, Rucki A (2013) A mechanism of solar cell degradation in high intensity, high temperature space missions. Prog Photovoltaics Res Appl 21(4):420–435

Chapter 13
Flight Apparatuses and Balloons in Venus Atmosphere

Alexander A. Bolonkin

13.1 Introduction

The Venus atmosphere is very different from the Earth atmosphere. One has different composition (CO_2), very high pressure (up 92 atm) and very high temperature (up 462 °C). The Earth standard atmosphere cannot be used for the estimation and computation of the air balloon, aircraft and re-entry space apparatus into Venus atmosphere. A model is proposed in this chapter for flight in the atmosphere of Venus.

13.2 Research the Venus by Space Apparatus

The former Soviet Union sent many space apparatuses to Venus (Venera 1–16). The USA, ESA, Japan also have sent some space apparatus.

The Soviet Venera 3 probe crash-landed on Venus on 1 March 1966. Its communication system failed before it was able to return any planetary data. On 18 October 1967, Venera 4 successfully entered the atmosphere and deployed science experiments. Venera 4 proved that the surface temperature was even hotter than Mariner 2 had measured, at almost 500 °C, and the atmosphere was 90 to 95 % carbon dioxide. The Venusian atmosphere was considerably denser than Venera 4's designers had anticipated, and slower than intended parachute descent meant its batteries ran down before the probe reached the surface. After returning descent data for 93 min, Venera 4's last pressure reading was 18 bar at an altitude of 24.96 km.

A.A. Bolonkin (✉)
C&R, Brooklyn, NY, USA
e-mail: aBolonkin@juno.com; aBolonkin@gmail.com

V. Badescu and K. Zacny (eds.), *Inner Solar System*,
DOI 10.1007/978-3-319-19569-8_13

One day later on October 19th, 1967, Mariner 5 conducted a fly-by at a distance of less than 4000 km above the cloud tops. Mariner 5 was originally built as backup for the Mars-bound Mariner 4; when that mission was successful, the probe was refitted for a Venus mission. A suite of instruments more sensitive than those on Mariner 2, in particular its radio occultation experiment, returned data on the composition, pressure and density of the Venusian atmosphere. The joint Venera 4 – Mariner 5 data was analyzed by a combined Soviet-American science team in a series of colloquia over the following year, in an early example of space exploration cooperation.

Armed with the lessons and data learned from Venera 4, the Soviet Union launched the twin probes Venera 5 and Venera 6 five days apart in January 1969; they encountered Venus a day apart on 16 and 17 May. The probes were strengthened to improve their crush depth to 25 bar and were equipped with smaller parachutes to achieve a faster descent. Because then-current atmospheric models of Venus suggested a surface pressure of between 75 and 100 bar, neither was expected to make it to the surface. After returning atmospheric data for little over 50 min, they both crushed at altitudes of approximately 20 km before hitting the surface on the night side of Venus.

The Vega program was a series of Venus missions which also took advantage of the appearance of Comet Halley in 1986. Vega 1 and Vega 2 were unmanned spacecraft launched in a cooperative effort among the Soviet Union (who provided the spacecraft and launch vehicle) and Austria, Bulgaria, Hungary, the German Democratic Republic, Poland, Czechoslovakia, France, and the Federal Republic of Germany in December 1984. They had a two-part mission to investigate Venus and also flyby Halley's Comet.

The flyby of Halley's Comet had been a late mission change in the Venera program following on from the cancellation of the US Halley mission in 1981. A later Venera mission was cancelled and the Venus part of the Vega 1 mission was reduced. Because of this, the craft was designated Vega, a contraction of "Venera" and "Gallei" (Russian words for "Venus" and "Halley", respectively). The spacecraft design was based on the previous Venera 9 and Venera 10 missions.

The two spacecrafts were launched on December 15 and 21, 1984, respectively. With their redesignated dual missions, Vega probes became part of the Halley Armada, a group of space probes that studied Halley's Comet during its 1985/86 perihelion.

13.3 Atmospheric Balloons on Venus

Vega 1 and 2 had air balloons. The balloons should have a mass of about 21.5 kg. The Vega 1 Lander/Balloon capsule entered the atmosphere of Venus (125 km altitude) at 2:06:10 UT (Earth received time.) on 11 June 1985 at roughly 11 km/s. At approximately 2:06:25 UT the parachute attached to the landing craft cap opened at an altitude of 64 km. The cap and parachute were released 15 s later at 63 km altitude. The balloon package was pulled out of its compartment by parachute 40 s later at 61 km altitude, at 8.1 degrees N, 176.9 degrees east. A second parachute opened at an altitude of 55 km, 200 s after entry, extracting the furled balloon. The balloon was inflated 100 s later at 54 km and the parachute and inflation systems were jettisoned. The ballast was jettisoned when the balloon reached roughly 50 km and floated back to a stable height between 53 and 54 km some 15 to 25 min after entry. The mean stable height was 53.6 km, with a pressure of 535 mbar and a temperature of 300–310 K in the middle, most active layer of the Venus three-tiered cloud system. The balloon drifted westward in the zonal wind flow with an average speed of about 69 m/s at nearly constant latitude. The probe crossed the terminator separating night-side from the day-side at 12:20 UT on 12 June after traversing 8500 km. The probe continued to operate in the daytime until the final transmission was received at 00:38 UT on 13 June from 8.1 N, 68.8 E after a total traversal distance of 11,600 km. It is not known how much farther the balloon travelled after that final telecommunication.

The Vega 2 Lander/Balloon capsule entered the Venus's atmosphere (125 km altitude) at 2:06:04 UT (Earth received time) on 15 June 1985 at roughly 11 km/s. At approximately 2:06:19 UT the parachute attached to the landing craft cap opened at an altitude of 64 km. The cap and parachute were released 15 s later at 63 km altitude. The balloon package was pulled out of its compartment by parachute 40 s later at 61 km altitude, at 7.45 degrees S, 179.8 degrees east. A second parachute opened at an altitude of 55 km, 200 s after entry, extracting the furled balloon. The balloon was inflated 100 s later at 54 km and the parachute and inflation systems were jettisoned. The ballast was jettisoned when the balloon reached roughly 50 km and floated back to a stable height between 53 and 54 km some 15 to 25 min after entry. The mean stable height was 53.6 km, with a pressure of 535 mbar and a temperature of 308-316 K in the middle, most active layer of the Venus three-tiered cloud system. The balloon drifted westward in the zonal wind flow with an average speed of about 66 m/s at nearly constant latitude. The probe crossed the terminator from night to day at 9:10 UT on 16 June after traversing 7400 km. The probe continued to operate in the daytime until the final transmission was received at 00:38 UT on 17 June from 7.5 S, 76.3 E after a total traverse distance of 11,100 km. It is not known how much further the balloon traveled after its final communication with Earth.

13.4 Current Missions to Venus

NASA's Messenger mission to Mercury performed two fly-bys of Venus in October 2006 and June 2007, to slow its trajectory for an eventual orbital insertion of Mercury in March 2011. It collected scientific data on Venus during both fly-bys.

The Venus Express probe was designed and built by the European Space Agency. Launched on 9 November 2005 by a Russian Soyuz-Fregat rocket procured through Starsem, it successfully assumed a polar orbit around Venus on 11 of April 2006. The probe is undertaking a detailed study of the Venusian atmosphere and clouds, including mapping of the planet's plasma environment and surface characteristics, particularly temperatures. One of the first results from Venus Express is the discovery that a huge double atmospheric vortex exists at the southern pole.

The Japan Aerospace Exploration Agency (JAXA) devised a Venus orbiter, Akatsuki (formerly "Planet-C"), which was launched on 20 May 2010, but the craft failed to enter orbit in December 2010. Hopes remain that the probe can successfully hibernate and make another insertion attempt in six years. Planned investigations included surface imaging with an infrared camera and experiments designed to confirm the presence of lightning, as well as the determination of the existence of current surface volcanism.

13.5 Projects of Future Missions and Explorations to Venus

The European Space Agency (ESA) hopes to launch a mission to Mercury in 2014, called BepiColombo, which will perform two fly-bys of Venus before it reaches Mercury orbit in 2020.

Under its New Frontiers Program, NASA has proposed a lander mission called the Venus In Situ Explorer to land on Venus to study surface conditions and investigate the elemental and mineralogical features of the regolith. The probe would be equipped with a core sampler to drill into the surface and study pristine rock samples not weathered by the harsh surface conditions. A Venus atmospheric and surface probe mission, "Surface and Atmosphere Geochemical Explorer" (SAGE), was selected by NASA as a candidate mission study in the 2009 New Frontiers selection, but the mission was not selected for flight.

The Venera-D (Russian: Венера-Д) probe is a proposed Russian space probe to Venus, to be launched around 2016, to make remote-sensing observations around the planet and deploy a lander, based on the Venera design, capable of surviving on the surface for a long time. Other proposed Venus exploration concepts include rovers, balloons, and aeroplanes (Fig. 13.1).

In late 2013 the Venus Spectral Rocket Experiment took place, which launched a sub-orbital space telescope.

A manned Venus fly-by mission, using Apollo program hardware, was proposed in the late 1960s. The mission was planned to launch in late October or early November 1973, and would have used a Saturn V to send three men to fly past Venus in a flight lasting approximately one year. The spacecraft would have passed approximately 5,000 km (3,100 mi) from the surface of Venus about

Fig. 13.1 Artist's impression of a stirling cooled Venus Rover devised by NASA (left); Venus aircraft concept (right)

four months later. Inspiration Mars includes a manned Venus flyby in their 2021 mission.

Various concepts for a Venus sample return include a high-speed upper atmosphere collection, an atmosphere sample return by slowing down and entering then returning, and a surface sample return.

The Venus Express spacecraft is now in orbit around the planet, probing deeper into the atmosphere using infrared imaging spectroscopy in the 1–5 μm spectral range. The JAXA probe Akatsuki which was launched in May 2010 was intended to study the planet for a period of two years, including the structure and activity of the atmosphere, but it failed to enter Venus's orbit during December 2010. A second attempt to achieve orbit will take place in 2015. One of its five cameras known as the "IR2" will be able to measure the atmosphere of the planet beneath its dense clouds, in addition to its movement and distribution of trace components. With a varied orbit from 300 to 60,000 km, it will be able to take close-up photographs of the planet, and should also confirm the presence of both active volcanoes as well as lightning.

The Venus In Situ Explorer, proposed by NASA's New Frontiers program is a proposed probe which would aid in understanding the processes on the planet that led to climate change and pave the way towards a later sample return mission.

Another craft called the Venus Mobile Explorer has been proposed by the Venus Exploration Analysis Group (VEXAG) to study the composition and isotopic measurements of the surface and the atmosphere, for about 90 days. A launch date has not yet been set.

After missions discovered the reality of the harsh nature of the planet's surface, attention shifted towards other targets such as Mars. There has been a number of proposed missions recently, many of these involve the little-known upper atmosphere. The Soviet Vega program in 1985 dropped two balloons into the atmosphere, but these were battery-powered and lasted for only about two Earth days each before running out of power, since then there has been no exploration of the upper atmosphere. In 2002 the NASA contractor Global Aerospace proposed a

Fig. 13.2 The USA project Venus' balloon

balloon (Fig. 13.2) that would be capable of staying in the upper atmosphere for hundreds of Earth days as opposed to two.

A solar flyer has also been proposed by Geoffrey A. Landis in place of a balloon, and the idea has been featured from time to time since the early 2000s. Venus has a high albedo, and reflects most of the sunlight that shines on it making the surface quite dark. The upper atmosphere, at 60 km, has an upward solar intensity of 90 %, meaning that solar panels on both the top and the bottom of a craft could be used with nearly equal efficiency. In addition to this, the slightly lower gravity, high air pressure and slow rotation allowing for perpetual solar power, make this part of the planet ideal for exploration. The proposed flyer would operate best at an altitude where sunlight, air pressure and wind speed would enable it to remain in the air perpetually, with slight dips down to lower altitudes for a few hours at a time before returning to higher altitudes. As sulfuric acid in the clouds at this height is not a threat for a properly shielded craft, this so-called "solar flyer" would be able to measure the area in between 45 km and 60 km indefinitely, as long as mechanical error or unforeseen problems do not cause it to fail. Landis also proposed that rovers similar to Spirit and Opportunity could possibly explore the surface, with the difference being that Venus surface rovers would be "dumb" rovers controlled by radio signals from computers located in the flyer above, only requiring parts such as motors and transistors to withstand the surface conditions, but not weaker parts involved in microelectronics that could not be made resistant to the heat, pressure and acidic conditions.

Russian space plan for 2006–2015 involves a launch of Venera-D (Venus-D) probe around 2016. The main scientific goals of the Venera-D mission are investigation of the structure and chemical composition of the atmosphere and investigation of the upper atmosphere, ionosphere, electrical activity, magnetosphere and gas escape rate.

13.6 Theory of Flight in Venus' Atmosphere

Below the author gives calculations of the density of carbon dioxide, helium and hydrogen in Venus's atmosphere. This data are necessary for Venus air balloons, Venus planes and Venus re-entry of outer and planetary space probes.

13.6.1 Density of Gases in Venus Atmosphere

The density of gases in Venus condition may be computed from general equation of gas condition

$$pV = \frac{m}{\mu} RT \tag{13.1}$$

where p is gas pressure, N/m^2; V is volume, m^3; R = 8314 J/(kmol·K) is universal gas constant; m is gas mass, kg; μ is molecular mass; T is gas temperature, K. We need the density of gas which we will use for computation in Venus atmosphere. We know the pressure, temperature and type of gases. From Eq. (13.1) we get the equation for computation density

$$\rho = \frac{m}{V} = \frac{\mu p}{RT} \tag{13.2}$$

Here ρ is gas density, kg/m^3. Let us the compute density gases: CO_2 – main gas of the Venus atmosphere ($\mu = 44$), helium He_2 ($\mu = 8$), H_2 ($\mu = 2$) – gases used for the air balloon. Results are shown in Table 13.1.

The carbon dioxide, CO_2 is a significantly heavier gas than N_2 – the main gas ($\approx$ 80 %) of the Earth's atmosphere. That means the lift force of the light gas for an air balloon produces more lift force than one inflated in the Earth atmosphere.

13.6.1.1 Air Balloon for Venus Atmosphere

Any air balloon in Venus's atmosphere allows scientists to study the atmosphere's flows, composition, pressure, temperature and to get detailed photographic pictures of Venus' surface. Unfortunately, the Venus's atmosphere has a high temperature and clouds which obstruct getting good picture. The high temperature does not allow for a low altitude photographic mission very easily.

The lift force of air balloon in the Venus's atmosphere is calculated by:

$$L = g(\rho_a - \rho_g)V \tag{13.3}$$

Table 13.1 Density of Carbon Dioxide, Helium and Hydrogen in conditions the Venus atmosphere

H km	0	10	20	30	40	50	60	70	80	90	100
T °C	462	385	306	222	143	110	−10	−43	−76	−104	−113
T K	745	658	579	495	416	383	263	230	197	169	159
p, atm	92.10	49.39	27.52	9.831	3.501	1.066	0.2357	0.1369	$4.76 \cdot 10^{3}$	$3.74 \cdot 10^{-4}$	$2.66 \cdot 10^{-5}$
CO_2, kg/m^3	65.4	38.12	20.57	10.5	6.286	1.472	0.474	0.0849	0.0128	$1.22 \cdot 10^{-3}$	$3.74 \cdot 10^{-5}$
He_2, kg/m^3	11.89	6.92	3.74	1.91	1.143	0.267	0.0862	0.0154	0.0023	$2.2 \cdot 10^{-4}$	$1.61 \cdot 10^{-5}$
H_2, kg/m^3	2.97	1.73	0.933	0.477	0.285	0.067	0.0215	0.0038	$5. \cdot 10^{-5}$	$5.53 \cdot 10^{-5}$	$0.4 \cdot 10^{-5}$

where L is list force, N; g = 8.69 m/s^2 is Venus gravitation; ρ_a - density of the Venus atmosphere kg/m^3; ρ_g – density of gas filled the air balloon, kg/m^3; V is volume of the air balloon, m^3. The need volume of the air balloon from Eq. (13.3) is

$$V = \frac{m}{\rho_a - \rho_g} \quad (13.4)$$

Here m is mass of the air balloon, kg.

Example of estimation of the Venus helium balloon. Let us take the mass of air balloon m = 21.5 kg, flight altitude H = 55 km. From Table 13.1 the density of Venus's atmosphere in given altitude are: CO_2 ρ = 1.067 kg/m^3, He_2 ρ = 0.17 kg/m^3. Substitute these data in Eq. (13.4) we find

$$V = \frac{m}{\rho_a - \rho_g} = \frac{21.5}{1.067 - 0.17} = 24.2 \quad \text{m}^3$$

Let us take the maximum volume equals V = 35 m^3. The requested maximal diameter is:

$$d = 2\sqrt[3]{\frac{3V}{4\pi}} = 2\sqrt[3]{\frac{3 \cdot 35}{4\pi}} \approx 4 \quad \text{m} \quad (13.6)$$

The calculated Venus air balloon is shown in Fig. 13.3.

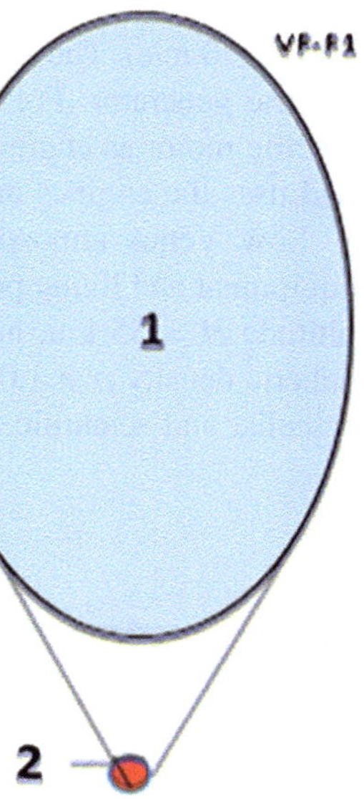

Fig. 13.3 Computed Venus air balloon. Notations: 1 – air balloon, 2 – scientific devices

13.6.1.2 Estimation and Computation the Aircraft for Venus

The project of a conventional aircraft for Venus is shown in Fig. 13.1 (right side), but possibilities are very limited. The flight time is short because the electric battery has not enough energy. The aircraft cannot fly for a long time over over a large are to permanently observe the interesting Venus location (for example an active volcano). One cannot recharge the battery.

The author offers the new idea – the temporary tethered aircraft which can fly for a longer time, like a helicopter and re-charge its vital battery many times. The aircraft clings to the surface and his propeller works as windmills and charge the battery. The sketch of the offered tethered aircraft is shown in Fig. 13.4.

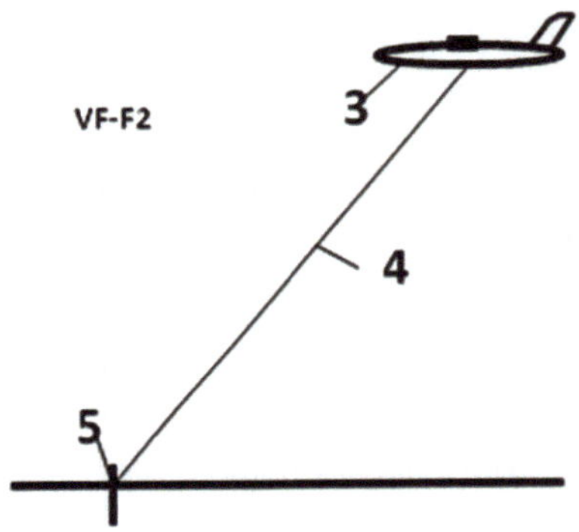

Fig. 13.4 Tethered aircraft used the wind. Notations: 3 is aircraft; 4 is a thin cable; 5 is an anchor

The Venus (like the Earth) has a permanently strong wind in a high altitude (up 100 ÷ 250 m/s). The aircraft electric propulsion has two regimes: electric motor and electric generator. For flight from one place to other the offered aircraft uses the electric motor as engine, for charging the battery it connects to surface (by anchor) and uses the engines and propellers as electric generator.

Low Venus atmosphere has very high temperature for a sensitive electronic equipment and living people. That way we make an estimation and computation for altitude $H = 55$ km, having temperature about 27 °C (see Table 13.1) and atmospheric density $\rho = 1.067$ kg/m^3 (see Table 13.1). The offerred project suitable for scientist and scientific devises is presented in Fig. 13.5.

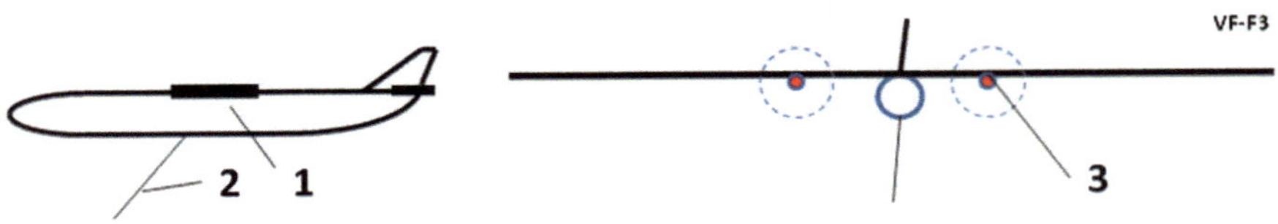

Fig. 13.5 Sketch of the Venus aircraft for scientists and scientific devices. Notations: 1 – aircraft, 2 – cable to anchor, 3 – engine: electric motor and electric generator

The lift force of the atmospheric flight apparatus is computed the equation:

$$L = C_L \frac{\rho V^2}{2} S \tag{13.7}$$

where C_L is lift force coefficient, ($C_L \approx 0 \div 2.5$); ρ is atmospheric density (see Table 13.1); V is speed (of flight or wind), m/s; S is wing area, m^2. The drag force is computed the equation:

$$D = C_D \frac{\rho V^2}{2} S \tag{13.8}$$

Here D is drag force, N; C_D is drag coefficient, ($C_D \approx 0.02 \div 0.5$).

Let us take the aircraft mass m = 100 tons = 10^5 kg, $C_L = 1$ and wind speed V = 100 m/s. The request wind area is (from Eq. 13.7)

$$S = \frac{2gm}{C_L \rho V^2} = \frac{2 \cdot 8.69 \cdot 10^5}{1 \cdot 1.067 \cdot 10^4} \approx 163 \quad \mathrm{m}^2 \tag{13.9}$$

Assume the typical wing aspect ratio is $\lambda = 10$. The average wing chord (width) equals:

$$b = \sqrt{\frac{S}{\lambda}} = \sqrt{\frac{163}{10}} = 4.1 \quad \mathrm{m} \tag{13.10}$$

where b is average wing chord, m. The wingspan equals

$$l = \frac{S}{b} = \frac{163}{4.1} \approx 40 \quad \mathrm{m} \tag{13.11}$$

Here l is the wingspan, m.

Windmills. The power of windmill may be calculated by equation

$$N = 0.5 \eta \rho A V^3 \tag{13.12}$$

where N is power, W; η is coefficient efficiency; ρ is atmospheric density, kg/m^3; A is propeller area, m^2; V is wind speed, m/s. The theoretical maximum of wind propeller efficiency is 0.67. In reality one is about 0.5. Drag and energy of windmill compute the equation

$$D = N/V, \quad N; \quad E = 8.33N, \quad kWh \tag{13.13}$$

Here D is wind propeller drag, N; E is wind energy, kWh.

Connection cable. The connection cable has a long length up 70 km. It must be made of a carbon artificial fiber having a tensile strength up 6000 MPa [1, Chap.1] and thermal resistance about 500 °C.

Let us to estimate the mass of connection cable having safe tensile stress $\sigma = 10^3$ N/mm^2 = 100 kg/mm^2 and maximal safe force $F = 200$ tons = 2×10^6 N. The gross section cable area is

$$s = \frac{F}{\delta} = \frac{2 \cdot 10^6}{10^4} \approx 200 \quad \text{mm}^2 \tag{13.14}$$

The volume of cable of length l_c = 70 km is

$$v = s \cdot l_c = 2 \cdot 10^{-4} \cdot 7 \cdot 10^4 = 14 \quad \text{m}^3 \tag{13.15}$$

Typical specific mass of artificial fiber is γ = 1800 kg/m^3. That means the mass of cable is

$$m_c = \gamma \cdot v = 1.8 \cdot 14 = 25.2 \quad \text{tons} \tag{13.16}$$

That is less the maximal mass of 200 tons.

Anchor of cable. Anchor is important element of tethered system. It must be strong enough to keep the cable and possibly release the cable if impossible to extract the anchor to next connection. Possible design of anchor is shown in Fig. 13.6.

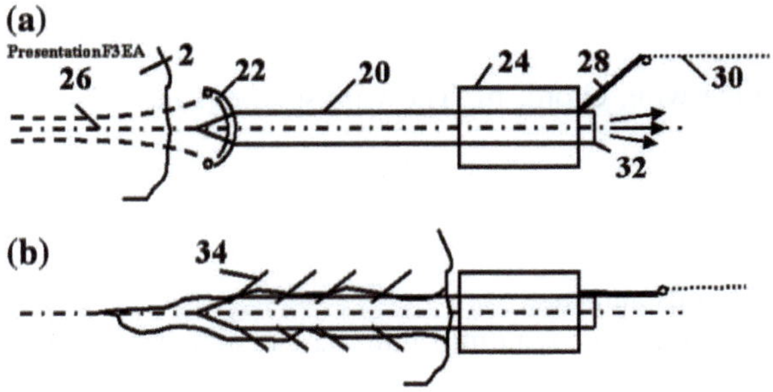

Fig. 13.6 (a) Anchor (harpoon fork). Notation: 2 – Venus surface, 20 – body of anchor, 22 – cumulative charge (shaped charge), 24 – rope spool, 26 – canal is made by shaped charge, 28 – rope keeper, 30 – rope, 32 – rocket impulse engine, which implants the anchor into the asteroid, 34 – anchor catchers. (b) Anchor connected to the asteroid

13.7 Conclusions

Offered equation allows to estimate the main parameters of atmospheric balloon, tethered aircraft and wind mill in Venus's atmosphere. They show that the tethered aircraft and air balloon are possible solutions and would allow us to get detail maps and information about Venus's surface, atmosphere and atmospheric flows.

Acknowledgments The author wishes to acknowledge Richard Cathcart (USA) for correcting the English and for offering useful advices and suggestions.

References

Bolonkin AA (2005) Non rocket space launch and flight. Elsevier, 488 pp. http://vixra.org/abs/1504.0011 v4, https://archive.org/details/Non-rocketSpaceLaunchAndFlightv.3

Bolonkin AA (2006) New concepts, ideas, innovations in aerospace, technology and the human sciences, nova, 2006, 510 pp. http://viXra.org/abs/1309.0193, https://independent.academia.edu/AlexanderBolonkin/Papers

Bolonkin AA (2011) Femtotechnologies and revolutionary projects. Lambert, USA, 538 pp. http://viXra.org/abs/1309.0191, http://www.archive.org/details/FemtotechnologiesAndRevolutionary-Projects

Bolonkin AA (2013) Innovations and new technologies. 30/7/2013, 309 pp. http://viXra.org/abs/1307.0169, https://www.academia.edu/11058394

Bolonkin AA (2009) Review of new ideas, innovations of non-rocket propulsion systems for space launch and flight (part 1). http://intellectualarchive.com #1368, http://www.archive.org/details/ReviewOfNewIdeasInnovationOfNonrocketPropulsionSystemsForSpace

Bolonkin AA (2009) Review of new ideas, innovations of non-rocket propulsion systems for space launch and flight (part 2). http://www.archive.org/details/Review2OfNewIdeasInnovations-OfNonrocketPropulsionSystemsForSpace

Bolonkin AA (2009) Review of f new ideas, innovations of non-rocket propulsion systems for space launch and flight (part 3). http://intellectualarchive.com #1369, http://www.archive.org/details/Review3OfNewIdeasInnovationsOfNonRocketPropulsionSystemsForSpace

Bolonkin AA (2011) Air catapult transportation. Scribd, NY, USA. http://viXra.org/abs/1310.0065

Bolonkin AA (2012) Re-entry space apparatus to earth. http://viXra.org/abs/1212.0003

Wikipedia (2012) Planets of solar system. http://wikipedia.org

Chapter 14
Mercury, Venus and Titan

Sushruth Kamath, Jullian Rivera, Michael Garcia and Haym Benaroya

14.1 Introduction

The exploration of space has been at the forefront of scientific thought and discovery for the last half century. Beginning with the first Sputnik satellite in 1957 to the Curiosity landing and the other numerous ongoing missions of today, mankind is setting its sights away from Earth.

As more information is gathered and the realm of understanding reaches farther into the cosmos, the important question of "where can humans land next?" inevitably arises. After conquering the Moon and exploring Mars, there is much debate on the future travel destinations. Therefore, it becomes imperative to fully catalog every possible option and ensure that all necessary data has been taken into account in defining the environment for which engineers must design habitats.

Speculation on futuristic and visionary possibilities for the exploration and settlement of the Solar System is a worthy intellectual exercise – it is fun, but it also helps us map current technological abilities into the near future. It is likely that when humanity decides to return in full force to space exploration and settlement it will be with a return to the Moon and then to Mars. This paper summarizes some key data that are needed by engineers on the environment of three bodies in the Solar System in order to place robots or humans on those bodies. The bodies

S. Kamath · J. Rivera · M. Garcia · H. Benaroya (✉)
Rutgers University, Piscataway, NJ, USA
e-mail: benaroya@rci.rutgers.edu

S. Kamath
e-mail: sushruth_kamath@comcast.net

J. Rivera
e-mail: nailluj93@gmail.com

M. Garcia
e-mail: michagar@scarletmail.rutgers.edu

V. Badescu and K. Zacny (eds.), *Inner Solar System*,
DOI 10.1007/978-3-319-19569-8_14

considered here are Mercury, with the orbit that is closest to the Sun, Venus, the planet that is closest in size to Earth, and Titan, Saturn's largest moon.

These three bodies in our Solar System are very different from each other as well as from Earth, its Moon and Mars, bodies with which we are most familiar. Mercury, Venus and Titan are three extremes in our Solar System. They offer challenges that we are not ready to currently undertake. Studying their environments and considering challenges that have to be overcome is necessary to explore them with robots and people. In essence, we are thinking out of the box with the hope that new ideas may be generated in efforts to explore and settle the Moon and Mars, and eventually the three bodies that are our focus here – Mercury, Venus and Titan.

14.2 Mercury

14.2.1 Introduction

One such, perhaps surprising, possibility is the planet Mercury. The small, first planet in the Solar System has largely been ignored by the countless missions and satellites from Earth. Yet collecting and analyzing existing and new data paints a picture of a planet that may be able to harbor a colonial base for future endeavors. As of now, it is incredibly difficult and dangerous to reach Mercury, but the plethora of resources and knowledge that can be gained from traveling to the small planet warrants its study and consideration for a manned mission.

The first step is to fully characterize Mercury's environment in order to understand its essence and nuances so that we can determine whether a settlement can be designed for and placed on the planet.

14.2.2 Missions to Mercury

Mercury has long been analyzed from Earth through various telescopes and ground based observation platforms. Being the smallest and fastest orbiting planet makes Mercury very difficult to study. Another challenge arises in the position of the planet's orbit compared to that of the Earth. Mercury's proximity to the Sun and distance from the Earth generally hampers man-made satellites and missions to orbit the planet and to observe it from close range.

So far, only two spacecraft have visited Mercury. The first was the limited Mariner 10. Launched on 3 November 1973, Mariner was designed to visit both the inner planets of Venus and Mercury. The vessel carried a basic array of instruments including TV and photographic cameras, airglow and occultation ultraviolet spectrometers, a charged particle telescope, an infrared radiometer, and magnetometers. Although the craft ran into technical difficulties reaching and passing Venus, Mariner 10 reached Mercury on March 29, 1974 for the first of three passes (*Mariner 10,* space.com). Mariner 10 and its orbital path are shown in Fig. 14.1.

Mariner 10 provided first images of Mercury and the on board equipment gave scientists much-needed data to begin to understand the planet. Mariner 10 carried very simple instruments and soon it ran out of fuel. On March 24, 1975, the craft turned off its transmitters and has not been heard from since. The data from the vessel was the only information on Mercury for more than three decades, until the planet was visited again in 2008.

On 3 August 2004, NASA launched the MErcury Surface, Space ENvironment, GEochemistry, and Ranging spacecraft. MESSENGER was a state-of-the-art vessel carrying an impressive payload of scientific equipment. The instruments included dual imaging systems, gamma-ray and neutron imaging systems, a laser altimeter, an atmospheric and surface composition spectrometer, and an energetic particle and plasma spectrometer. After using Venus as a gravitational slingshot, MESSENGER arrived at Mercury on 14 January 2008. The spacecraft and its orbital path are shown in Fig. 14.2.

Within days of arriving, MESSENGER had mapped 99 % of the planet's surface and had eclipsed the data total from Mariner 10. MESSENGER was inserted into Mercurian orbit on 18 March 2011 and continues to beam back torrents of data. The majority of what we know about Mercury today comes from the work MESSENGER has done (Bedini et al. 2012).

14.2.3 Interior Planet Structure

The European Space Agency (ESA) and the Japan Aerospace Exploration Agency (JAXA) are currently in the process of planning and building the third mission to Mercury. Named BepiColombo and set for launch in mid-2016, this craft will add significantly to our understanding of the innermost planet of the Solar System (*BepiColombo,* European Space Agency). Next, we review all the newly MESSENGER-discovered facts and examine whether or not a landing on Mercury is feasible.

In order to characterize the first planet fully, a suitable starting point is Mercury's interior and core. Using basic readings from MESSENGER and from ground based observations of rotational time and speed, the average density of Mercury has been found to be 5.427 g/cm^3. This means that Mercury is the second densest planet, second to Earth whose density is 5.51 g/cm^3.

The reason for the high density of the smallest planet perplexed scientists who were modeling its interior (Cameron et al. 1988). However, after analyzing MESSENGER's gamma-ray spectrometers, magnetometers and x-rays, it was determined with a high certainty that Mercury is composed of a large molten iron core that is nearly 3/4 the diameter of the planet. The rest of the interior is composed of a 600 km thick silicate and basalt mantle and a 200 km thick crust made of the same composition (Rivoldini and Van Hoolst 2013). In addition, there appears to be a layer of solid iron sulfide lying between the core and the mantle (*Lunar and Planetary Science Institute*, NASA).

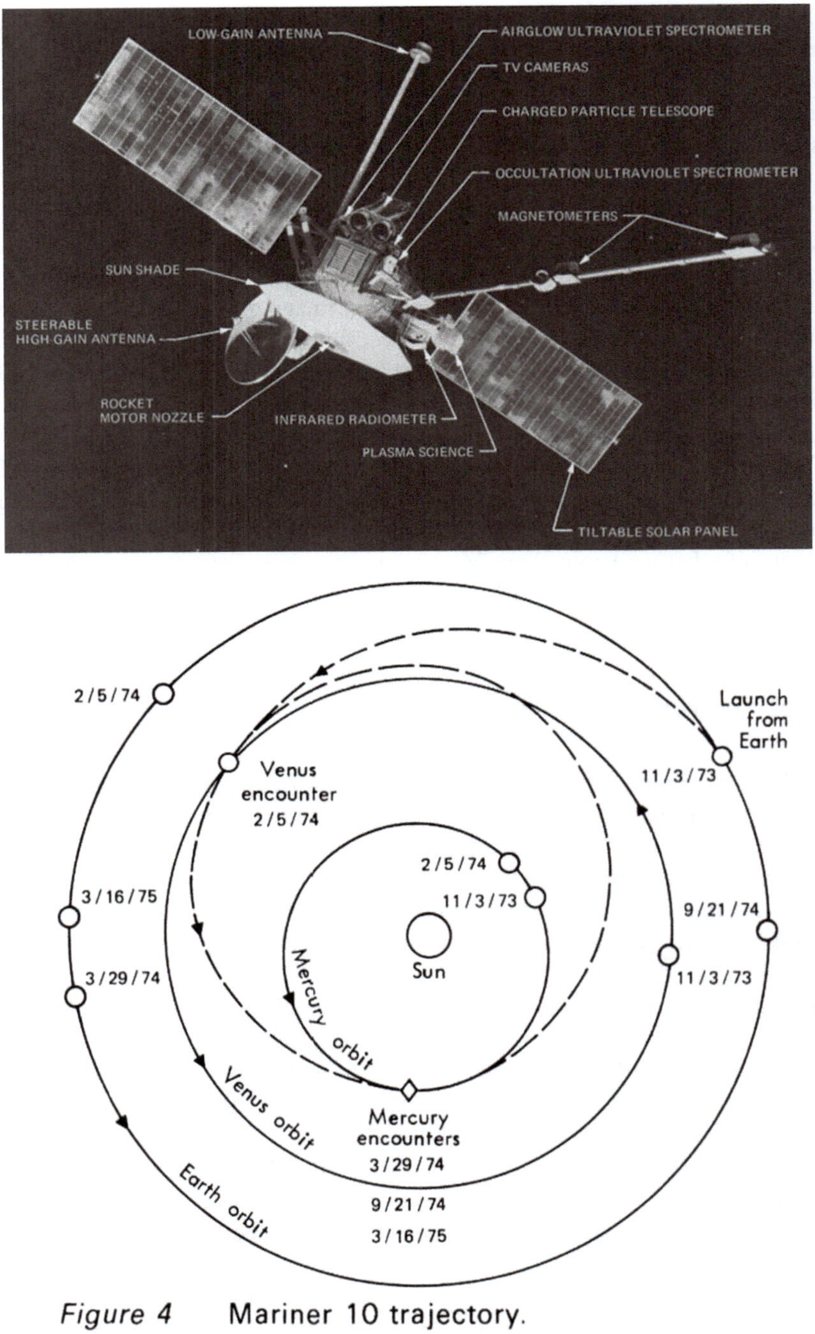

Fig. 14.1 Mariner 10 and its trajectory. NASA images

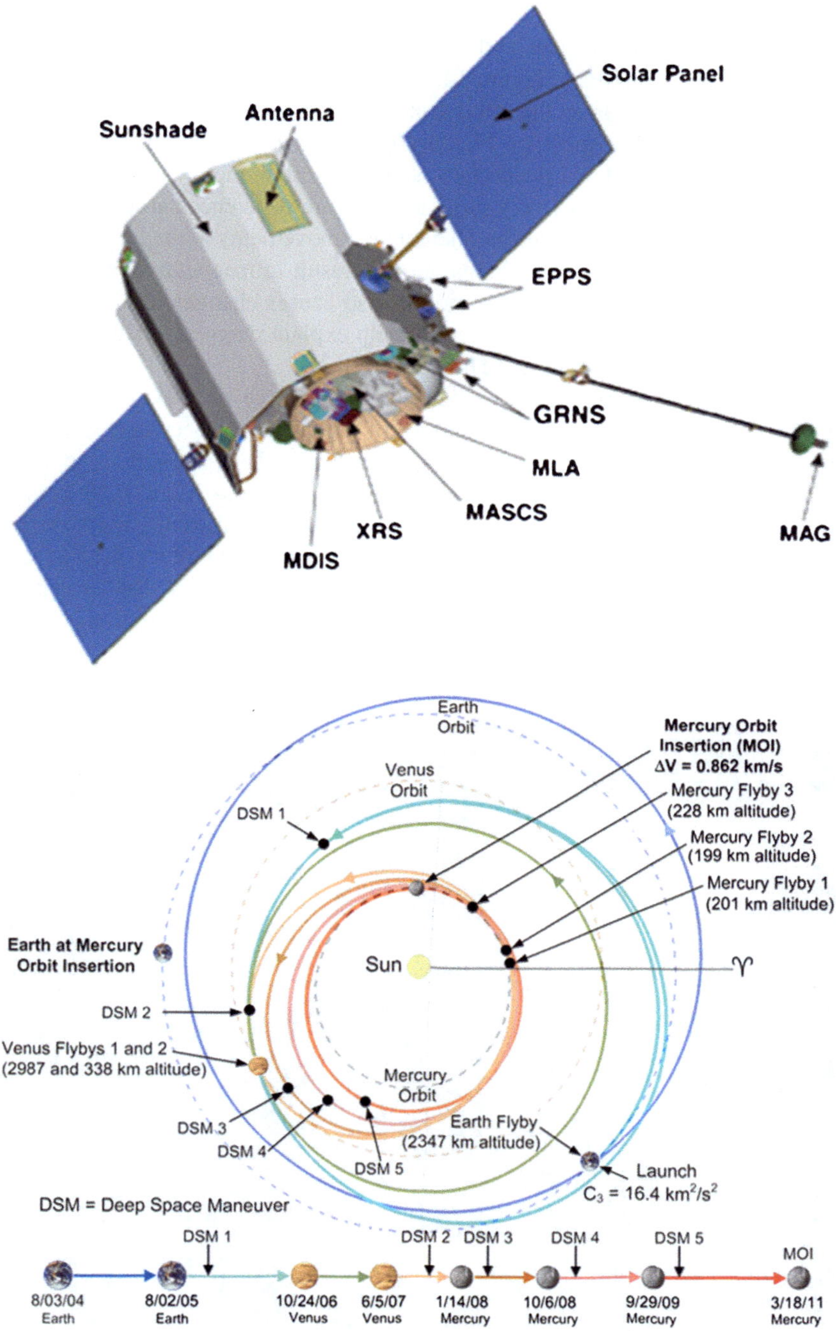

Fig. 14.2 MESSENGER and its trajectory. NASA images

The composition and structure of the inside of Mercury has a significant impact on the way the planet behaves and also dictates the planning and preparations of any future landing missions. Figure 14.3 shows the structure of Mercury's interior along with a comparison to the cross section of the Earth.

Another important aspect of Mercury's core that must be understood is its proclivity to be affected by outside factors. The first such factor is temperature. The proximity to the Sun causes very large temperature differences around the planet. This in turn causes heat currents and heightened movements in the large liquid core that not only affect surface conditions but also result in magnetic field fluctuations (Solomon 2003). Increased internal motion also causes changes in planet libration dynamics, which, if studied further, could help explain magnetic field strength and rotational properties (Koning and Dumberry 2013).

Another influence on Mercury's interior is the force of gravity (Smith et al. 2010). The gravitational pull on the surface of the planet has been determined to be around 0.38 g. The reason the planet's gravitational field strength is more than double that of the Moon's 0.16 g, and roughly the same as Mars' 0.38 g, is that, despite its small size, Mercury's high density iron core is inherently massive (*Solar System Exploration,* NASA). How this relatively high gravity affects the liquid core of a rocky inner planet is not entirely known. Scientists expect that the data sent daily by MESSENGER (Genova et al. 2013) will address this question.

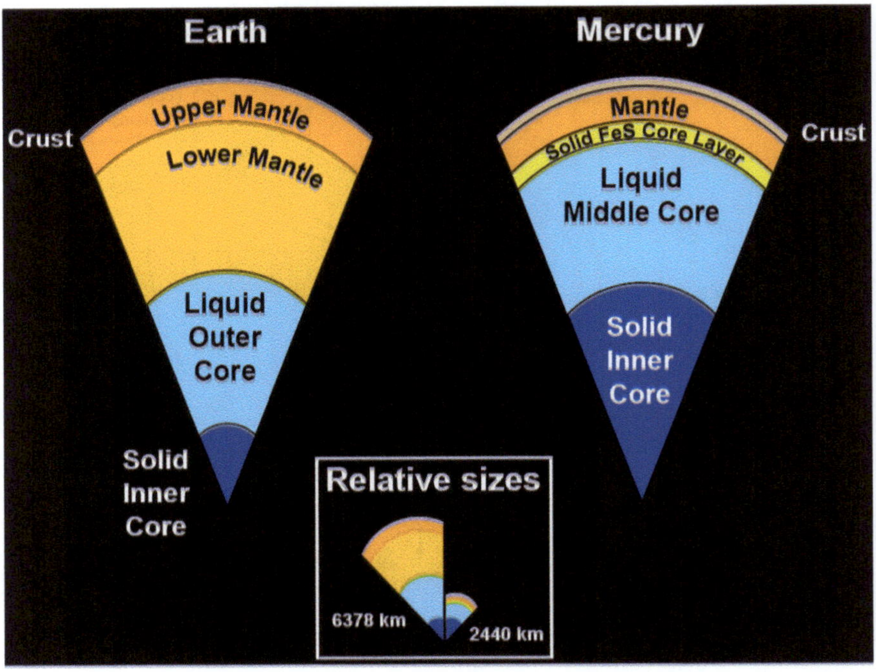

Fig. 14.3 Comparison of Earth and Mercury interiors. NASA image

Due to its close proximity to the Sun, less than 0.47 AU, Mercury experiences some of the strongest tidal forces in the Solar System (Solomon 2003). As opposed to oceanic bulges caused by the Moon on Earth, the core and surface of Mercury show significant protuberance due to solar gravity. It was long thought that Mercury was tidally locked with the Sun; however, careful ground based observations and calculations led to the discovery and confirmation of the planet's stable 3:2 spin-orbit resonance and tidal bulge properties (Solomon 2003).

Another external factor that may have had serious effects on the planet's core is debris impact. It has been speculated that medium to large asteroid impacts on the inner planets could potentially cripple core dynamos (Arkani-Hamed and Ghods 2011). A dynamo is a mechanism by which a celestial body such as Earth or a star generates a magnetic field. The asteroid impact theory applies particularly to Mercury since the planet has a small crust/mantle and a reasonably large liquid core. Analyzing the antipodal region of the large Caloris Basin on Mercury illustrates that severe impacts may affect both the core and the rest of the planet's surface. MESSENGER's images of the Caloris Basin (left) and the region known as the Weird Terrain (right), antipodal to the basin, are shown in Fig. 14.4.

An advanced mission to land on Mercury may be able to tap the core of the planet for resources, but also exposure to hazards. The character of the interior of Mercury plays a large role in how we might plan a landing on the planet. One could speculate that in a not-near future the core could be mined for metals provided the crust is thin enough for our drilling technologies. The vast iron and sulfur sources could be a valuable asset to settlement and exploration. In addition, the core's intrinsic heat currents could provide geothermal power for a base on the surface. The motion of the liquid core ensures a magnetic field around the planet, which provides minimal safety from the extreme radiation delivered via the solar wind.

However, as discussed earlier, the core may not be stable since it is possible that tidal forces and repeated impacts may disrupt the core significantly enough to have destructive consequences.

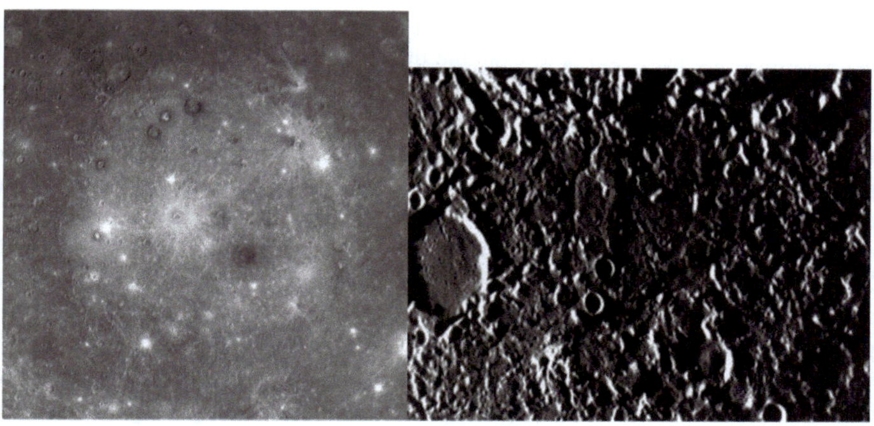

Fig. 14.4 (L) Caloris basin and (R) weird terrain. NASA images

14.2.4 Planetary Surface Structure

In order to have a successful touchdown on Mercury, a thorough and exhaustive analysis of the planet's surface must be done. Before Mariner 10, countless ground based systems had predicted that the surface of Mercury would be very similar to that of the Moon. In 1974, this notion was proven right. Early images from Mariner 10 showed similar characteristics between the Mercurian landscape and that of the Moon (*Mariner 10,* space.com).

Following the more advanced and careful screening done by MESSENGER, the composition and structures of Mercury were identified and explained (*MESSENGER,* NASA/JHUAPL). The surface of the first planet is composed of silicates and basalts similar to those found on other terrestrial planets (Carli and Sgavetti 2011). However, the notable difference arises in the formation and maturity of this sand-like substance.

Through MESSENGER's spectrometers, the color of Mercury's surface seems to be more uniform and not as dark as the surface on Mars or the Moon. This indicates an absence of iron and titanium rich silicates, as these materials would have created much darker surface planes and crevices (Blewett et al. 2009). The data received also indicates low quantities of iron oxides, which presents the theory that Mercury was formed in conditions without oxygen, or conditions in which oxygen was stripped away (Elser et al. 2012).

The surprising conclusion is that Mercury's surface contains very little iron-based material when the planet has a large iron core (Izenberg et al. 2014). This may be due to the fact that the Mercurian surface is more "mature" than those found elsewhere (Wasson 1988). Since the crust of the planet is exposed to multiple harsh influences, including continual extreme radiation, superheating, and supercooling, the composition is considered to have numerous unknown and unique characteristics, hence the term "mature" (Rhodes et al. 2011). MESSENGER's multi-spectral images of planet Mercury's surface were used to indicate possible compositions as shown in Fig. 14.5.

After understanding the current data on the composition of the innermost planet's surface, the next step is to examine the geologic structures present. Much of the knowledge about the formations on Mercury comes from carefully analyzing the images taken not only from Mariner 10 and MESSENGER, but also from Earth-based instruments (Ksanfomality 2008). The most obvious features on the planet's exterior are the lunar mare-like planes and impact craters. Photographic evidence also shows volcanic activity and vents on the surface of the planet.

The mares were likely formed by cooling volcanic eruptions (Rothery et al. 2014). The craters serve as an historic record of numerous debris impacts. As there are very limited atmospheric effects on Mercury, asteroids do not get destroyed prior to impact, as on Earth, and craters are not erased by fluid dynamics. Thus, if volcanic ejecta cools on the surface and is not subsequently hit, the only means of affecting its composition is through solar heat, radiation and gravity (Xiao et al. 2014). In addition, the vent-like structures give rise to questions about potential gas

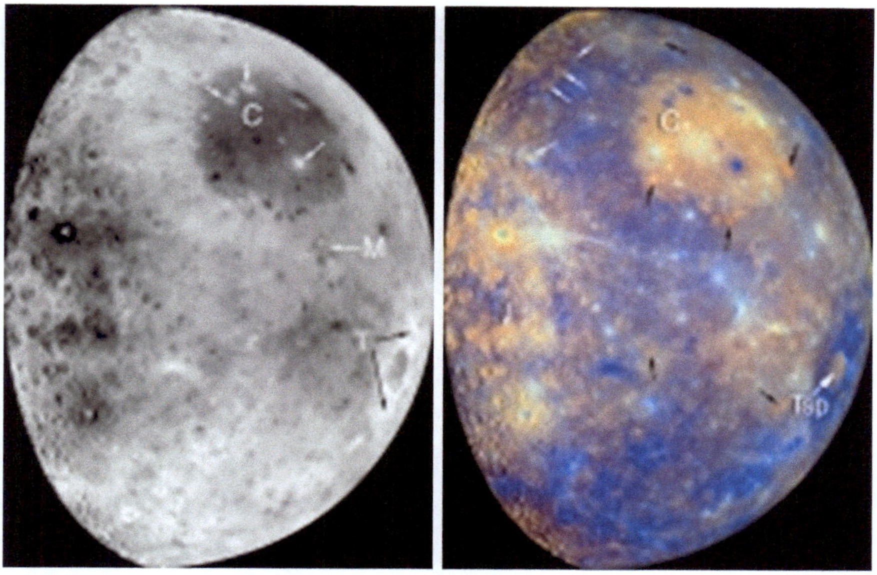

Fig. 14.5 MESSENGER's multi-spectral images. NASA/JHUAPL images

deposits and the existence of a significant pseudo-atmosphere (Xiao and Komatsu 2013).

MESSENGER's images of the planes and craters on Mercury are shown in Fig. 14.6, and of possible gaseous vents on the surface in Fig. 14.7.

The most surprising surface structures found on Mercury were several lobate (curved scalloped edges) scarps (significant vertical landform), or rupes. These winding cracks in the facade of the planet strongly indicate the presence of fault lines, and consequently, tectonic activity and movement (Watters et al. 2009). Therefore, the possibilities of an unstable crust and earthquake-like phenomena are heightened. In addition, the scarps indicate areas of high fault pressure where surface structures are destroyed or fused depending on motion. Many rupes were only recently discovered in images from MESSENGER and, as more information is gathered, will better understood (Ruiz et al. 2012). The image shown in Fig. 14.8 is of a lobate scarp on the surface of Mercury.

The final and most important characteristic of the planet's exterior is the extreme temperature differences. Mercury was wrongly assumed to be tidally locked with the Sun for a long time, leading to the belief that only one side of the planet endured the intense solar heat, while the other 'dark' side would only be exposed to the bitter cold of space. The Sun's close proximity to Mercury and the planet's 3:2 spin-orbit resonance means that the side facing away from the Sun can experience temperatures as low as 80 K (−316 °F or −193 °C). Then, as the planet rotates, that side faces the Sun and is subjected to temperatures as high as 700 K (800 °F or 427 °C) (*Solar System Exploration*, NASA). Thus, surface materials face conditions

Fig. 14.6 Mercury planes. NASA image

Fig. 14.7 Mercury vents. NASA image

cold enough to almost freeze nitrogen and hot enough to melt lead in just a few rotations of the planet (Helbert et al. 2013).

The poles, however, experience a far less dramatic temperature differential due to their positions relative to the Sun. Thus, areas of relatively stable conditions exist around each pole. The temperatures in these zones are near the surface temperatures

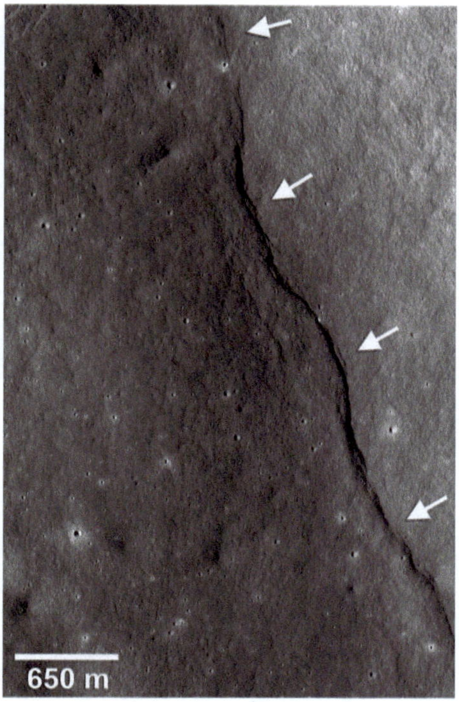

Fig. 14.8 Lobate scarp on Mercury. NASA image

found on Earth, albeit much less stable. Figure 14.9 shows the biannual maximum temperature (left) and average temperature (right) around the North Pole.

When planning travel to first planet, we should consider that the surface of Mercury can provide many natural resources as well as several potent dangers. One of the most significant finds on the planet's exterior is a substantial deposit of ice located just under the North Pole (Prockter 2005). Analyzing MESSENGER's x-ray and spectrometer data along with Earth-based radar readings and outputs revealed that large quantities of ice were present in Mercury's polar regions.

Evaluating surface maps showed how large, deep craters had provided the necessary protection from the heat of the Sun to harbor pockets of ice (*Lunar and Planetary Science*, NASA). Figure 14.10 shows the surface map of the North Pole with the red overlay indicating areas of lowest illumination, that is, most shadowed, and the yellow overlay indicating detected pockets of ice.

Although the surveying of the pole is far from complete, if the ice there is deemed usable and harvestable, it would be an enormously crucial asset for manned missions to Mercury (Vasavada et al. 1999). In addition, the regolith can be used as building blocks for construction or as protection from destructive radiation and from meteorites.

The heat and light from the Sun can provide unlimited solar and thermal power for the entire planet, assuming the solar panels can be prevented from vaporizing almost instantaneously. On the other hand, the surface poses its own set of dangers,

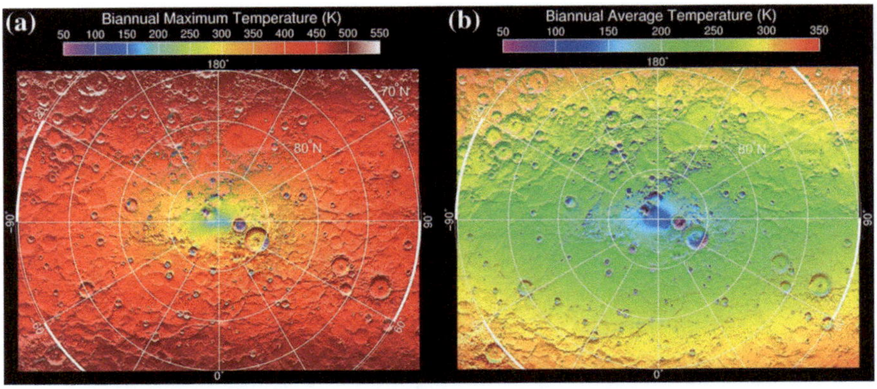

Fig. 14.9 Biannual Mercurian pole temperature. NASA/JHUAPL images

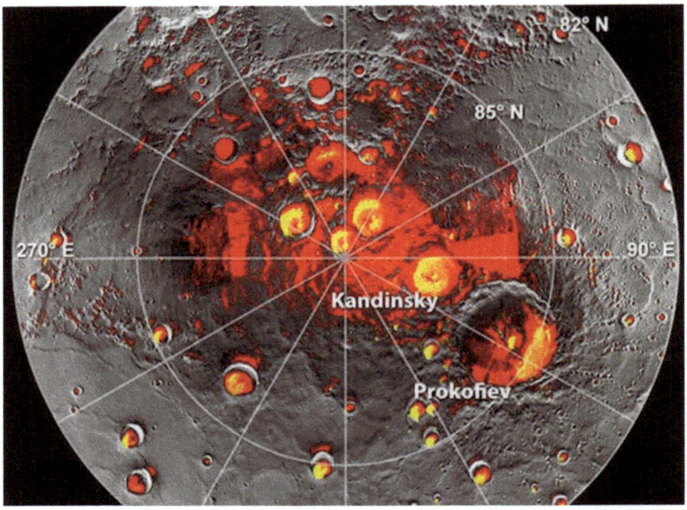

Fig. 14.10 Mercury polar ice overlay. NASA image

as well. The vast temperature differences that are either too hot or too cold for habitation must be quarantined from humans or robots. Secondly, asteroid impacts pose a major threat to colonies or any structures on the planet.

14.2.5 Planetary Atmospheric Conditions

Mercury hosts a thin atmosphere and a unique magnetic field, and understanding both is critical to any mission to visit the planet. The proportionally large molten iron core is strong enough to generate a magnetic field around the planet (Grosser et al. 2004). Measured to peak at approximately 300 nT, Mercury's field is only approximately 1.1 % the strength of that of the Earth (*Lunar and Planetary Science Institute*, NASA). However, Mercury's innermost position in the Solar System means that in order to protect the planet's surface from the harsh solar wind, it would need a magnetic field immensely more powerful.

Mercury's magnetic field has been carefully documented by MESSENGER's sensors (McNutt et al. 2010). After passing through the bow shock, particles hit Mercury's magneto sheath. Particles not deflected away are then met with the planet's magnetopause, which is structured much like that of the Earth's, with polar cusps, lobes, and an intermediate plasma sheet (Hiremath 2012). Since the solar wind is very strong in the orbit of Mercury, most of the Sun's radiation passes through the magnetosphere and bombards the surface, possibly changing surface composition characteristics.

There is much to be learned from Mercury's electromagnetic dynamics (Anderson et al. 2011). Mercury is the only inner planet, other than Earth, to have a substantial inherent magnetosphere. In addition, it is imperative to uncover how the instability of the core and the intense heat currents affect the production and strength of the magnetic field. The plasma environment near the planet may provide valuable insights on the effects of solar wind (Raines et al. 2011).

Figure 14.11 shows a concise diagram of Mercury's magnetic field components. Figure 14.12 shows the measured strength of the magnetosphere of Mercury from the proximity of the northern half of the planet.

Within Mercury's magnetic field is a tenuous and little studied atmosphere (Baker et al. 2011). Mercury might have an exceptionally high density, but this does not make up for its small volumetric size. As such, Mercury is simply not massive enough to harbor a large, developed atmosphere. In addition, the high temperatures on the planet expel gasses and prevent the planet from retaining an atmosphere (Wasson, *Building Stones*).

It was long thought that Mercury had no gaseous upper layer at all; however, data from Mariner 10 and also from MESSENGER proved the existence of both a weak, unstable atmosphere and metal rich "tails" (McNutt et al. 2014).

The planet's atmosphere, barely more than a flimsy sheet around the surface, is composed of 42 % molecular oxygen, 29 % sodium, 22 % hydrogen, 6 % Helium, and 0.5 % potassium. The other 0.5 % is comprised of trace amounts of argon, nitrogen, and various other elements as well as carbon dioxide and water vapor (*Lunar and Planetary Science*, NASA). This gaseous layer is not considered to be stable. In other words, the various molecules in the atmosphere are continually lost and replaced via various processes.

For example, hydrogen and helium are thought to be remnants of solar wind impacts on the magnetosphere. The atoms are trapped for a short while by the magnetic field before escaping and being substituted by newer arrivals. Sodium, potassium, and other metals are most likely emitted from the surface through the effects of radiation on the planetary regolith, through the ejecta of unending asteroid

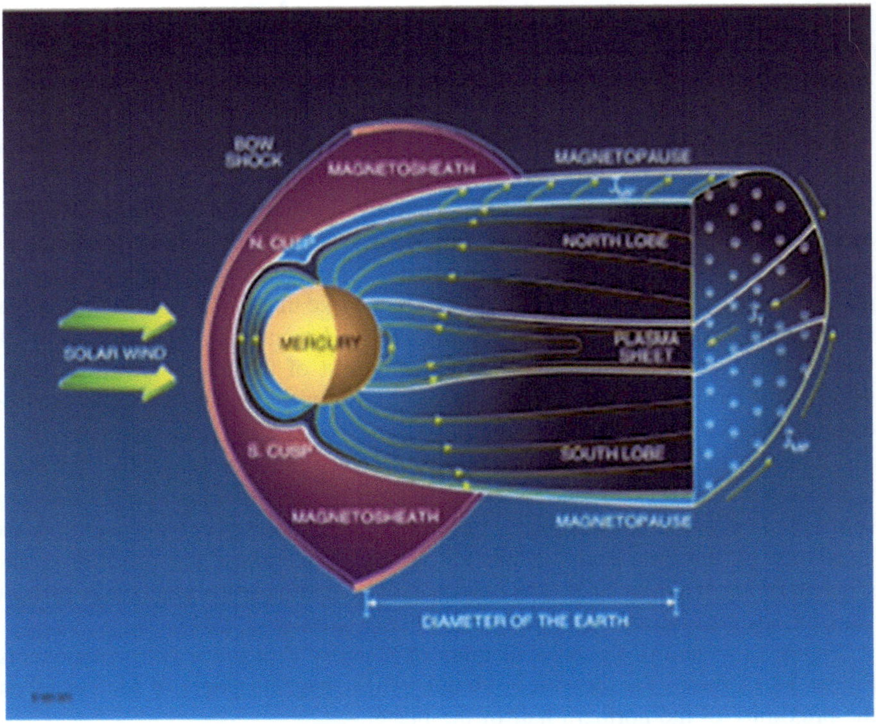

Fig. 14.11 Magnetic field structure. NASA/JHUAPL images

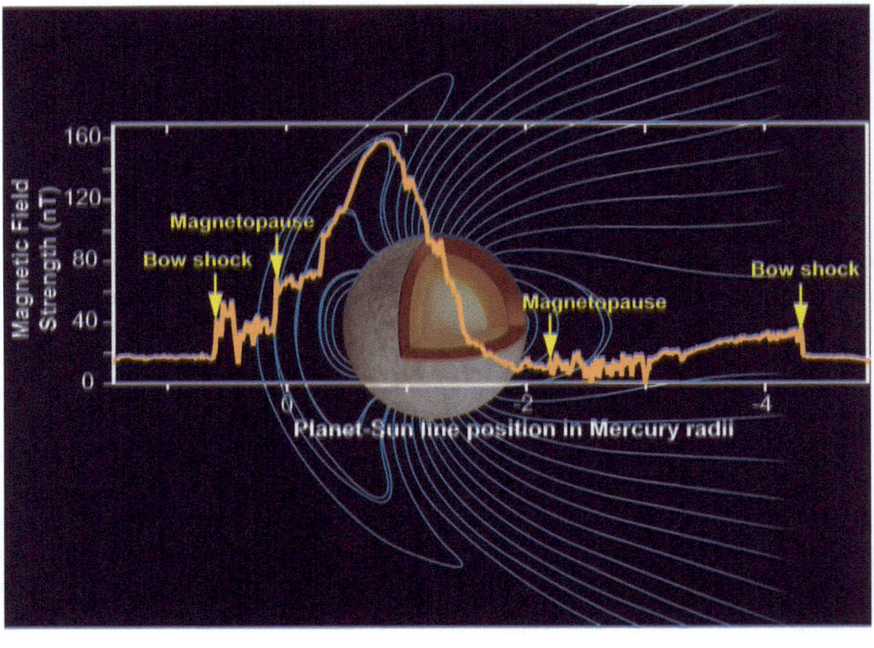

Fig. 14.12 Magnetic field strength. NASA/JHUAPL images

impacts, or through the photographed vents on the surface (Wasson, *Building Stones*).

Thus, as opposed to the stable and contained atmosphere on Earth, Mercury does not have a strong enough gravitational force to retain gasses. Instead, it is always replenishing its dwindling gaseous layer. In other words, Mercury possesses an exosphere rather than an atmosphere (*Solar System Exploration*, NASA).

MESSENGER and ground based systems also discovered "tails" behind the planet. As solar wind passes any planet, it picks up elements of the surface, atmosphere, and other loose components; these can be detected and analyzed to better understand planetary characteristics. The most important and valuable tails are those of the element sodium (Leblanc and Johnson 2003). Thus, the detected sodium tail, along with the notable magnesium tail and calcium tail, from MESSENGER's on board spectrometers are shown below in Fig. 14.13.

While some of Mercury's resources lie in the magnetosphere and the atmosphere/exosphere, these areas are also the most hazardous. The magnetic field of the planet provides a layer of security for the surface by deflecting some harmful radiation away. In addition, the atmosphere provides the natural gas resources of sodium, oxygen, and hydrogen.

More research must be done in order to determine whether all the detected elements are from the solar wind or from the surface via asteroids and vents. Also, more data and information is needed on possible ways of harvesting these transient gasses (Schmidt et al. 2009).

Unfortunately, Mercury's magnetic field is extremely leaky. The sheer amount of solar emission affecting the planet causes much radiation to penetrate the field and directly impact the surface. In many cases, vortices of radioactivity, almost like tornadoes of radiation, have been detected ravaging Mercury's surface. Further investigation into the full effects of the continual solar wind battering is necessary to better understand this phenomenon.

As for the atmosphere, while it does provide gaseous elemental resources and allows for much of the planet's heat to dissipate without being trapped by greenhouse effects, it is highly unstable and does little to protect the surface from asteroid impacts. On Earth, much of the debris heading for the planet is burned up in the upper atmosphere. The thin gaseous layer in Mercury does not provide any protection from incoming asteroids. Hence, strikes on the surface of Mercury are a very common and dangerous occurrence.

In summary, although Mercury is one of the few reachable entities with its own magnetic field and tenuous, but existent, atmosphere, landing on the planet does entail serious hazards from the dearth of intrinsic planetary surface protection.

14.2.6 Discussion and Conclusions - Mercury

Mercury is an extraordinary planet. As the innermost solid body orbiting the Sun, Mercury experiences the harshest conditions. From extreme cold to incredible heat

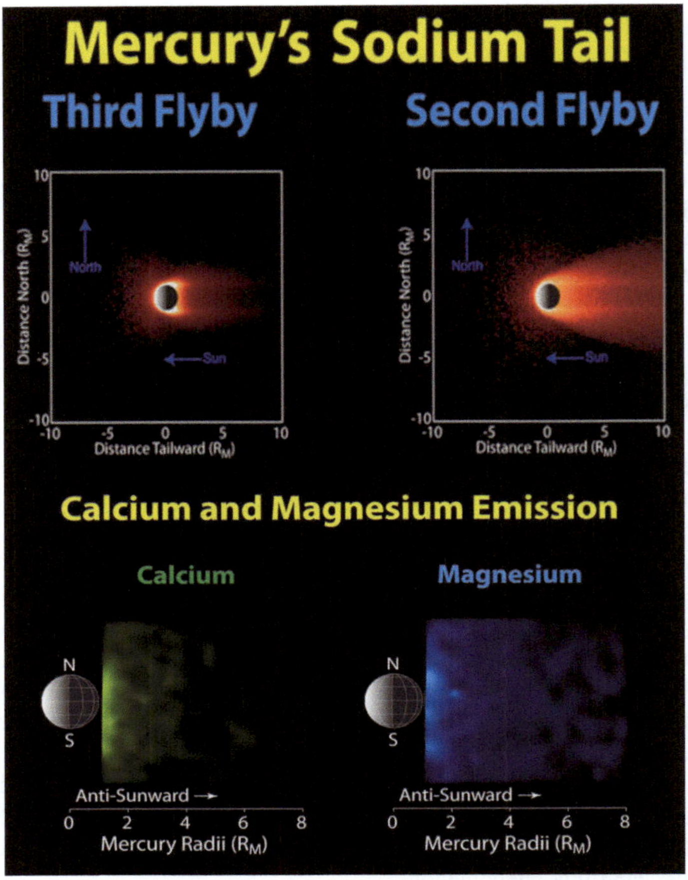

Fig. 14.13 Mercury tails and emissions. NASA images

(80–700 K), a consistent and severe solar radiation and unending asteroid impacts, the small planet weathers it all.

From a scientific standpoint, Mercury presents the perfect example of an accessible planet from which we can learn much about the creation of our Solar System.

It is clear that more research is needed to better understand the numerous hazards of travelling so near to the Sun and to such a barren land. The upcoming ESA and JAXA joint mission of BepiColombo likely will shed more light on these mysteries. BepiColombo, set for arrival at Mercury in early 2024, will consist of two spacecraft: the Mercury Planet Orbiter (MPO) and the Mercury Magnetospheric Orbiter (MMO). The MPO will scrutinize the surface and crust of the planet in order to better define its properties (Rothery et al. 2010). The MMO will study the magnetic fields and atmosphere of Mercury in order to analyze the properties of such a field and its interactions with solar wind. With the completion of BepiColombo's one-year mission, Mercury will be much better understood, and one day this knowledge will allow us to plan robotic and human travel to it, possibly with even a manned landing.

There would be many unique advantages of creating a base on Mercury. Proximity to the Sun is the most significant advantage. The technology to reach the planet exists and has been implemented. As advances in low thrust propulsion methods are made, novel orbits of the planet and missions carrying more loads and heavier cargo can be designed for landings on the planet (Anderson et al. 2014).

Once on the planet's surface, the regolith could be used to build shelters and protective shells for permanent structures. Another option is to construct underground laboratories. These would use the natural defenses of the regolith and also potentially enable access to not only the aforementioned ice, but also to metals and various resources of the core. In addition, a successful base on Mercury would allow departing vessels to use the enormous solar gravity well as a slingshot to further destinations. Clearly, vast arrays of incredible assets are untapped.

As with any planetary body, Mercury presents serious hazards. The instability of the core and crust due to asteroid impacts, tidal bulging, and even heat currents must be fully understood and prepared for. The surface temperature variations pose many challenges to survival. Furthermore, the possible tectonic activity must be studied and then mapped, and the mountains and vents thoroughly explored to ensure no volatile ejections.

Most importantly, the leaky magnetic field that allows violent radioactive storms on the planet's surface must be mitigated, as any landing would need sufficient protection from the Sun's deadly rays. If a manned mission is to be made, essentials for life must be accounted for. Food and water must be transported and stored on the planet. Oxygen for respiration must be harvested or brought in. Pressurized capsules must be brought in for the initial habitats.

Taking all of these issues into consideration, it seems that the most plausible next step after BepiColombo would be to send a landing craft or rover onto the surface of Mercury. Originally, the BepiColombo craft was to have carried a lander element. It was proposed that this "self-inserting 'mole' device" would execute an airbag-assisted soft landing before entering the Hermian regolith and being hammered down to a depth of 2–5 m (Spohn et al. 2001). The lander craft was eventually cut from the mission due to budget constraints; however, the ideas can still be utilized in the future.

A full topographical map of the surface must be constructed in order to visualize and comprehend the intricacies of a possible landing site (Oberst et al. 2011). Lastly, and most importantly, Mercury's crust may be unstable due to the effects of tidal bulging and the unknowns regarding the lobate scarps. Luckily, these mysteries might be solved with the ongoing and future research, eventually enabling access to the surface of the innermost planet.

Similar to the Mars Curiosity rover, a Mercurian unmanned exploration craft could be well protected from the harsh environment, could land using an airbag deployment method and could help us better understand and characterize surface conditions, paving the way for other landing missions and even human settlements.

It may be quite a few decades before a human sets foot on another planet, but the building blocks are being created upon the knowledge that MESSENGER is providing. With research and continued exploration, the next giant leap for mankind might be one small step onto the first planet, Mercury.

14.3 Venus

14.3.1 Introduction

Venus can be considered to be Earth's twin in the Solar System. However, the divergent evolution of the Venus environment has made it a scorched, lifeless rock with a toxic atmosphere. These harsh conditions pose many challenges for the planet's exploration, including temperatures (740 K) that melt electronic systems on the surface, the crushing pressures of 100 times those on Earth, and clouds of corrosive and poisonous sulfuric acid (Gao et al. 2014; Landis 2003; Williams 2014).

Here, we seek to characterize the Venus environment and outline the challenges that faced past missions to the planet and threaten the success of future endeavors. We also speculate on technological breakthroughs that could make the in-depth exploration of the planet more feasible. Lastly, we identify a region in the Venus environment that may be sufficiently Earth-like, to facilitate the construction and survival of a human settlement.

Before the beginning of the space exploration era, mankind observed the planets through ground-based telescopes, using analysis techniques to draw conclusions where raw data was unobtainable. From these observations, scientists discovered a world that outwardly looked very much like our own: Venus. Closest to Earth in orbital distance from the Sun, in radial size, in bulk mass, in bulk composition, and in gravitational pull, many considered this yellow planet covered in a thick carbon dioxide atmosphere to be Earth's twin sister. See Table 14.1. Naturally, there were those who wondered if the conditions were similar enough to also harbor life, but a cloud layer that covered the entire surface from view prevented Earth observers from finding an answer.

Table 14.1 Side by side comparison of bulk characteristics of the inner planets

	Earth	Venus	Moon	Mars
Mean distance from Sun	149,598,261 km	108,208,000 km	Same as Earth	227,939,100 km
Mean radial size	6371.0 km	6051.8 km	1737.1 km	3389.5 km
Gravity in Earth g's	1 g	9/10 g	1/6 g	1/3 g

14.3.2 Missions to Venus and Their Discoveries

To confirm the presence of other life in the Solar System, in situ missions to Venus are required. As the Space Race picked up speed and mankind forged its way into space, the Soviet Union selected Venus to be the site of humanity's first expedition to another planet.

Starting in 1961, the Soviet Space Program launched two attempts at flybys of Venus with the unmanned Sputnik 7 and Venera 1 spacecraft. Both missions failed, with the Venera 1 coming closest to Venus but losing communications with the Earth along the way (Williams 2014).

The second attempt by NASA in 1962 made its way successfully to Venus, communications intact. The unmanned Mariner 2 accomplished this feat, performing a flyby of the planet within 35,000 km (Williams 2014). While passing Venus on the way to Mars, Mariner 2 took several measurements of the conditions on Venus, confirming a carbon dioxide atmosphere, a thick cloud layer, and also a surface temperature reading of over 400 °C (Williams 2014). This last discovery of temperatures high enough to vaporize water and to melt lead killed hopes for life on Venus. Figure 14.14 shows some of the spacecraft that made it to Venus.

Exploration of the planet continued and new questions arose. Chief among these questions was how Venus evolved so differently from Earth over its lifetime and if knowledge of its past could give insights to the Earth's own future and planetary mechanisms (Squyres et al. 2011).

Soviets conducted most of the early exploration, sending dozens of flyby probes, orbiters, and landers to Venus over a 20-year period. It became increasingly apparent that Venus not only was lifeless but also bore extremely hostile conditions.

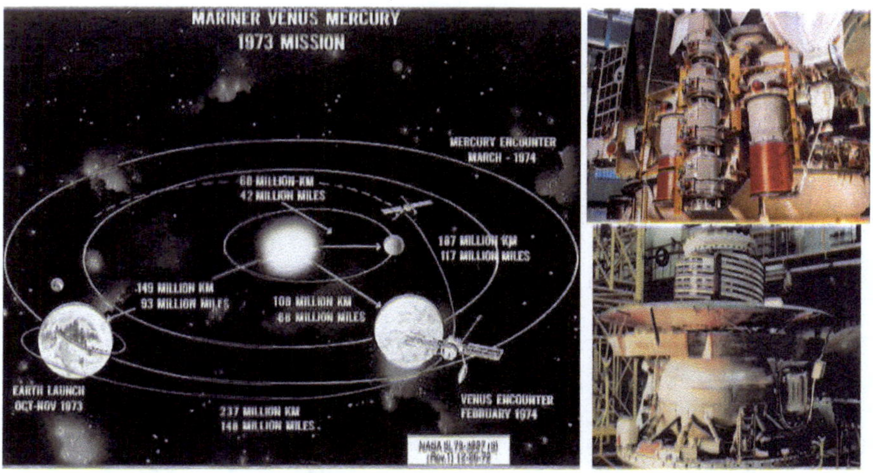

Fig. 14.14 (Left) 1973 Mariner mission to Venus, (Top Right) Soviet Vega probe, (Bottom Right) Soviet Vega lander. NASA images

Hard lessons were learned by surface probes that were destroyed, while descending through the atmosphere, due to the rising temperatures and pressures.

Even when probes survived the descent through the atmosphere, via engineering solutions such as the cooling of their interiors to sub-zero temperatures and their containment in pressurized vessels, each probe barely lasted an hour or two before all systems failed and communication was lost.

The Soviets eventually ended their Venera and Vega mission programs. Their last probes, and the last probe to touch the soil of Venus, were the Vega 1 and 2 in 1984 (Williams 2014). Venus has long been placed on the backburner of space exploration priorities. Only a few key missions have been planned for the next 30 years. These include the Magellan orbiter, the Pioneer Venus orbiter, the MESSENGER flyby from NASA, and the Venus Express orbiter from the European Space Agency. The Japanese space program JAXA will attempt an orbit of its AKATSUKI probe around Venus in 2015 (Williams 2014).

While those missions have returned most of the data we have today about Venus, they are limited by the lack of direct physical measurement of the environment. In order to fully understand the planet, a series of new in situ missions are needed.

Here, we summarize some of the information collected over the last 50 years of the direct exploration of Venus and from observations using Earth-based telescopes, to provide a characterization of the Venus environment for purposes of engineering a habitat.

We focus on defining the harsh conditions of the planet, as illustrated in Fig. 14.15, and the threats these pose to future missions. Ultimately, we speculate on future technology that might make overcoming the environment possible as well as a location in the Venusian atmosphere that might be suitable for human habitation.

Fig. 14.15 Artist's representation of the Venus environment. NASA image

14.3.3 The Surface of Venus

The surface of Venus has been described as a "hellish land" riddled with volcanoes and gigantic plains of lava rock. It experiences vaporizing temperatures of around 740 K (around 470 °C or 870 °F), much higher than Earth's hottest regions (Williams 2014). These temperatures vary little over time due to the oven-like effect created by Venus' cloud layer above, remaining intense regardless of night or day or season, depending more on elevation and surface composition than other factors (Cañon-Tapia 2014). Likewise, pressures of nearly 90 atm have been recorded (Williams 2014).

Past missions to the surface of Venus focused heavily on analyzing its composition and its geological history (Squyres et al. 2011). Primary issues that limited the amount of data that could be gathered included a lack of well-defined target locations, deployment and landing difficulties damaging sensors and drilling equipment, poor communications with the landers, and the limited time of operation associated with an environment blocked from the Sun and experiencing extreme temperatures/pressures (Williams et al. 2014). Measurements from different regions of the planet were recovered by the Russian Venera and Vega probes (Williams 2014).

The surface of Venus is composed primarily of mafic, or basaltic, rock, a result of millions of years of gradual lava flow (Basilevski et al. 2012). Volcanic plains of up to 400–500 m thickness produced by enormous, shield-like volcanoes that dwarf those on Earth cover about 70 % of the planet. The plains and volcanoes are categorized extensively based on formation, structure, and level of deformation by tectonics (Ivanov and Head 2013). However, though vast and diverse, these structures offer very little information about the underlying composition of Venus and its history, merely representing the years of covering up and erasing the planet experienced (Basilevski et al. 2012).

The remaining 30 % of the surface consists of either heavily tectonized volcanic plains or, more promisingly, areas of complex rock and lava flows that may hold a key to understanding the structure and evolution of Venus below the plains (Ivanov and Head 2013). Infrared and ultraviolet emissivity readings were used to identify these regions, which differ from those indicative of mafic plains. It was discovered that Venus possesses another type of volcanic structure other than shield-like domes. Steep slopes were found instead of gentle domes. This suggests slower, more viscous or foamy lava flows that could be attributed to more enriched lava containing large amounts of dissolved water or other chemical/mineral components.

Plateau-like formations called Tessera terrain were also found in these regions, believed to be composed of a material more geochemically differentiated from the mafic rocks, perhaps feldspathic silicates, intermediate rocks, or anorthosites (Basilevski et al. 2012). Some suggest though, that tesserae may just be more tectonically deformed rock made up of older materials possibly of volcanic origin (Ivanov and Head 2013).

Figure 14.16 is a map based on data from the Pioneer Venus Orbiter.

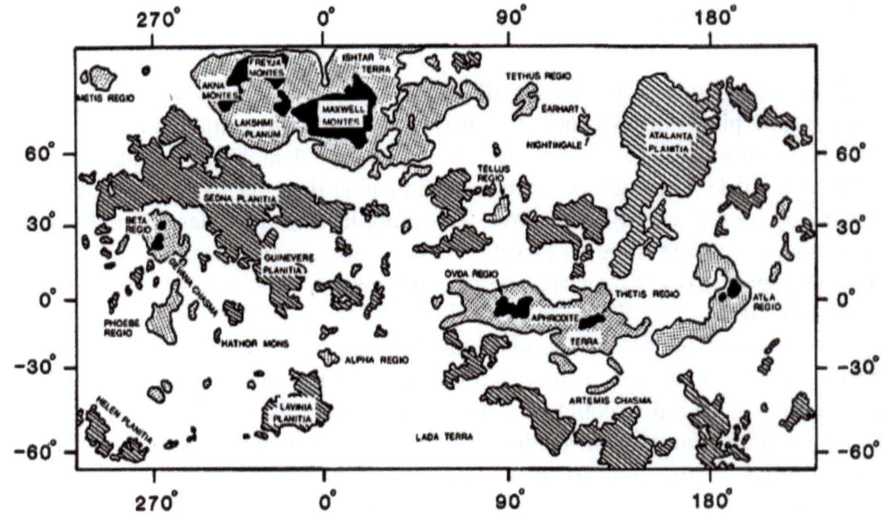

Fig. 14.16 Notable mountains, plateaus and other land formations as mapped by the Pioneer Venus orbiter. NASA image

In the early years of Venus exploration, it was speculated that Venus was a planet of extreme volcanic and tectonic activity due to the vastness of the volcanic plains. As more data was assembled and further studies were conducted, many clues were brought to light that allowed scientists to begin to piece together the geological and evolutionary history of Venus, clues that include relatively recent but minimal volcanic activity and a low number of impact craters visible on the surface (Romeo 2013).

Two theories for the geological history of the planet have been proposed based on the above information and that a large percentage of the surface is covered in lava formations. The first suggested that Venus underwent extensive volcanic activity over the years that both covered the surface and erased a large number of impact craters that otherwise would be visible. Moving forward in time, this activity gradually decreased to the minimal activity we see today.

The second theory proposes a more dynamic history that involved a "catastrophic resurfacing" of Venus several million years ago when a violent volcanic event covered the planet in lava flows (Romeo 2013). Following that event a sharp decline in volcanic activity ensued, a calming period that we experience still today. All of the geological landmarks and impact craters of the past were erased by the resurfacing and what we see today are only the most recent of markings (Romeo 2013). Many models were generated to simulate these two theories in an attempt to reproduce a scenario and surface similar to the one we see today of Venus. The present consensus based on all this information, supports the resurfacing theory.

Interestingly, it has been observed that the majority of active volcanoes are heavily concentrated on one hemisphere of the planet, suggesting that there may

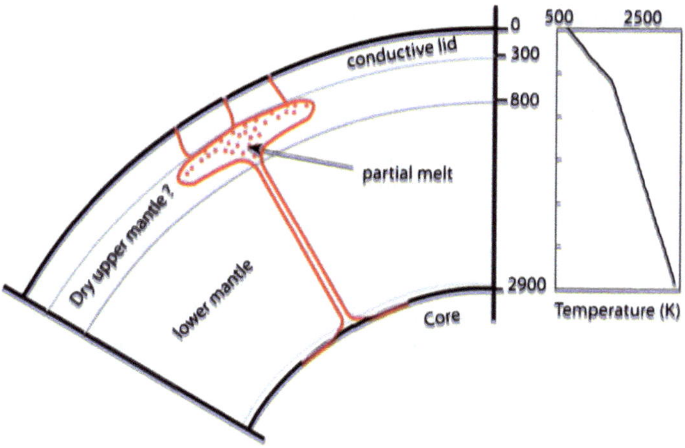

Fig. 14.17 Representation of Magma plume theory and inner structure of planet. NASA image

exist within the inner structure of Venus a plume-like magma process on this side of the planet that has found its way to the surface through a thinner layer of the lithosphere (Cañon-Tapia 2014). Figure 14.17 is a schematic of this possibility.

Besides volcanic activity, Venus also possesses signs of tectonic activity that shapes its surface. The processes by which this occurs remain largely unknown. There is no evidence of the plate mechanics that occur on Earth and it is believed that subterranean Venus is far less dynamic than subterranean Earth. The weak magnetic field of Venus is considered to be supporting this theory by comparison with Earth's strong magnetic field that generated by the polarization of its moving liquid mantle and core.

It is theorized that Venus once existed in an era dominated by tectonic activity much like that of Earth. This then subsided and the dynamics of the planetary insides slowed, giving way to a volcanism-dominated era (Ivanov and Head 2013). Mantle degassing and the disappearance of a magnetic field contribute to the "runaway" greenhouse conditions that exist today. Sulfur and carbon were spewed into the atmosphere, water was consumed by lava flows or evaporated away, the surface was covered, radiation bombarded the planet, hydrogen and oxygen were lost to the atmosphere, temperatures and pressures rose, and the clouds formed (Driscoll and Bercovici 2013). At the end of this era of great change, the final catastrophic resurfacing event occurred and reshaped the planet permanently, leaving behind a new, volcanic plain-covered, crater-free surface. This was the beginning of the era that continues today, of continuous but diminished volcanic and tectonic activity with a scorched surface and a highly toxic atmosphere (Ivanov and Head 2013).

14.3.4 The Lower Atmosphere of Venus (0–45 km Altitude)

The lower atmosphere has been poorly characterized and might be one of the biggest mysteries of Venus. Other than its general transition from surface to cloud level temperatures and pressures, we know very little of its composition and dynamics due to being masked from view by the thickness of the sulfur clouds above and the intensity of conditions at the surface (Squyres et al. 2011). Balloon probes have entered the clouds and landers have reached the surface. Orbiters and ground-based telescopes have observed the upper atmosphere and cloud tops. None have delved into the tiny region tucked between all those areas. Thus, the lower atmosphere of Venus is a high priority target for future in situ missions to Venus (Squyres et al. 2011).

What is known is that the lower atmosphere of Venus transitions quickly from temperatures of up to 750 K at the surface to about 400 K at the cloud bottoms at an altitude of 50 km. There is also a dip in pressure from under 100 atm at the surface to 1–2 atm at 50 km. The lower atmosphere is presumed to be windy, though it does not possess the super-rotational wind flows of the upper atmosphere, as will be discussed later. On the other hand, powerful bursts of wind are generated by the convection currents as air flows between regions differing in temperature by up to 350 K. These upwelling and down-welling gusts are largely responsible for volcanic outgassing components rising into the atmosphere and into the clouds (Imamura et al. 2014).

Compositionally, we know that over 95 % of the atmosphere is CO_2 and there is evidence of high CO concentrations (Cotton et al. 2012) and of a sulfur particle haze below the clouds at altitudes of around 45 km (Gao et al. 2014). Water vapor traces have been observed in the atmosphere, concentrations of about 30 ppm at 30 km altitude (Chamberlain et al. 2013; Cottini et al. 2012). Though most water has been lost to space over time, water is still a key source of hydrogen necessary for the formation of acids, especially the sulfuric acid in Venus' clouds, suggesting that Venus is presently more hydrated than oxidized and that hydrogen may be dissolved and outgassed by volcanoes or deposited in the past by comets (Cottini et al. 2012).

14.3.5 The Sulfur Clouds of Venus (45–70 km Altitude)

The most iconic features of Earth's "evil twin" are no doubt her enormous super-rotating clouds. These clouds generally hover at heights of about 45–70 km (Khatuntsev et al. 2013; Stoddard and Jurdy 2012). The clouds consist mainly of sulfuric acid, which was spewed into the atmosphere over time from volcanic out-gassing and upwelling convection currents at the surface (Driscoll and Bercovici 2013). Figure 14.18 is an artist's rendering of these clouds.

Fig. 14.18 Artist's representation of the Venus clouds and the presence of lightning. NASA image

Once inside the atmosphere, the sulfuric acid coagulates and condenses in the middle region of the cloud layer, initially forming tiny micrometers particles (Gao et al. 2014). Gradually, these particles combine and form even larger, heavier particles that then descend and form a lower haze at the bottom of the clouds while displacing smaller particles upwards. Strong updrafts from the much hotter Venus surface blow small midlevel particles higher where they can nucleate with meteoric dust in the upper atmosphere and form an upper haze of sulfur (Gao et al. 2014). Average particle size in the clouds increases from mid latitudes toward the poles and toward the equator. Thus, thicker clouds are found around those regions than in latitudes between 30^0 and 60^0 in both hemispheres (Haus et al. 2014).

It has been speculated that another key chemical component other than sulfuric acid exists in the clouds. From ground-based observations using emission spectrum analysis of Venus' atmosphere, it has been noted that the clouds are absorbing a higher amount of UV radiation than sulfuric acid typically would on its own (Markiewicz et al. 2014). Studies of an optical phenomenon known as "Glory" suggest that the unknown component may be $FeCl_3$ or elemental sulfur either in the core of particles or as a coating (Markiewicz et al. 2014). Similar to traces of water vapor found in the lower atmosphere, notable concentrations of water vapor have been observed at the cloud top layer as well (Cottini et al. 2012).

Khatuntsev et al. (2013) describes clouds moving at high speeds under high winds, orbiting the planet in approximately three days at the mid-latitudes to the poles, and in about five days at the equator. Due to having constant periods of rotation of around three days in such a large continuous region, it has been theorized that the clouds near the poles behave as a quasi-solid body with a much denser consistency than the more fluid flowing clouds at low latitudes. Mean velocities in the upper clouds peak at 80–110 m/s with daily oscillations due to up-welling and down-welling currents from the lower atmosphere.

This super-rotational flow is directed retrograde with a polar flow component of around 10 m/s. In the mid-latitude regions, there is a strong variability in wind speeds over short periods of time of up to 35 m/s. This "mid-latitude jet" indicates that this region vacillates between the jet-like behavior of the low latitudes and the quasi-solid body behavior of the high latitudes. The lower layer winds from the middle cloud deck downward peak at 70–80 m/s.

In normal Earth conditions, the day typically produces stronger winds than during the night due to sunlight heating up the surface air. Convection currents are created as hot air rises and cools in the upper atmosphere. The opposite cycle occurs on Venus (Imamura et al. 2014). The high temperatures on the surface of Venus remain relatively constant while the sulfur clouds block and absorb most of the solar heat that reaches the planet, resulting in calmer convection currents in the lower atmosphere that become even calmer during the day as the clouds heat up in the Sun resulting in more uniform atmospheric temperatures (Imamura et al. 2014).

The average temperatures for the cloud region range from a cold 240 K in the top layers to as high as 380 K in the bottom layers (around −30 to 110 °C). Cloud temperatures can vary 10 K in a daily cycle but the general thermal structure of the planet does not vary with time. Over 23 years of observations, the average temperatures of the planet have not changed significantly (Haus et al. 2014).

14.3.6 The Upper Atmosphere of Venus (70+ Km Altitude)

As first surmised by Landis (2003), despite the harshness of the surface and lower atmospheres of Venus, the upper atmosphere of Earth's twin has relatively tame Earth-like conditions of temperature, pressure, and chemical composition. This is at altitudes of 65–120 km, above the toxic sulfur cloud layer and under the radiation protection of a magnetized ionosphere.

As discussed in the previous sections, the sulfur clouds are a dense area of high pressure and high temperature. At higher altitudes, sulfuric acid particulates fail to form and atmospheric pressures plummet from peaks of 90 atm at the surface to 5 atm in the clouds, to less than 0.1 atm at an altitude of 65 km (Migliorini et al. 2012). At greater altitudes, the atmosphere becomes extremely thin and pressures drop further, in part due to the loss of atoms and molecules stripped away in the outward flux of particles into space (Lundin et al. 2011). Figure 14.19, based on the data of Lundin, shows these variations of pressure with altitude.

Temperatures also decrease significantly at higher altitudes. The sulfur clouds absorb and entrap solar radiation and heat. Thus, while the lower atmosphere and cloud layer remain at superheated temperatures of several hundred Kelvin, at altitudes just above the clouds of 55–65 km, temperatures mellow out to Earth-like, livable conditions of −243 K to 303 K (−30 to 30 °C). At altitudes of 100 km and greater temperatures begin to drop to less than 150 K (−123 °C) (Haus et al. 2014). Figure 14.20, based on the data of Haus, shows these variations in temperature with altitude.

Like the rest of Venus, the upper atmosphere is a CO_2 based greenhouse (Driscoll and Bercovici 2013). However, it does contain traces of N, O_2 and CO,

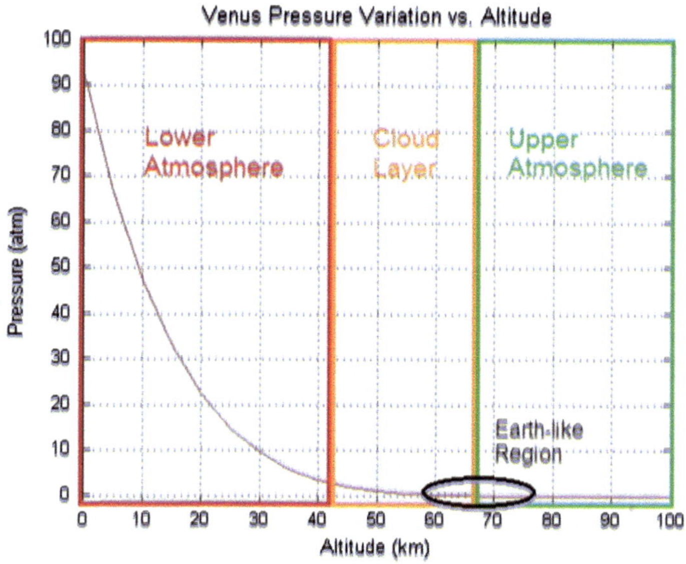

Fig. 14.19 Venus average pressure vs. altitude

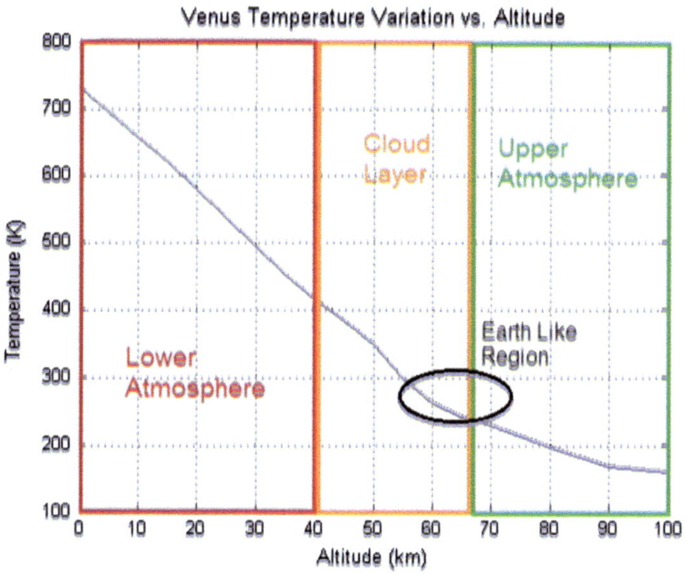

Fig. 14.20 Venus temperature vs. altitude

which are the primary constituents of our atmosphere on Earth (Landis 2003). There is evidence indicating that water vapor can be found in low concentrations at the cloud tops in certain locations around the planet (Lundin et al. 2011). Concentrations of CO are quite high above the clouds, though the values can vary greatly over periods of 20 or more days (Cotton et al. 2012). Figure 14.21 compares such concentrations.

While this benign region of Venus at altitudes of 65–90 km has potential for human and robotic use, high winds in the range 80–110 m/s are present. At higher elevations, around 90–120 km, wind characteristics are poorly defined but have strong variability in wind velocity. Beyond 120 km, winds are generated by the mix of solar generated heat and the cold air, resulting in peak speeds of 130 m/s (Sornig et al. 2013).

Many studies have attempted to create accurate models of the wind flows at different altitudes, latitudes, and times, but they still fail to encompass the complexity of the dynamics observed in all regions and altitudes (Sornig et al. 2013). Overall, there is a higher average velocity at the equator and lower average near the poles with temporal variations resulting in peak gradients of ±50 m/s in certain areas over several days (Sornig et al. 2013). Long-term data shows that average wind speeds consistently are increasing over time (Khatuntsev et al. 2013). Figure 14.22 shows this trend.

14.3.7 The Ionosphere of Venus

Unlike Earth, Venus generates a very weak magnetic field. This in turn allows much of the Sun's more intense electromagnetic radiation to penetrate the planet instead

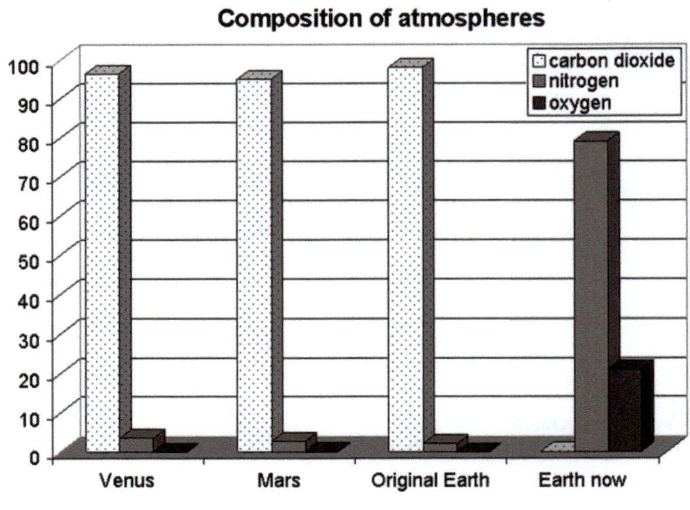

Fig. 14.21 Atmospheric composition of Venus (Evans et al. 2012)

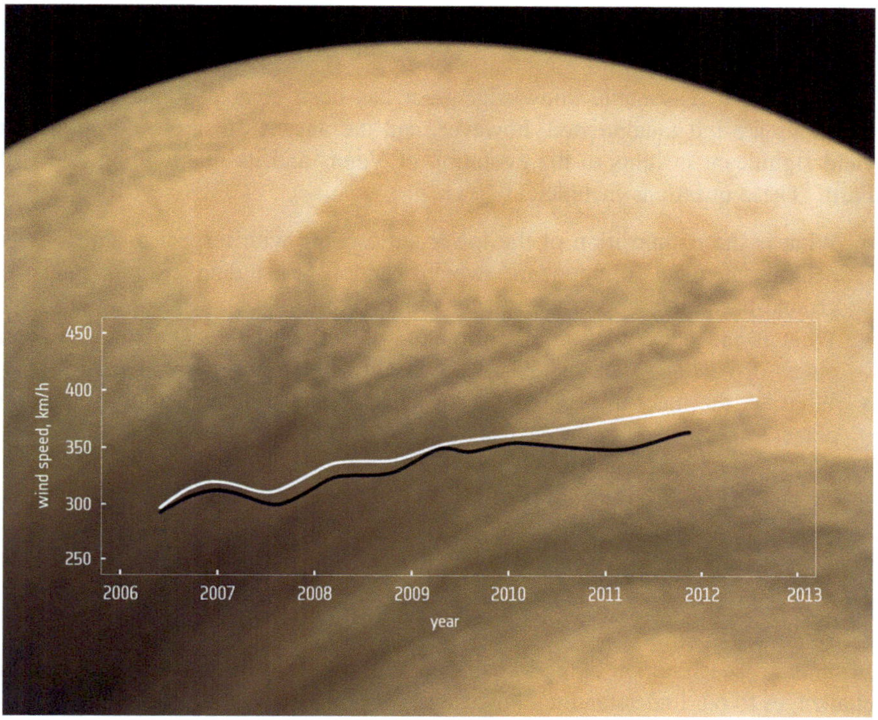

Fig. 14.22 Venus cloud level wind speed trends over recent years. NASA image

of being deflected or neutralized (Driscoll and Bercovici 2013). As a result of this high solar insolation, the particles at the very edge of the atmosphere become highly energized to the point of ionization, creating what is known as Venus' upper ionosphere, which becomes magnetized and deflects (Zhang et al. 2006). This barrier consists of mainly O^+ and H^+ ions (Lundin et al. 2011).

Due to continuous solar wind bombardment, Venus' upper atmosphere ions and particles are stripped away under the drag and momentum exchange between the wind and the ionosphere (Lundin et al. 2011). This loss is theorized to have contributed to the depletion of hydrogen, ozone, oxygen, and other chemical substances in the atmosphere, leading to the runaway greenhouse evolution of the planet (Ruess et al. 2006). Studies suggest that the flux loss occurs at a rate of 10^{25} ions or 250 g per second (Ruess et al. 2006). Comparatively, this is far less than the 3 ktons per second lost by Haley's Comet and the 1 ton per second lost by Io, but it has been noted that the acceleration and loss of these ionized and plasma particles forms a tail effect around Venus similar to that of a comet (Ruess et al. 2006). Another subsequent result of the drag interaction between the ionosphere and solar wind is the addition of neutral particle winds in the upper atmosphere (Lundin et al. 2011).

14.3.8 Venus: The Remaining Questions

We see that a reasonable knowledge base has been created about Venus. Key questions are left unaddressed, however, and the answers to these questions can yield significant insights to the evolution of Venus and its intertwining fate with Earth. These questions include:

- What is the composition of the lower atmosphere with respect to volatile and neutral chemical components, which cannot be studied from orbiters or ground-based telescopes beneath the cloud layer?
- What is the underlying composition of the surface below the thick regional plains of volcanic rock, with Tessera terrain being chief target areas?
- What are the mechanisms behind the volcanic and tectonic activity within the planet?
- What is the inner structure of the planet?

14.3.9 Venus: Summary of Engineering Challenges

In summary, each region of the Venusian environment poses different extremes and challenges to our understanding, but also to our ability to place instrumented rovers on, or in orbit around the planet.

14.3.9.1 Surface and Lower Atmosphere

- Constant high temperatures of 740 K (Cañon-Tapia 2014)
- High pressures in the range 90–100 bar (Williams 2014)
- Volcanic and tectonic activity, active lava flows (Ivanov and Head 2013)
- Sulfuric acid out-gassing and upwelling into clouds (Driscoll and Bercovici 2013)
- Sulfuric acid haze and possible acid rains (Gao et al. 2014)
- Limited sunlight, like very cloudy Earth days (Williams 2014)
- Poor signal reception, high interference from cloud layer and ionosphere

14.3.9.2 Cloud Layer and Upper Atmosphere

- High global wind speeds upwards of 80 m/s, retrograde super-rotational flow (Khatuntsev et al. 2013; Sornig et al. 2013)
- Dense clusters of sulfuric acid particles and haze (Gao et al. 2014)

- Ionized upper atmosphere due to direct exposure to solar radiation, may cause communications interference and with probe materials (Zhang et al. 2006)
- "Cold" temperatures (Haus et al. 2014)
- Nightside and dayside cycles of about 117 Earth days (Landis 2003)

With respect to landing probes on the surface, measures should be taken to make sure that the probes would endure the very high temperatures that exist. Unlike the Moon and Mars, which fluctuate between lesser extremes of cold and heat, the surface of Venus remains over 700 K at all times due to the oven-like conditions the cloud layer creates over the lower atmosphere and the planet's surface. These temperatures are hot enough to melt the lead solder and circuit boards of the current electronic systems on Mars probes (Landis 2003; Williams 2014). High Temperature Electronics or HTEs are a possible solution. While current technology falls short of operating at Venus conditions, HTEs developed for Earth-application in oil well drills, avionics, and automobiles have been proven to operate at up to 200 °C (Watson and Castro 2012).

If an extended presence is demanded of a surface probe similar to the operating lifetimes of Mars probes, methods for renewable energy will be needed as little sunlight penetrates the dense cloud layer above and Venus' nights are many Earth days long. Possible solutions are harnessing wind or thermal energy instead or deriving fuel in situ from Venus' surface. Protecting wind turbines and airfoils from dust and corrosion will be difficult challenges, however.

With sulfuric acid and other particulates outgassed into the atmosphere through volcanic events, anti-corrosion materials and systems should be considered for longer missions on the surface. There are materials with high sulfuric acid exposure tolerance, but they have never been used in planetary probes and at Venus temperatures.

For probes that intend to operate in and study the atmosphere of Venus, depending on the specific altitude, they will need several specialized systems as well. In the final surface and atmospheric missions to Venus by the Soviet Vega 1 and 2 craft, balloon probes were launched into the atmosphere along with the landing probes that were sent to the surface (Williams 2014). These balloons were made of lightweight material protected from sulfuric acid corrosion by a special coating. While this method protected the balloons over their two-day long flights, it is questionable whether such light coating is enough to protect a probe for an extended period of time (Williams 2014). Because Venus' atmosphere is so dense, the buoyancies of helium and even breathable air is higher (Landis 2003). Therefore, atmospheric probes can be made of heavier materials if they can provide sufficient protection from corrosion while also making them more structurally sound against high winds.

If these balloon probes are designed for the lower atmosphere or within the cloud layer, similar energy supply considerations should be taken as with the surface location. With limited exposure to sunlight, harnessing wind energy is an option, though this adds much more complexity to the probes. Likewise, while temperatures and pressures do decrease at higher altitudes on Venus, temperatures and pressures in the lower atmosphere and clouds are still at values much higher than conditions on Earth, Mars, or the Moon. While high outer pressure means that the threat of leaking lifting gas from balloons is reduced, high temperatures means that HTEs will need to be applied to the atmospheric probes as well.

Landis (2003) makes useful suggestions in order to mitigate potential communications problems between landing probes and lower atmospheric probes with Earth. Serious additional challenges exist if samples are to be taken from the atmosphere and soil of Venus and brought back to Earth for extensive analysis; problems arise with the reclamation of probes since they will need to be designed not only for survival but also for powered flight out of the Venus gravity well.

14.3.9.3 Establishing a Settlement on Venus for Future Missions

While there are many difficulties facing stand-alone probes operating in the Venus environment for extended periods of time, Landis suggested that there was a region on the planet that has almost Earth-like conditions. By creating a base of operations in this region with the focus on possible human habitation, probes can be constructed and deployed on smaller scale missions with more feasible prospects for data retrieval, sample reclamation, and extensive study by scientists, in situ. This region is within the upper atmosphere of Venus just above the cloud tops and between 65–70 km altitudes and is characterized by:

- Mild temperatures varying from 240–300 K (Haus et al. 2014)
- Low pressures from 0.2–1 bar (Williams 2014)
- High winds ranging from 80 to 100 m/s (Sornig et al. 2013)
- Ample sunlight access with super-rotational days of 50 h (Landis 2003)
- Low radiation exposure due to protection by the ionosphere and upper atmosphere (Landis 2003)
- Lower dangers of sulfuric acid corrosion above the cloud tops (Gao et al. 2014)
- Thick carbon dioxide atmosphere with traces of oxygen, nitrogen, hydrogen, and water vapor
- 90 % Earth's gravity (Landis 2003)

Although the time when mankind can realistically consider constructing a settlement on Venus or on any other body in the Solar System remains far in the future due to limits on technology, infrastructure, costs, and economic and other incentives, an analysis of a settlement in a future scenario where it is of consideration can still be made.

The conditions mentioned above, at that "sweet-spot" within Venus' atmosphere, offer a unique opportunity for human habitation unlike any concepts considered for settlements on the Moon, Mars, or any other planet, that is, a floating base in the atmosphere, an "aerostat base" (Landis 2003). Figure 14.23 shows such a hypothetical base.

Because of benign pressures at the aerostat elevation, Landis considers risks associated with gas losses as manageable. Landis further points out that an aerostat

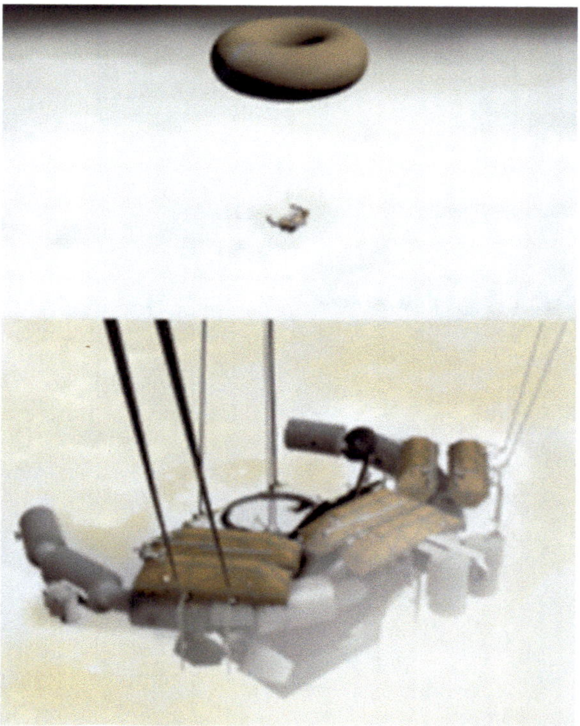

Fig. 14.23 "Venus Balloon Outpost" – hypothetical aerostat habitation on Venus with torus shaped balloon. Wikimedia image

habitation that is free to drift in the atmosphere will experience pseudo-days of only 50 h. Furthermore, Venus' thick atmosphere and magnetized ionosphere will block much of the radiation reaching the base (Landis 2003).

Of course, not only must this settlement survive in the environment, but it must also be able to generate the vital resources necessary for self-sufficiency. It has been reported that the atmosphere at the top cloud layer is predominantly carbon dioxide but contains traces of other elements like oxygen, nitrogen, and hydrogen. While these trace elements can be harvested in small quantities, more oxygen and hydrogen can be obtained through chemical processing of the carbon dioxide and sulfuric acid readily available in large quantities (Landis 2003). These can then be combined into water through known processes.

One of the great unknowns and dangers regarding manned space activities is the disparity in local gravitational fields as compared to that of Earth. Venus, with 90 % Earth gravity, removes one troublesome problem that exists for manned operations on the Moon and Mars.

As with the manned settlement of any body in the Solar System, settling Venus will require a significant transport of mass from Earth (or the Moon more likely at the time when we consider settling Venus) for the initial infrastructure. It will take time to create an in situ resource utilization infrastructure.

Fig. 14.24 Russian concept for a floating settlement on Venus. OrbitalSwap 2013

Regardless, should these challenges be surmounted, eventually more bases can be deployed to Venus while older settlements can grow in size, and become more self-sufficient. Permanent settlements can be created on Venus. Figure 14.24 shows an artist's rendering of a significantly evolved base floating and orbiting in the Venusian atmosphere.

14.3.9.4 Discussions and Conclusions – Venus

Venus is a unique planet in the Solar System, outwardly very similar to Earth but having diverged in evolution to the point of being hostile to all life on its surface and in its lower atmosphere. These harsh conditions make exploring the Venus environment very difficult and leave scientists with many unanswered questions about the underlying structure of the planet and how it came to be. To answer these questions, future missions are needed, but manned missions to the surface and atmosphere are out of the question at this time for life-threatening reasons. Similarly, unmanned probes face challenges unlike those faced on other bodies, like the Moon and Mars, which prevent them from operating for extended periods of time. While orbiters collect large amounts of data and can operate for many years around Venus, they are limited due to interference and lack of visibility through the planet's thick atmosphere.

In order to fully explore Venus, technological breakthroughs must be made that allow probes to cope and function in the Venusian environment for more than the few hours they survived previously. Additionally, at a certain altitude in the upper atmosphere of Venus, there may be conditions suitable for the construction of a floating human habitation that will give mankind a permanent foothold on (in low orbit over) the planet. There, human explorers can live and conduct more feasible exploration of the planet. These settlements can eventually grow and become more self-sufficient, a human colony independent of outside resources for its survival.

This poses many advantages for the evolution and expansion of mankind beyond Earth as well as economic advantages of a better location with which to reach and mine asteroids.

14.4 Titan

14.4.1 Introduction

We are interested in Titan because it has drawn the interest of the scientific and engineering communities as more data has been gathered about its composition and environment. Understanding particular planetary bodies can provide a framework for understanding all the planetary bodies, including Earth. From the perspective of engineers who will be called upon to design and construct facilities and habitats on our future homes in the Solar System, it is clear that the technologies will have overlapping capabilities, even though all these bodies have dissimilar environments. It is in this spirit that we include Titan in this paper on Mercury and Venus.

Here we focus on Titan's environmental characteristics that would affect a structure on its surface. Important factors for structural integrity include temperature, atmospheric pressure, troposphere density, Titan's regolith composition, wind velocity, seismic activity, and the surface liquid density. These factors will have to be taken into consideration when designing a surface structure and constructing it, either using available surface materials or using materials brought to the surface from Earth (or more likely, from the Moon). We will also mention a likely location for such a settlement.

Titan's environment is unique in our Solar System in the sense that it is very similar to Earth. The atmospheric composition on Titan is mostly nitrogen, like on Earth, and the atmospheric pressure is only 0.467 atm (Harria et al. 2006). The atmospheric density in the troposphere is roughly 4 times greater than Earth's surface air density (Lorenza et al. 2012). The oceans on Titan, near the equator, have a density of 614 kg/m^3, which is relatively close to Earth's 1025 kg/m^3 oceans (Tan et al. 2013). These characteristics, along with a few other examples, are why Titan will eventually be a suitable candidate for building a surface habitat.

14.4.2 Atmospheric Composition

Titan's atmosphere has components that are similar to our own on Earth. Titan is one of the only places in our Solar System where there is a surface pressure similar to Earth's atmospheric pressure at sea level. Titan's atmospheric pressure is 146.7 kPa (+/−0.1 kPa) while Earth's sea level pressure is 101 kPa (Harria et al. 2006). The gases present in the atmosphere are 98.5–98.6 % N_2 and 1.4–1.5 % CH_4 (Tokano 2014). Earth's atmosphere is 78.09 % N_2 and 20.95 % O_2. N_2 can be used for plant fertilizer, allowing plants to be grown inside a surface structure. N_2 is a macronutrient that is a key component in chlorophyll. The atmospheric density was found using the nominal Yelle model (Yelle et al.). The troposphere density was reported to be 5.24 kg/m^3 at the surface using the Yelle model. Saturn induces a magnetic field on Titan, although it does occasionally leave this magnetic field (Simon and Motschmann 2009).

14.4.3 Temperature Profile

The temperature profile of Titan is one of the major features of this moon that is very unlike our own planet. The surface temperature has been recorded by the Cassini spacecraft to be 92.5 K +/−2.5 K close to the equator. Tropospheric temperatures become cooler as we go from the equator to the poles by approximately 15 K in contrast to the equator (Schinder et al. 2012). Seasons on Titan last about 30 times longer than seasons on Earth. During northern winter the South Pole is always illuminated, which causes a temperature difference that allows a Hadley cell (a tropical atmospheric circulation) to form in the troposphere (Rodriguez et al. 2011). Data concerning the temperature profile was gathered in order to understand how fluctuations and temperature gradients could affect a structure on Titan.

14.4.4 Orbit

Titan has an orbital period of 15.95 Earth days. The length of a Titan "day" is the same as its orbital period. The surface gravity is 1.354 m/s^2 and Titan has a mean radius of 2,574.7 km.

14.4.5 Wind

The most probable wind speeds have been estimated to be 0.3 m/s at 10 m from the surface. The surface wind can go as high as 0.9 m/s. Greater wind speeds are seen as we begin to ascend in altitude – roughly 10–60 m/s from 30–50 km above the surface (Lorenza et al. 2012). It may be possible for energy to be harvested at higher altitudes on Titan using airborne turbines.

14.4.6 Surface Liquids

Surface liquids vary slightly from the equator to the poles in density and composition because of temperature differences. Surface liquid density at the equator is 614 kg/m^3 with a composition of C_2H_6 (53 %), CH_4 (32 %), C_3H_8 (7 %), and N_2 (7 %) with the given mole percent. Toward the poles the density is 551 kg/m^3 with compositions of 68 % CH_4, 22 % N_2, and 8 % C_2H_6 (Tan et al. 2013).

14.4.7 Titan Regolith

The ESA-build spacecraft, the Huygens probe, landed on Titan in 2005. It was the first spacecraft to land on Titan. The regolith near the probe-landing site, the Xanadu region, was determined to have a muddy consistency. It was determined that the regolith may consist of a damp and cohesive material with interstitial liquid contained between its grains (Atkinson et al. 2010). The liquid could possibly be methane or another surface liquid that is present. On the surface there are certain areas that have concentrations of water ice. There are also deposits of what could be CO_2 ice or HC_3N (Soderblom et al. 2009). The density of the liquid in between the grains is important to consider because its density changes as we approach the poles.

14.4.8 Seismic Activity

Titan may have cryovolcanoes on its surface. The surface could be experiencing a few types of cryovolcanism, such as mud volcanism, which involves acetylene. Another type of volcanism that could be occurring is methane-clathrate-hydrate volcanism. Finally a type of cryovolcanism involving ammonia could be occurring on Titan. The magnitude of these eruptions is not known but it is believed that one of these processes is currently happening and is depositing material on the surface (Soderblom et al. 2009).

14.4.9 Discussion and Conclusions – Titan

Titan has some Earth-like traits although there are some very important exceptions. Titan lacks numerous materials needed for long-term settlements, such as metals and certain macronutrients for plants. Materials needed for structures and crops may have to be brought from Earth or even mined from nearby moons. Saturn's rings are thought to have concentrations of Fe_2O_3 that could be mined for the iron and oxygen content (Filacchione et al. 2012). This may be a cost effective way to obtain iron and oxygen.

A suitable location for a structure could be along the equator and away from the Hotei region, which is a volcanically active area of Titan. The equator experiences less temperature fluctuations and has a greater average temperature. The temperature affects the density of the liquid in some areas with muddy regolith, which would affect a surface structure.

Deposits of water ice on the surface will also be an added benefit to future settlers. Using electrolysis water could be separated into oxygen and hydrogen. The hydrogen then could be used alongside nitrogen in the atmosphere in a Haber process to produce ammonia for crops and oxygen for breathing. Ammonia may be present in some volcanically active areas of Titan. Mining it may be an alternative.

Few possible methods exist for powering such a structure. Solar panels would not work efficiently because of the distance from the Sun and also due the smog present in Titan's atmosphere. Nuclear reactors or some type of reaction involving radioactive material would also be a plausible way to generate power. Another method involving airborne wind turbines could be a possibility. Titan has a very thick atmosphere and low gravitational pull. In the stratosphere there are large wind speeds that could be taken advantage of in order to harvest energy.

14.5 Robotic Outpost and Human Habitation – The Lunar Experience

We appear to have a good bit of information, at least enough to be able to perform preliminary analyses and designs for structures that will safeguard robots and humans who are sent to Mercury, Venus, and Titan to explore and, in the case of humans, life for a time. We will call these structures, whether for machine or man, habitats. In this section, we will summarize the key considerations for habitat structural analysis in extraterrestrial environments.

It is fair to assume that the first habitats on Titan and Mercury will be placed on the surface in whole. Without any infrastructure, the habitats cannot be erected locally. On Venus, given its unique atmosphere, it has been suggested that an orbital/floating habitat is not only feasible, but also optimal. It certainly will be a greater challenge to place a habitat in orbit rather than on a planetary surface, but perhaps not.

Some of the discussion below is based on Ruess et al. (2006) and Benaroya and Bernold (2008).

14.5.1 Lessons from Our Studies of Lunar Habitats

Any structure outside the safe Earth environment will need to be designed for and built with the following prime considerations:

Safety and Reliability. Human safety and the minimization of risk to "acceptable" levels are always at the top of the list of considerations for any engineering project. Extraterrestrial sites offer new challenges to the engineering designer, some of which are still problematic to resolution today. Minimization of risk implies in particular structural redundancy, and when all else fails, easy escape for the inhabitants. The key word is "acceptable." It is a subjective consideration, deeply rooted in economics. What is an acceptable level of safety and reliability for an extraterrestrial site, one that must be considered highly hazardous? Such questions go beyond engineering considerations and must include policy considerations: Can we afford to fail?

Other than one-g gravity. For example, on the Moon traditional structures will have, in gross terms, six times the weight bearing capacity as it does on the Earth. (Cable structures, such as cableways, will benefit even further from the reduced gravity.) Or, to support a certain loading condition, one-sixth the load bearing strength is required on the Moon as on the Earth. In order to maximize the utility of concepts developed for extraterrestrial structural design, mass-based rather than weight-based criteria should be the approach of extraterrestrial structural engineers. All of NASA's calculations have been done in *kg-force* rather than *Newtons*. Calculations are always without the gravity component; use kg_f/cm^2 as pressure, for example.

In the area of foundation design, most classical analytical approaches are based on the limit state condition. That means that the design is based on the limit of loading on a wall or footing at the point when a total collapse occurs, that is, the plastic limit. Since extraterrestrial structures require accurate pointing capabilities for astronomy and communications, for example, a settlement based design method would be more useful.

A note against assuming that less gravity means a footing can support more load: if soil can be assumed to be linearly elastic material, then the elastic modulus is not affected by gravity. However, the load bearing capacity of a real soil depends on the confining stress around it. If the soil surrounding the point of interest is heavier because of a larger gravity, the confining stress would be higher and the soil at the point of interest can support a higher load without collapsing. On the Moon, the lunar regolith has very little cohesion, and the undisturbed regolith is very dense. These facts will affect how the regolith is used and controlled in a construction setting.

Internal Air Pressurization. The extraterrestrial structure will be a life-supporting closed environment. It will be a pressurized enclosed volume with an internal pressure of 6.9×10^4 to 10.3×10^4 Pa. The enclosure structure must contain this pressure, and must be designed to be "fail-safe" against catastrophic and other decompression caused by accidental and natural impacts.

Shielding. A prime design consideration is that the structure be able to shield against the types of hazards found extra-terrestrially: continuous solar/cosmic radiation, meteorite impacts, and extreme variations in temperature and radiation. On the Moon, a layer of regolith (lunar soil) is placed atop the structure is stipulated to be adequate for shielding, where the added weight would only partially (in the range of 10–20 %) balance the forces on the structure due to internal pressurization mentioned above. In addition to general shielding, special radiation shelters will be needed for periods of increased solar activity. Extreme care would have to be taken not to disturb the regolith due to its damaging and carcinogenic nature.

Shielding against micrometeorite impacts is accomplished in one way by covering the outpost with dense and heavy materials, in this case compacted regolith, to absorb the kinetic energy. Some suggest that for shielding purposes alone, it is better to design and place human rated structures underground. This may be so, but it is then necessary to factor in the added costs and difficulties of subsurface work.

Long-term sustained low-level radiation effects, such as those that would be encountered on the Moon, lead to an annual dose-equivalent on humans on the exposed lunar surface of about 0.3 Sv and the dose-equivalent over an 11-year solar cycle is about 10 Sv, with most of the particles arriving in one or two gigantic flares lasting one to two days. It is estimated that at least 2.5 m of regolith cover would be required to keep the annual dose of radiation at 0.05 Sv, which is the allowable level for radiation workers (0.005 Sv for the general public). A shallower cover may be inadequate to protect against the primary radiation and a thicker cover may cause the secondary radiation, which consists of electrons and other radiation as a result of the primary radiation hitting atoms along its path.

In recent years, there is a move away from silicon- and germanium-based electronic components towards the use of gallium arsenide. Lower current and voltage demand, and miniaturization of electronic components and machines would make devices more radiation hardened.

Vacuum. A hard vacuum surrounds the Moon. This will preclude the use of certain materials that may not be chemically or molecularly stable under such conditions. This issue warrants further research.

Construction in a vacuum has several problems. One would be the possibility of out-gassing of oil, vapors, and lubricants from pneumatic systems. Hydraulic systems are not used in space for this reason. The out-gassing is detrimental to astronomical mirrors, solar panels, and any other moving machine parts because they tend to cause dust particles to forms pods. Another problem is that surface-to-surface contact becomes much more abrasive in the absence of an air layer. The increase in dynamic friction can cause fusion at the interfaces, such as a drill bit fusing with rock. This is of course aggravated by the fact that the vacuum is a bad conductor of heat. The increase in abrasiveness at interfaces also increases wear-and-tear on all moving parts, for example, railways and wheels.

Blasting in a vacuum is another interesting problem to consider. When the explosive in a blast hole is fired, it is transformed into a gas, the pressure of which may sometimes exceed 100,000 terrestrial atmospheres. How this would affect the area around the blast and the impact of ejecta resulting from the blast is difficult to predict. Keeping in mind that a particle set in motion from the firing of a rocket from a lander could theoretically travel half way around the Moon, the effects of surface blasting on the Moon, for example, would be something to be concerned about.

Dust. The lunar surface has a layer of fine particles that are disturbed and placed into suspension easily. These particles cling to all surfaces and pose serious challenges for the utility of construction equipment, air locks, and all exposed surfaces.

Lunar dust consists of pulverized regolith and appears to be charged. The charge may be from the fractured crystalline structure of the material or it may be of a surficial nature, for example, charged particles from the solar wind attaching themselves to the dust particles. It was reported that the dust particles levitated at the lunar terminator (line between lunar day and lunar night) may be due to a change in polarity of the surficial materials.

Ease of Construction. The remoteness of the lunar site, in conjunction with the high costs associated with launches from Earth, suggests that lunar structures be designed for ease of construction so that the extra-vehicular activity of the astronaut construction team is minimized. Construction components must be practical and, in a sense, modular, in order to minimize local fabrication for initial structural outposts.

Simple and conventional devices with no moving parts are preferred in the extraterrestrial environment over ones that involves multiple degrees of freedom in an exotic configuration involving a yet to be developed artificial intelligence control. Another misconception is that construction on the Moon, for example, is simply a scaling of the effects of similar operations on Earth and that theoretical predictive tools, especially those performed with computers, can accurately predict events. It is also critical to realize that astronauts do not make for efficient long-term construction crews. Automated construction is critical.

Use of Local Materials. This is to be viewed as extremely important in the long-term view of extraterrestrial habitation. But feasibility will have to wait until a minimal presence has been established on the Moon. Initial lunar structures will be transported for the most part in components from the Earth.

The use of local resources, normally referred to as ISRU (in situ resource utilization), is a topic that has been studied, now more intensely because of the possibility of actually establishing human presence on the Moon, Near-earth-orbit [NEO] and Mars. We can expect that ISRU will be the foundation for a long-term robotic and manned presence in the Solar System.

Water. The single most critical resource is water. As a need for human and plant life, it is without substitute. Its elementary components, hydrogen and oxygen, are fundamental to many processes that are needed in an industrial infrastructure. Water in some form has been identified on the Moon, and, perhaps remarkably – or not – on many of the planetary bodies in the Solar System. Water, in addition to other local elements, will set the parameters of the local infrastructure.

14.5.2 Discussion and Conclusions – Habitats

There is a vast literature on the exploration of the Moon and Mars, along with research on what eventual habitats will look like, especially on the Moon. While the Moon and Mars have quite different environments than do Mercury, Venus or Titan, there are certain engineering criteria that all habitats must fulfill so that robots and humans can survive on any extraterrestrial body. The formalism of engineering design is valid regardless of the target for that design. The physical laws are the same and, as long as the environment is fully defined, engineering can address the need.

14.6 Discussion and Conclusions - Overall

Of course, there are several additional issues that need to be considered. There are the costs associated with long space flights and the sustenance of humans and machines far from home. There are the logistics of supplies. Physiological constraints of humans and plants require further understanding. We are as yet unsure of the robustness of the human body in any other than full g. Even small radiation dosages can prove fatal when applied over the years that space activity requires of individuals. Psychological limitations can be challenging. The closeness required for long space flights cannot be tolerated by any but the very few who can endure small spaces for long periods of time. This is coupled with the risks of such voyages and distance from home and family.

It is almost certain that the first human settlements will be on the Moon and Mars, preferably in that order since the Moon's proximity to Earth gives it a special role for our development of survival techniques and technologies that will serve us when we choose to go further to Mars. It is also near certain that once we are able to survive in "Earth-Moon-Mars space" that the outer and inner Solar System will relatively rapidly become our proverbial backyard.

The foundation infrastructure that we will have to develop in order to settle the Moon and Mars will likely be sufficient for entrepreneurs and pioneers to do the same for Mercury, Venus and Titan. These are three locations that are ours to settle once we activate our spacefaring abilities with the Moon and Mars. We just have to decide to do it.

Acknowledgments It is a pleasure to thank the editors of this volume for organizing and finalizing this book of chapters on the exciting planets Mercury and Venus. We appreciate being included. We are also grateful to the two reviewers who took their reviewing tasks seriously enough to read this long chapter and provide us with useful suggestions for clarification and completeness. Thank you both.

References

Mercury

Anderson BJ, Slavin JA, Korth H, Boardsen SA, Zurbuchen TH, Raines JM, Gloeckler G, McNutt RL Jr, Solomon SC (2011) The dayside magnetospheric boundary layer at Mercury. Planet Space Sci 59:2037–2050

Anderson P, Macdonald M, Yen CW (2014) Novel orbits of Mercury, Venus, and Mars enabled using low thrust propulsion. Acta Astronaut 94:634–645

Arkani-Hamed J, Ghods A (2011) Could giant impacts cripple core dynamos of small terrestrial planets? Icarus 212:920–934

Baker DN, Odstrcil D, Anderson BJ, Arge CN, Benna M, Gloeckler G, Korth H, Mayer LR, Raines JM, Schriver D, Slavin JA, Solomon SC, Travnicek PM, Zurbuchen TH (2011) The space environments of Mercury at the times of the second and third MESSENGER flybys. Planet Space Sci 59:2066–2074

Bedini PD, Solomon SC, Finnegan EJ, Calloway AB, Ensor SL, McNutt RL Jr, Anderson BJ, Prockter LM (2012) MESSENGER at Mercury: a mid-term report. Acta Astronaut 81:369–379

BepiColombo (2014) European Space Agency. http://sci.esa.int/bepicolombo/. Accessed 25 July 2014

Blewett DT, Robinson MS, Denevi BW, Gillis-Davis JJ, Head JW, Solomon SC, Holsclaw GM, McClintock WE (2009) Multispectral images of Mercury from the first MESSENGER Flyby: analysis of global and regional color trends. Earth Planet Sci Lett 285:272–282

Cameron AGW, Fegley B Jr, Benz W, Slattery WL (1988) The strange density of Mercury: theoretical considerations. Mercury. In: Matthews MS, Chapman C, Vilas F (eds) The University of Arizona Press, Tucson

Carli C, Sgavetti M (2011) Spectral characteristics of rocks: effects of composition and texture and implications for the interpretation of planet surface compositions. Icarus 211:1034–1048

Elser S, Meyer MR, Moore B (2012) On the origin of elemental abundances in the terrestrial planets. Icarus 221:859–874

Genova A, Iess L, Marabucci M (2013) Mercury's gravity field from the first six months of MESSENGER data. Planet Space Sci 81:55–64

Grosser J, Glassmeier KH, Stadelmann A (2004) Induced magnetic field effects at planet Mercury. Planet Space Sci 52:1251–1260

Helbert J, Nestola F, Ferrari S, Maturilli A, Massironi M, Redhammer GJ, Capria MT, Carli C, Capaccioni F, Bruno M (2013) Olivine thermal emissivity under extreme temperature ranges: implication for Mercury surface. Earth Planet Sci Lett 371–372:252–257

Hiremath KH (2012) Magnetic field structure of Mercury. Planet Space Sci 63–64:8–14

HORIZONS Web-Interface. (2014) NASA JPL. http://ssd.jpl.nasa.gov/horizons.cgi. Accessed 3 Aug 2014

Izenberg NR, Klima RL, Murchie SL, Blewett DT, Holsclaw GM, McClintock WE, Malaret E, Mauceri C, Vilas F, Sprague AL, Helbert J, Domingue DL, Head JW III, Goudge TA, Solomon SC, Hibbitts CA, Dyar MD (2014) The low-iron, reduced surface of Mercury as seen in spectral reflectance by MESSENGER. Icarus 228:364–374

Koning A, Dumberry M (2013) Internal forcing of Mercury's long period free librations. Icarus 223:40–47

Ksanfomality LV (2008) The surface of Mercury in the 210–350 W longitude range. Icarus 200:367–373

Leblanc F, Johnson RE (2003) Mercury's sodium exosphere. Icarus 164:261–281

Lunar and Planetary Science (2014) NASA. http://nssdc.gsfc.nasa.gov/planetary/. Accessed 25 July 2014

Mariner 10 (2014) Space.com. http://www.space.com/18301-mariner-10.html. Accessed 25 July 2014

McNutt RL Jr, Solomon SC, Bedini PD, Finnegan EJ, Grant DG, The MESSENGER Team (2010) The MESSENGER mission: results from the first two Mercury flybys. Acta Astronaut 67:681–687

McNutt RL Jr, Solomon SC, Bedini PD, Anderson BJ, Blewett DT, Evans LG, Gold RE, Krimigis SM, Murchie SL, Nittler LR, Phillips RJ, Prockter LM, Slavin JA, Zuber MT, Finnegan EJ, Grant DG, The MESSENGER Team (2014) MESSENGER at Mercury: early orbital operations. Acta Astronaut 93:509–515

MESSENGER (2014) NASA/JHUAPL. http://messenger.jhuapl.edu/. Accessed 25 July 2014

Oberst J, Elgner S, Turner FS, Perry ME, Gaskell RW, Zuber MT, Robinson MS, Solomon SC (2011) Radius and limb topography of Mercury obtained from images acquired during the MESSENGER flybys. Planet Space Sci 59:1918–1924

Prockter LM (2005) Ice in the solar system. Johns Hopkins APL Technical Digest, vol 26, Num 2, pp 175–188

Raines JM, Slavin JA, Zurbuchen TH, Gloeckler G, Anderson BJ, Baker DN, Korth H, Krimigis SM, McNutt RL Jr (2011) MESSENGER observations of the plasma environment near Mercury. Planet Space Sci 59:2004–2015

Rhodes EA, Evans LG, Nittler LR, Starr RD, Sprague AL, Lawrence DJ, McCoy TJ, Stockstill-Cahill KR, Goldsten JO, Peplowski PN, Hamara DK, Boynton WV, Solomon SC (2011) Analysis of MESSENGER gamma ray spectrometer data from the Mercury flybys. Planet Space Sci 59:1829–1841

Rivoldini A, Van Hoolst T (2013) The interior structure of Mercury constrained by the low degree gravity field and the rotation of Mercury. Earth Planet Sci Lett 377–378:62–72

Rothery DA, Marinangeli L, Anand M, Carpenter J, Christensen U, Crawford IA, DeSanctis MC, Epifani EM, Erard S, Frigeri A, Fraser G, Hauber E, Helbert J, Hiesinger H, Joy K, Langevin Y, Massironi M, Milillo A, Mitrofanov I, Muinonen K, Naranen J, Pauselli C, Potts P, Warell J, Wurz P (2010) Mercury's surface and composition to be studied by BepiColombo. Planet Space Sci 58:21–39

Rothery DA, Thomas RJ, Kerber L (2014) Prolonged eruptive history of a compound volcano on mercury: volcanic and tectonic implications. Earth Planet Sci Lett 385:59–67

Ruiz J, López V, Dohm JM, Fernández C (2012) Structural control of scarps in the Rembrandt region of Mercury. Icarus 219:511–514

Schmidt CA, Wilson JK, Baumgardner J, Mendillo M (2010) Orbital effects on Mercury's escaping sodium exosphere. Icarus 207:9–16

Smith DE, Zuber MT, Phillips RJ, Solomon SC, Neumann GA, Lemoine FG, Peale SJ, Margot J, Torrence MH, Talpe MJ, Head JW III, Hauck SA II, Johnson CL, Perry ME, Barnouin OS, McNutt RL Jr, Oberst J (2010) The equatorial shape and gravity field of Mercury from MESSENGER flybys 1 and 2. Icarus 209:88–100

Solar System Exploration: Mercury (2014) NASA. http://solarsystem.nasa.gov/planets/profile.cfm?Object=Mercury. Accessed 25 July 2014

Solomon SC (2003) Mercury: the enigmatic innermost planet. Earth Planet Sci Lett 216:441–455

Spohn T, Ball AJ, Seiferlin K, Conzelmann V, Hagermann A, Komlec NI, Kargl G (2001) A heat flow and physical properties package for the surface of Mercury. Planet Space Sci 49:1571–1577

Vasavada AR, Paige DA, Wood SE (1999) Near-surface temperatures on Mercury and the moon and the stability of polar ice deposits. Icarus 141:179–193

Wasson JT (1988) The building stones of the planets. In: Matthews MS, Chapman C, Vilas F (eds) Planetary Research. Mercury. The University of Arizona Press, Tucson, pp. 622–650

Watters TR, Solomon SC, Robinson MS, Head JW, André SL, Hauck SA II, Murchie SL (2009) The tectonics of Mercury: the view after MESSENGER's first flyby. Earth Planet Sci Lett 285:283–296

Xiao Z, Komatsu G (2013) Reprint of: impact craters with ejecta flows and central pits on Mercury. Sci Direct. doi:10.1016/j.pss.2013.07.001i

Xiao Z, Strom RG, Chapman CR, Head JW, Klimczak C, Ostrach LR, Helbert J, D'Incecco P (2014) Comparisons of fresh complex impact craters on mercury and the moon: implications for controlling factors in impact excavation processes. Icarus 228:260–275

Venus

Basilevski AT, Shalygin EV, Titov DV, Markiewicz WJ, Scholten F, Roatsch TH, Kreslavsky MA, Moroz LV, Ignatiev NI, Fiethe B, Osterloh B, Michalik H (2012) Geologic interpretation of the near-infrared images of the surface taken by the Venus Monitoring Camera. Venus Expr Icarus 217:434–450

Cañon-Tapia E (2014) Insights into the dynamics of planetary interiors obtained through the study of global distribution of volcanoes II: tectonic implications from Venus. J Volcanol Geotherm Res. doi:10.1016/j.jvolgeores.2014.05.013

Chamberlain S, Bailey J, Crisp D, Meadows V (2013) Ground-based near-infrared observations of water vapor in the Venus troposphere. Icarus 222(1):364–378

Cottini V, Ignatiev NI, Piccioni G, Drossart P, Grassi D, Markiewicz WJ (2012) Water vapor near the cloud tops of Venus from Venus Express/VIRTIS dayside data. Icarus 217(2):561–569

Cotton DV, Bailey J, Crisp D, Meadows VS (2012) The distribution of carbon monoxide in the lower atmosphere of Venus. Icarus 217:570–584

Driscoll P, Bercovici D (2013) Divergent evolution of Earth and Venus: influence of degassing, tectonics, and magnetic fields. Icarus 226(2):1447–1464

Evans C, Peters R, Thompson M, Gadsby C, Partridge K, Mylan R, Sivan Y, Nation T, Follows D, Seeboo VH (2012) Atmosphere composition—D:\My Webs\index.htm. Licensed under Creative Commons Zero, Public Domain Dedication via Wikimedia Commons

Gao P, Zhang X, Crisp D, Bardeen CG, Yung YL (2014) Bimodal distribution of sulfuric acid aerosols in the upper haze of Venus. Icarus 231:83–98

Haus R, Kappel D, Arnold G (2013) Self-consistent retrieval of temperature profiles and cloud structure in the northern hemisphere of Venus using VIRTIS/VEX and PMV/VENERA-15 radiation measurements. Planet Space Sci 89:77–101

Haus R, Kappel D, Arnold G (2014) Atmospheric thermal structure and cloud features in the southern hemisphere of Venus as retrieved from VIRTIS/VEX radiation measurements. Icarus 232:232–248

Imamura T, Higuchi T, Maejima Y, Takagi M, Sugimoto N, Ikeda K, Ando H (2014) Inverse insolation dependence of Venus' cloud-level convection. Icarus 228:181–188

Ivanov MA, Head JW (2013) The history of volcanism on Venus. Planet Space Sci 84:66–92

Khatuntsev IV, Patsaeva MV, Titov DV, Ignatiev NI, Turin AV, Limaye SS, Markiewicz WJ, Almeida M, Roatsch T, Moissl R (2013) Cloud level winds from the Venus express monitoring camera imaging. Icarus 226(1):140–158

Landis GA (2003) Colonization of Venus. In: AIP conference proceedings, vol 654, pp 1193–1198. doi:http://dx.doi.org/10.1063/1.1541418

Lundin R, Barabash S, Futaana Y, Sauvaud JA, Fedorov A, Perez-de-Tejada H (2011) Ion flow and momentum transfer in the Venus plasma environment. Icarus 215(2):751–758

Markiewicz WJ, Petrova E, Shalygina O, Almeida M, Titov DV, Limaye SS, Ignatiev N, Roatsch T, Matz KD (2014) Glory on Venus cloud tops and the unknown UV absorber. Icarus 234:200–203

Migliorini A, Grassi D, Montabone L, Lebonnois S, Drossart P, Piccioni G (2012) Investigation of air temperature on the nightside of Venus derived from VIRTIS-H on board Venus-Express. Icarus 217(2):640–647

Romeo I (2013) Monte Carlo models of the interaction between impact cratering and volcanic resurfacing on Venus: the effect of the Beta-Atla-Themis anomaly. Planet Space Sci 87:157–172

Russell CT, Luhmann JG, Strangeway RJ (2006) The solar wind interaction with Venus through the eyes of the Pioneer Venus Orbiter. Planet Space Sci 54(13–14):1482–1495

RUSSIAN FLOATING BASE FOR VENUS (2014) OrbitalSwap collective. Orbital swap, 9 Feb 2013. http://orbitalswap.com/?p=52. Accessed 25 July 2014

Sornig M, Sonnabend G, Stupar D, Kroetz P, Nakagawa H, Mueller-Wodarg I (2013) Venus' upper atmospheric dynamical structure from ground-based observations shortly before and after Venus' inferior conjunction 2009. Icarus 225(1):828–839

Squyres S, Soderblom LA, Calvin WM, Cruikshank D, Ehrenfreund P, Hubbard G S, Huntress WT, Kivelson MG, Lee G, Luu J, Mackwell S, McNutt RL, McSween HY, Paulikas GA, Simon-Miller A, Stevenson D, Young AT (2011) Planetary Decadal Survey 2013–2022: the future of planetary science. Solar System Exploration. NASA, 1st July 2014

Stoddard PR, Jurdy DM (2012) Topographic comparisons of uplift features on Venus and Earth: implications for Venus tectonics. Icarus 217(2):524–533

VenusBalloonOutpost by Anynobody—Original work—This file was created with Blender. Licensed under Creative Commons Attribution-Share Alike 3.0 via Wikimedia_Commons http://commons.wikimedia.org/wiki/File:Venusballoonoutpost.png#mediaviewer/File: Venusballoonoutpost.png

Watson J, Castro G (2012) High-temperature electronics pose design and reliability challenges. Analog dialogue. Analog Devices, Inc., 4 Aug 2014

Williams DR (2014) Chronology of Venus exploration. Venus exploration timeline. NASA, 29 May 2014. http://nssdc.gsfc.nasa.gov/planetary/chronology_venus.html. Accessed 15 July 2014

Zhang TL, Baumjohann W, Delva M, Auster HU, Balogh A, Russell CT, Barabash S, Balikhin M, Berghofer G, Biernat HK, Lammer H, Lichtenegger H, Magnes W, Nakamura R, Penz T, Schwingenschuh K, Vörös Z, Zambelli W, Fornacon KH, Glassmeier KH, Richter I, Carr C, Kudela K, Shi JK, Zhao H, Motschmann U, Lebreton JP (2006) Magnetic field investigation of the Venus plasma environment: expected new results from Venus Express. Planet Space Sci 54 (13–14):1336–1343

Titan

Atkinson KR, Zarnecki JC, Towner MC, Ringrose TJ, Hagermann A, Ball AJ, Leese MR, Kargl G, Paton MD, Lorenz RD, Green Simon F (2010) Penetrometry of granular and moist planetary surface materials: application to the Huygens landing site on Titan. Icarus 210(2):843–851

Filacchione G, Capaccioni F, Ciarniello M, Clark R, Cuzzi J, Nicholson PD, Cruikshank DP, Hedman MM, Buratti BJ, Lunine JI, Soderblom LA, Tosi F, Cerroni P, Brown RH, McCord TB, Jaumann R, Stephan K, Baines KH, Flamini E (2012) Saturn's icy satellites and rings investigated by Cassini–VIMS: III—radial compositional variability. Icarus 220 (2):1064–1096

Harria AM, Mäkinen T, Lehto A, Kahanpää H, Siili T (2006) Vertical pressure profile of Titan—observations of the PPI/HASI instrument. Planet Space Sci 54(12):1117–1123

Lorenza RD, Newman CE, Tokano T, Mitchell JL, Charnay B, Lebonnois S, Achterbergf RK (2012) Formulation of a wind specification for Titan late polar summer exploration. Planet Space Sci 70(1):73–83

Rodriguez S, Le Mouélic S, Rannou P, Sotin C, Brown RH, Barnes JW, Griffith CA, Burgalat J, Baines KH, Buratti BJ, Clark RN, Nicholson PD (2011) Titan's cloud seasonal activity from winter to spring with Cassini/VIMS. Icarus 216(1):89–110

Schinder PJ, Flasar FM, Marouf EA, French RG, McGhee CA, Kliore AJ, Rappaport NJ, Barbinis E, Fleischman D, Anabtawi A (2012) The structure of Titan's atmosphere from Cassini radio occultations: occultations from the Prime and Equinox missions. Icarus 221 (2):1020–1031

Simon S, Motschmann U (2009). Titan's induced magnetosphere under non-ideal upstream conditions: 3D multi-species hybrid simulations. Planet Space Sci 57(14–15):2001–2015

Soderblom LA, Brown RH, Soderblom JM, Barnes Jason WK, Randolph L, Sotind C, Jaumann R, Mackinnon D (2009) The geology of Hotei Regio, Titan: correlation of Cassini VIMS and RADAR. Icarus 204(2):610–618

Tan SP, Kargel JS, Marion GM (2013) Titan's atmosphere and surface liquid: new calculation using Statistical Associating Fluid Theory. Icarus 222(1):53–72

Tokano T (2014) Non-uniform global methane distribution in Titan's troposphere evidenced by Cassini radio occultations. Icarus 231(1):1–12

Yelle R, Strobell D, Lellouch E, Gautier D (n.d.) From Lunar and Planetary Laboratory, University of Arizona. http://www.lpl.arizona.edu/~yelle/eprints/Yelle97b.pdf

Robotic Outpost and Human Habitation

Ruess F, Schaenzlin J, Benaroya H (2006) Structural design of a lunar habitat. ASCE J Aerosp Eng 19(3):133–157

Benaroya H, Bernold L (2008) Engineering of lunar bases. Acta Astronaut 62(4–5):277–299

Bibliography

Spohn T, Breuer D, Johnson TV (2014) Encyclopedia of the solar system, 3rd edn. Elsevier, Amsterdam

Chapter 15
Deployable Structures for Venus Surface and Atmospheric Missions

K. Ozdemir

15.1 Introduction

"Venus, the "greenhouse planet", is scientifically fascinating place." (Landis G, NASA). Based on the fact of growing interest in Venus exploration, backed by the US National Academies of Sciences' listing of the Earth's "hellish twin" as one of the mission destinations with high priority, exploration mission architectures for Venus are expected to respond to operational expectations that are higher than the early period of Venus missions. High profile missions point out the need for more capable, more flexible, better designs.

Deployable Structures are structures that can change their shape so that they have a compact form for, e.g., transportation or storage, but can then be expanded for their final use (University of Cambridge 2014). These type of structures have been used in space missions successfully ranging from one-dimensional (e.g. masts) to two (e.g. solar arrays) and three dimensional systems (e.g. compartments). The deployable structure designs provide the space mission planners the options to convey relatively large-sized elements to their operation environment in their available envelopes of transport. As observed in the hardware assets of the earlier Venus mission architectures, the extreme conditions of Venus atmosphere and surface, in regard to acidity, temperature and pressure levels, have dictated robust designs leaving very limited room for deployable structures. This, apparently, is about to change. Almost all of the future mission designs for Venus integrate deployable components into their architecture. In order to cope with the extreme conditions of Venus, advances in material technology as well as improved designs play a crucial role.

K. Ozdemir (✉)
Yeditepe University, Istanbul, Turkey
e-mail: office@fezai.com

V. Badescu and K. Zacny (eds.), *Inner Solar System*,
DOI 10.1007/978-3-319-19569-8_15

The goal of this chapter is to study the existing designs, developed for Venus atmospheric and surface missions. Furthermore, novel concepts of Venus "deployables" are reflected with their background of development schemes.

15.2 Deployable Space Structure Design: An Overview

In this section, an overview to the background of research on deployable structures for space flight (Haeuplik and Ozdemir 2012) is provided. The mentioned examples of "deployables" are mainly from habitation segment.

15.2.1 Up to 1960: Beginnings

The origins of research on deployable structures lead back to the work of an English engineer F.W. Lancaster in 1917 (Kuhn 1991). The first inflatable structure built with an engineering process was the "radome", designed by Walter Bird. The use of Nylon as a novel material made the construction of the design with a diameter of 15 m possible. Apart from inflatable structures built by Goodyear Aerospace Corporation for radar and communication infrastructure, Wernher von Braun's conceptual design for a wheel-shaped inflatable space station is among the advances in deployable aerospace design in the 1950s. As an example of inflatable structures, the Echo 1 balloon demonstrated a remarkable packaging ratio of 30:1 in 1960 (Zadunaisky et al. 1961).

15.2.2 1960–1980: Man in a Can

The advances in the human spaceflight during the 1960s triggered also the deployable space structure development beside many other space technologies. Ranging from the initial inflatable space station designs of NASA to the inflatable airlock of Voskhod that made the first spacewalk of the history possible, the progress in deployable structure design was taken beyond the Low Earth Orbit to lunar environment (Haeuplik and Ozdemir 2012). The 'Stay Time Extension Module' of Grumman and NASA was intended to be an inflatable temporary shelter for the Lunar Module of Apollo program. During the 1970s inflatable decelerators for (re) entry were studied by American and Russian experts. Meanwhile, the first European inflatable reflector antennas were developed by ESA and partners.

15.2.3 1980-Present: Progress

1982 saw the first successful operation of Venera surface lander mission series, by the missions Venera 13 and 14 (Table 15.1), thus the first deployment of a surface lander component on Venus, the deployable instrument arm on both landers (Lorenz et al. 2011). In the habitation segment of the deployable space structures, apart from prototypes, that never saw action, such as NASA Personal Rescue Enclosure from 1984, there are design studies for LEO and Lunar stations, including the Lawrence Livermore Lunar Habitat. The development of the 'Transhab', a long-duration space habitat with a hybrid structure of a rigid core and an inflatable body, marks the beginning of the 2000s. The transhab design was intended to be integrated into the International Space Station as crew quarters. The inflatable landing cushions of the NASA Pathfinder Rover Mission in 1997 and Mars Exploration Rover Missions in 2004 are among the deployable space structures that kept progress in space exploration in motion. The InFlex Lunar Habitat prototype, an inflatable habitat for a crew of two was built in 2007. Furthermore, Bigelow Aerospace demonstrated its focus on inflatable space structure development by Genesis I and II prototypes that were launched in 2006 and 2007.

Table 15.1 Existing deployable structure concepts for Venus surface and atmospheric missions (*Data source* NASA 2008, 2013a, b; Landis 2006, 2010)

Structure	Decelerator	Aircraft	Balloon	Aerostat	Airship	Bellow	Rover	Lander
Deployment environment	Orbit	Orbit/mid-air	Mid-air	Mid-air	Mid-air	Surface	Surface	Surface
Deployed component	Ablator surface	Wings, fuselage	Gas envelope	Gas envelope	Gas envelope	Gas envelope	Legs, chassis, sail	Inst. arm, outriggers
Deployment character	Inflation Mechanical	Inflation Mechanical	Inflation	Inflation	Inflation	Inflation	Mechanical	Mechanical
Mission/study origin	NASA VISE, VITAL	VAMP, NASA Venus exploration	VEGA, NASA Venus exploration	Future Venus exploration	NASA Venus exploration	NASA Venus mobile explorer	NASA Venus in-situ explorer	Venera 13-14, SAGE

There are plans to integrate and test an inflatable module built by Bigelow at the ISS in 2015. NASA tested its LDSD (Low Density Supersonic Decelerator) system, including an inflatable kevlar ring and a large-sized parachute, which is intended to be used in future planetary missions (Grossman 2014), successfully using a balloon launch in 2014.

15.3 Existing Deployable Structure Concepts for Venus

The existing concepts for deployable structures in Venus missions extend in a range of surface mobility platforms to floating units (Table 15.1). In the following sections, the existing concepts are assessed in three groups, according to their environment of deployment, namely:

- Orbital Deployments
- Mid-air Deployments
- Surface Deployments

15.3.1 Orbital Deployments

As the initial environment of Venus surface and atmospheric mission operations, Venus orbit is home to the lightweight deployable structures of aero-capture & entry systems and an aerial platform design. Since these structures will be subject to atmospheric entry loads, low ballistic co-efficiency is a key factor in their designs.

Table 15.2 ADEPT identification chart (*Data source* Glaze 2013)

ADEPT Adaptive deployable entry and placement technology (NASA 2011)	
Design objective	Integrating a flexible, light-weight, low ballistic coefficiency decelerator
Structural features	Lightweight metal structure and carbon textile in folded stowage configuration
Deployment character	Umbrella-like mechanical deployment
Mission – research frame	NASA studies: Venus in-situ explorer mission, Venus intrepid tessera lander

15.3.1.1 ADEPT: Mechanically Deployed Soft-Shell Decelerator

The Adaptive Deployable Entry and Placement Technology (ADEPT), is an entry system that can provide a lighter system alternative key operational benefits and risk reduction compared to a rigid aeroshell (Glaze et al. 2013). The ADEPT structure is comprised of a ribbed metal framework, covered with a tensioned membrane in deployed layout. The system is deployed in a similar way as an umbrella. The design provides a relatively large and high drag decelerator surface with a low ballistic co-efficiency, compared to conventional rigid aeroshell systems. Covering the leading side, ADEPT does not create a full envelope around the payload, decreasing the total mass of the system compared to rigid-body aeroshells (Table 15.2).

The ADEPT structural system is composed of four main components:

- main body
- nose cap
- ribs
- struts

The titanium upper and lower rings are connected to each other by titanium struts, forming a truss system that encloses the payload compartment (Fig. 15.1).

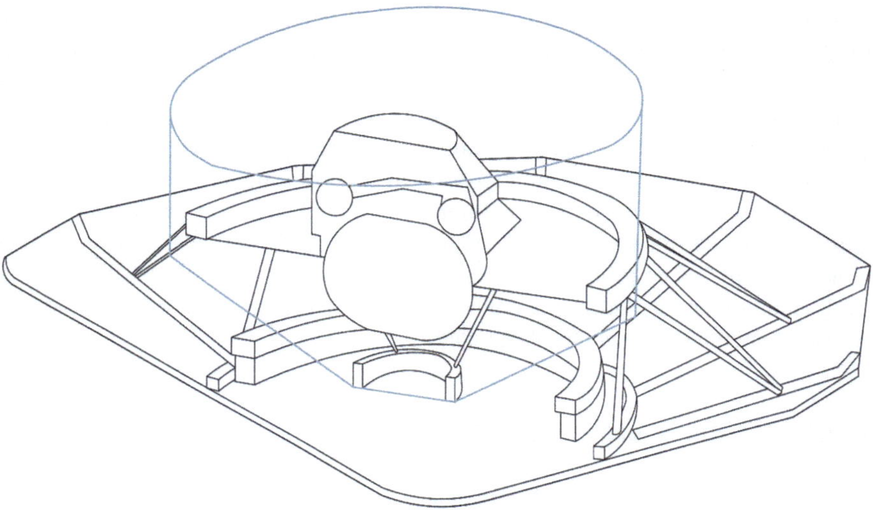

Fig. 15.1 ADEPT deployable decelerator cutout view (*Credit* Erdogdu)

The lower ring of the main body has a box-form profile and provides the attachment element to the cruise stage of the spacecraft, besides supporting the struts. The upper ring includes the interface to the landing ring of the lander and is the attachment location of the nose cap.

The nose cap forms the leading edge of the entry vehicle and is structured similar to a conventional rigid aeroshell. It is constructed using a titanium honeycomb core sandwich and Graphite and uses a light-weight TPS. The titanium ribs, when deployed similarly to an umbrella, are supported by the struts to tension the 3D woven carbon cloth, creating the aeroshell surface together with the nose cap. The deployment occurs as the nose cap is pulled aft by a cable-winch system, forcing the struts to push the ribs outwards. This results in the umbrella-like opening of the decelerator.

Design and development work on ADEPT continues. The feasibility of the system for a Venus mission is established at preliminary design level.

15.3.1.2 BALLUTES: A Special Breed

Another light-weight option in the deployable decelerator design segment is the ballute, a hybrid structure of balloon and parachute. Ballutes have been used since 1950s as ground impact speed reduction devices for empty sounding rockets. Assessed by NASA as an auxiliary decelerator unit for Viking Landers in the early 1970s, ballutes began to appear in the planetary mission architectures. These initial assessments indicated a mushroom-like shape, rather than a spherical form. The ballutes were proposed as an entry decelerator for a Venus mission first in 1976 (Akiba et al. 1976). The material assessments, as well as entry computations were presented in this work. The results of a more recent study were presented by McDonald (McRonald 1999) indicate that decelerator ballutes can be designed to operate at temperature and g-load conditions of Venus exo-atmospheric environment. Several configurations appear in the work, including spherical, lenticular and annular ballute forms. In order to become ready for project use, the study suggest to establish a development program to assess concepts of long term stowage and rapid & fail-safe deployment of the system, including the stable and predictable behaviour of the structure under extreme conditions of entry. Nevertheless, the possibility of aerocapture and entry using ballutes enables a payload to be placed in orbit or to land without being included in a conventional entry vehicle with heat shield can be seen as a promising outcome of the study.

15.3.1.3 VAMP: Directly from the Orbit

VAMP (Venus Atmospheric Maneuverable Platform) is a semi-buoyant propelled aerial vehicle concept, proposed by Northrop Grumman Aerospace and L'Garde Inc. for NASA Venus exploration study (Griffin 2013). Designed to cruise at ~70 km of altitude, the deployable aerial platform will house a science package for experiments and data collection. One eccentric feature of VAMP is the deployment in orbit, which enables the system to eliminate the aeroshell component and its mass, leaving more room for the scientific load onboard (Table 15.3). The design integrates diverse know-how from related background of project partners:

- Stowage, deployment, inflatable entry (Northrop Grumman and L'Garde)
- Venus Exploration Technology (NASA Glenn, JPL and others)
- LEM-V Semi-Buoyant Vehicle Design & Technology (Northrop Grumman)
- Global Hawk UAV Design and Technology, especially in long duration flight (Northrop Grumman)

VAMP design features a propeller system for propulsion at cruise altitude of 70 km, while the vehicle can use its buoyancy to float at 55 km for passive flight. The orbital deployment of the platform provides an atmospheric entry shield with low ballistic coefficient, in regard to the total surface area of the deployed system and reduced mass. Primarily, two shape options for the vehicle are being assessed. A trade-off comparison between trapezoidal flying wing and a lenticular shape of conventional reentry ablation surfaces, the first one appears to be the favorable one, though further assessment work is in progress. The present study on configuration and deployment scheme proposes a non-deployable core unit, housing the propeller system and scientific hardware and inflatable wings with rigid ribs and linearly deployed spars (Fig. 15.2).

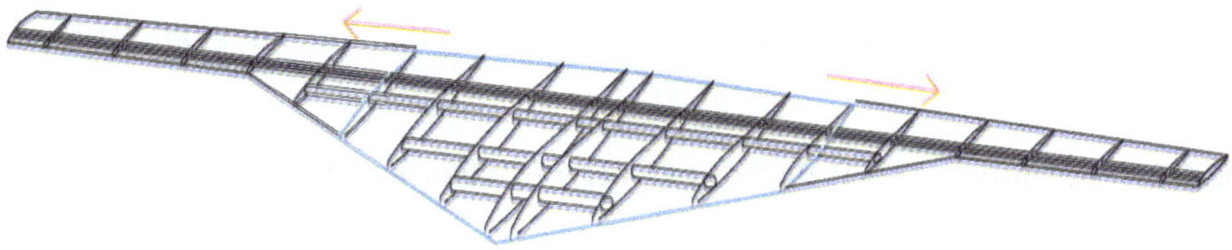

Fig. 15.2 Depiction of VAMP deployable aircraft structure (*Credit* Erdogdu)

Table 15.3 VAMP identification chart (*Data source* Griffin K 2013)

VAMP_Venus atmospheric maneuverable platform (Northrop Grumman and L'Garde 2012)	
Design objective	Design of a long-duration aerial platform
Structural features	Whole-wing aircraft w. pneumatic wing body, rigid rib structure
Deployment character	Stacked ribs, extending spars, inflatable wings
Mission – research frame	Aerial platform study by Northrop Grumman and L'Garde

In stowage configuration, the ribs are stacked side by side to be pushed along the wing axis by spars for full deployment. During the in-orbit deployment, the rudders attached to deployable booms are positioned first, followed by full wing-body extension.

The next steps in the project are:

- Laboratory Testing with Sub-scale Engineering Model, including wind tunnel and material tests
- Balloon Release of Demo Vehicle at Altitude, to test primarily the in-flight behavior
- Sounding Rocket Release of Demo Vehicle Exo-atmospherically, for entry tests

VAMP, in its current state, appears to be a reliable and capable aerial platform design with numerous advantages, sourced by its deployment concept. The field experience of its project partners is a key factor in the development of the design.

15.3.2 Mid-air Deployments

The benign conditions of the Venus atmosphere compared to the extreme environment if the surface for the deployment of light structures motivate the designers to develop systems with "mid-air" deployment capabilities. Including, but not limited to dirigibles mid-air deployed mission assets provide a set of aerial capabilities for research and communication functions. The success of VEGA balloon missions of 1984 indicates the potential of aerial deployed exploration platforms.

15.3.2.1 Sustained Aerial Platforms: Vega Mission and Beyond

Sustained air platforms, or "balloons" in short, appear among the major elements of Venus exploration, especially since the success of two Russian-led VEGA missions. These exploration platforms can be designed to operate at one or more vertical levels in the Venus atmosphere, studying the cloud regions to understand the chemical processes, the atmospheric stability, structure and evolution of the atmosphere and to constrain bio-signatures. Concerning the balloon design for cloud top level in Venus atmosphere, Venus presents a benign environment for the deployment and inflation of a balloon (Hall 2013). The atmosphere is dense, allowing very slow parachute descent. The cloud-top level balloons can finish inflation at a level below the equilibrium float altitude and rises to that point after inflation. This feature avoids the over-pressurizing of the gas envelope in higher altitudes and was also used in VEGA missions.

Consolidated by VEGA success, sustained air platform technology continues to be a field with significant potential. NASA conducts a development study on superpressure balloons for Venus exploration, where possibilities of advancements in balloon technologies are assessed.

15.3.2.2 VEGA: The Vanguard

As the vanguard of Venus ballooning, the primary scientific objective of Vega balloon mission was to obtain information about the large- and small-scale motions, structure, and cloud properties of the Venus atmosphere at the float altitude of 54 km (Table 15.4). The balloons floated over a period of ~46 h in the day- and night-side, performing measurements of pressure, temperature, lightning, illumination levels and cloud properties (Sagdeev 1986).

The Vega balloon system was comprised of a 3.4 m diameter superpressure balloon and a gondola housing the scientific pack and vehicle control unit. The gondola was suspended below the balloon by a 13 m long tether strap. The balloon membrane material was Teflon film coated Teflon cloth matrix. The gas envelope held Helium as lifting gas at 30 mbar overpressure. The helium gas diffused out of the envelope slow enough to keep the system afloat in the battery life-time frame.

Table 15.4 VEGA balloon identification chart (*Data source* NASA 2013)

VEGA_Venus-Halley mission (IKI RAN and CNES 1984) Venus atmospheric balloon	
Design objective	Design of an aerial scientific platform for Venus Atmosphere
Structural features	Superpressure, teflon membrane gas envelope balloon
Deployment character	Parachute extraction + inflation deployment
Mission – research frame	VEGA_Venus-Halley mission

The whole deployed Vega balloon system summed up to a mass of 21.5 kg. The vehicle, including the balloon, gondola, parachute, ballast, tanks of helium and timing electronics was stowed in a toroidal-shaped compartment around the lander antenna cylinder in the spherical aeroshell.

The deployment of the Vega balloon probe progressed as follows: Shortly after the entry into the Venus atmosphere, the parachute attached to the aeroshell opened to detach the upper aeroshell hemisphere. After the upper cap release a main extraction parachute is deployed to pull the balloon package out of its compartment. A second parachute opened at an altitude of 55 km to extract the furled balloon from the package. The balloon inflation is sustained till 54 km altitude and the parachute and inflation system were jettisoned. Upon descent to ~50 km of altitude, the ballast is released to let the balloon system stabilize back to 54 km of altitude ~20 min after atmospheric entry.

15.3.2.3 Venus Superpressure Balloon Development: The Next Balloons

Designed to operate at the same altitude region as VEGA balloon probes, two balloons were fabricated by JPL and project partners to be tested as technology demonstrator samples (Table 15.5). The superpressure balloons with 5.5 m diameter have a payload capacity of 45 kg and a lifetime of 30 earth days at 55 km of altitude. The demonstrators were tested for buoyancy, leakage, sulfuric acid resistance, aerial deployment/inflation and stowage folding robustness. Within the study, VEGA balloon material, the Teflon-coated cloth matrix, is assessed to be traded for better performing Vectran fabric based materials through pinhole effect tests under exposure to sulfuric acid. Furthermore, the Vectran fabric provides improved resistance to transient loads during the deployment and inflation process (Hall 2013). Such improvements in material technology can also enable longer duration performance of the balloons under acidic conditions of the upper Venus atmosphere.

The Superpressure Balloon Development reveals the challenges and the innovation potential of the VEGA balloon probe systems' heritage, especially in regard to the scalability of the designs. The poor scalability aspect of the VEGA balloon design elements (e.g. membrane material) is among the challenges to be addressed in further advancements. The NASA program sets a 7 m diameter balloon with 100 kg payload capacity as an engineering development goal.

Table 15.5 NASA Venus superpressure balloon design identification chart (*Data source* Hall 2013)

NASA Venus superpressure balloon design (NASA 2011)	
Design objective	Design of an aerial platform for Venus atmosphere
Structural features	Superpressure, membrane gas envelope balloon
Deployment character	Parachute extraction + inflation deployment
Mission – research frame	NASA Venus superpressure balloon development

15.3.2.4 Venus In Situ Explorer: A Set of Aerial Platforms (Balloon, Airship, Aircraft)

Venus, with her greenhouse-like environment is a fruitful place for scientific studies. Listed with a high priority level in planetary science by The US National Academies of Sciences, a Venus surface in situ explorer mission requires assets in different categories. Designed by NASA, a mission concept for a robotic mission to study the surface and atmosphere of Venus assessed a variety of options for aerial and surface exploration platforms (NRC 2003). The mission includes both surface robots, designed with an operational lifetime of 50 days on the surface of Venus, and also solar-powered airplanes to probe the middle atmosphere (Table 15.6). This set of global Venus mission comprises the four surface rovers, each with a dedicated aerial platform hosting the computer and control electronics, four additional aerobots dedicated to the atmospheric survey, a communications relay satellite and a number of auxiliary research elements.

Table 15.6 NASA VISE aerial platforms trade-off chart (*Data source* NRC 2003)

Platform	Balloon	Airship	Aircraft
Storage & deployment	Compact/simple	Bulky/complicated	Compact/simple (ARES)
Altitude manipulation	Powerless/complicated	Powered/complicated	Powered/simple
Location manipulation	Drifter	No station-keeping	Possible
Technology heritage	VEGA	–	Mars ARES
Summary	Not flexible/practical	Complex & inflexible	Chosen as aerial platform

Concerning the atmospheric research platform, three options were taken into consideration. Based on their applicability, these existing concepts of aerial platforms were cross-assessed, within a trade-off process. The deployable aerobot design candidates for in situ Venus exploration are as follows:

- Balloon
- Airship
- Aircraft

Balloon

Built on the VEGA Balloon Mission and Venus Superpressure Balloon development heritage (Hall 2013), a balloon design to act as a Venus in situ explorer aerial platform make use of certain advantages. The fact that VEGA balloon flight took place above the cloud layer at about 54 km for up to 50 h, provides a probed-high level of practicability in terms of stowage/deployment, material and cruise operation. Nevertheless, as seen in the trade-off chart (Table 15.6), hard limitations in altitude and location control resulted in preference of a more flexible aerobot option, the solar aircraft.

Airship

According to a late study on a dirigible aerial Venus platform, the configuration would be similar to a standard airship that would operate on Earth. The float-envelope of the platform would be ellipsoidal with rear fins for stability and control. A spherical pressure vessel encapsulates the operational equipment and the payload. An electric motor and propeller are used for propulsion, driven by a radioisotope power Stirling engine. The radioisotope reactor provides the system with propulsion power as well as cooling power in order to maintain the interior of the pressure vessel at a temperature in which the electronics and payload can operate (Colozza 2012).

The Venus in situ explorer aerial platform trade-off indicates the drawbacks of a dirigible system in regard to its limited capabilities in altitude and location control, as well as difficulties in stowage and deployment of the system (Table 15.6).

Aircraft

Various studies (Landis 2001, 2003, 2006) on Venus aircraft systems indicate a solar-powered Venus aircraft can cruise in higher velocities than the local wind. This feature can enable the vehicle to station-keep at a sub-solar coordinate or hover over a specific location. Considering the encapsulated entry into the Venus atmosphere in an aeroshell and overall transport configuration a folding wing and fuselage design (Table 15.7) is essential for this type of aircraft. The wings of the candidate design are shown to be folded into halfway of their length to be deployed from the aeroshell upon entry into the Venus atmosphere. The wings are designed to be equipped with solar cell panels on top and bottom surfaces (Fig. 15.3).

The deployment process of folded wings has been studied in the proposed ARES project for a Mars exploration mission, through a demonstrator. Considering the similarity in size and the deployment sequence, ARES demonstrator appears to be a valuable precursor for the Venus solar aircraft (Guynn et al. 2003). As observed at a video of the proposed Venus in situ explorer mission, published by NASA (NASA 2008), the solar aircraft can be seen deployed from its aeroshell simply by being dropped-open by its own weight.

Built on the mentioned basic configuration, different designs in different sizes can be seen in NASA documents. These include single unfolding fuselage and twin extending fuselage types, fitted into 1.2 m (Pioneer-Venus small probe-type) and 3.7 m (Viking lander-type) aeroshells.

The variety of final configurations present flexibility for different uses (e.g. rover systems component vs. atmospheric survey) Nominal flight altitude is slightly above the cloud level, 75 km above the surface. Though, the effects of highly corrosive atmosphere of Venus on the vehicle components present challenges that can be overcome with progress in material technology based on Vega Balloon mission experience that took place in similar altitudes. The highly corrosive character of Venus atmosphere, including clouds of sulfuric acid droplets, points out to the necessity to protect the exposed surfaces of the aircraft. The human experience in dealing with sulfuric acid, in terms of material technology, provides valuable know-how to construct sturdy designs like the VEGA balloons that flew for relatively long durations (~50 h) at cloud level in the Venus atmosphere.

Table 15.7 NASA Venus aircraft design identification chart (*Data source* Landis 2003, 2006)

NASA Venus exploration aircraft (NASA 2003)	
Design objective	Integration of an aerial exploration and relay platform
Structural features	Lightweight rigid structure in folded stowage configuration
Deployment character	Unfolding initiated by free-fall detachment from entry capsule
Mission – research frame	NASA Venus (in-situ) exploration mission study

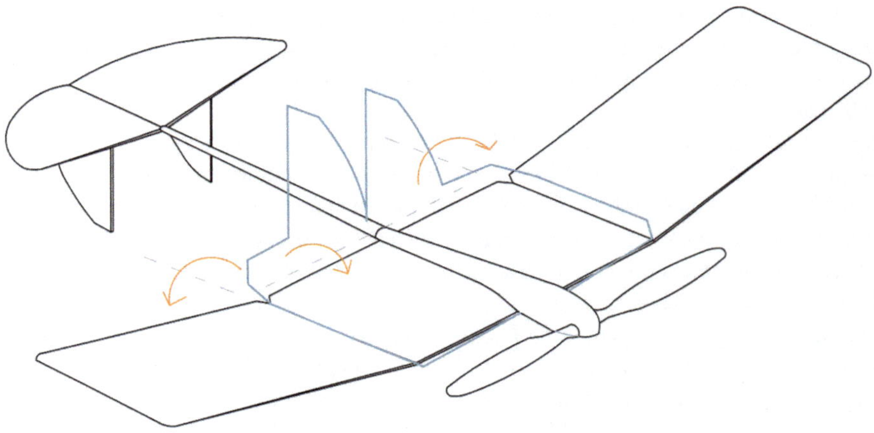

Fig. 15.3 NASA Venus exploration aircraft in deployed configuration (*Credit* Erdogdu)

15.3.3 Surface Deployments: Surviving the Extreme

Due to extreme conditions on the Venus surface, ground systems are envisioned with high levels of robustness, limiting the integration of deployable components within. The Venera missions of 1970s demonstrate the use of rigid lander components with reduced complexity for a fail-safe mission progress under extremities.

15.3.3.1 BELLOWS: Robust and Flexible

The semi-rigid inflatable air platforms, bellows in short, are among the considered options for the concept of deployable mobile platforms on Venus. Stowed and deployed using the typical bellows layout, these systems provide an alternative to membrane inflatable bodies that are not suitable for surface conditions of Venus (Baker et al. 2010a, b). The bellows design has been studied for Venus surface missions (e.g. VESSR) that require mobility. The NASA Venus Surface Sample Return (VESSR) Mission concept features the collection of surface material from Venus and transport to our planet.

Table 15.8 Bellows identification chart (*Data source* Baker et al. 2010a, b; Sweetster 2003)

Bellows semi-rigid inflatable air platform (NASA 2010)	
Design objective	Integration of an surface mobility/aerial transport platform
Structural features	Lightweight metal skin structure in bellow-folded stowage configuration
Deployment character	Inflation to extended bellow configuration
Mission – research frame	NASA VME (Venus mobile explorer)/VESSR (Venus surface sample return)

The whole setting, with its components is technically challenging. Lighter than air vehicles are the only practical means of raising samples from the surface of Venus to where the atmosphere is thin enough for rocket injection into space.

Two high-rated Venus mission concepts are proposed in the National Science Foundation Decadal Survey feature lighter-than-air vehicles to lift payloads from Venusian surface to different altitudes: Venus Surface Sample Return (VESSR) and Venus In Situ Explorer (VISE). Both mission proposals include deployable bellows inflatables. In regard to high pressure and high temperature environment of Venus surface, metallic bellows structure (Table 15.8) appears as a feasible solution for a deployable balloon system. The simple stowage and deployment character besides the potential for fabrication from a suitable metal (e.g. stainless steel) renders bellows structures as a promising candidate for deployable mobility systems.

Venus Surface Sample Return Mission (VESSR): Two Balloons of Different Kind

As one of the most challenging segments of the proposed Venus Surface Sample Return (VSSR) missions is the transport of the sample from the surface to the Venus orbit, different approaches can be observed in mission designs (Sweetster 2003). Due to high atmospheric density, thus high drag at the Venus surface, a direct rocket ascent is avoided in the mission designs. One of the mission layouts features a two step ascent to the specified altitude using a multi balloon system. High pressure and high temperature atmospheric environment also adds a constraint concerning the balloon material. Therefore, an inflatable system featuring a semi-rigid gas envelope is proposed as a first stage ascent platform (Cutts et al. 1999). The gas envelope is designed to deploy like a bellows body and can be made of stainless steel or a suitable alloy. The relatively high mass of the metal structure can be tolerated for the first stage balloon, which will be smaller in size than the second stage, considering the high density of the atmosphere at lower levels. The inflation of the metal bellows with helium or hydrogen gas is started during the descent of the combined vehicle to the Venus surface. During the descent and short stay of sample collection on the surface, the second balloon stays in the thermally insulated compartment (Fig. 15.4). The metal "balloon" is designed to lift the sample container from the surface to ~10 km altitude, where the second balloon, made of Teflon-coated Kapton film, deploys for ascent to ~60 km altitude, where the rocket transit initiates. The temperature at the cross-over altitude of ~10 km is suitable for the membrane balloon deployment. Since the first stage will be discarded at the cross-over altitude, the bellows deployment is a plastic expansion, thus not retractable.

A demonstrator prototype of metal bellows has successfully been tested by Cardner Bellows Corporation and the Jet Propulsion Laboratories of NASA for inflation and leakage rating. The demonstrator, fabricated from stainless steel is deployed from the initial height of 0.19 m to final height of 2.16 m. The demonstrator had a diameter of 0.53 m and 70 convolutions. The challenges in optimization of deployment, leak-free operation and altitude control remain to be studied.

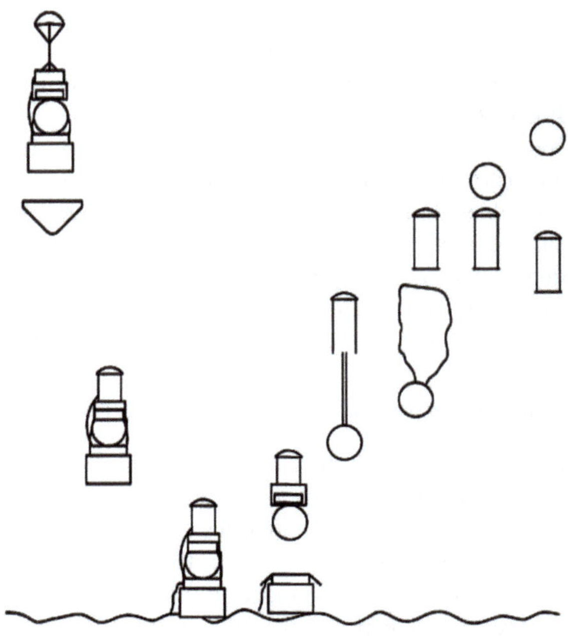

Fig. 15.4 NASA Venus surface sample return mission layout featuring bellows (*Credit* Erdogdu)

Venus Mobile Explorer Mission (VME): Two Locations, One Lander

The Venus Mobile Explorer (VME) mission proposal of NASA also includes a buoyancy envelope, deployed in bellows form (Baker et al. 2010a, b). The goal of access two locations dictates the requirement of a mobility system for the surface component of the mission. The mission design indicates a lander, capable of traversing between two measurement stations by ascending to a specific altitude by a buoyancy system and drifting with the local winds (Table 15.8). The bellows system uses Helium gas that is lighter than the ambient Carbon dioxide gas and the internal pressure will be 0.5 bar above the ambient atmospheric pressure. The bellows system, made of stainless steel, is designed to lift the lander from the first location and keep it afloat for 20 min at the specified altitude for drifting with the wind. At the end of the transit period, the bellows envelope is detached from the lander, letting it descent to the second location of exploration. Due to the plastic deformation, the deployed system can not be retracted, therefore used just for one transit.

15.3.3.2 Venus Surface Rovers: Robots with Character

In planetary exploration studies, surface mobility is a key issue. Remotely-operated or autonomous, robotic mobile vehicles are valuable assets of operations in extreme environments. In Venus surface exploration, rovers are among the major mission architecture components. Due to harsh environmental constraints, Venus rover designs, few in numbers, present robust configurations in order to minimise the risk of mechanism failures under extreme temperature, pressure and corrosion conditions. This design approach limits the use of deployable components in rover structures. Nevertheless, rover designs featuring simple-deployment components can be seen in the overall exploration landscape of Venus.

As a repercussion of The US National Academies of Science Space Studies Board decadal study, (NRC 2003), *New Frontiers in the Solar System*, robotic Venus mission design options were developed by NASA, incorporating different assets of planetary exploration. Among these elements, two different rover designs appear to have deployable components.

Venus Exploration Rover

The NASA in situ exploration mission design for Venus (Landis 2003) reveals an integrated approach, in regard to the surface, aerial and orbital assets (Table 15.9). Using the orbital unit as relay platform, surface rovers are envisaged to perform collaboratively with air vehicles. Both units are delivered into Venus atmosphere in similar-type aeroshells (NASA 2008), then deployed in mid-air and on the surface respectively.

The surface rover, after the aeroshell detachment, is landed directly on a parachute, without using a separate lander vehicle.

Table 15.9 Venus surface rover identification chart (*Data source* Landis 2003)

NASA Venus exploration rover (NASA 2003)	
Design objective	Design of a mobile surface exploration platform
Structural features	Lightweight metal structure in folded stowage configuration
Deployment character	Mechanical parallelogram deployment of wheel struts
Mission – research frame	NASA Venus exploration studies

This landing on wheels is featured also in MSL mission on Mars, providing a valuable technology heritage. The rover is designed to be stowed in the aeroshell, wheels retracted on parallelogram struts. The leg system deploys like a pantograph, prior to contact with the surface. The scissor-like parallelogram geometry of the legs provides a shock-absorbent effect at touch-down, in addition to the rotating wheels that avoid tipping over. The unit has a deployment mass of 330 kg and has a radioisotope power system, as stated in the study.

Zephyr: A Landsailing Rover

Another rover design, developed by NASA is named Zephyr, a surface vehicle with land-sailing mobility concept. Zephyr rover utilize the local winds on the surface of Venus for mobility (Fig. 15.5). These winds are low in speed but produce enough force to sail in regard to the high atmospheric density. As observed in documents, issued by NASA, there are two Zephyr rover configurations.

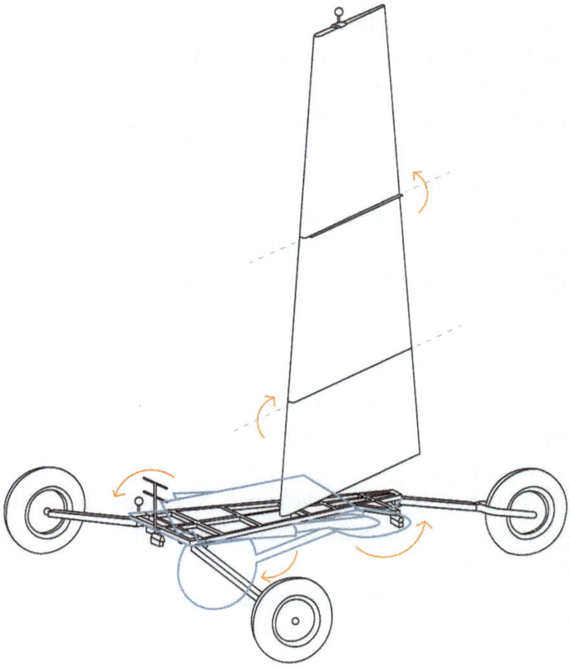

Fig. 15.5 NASA Zephyr landsailing Rover design for Venus (*Credit* Erdogdu)

The major configuration (Table 15.10) sports a 6.6 m tall terrestrial landsailer-like design, with a windsurf-type sail profile and a light, three-wheel chassis. The stowed vehicle fits into a 3.6 m diameter aeroshell. The vehicle chassis appears to be folded into two parts, with wheel struts retracted. The sail is shown to be folded in three layers. The NASA presentation on this concept (NASA GRC 2012) indicates an aerial deployment of the chassis on parachute descent following the aeroshell separation. The sail unit, deployed on surface is pivoted on the chassis for trimming into the suitable angle for sailing.

The second Zephyr design configuration reflects a simple and robust layout. The triangular plan vehicle has a low profile with three wheels can deploy its sail that covers the half of its top surface plate by tilting upwards. The sail in a triangular shape and covered with solar cells, is pivoted on the rover body, allowing it to be "trimmed" in accordance with the surface winds of Venus. This Zephyr version reveals a simple design with a basic deployment system of its novel mobility system, and can be a demonstrator version of the design concept.

The designer team also assessed the potential of wind turbines for another way of in situ resource utilization, which can have a potential for deployable design. However, the assessments focus on a low cost mission and apparently avoid the risk factor of deployable designs (Landis et al. 2013).

Table 15.10 Venus zephyr rover identification chart (*Data source* NASA GRC 2012)

NASA Venus zephyr rover (NASA 2012)	
Design objective	Design of a wind-driven mobile surface exploration platform
Structural features	Lightweight metal and membrane structure in folded stowage layout
Deployment character	Mechanical deployment of folded chassis and sail
Mission – research frame	NASA Venus exploration studies

15.3.3.3 Venera Lander Deployable Arm: Stinging the Surface

The landers of Venera and Vega program, as well as the landers of upcoming missions reflect, understandably, quite simple and robust configurations, considering the harsh surface conditions of Venus. The common layout is comprised of a pressure vessel, housing the electronics, an external hardware unit and a fixed landing ring (Lorenz et al. 2011). Each component typically is rigid and has minimum number of elements. However, even few, there are deployable components, including an instrument arm and landing legs/outriggers.

The Soviet Venera missions form the main body of realized surface exploration on Venus. Through these missions valuable information on the atmosphere and soil properties of Venus have been acquired. The collection of the data was performed by different measurement devices, including a dynamic penetrometer (Table 15.11) in Venera missions 13 and 14, which took place in 1981. The goal of the dynamic penetrometer experiment was to determine the soil properties of Venus surface. The measurement device, the penetrometer was housed in a deployable arm. The deployable arm was positioned on the lower edge of the lander and designed to impact the soil far enough from the lander that there would be no effects from deformation of soil due to landing. The measurement device was attached at the end of the deployable arm that is hinged at the other end to the lander. The penetrometer featured a cone-shaped die that would penetrate the soil mechanically and electrically.

Table 15.11 Venera lander deployable unit identification chart (*Data Source* Lorenz et al. 2011)

Venera lander (Russian space agency 1981) deployable instrument arm	
Design objective	Design of a deployable arm housing an impact measurement device
Structural features	Metal structure in folded stowage configuration
Deployment character	Mechanical deployment
Mission – research frame	Venera 13 & 14

The Venera 13 penetrometer showed the load-carrying capacity of the soil to be 2.6–10 kg/cm^2, analogous to heavy clays or compacted dust-like sand. The Venera 14 instrument showed load carrying ability of 65–250 kg/cm^2, similar to volcanic tuffs or fissured rocks. However, it is possible that the penetrometer landed on top of the ejected lens cap, and actually measured the lens cap on top of the rock rather than the rock itself. This has never been confirmed officially (NASA 2014).

15.3.3.4 Sage Lander Outriggers: Five Landing Canes

Proposed by Colorado University and selected by NASA in 2009 as one of the agency's next space venture to another celestial body in our solar system, The Surface and Atmosphere Geochemical Explorer mission, features a Venus lander that would conduct measurements on the surface (LASP 2014). The lander (Table 15.12) is composed of a landing gear (i.e. pad & outriggers), a pressure vessel topped by a rigid aluminum drag plate for terminal descent deceleration and an external surface excavation subsystem. The vehicle is designed to be landed on a honeycomb crushable pad for landing load attenuation, balanced by deployable outriggers. The outriggers are hinged at one end to the lander and kept folded into two struts in stowed configuration. They are supposed to deploy mechanically during the descent to the surface right after the heat shield separation to ensure a balanced touch-down on the surface (Fig. 15.6).

In the SAGE joint effort, University of Colorado at Boulder's Laboratory of Atmospheric and Space Physics (LASP 2014) would provide the science leadership, data archiving, education and public outreach effort. Partners include NASA JPL (project management); Lockheed Martin (construction of the carrier spacecraft); ARC.; Goddard; Langley Research Center; and Canadian Space Agency (robotic arm).

Table 15.12 SAGE Lander outriggers identification chart (*Data source* LASP 2014)

SAGE Lander (University of Colorado LASP 2014) deployable outriggers	
Design objective	Design of a landing component for SAGE lander
Structural features	Metal structure in folded stowage configuration
Deployment character	Mechanical deployment
Mission - research frame	Surface and atmosphere geochemical explorer (SAGE)

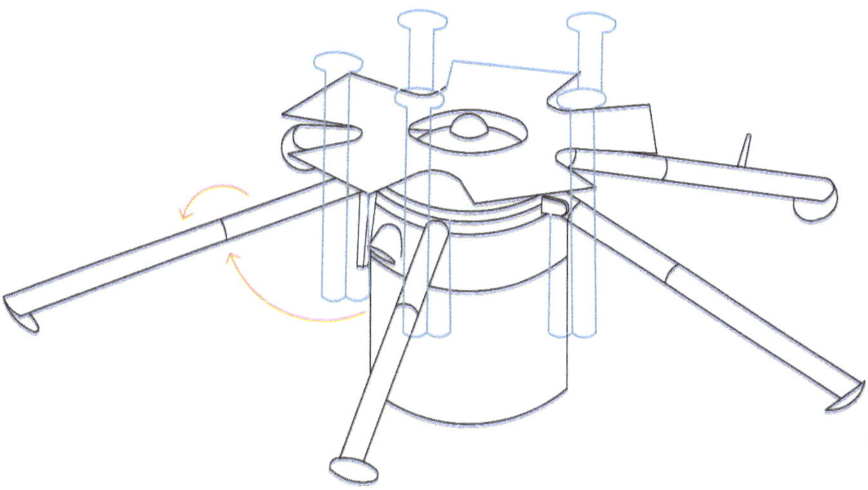

Fig. 15.6 SAGE Lander with deployable outriggers (*Credit* Erdogdu)

15.3.4 Summary: Overall Evaluation of the Existing Concepts

The presented existing deployable structure concepts for Venus sub orbital missions are based on the progress line of gradual hardware and material development, initiated by Venus missions of 1960s–1980s period of space exploration (Table 15.13). Through these missions not only the main environmental factors are understood, but a heritage of probed materials and structures is accumulated. Mainly, the major challenges in the future of these systems lie in material development and establishment of test & enhancement programs. Though limited in capabilities and configurationally inflexible, the deployable systems of the intermediate Venus exploration era heritage provide the designers with a ground of solid experience for further development. Taking the progress of material technology and the level of mission architectures so far into consideration, these space-flown structures can be re-integrated (Venera-D 2014) relatively easy with enhancements into upcoming missions of the present Venus exploration period. Mostly presented in a robust stowage, deployment and operational configuration, wheeled roving vehicle designs (Fig. 15.7) are among the essential components of the existing surface mission architectures. The deployable aerial platform segment requires material innovation for longer & better resistance to acidic environment of Venus atmosphere. The deployable aerial vehicles appear to be a crucial component of the emerging Venus exploration period, presenting a layer of transport, observation, command and communication assets.

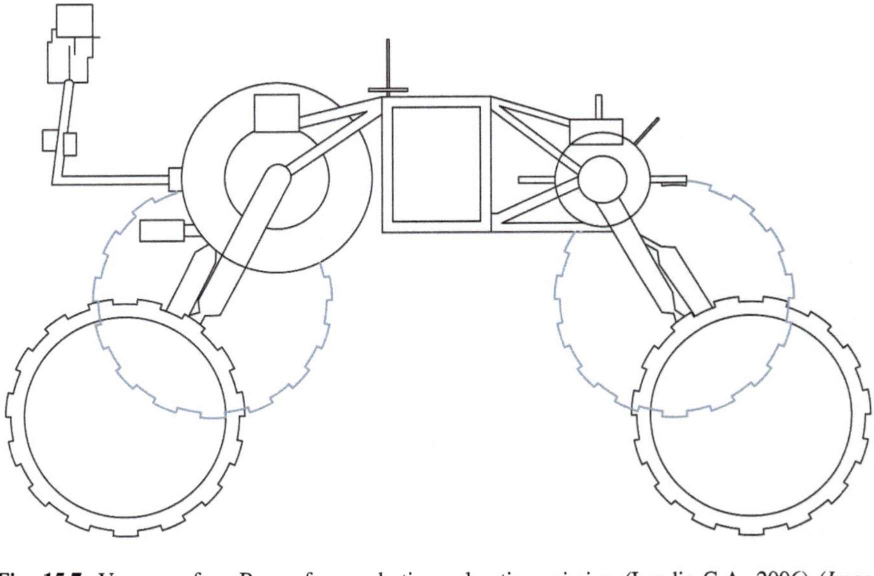

Fig. 15.7 Venus surface Rover for a robotic exploration mission (Landis G.A. 2006) (*Image credit* Erdogdu)

Furthermore, VEGA mission as a whole presents a good example of international cooperation, where agencies with relevant expertise can share the responsibility of mission components, as French space experts did for the deployable balloon system. Bellows and sail systems, though conceptual so far, have the potential to establish ever-growing mobility capabilities on Venus surface, once test & development programs in material and structural research are set up. The lander component deployment experience, such as Venera lander instrument arm indicates the need for "smart" deployment, with better control in accordance with the mission environment and conditions. Nevertheless, deployable components can arm the landers that have simple/limited layouts by nature, with enhanced operational capabilities (e.g. landing and sampling on uneven terrain).

15.4 Future Deployable Structure Concepts for Venus Exploration

Beyond the existing concepts for deployable structures for Venus surface and atmospheric exploration, there are a number of likely-to-appear design candidates in the given short- and long-term research vision. One of the short-term concept candidates for Venus deployables can be the segment of deployable aerodynamic units. These include:

- wind turbine systems
- sail systems
- kite structures

Wind turbines are featured in the frame of surface missions to harvest the local source of winds that are low in speed, but produce sufficient force on the surfaces, considering the high density of the Venus atmosphere. The upcoming Venera D mission is likely to include a wind turbine to generate power for the lander (Venera-D 2014). There are also proposals (Landis and Colozza 2013) to integrate wind turbines and sail structures into surface rovers in order to simplify the power generation system, with impacts on the mass and configuration of the rovers.

As for kite structures, a novel concept for a wind-flyer housing lightweight instruments and transmitters is proposed by Russian NPO Lavochkin for the Venera-D mission (Venera-D 2014). Lavochkin suggests that the idea is applicable leaning on experiments conducted. The device is envisaged to be deployed from the lander to stay aloft using the winds for long durations at ~50 km altitude. The platform is referred to as a "vetrolet" (Russian term for wind-flyer).

In regard to the stowage and deployment patterns, biomimetical transfer principles can improve the overall quality of the designs. Ranging from favorable packaging ratios to deployment efficiency, biomimetical designs can enhance the deployable structure concepts for Venus exploration, as it has been doing for other segments of spaceflight. Beside bio-inspired designs, terrestrial extreme environment architecture can also supply relevant role-models for future deployable structures in Venus environment (Ozdemir 2009).

Concerning the long-term Venus exploration vision, long duration human presence factor can play an important role in the way leading to the colonization of the planet. Recent perspectives on colonization (Landis 2003) point out that the cloud-top level of Venusian atmosphere present suitable conditions for human missions and stay. The Venus environment at this level is considered to be benign, holding

Table 15.13 Existing deployable structure concepts evaluation chart (*Data source* see Tables 15.1–15.11)

Deployed structure	Decelerator	Aircraft	Balloon
Deployment environment	Orbit/mid-air	Orbit/mid-air	Mid-air
Challenges	Activation after long term stowage	Sulfuric acid proof materials	Sulfuric acid proof materials
Potential for development	Elimination of entry vehicle & heat shield	Versatile expl. platform	flotilla build-up, aerial platform
Mission/study origin	NASA VISE /VITAL	VAMP NASA Venus exploration	VEGA NASA Venus exploration

Bellow	Rover	Lander component
Surface	Surface	Surface
Leak-free operation, altitude control	Venus-rated materials	Smart deployment
Surface mobility, sample return	Low-cost hi-eff missions	Lander performance enhancements
VESSR NASA Venus Mobile Explorer	NASA Venus in-situ explorer	Venera 13-14 SAGE

- atmospheric pressure of 1 atm
- abundant solar energy
- suitable ambient temperature in "liquid water range"
- primary volatiles required for life (Carbon, Hydrogen, Oxygen, Nitrogen, and Sulfur)
- 90 % of the gravity at the surface of Earth

to provide suitable circumstances compared to the environment on the planet's surface. Thus, a progress line of human habitation systems at cloud-top level, ranging from short-duration mission outposts to larger permanent settlements can be envisaged with appropriate structural systems, the Aerostats. In order to keep the whole system afloat, theoretically, breathing air contained in an envelope would be sufficient, since our good old oxygen-nitrogen mixture is a lifting-gas on Venus. Thus, the internal volume of the habitat is included in the gas envelope of the aerostat. This can have direct impacts on the configuration of the outpost, concerning the geometry of the interior, stowage and deployment pattern of the envelope membrane. Nevertheless, using the given altitude as a tele-operation command post for robotic surface missions, rather than staying in orbit can have certain advantages, in regard to communication quality with the surface, presence of gravity and scientific research possibilities.

Besides, in the long term Venus vision, floating settlements in the size of a modest city are also envisaged (Landis 2003), yet as idea sketches. Rough estimations of a gas envelope of few-kilometer diameter indicate the need for yet-to-come novelties in material, structural and other technologies. These systems will possibly present a significant amount of opportunities to integrate deployable sub-structures.

Furthermore in deployable structure design field, sampling hardware sets of landers is another segment where advances in material technology and design can have solid impacts on the mission outcomes. Extended drilling equipment, including deployable optical covers to enhance the optical properties of the drill surface or deployable jackstands to stabilize the platform indicate the room for development in this hardware environment (Zacny 2009).

15.5 Resume & Outlook

The applied deployable structures technology of Venus sub-orbital exploration so far has projected a steady and conservative progress line. This gradual and rather slow development of Venus "deployables" can be related to the dominance of Venera and Vega mission technologies. These limited number of missions (including the NASA Pioneer Multiprobe Mission, provide the fire-baptized experience for the research & development of these structures. Besides exploitation of this valuable heritage, technological investments in innovative concepts such as wind-driven mobility can prove to have game-changing effects in Venus exploration endeavor.

The most innovative deployable concepts (e.g. solar aircraft) utilize reciprocal know-how from other planetary exploration architectures. Therefore, research and development for planetary exploration missions will continue to produce common knowhow to be exploited by deployable structure research & development of Venus exploration programs. Especially, in the segment of aerial vehicles of Mars and Venus sub-orbital exploration, a cross innovation of concepts, materials and integration can be expected (Smith et al. 2010).

Concerning the future work on compiling the deployable structure concepts for Venus exploration, the deployable systems of fly-by or orbital missions can also be studied, beside the surface and atmospheric missions. An improvement can also be made in the assessment structure, where scales of deployable systems form the main sections of the article. This can enable the researchers to view the concepts in a spectrum of sizes for extensive evaluation.

Acknowledgments The author would like to thank Elif Erdogdu, Sandra Häuplik-Meusburger and Stephen Ransom for providing the visuals and conducting reviews respectively.

References

Akiba H et al (1976) Feasibility study of buoyant venus station placed by inflated balloon entry, 27th international conference of astronomy. Anaheim, CA

Baker C, Glaze L et al (2010a) Venus mobile explorer (VME): a mission concept study for the National Research Council Planetary Decadal Survey, 7th international planetary probe workshop, Barcelona

Baker C, Glaze L et al (2010b) Venus intrepid tessera lander (VITAL): a mission concept study for The National Research Council Planetary Decadal Survey, NASA GSFC

Baker C, Smith B, Prabhu D, Gage P, Glaze L (2013) Venus in situ explorer mission design using a mechanically deployed aerodynamic decelerator, IEEE explore, Accessed Jan 2014

Ball AJ, Garry J, Lorenz R, Kerzhanovich V (2007) Planetary landers and entry probes. Cambridge University Press, Cambridge, UK

Balint T et al (2007) Extreme environment technologies for future space science missions, Technical Report JPL D–32832. National Aeronautical and Space Administration, Washington, D.C

Balint T, Cutts J (2009) Technologies for future venus exploration, White Paper to the NRC

Basilevsky A et al (2007) Landing on Venus: past and future. J Planet Space Sci 55:2097–2112

BBC (2009) Venera-D Mission Report "Russia plots return to Venus". http://news.bbc.co.uk/2/hi/science/nature/8294925.stm (May 2014)

Bienstock B (2003) Pioneer Venus and Galileo entry probe heritage. In: Proceedings of international workshop "Planetary probe atmospheric entry and descent trajectory analysis and science", Lisbon

Chassefière E et al (2002) The Lavoisier mission : a system of descent probe and balloon flotilla for geochemical investigation of the deep atmosphere and surface of Venus. Adv Space Res 29 (2):255–264, Elsevier UK

Chassefière E et al (2007) European Venus explorer (EVE), An in-situ mission to Venus, ESA cosmic vision proposal

Clark IG et al (2009) Supersonic inflatable aerodynamic decelerators for use on future robotic missions to Mars. J Spacecraft Rockets 46(2)
Colozza A (2004) Evaluation of solar powered flight on Venus, 2nd international energy conversion engineering conference, RI
Colozza A (2012) Radioisotope stirling engine powered airship for low altitude operation on Venus, NASA/CR—2012-217665
Cutts J et al (1999) Venus surface sample return: role of balloon technology, AIAA balloon technology conference, Norfolk, VA
Cutts J, Balint T, Chassefière E, Kolawa E (2007) Technology perspectives in the future exploration of Venus, AGU
Decadal survey inner planets sub-panel, VEXAG_Venus exploration analysis group. http://www.lpi.usra.edu/vexag/
Dyson R, Schmitz P et al (2009) Long-lived Venus conceptual lander design. The international energy conversion engineering conference, Denver, CO
Esposito LW, Stofan E, Cravens T (2013) Exploring Venus as a terrestrial planet. Wiley, NJ, USA
Evans M et al (2005) A Venus Rover capable of long life surface operations, 2005 AGU fall meeting, San Francisco, CA
Ford PG, Pettengill GH (1992) Venus topography and kilometer-scale slopes. J Geophys Res 97:13103–13114
Griffin K (2013) Venus atmospheric maneuverable platform (VAMP): a concept for a long-lived airship at Venus, Northrop grumman aerospace public release, NGAS Case 13-1131
Grossman L (2014) NASA 'flying saucer' for Mars to land in Hawaii. New Scientist, Elsevier, 222 (2965)
Guest SD (1994) Deployable structures: concepts and analysis, Dissertation submitted to University of Cambridge, UK
Guynn MD et al (2003) Evolution of a Mars airplane concept for the ARES Mars scout mission, Paper AIAA 2003-6578, Second AIAA unmanned unlimited systems, technologies, and operations conference, 15–18 Sept 2003, San Diego, CA
Hajos G et al (2005) An overview of wind-driven rovers for planetary exploration, 43rd AIAA aerospace sciences meeting, Reno, NV
Hall J et al (2008) Prototype design and testing of a Venus long duration, high altitude balloon, Adv Space Res 42
Hall J, Cutts J (2013) Venus aerial platform technology. Presentation at the Venus technology forum, JPL, CA
Hall J, Yavrouian A (2013) Pinhole effects on Venus superpressure balloon lifetime. AIAA balloon systems conference, Daytona, FL
Harvey B, Zakutnyaya O (2011) Russian space probes, Springer Praxis Books
Haeuplik-Meusburger S, Ozdemir K (2012) Deployable lunar habitation design. In: Badescu V (ed) MOON: Prospective energy and material resources. Springer
Ivanov MA (2001) Tessera terrains on Venus: a summary of the global distribution, characteristics and relation to surrounding units from Magellan data, J Geophys Res 101
JPL NASA (2010) Surface and atmosphere geochemical explorer, facts sheet, online May 2014
JPL NASA (2013) Low density supersonic decelerators, NASA fact sheet JPL 400-1530 6/13
Kerzhanovic V et al (2000) Venus aerobot multisonde mission: atmospheric relay for imaging the surface of venus, IEEE
Kerzhanovic V, Yavrounian A, Hall J, Cutts J (2005) Dual balloon concept for lifting payloads from the surface of Venus, AIAA 5th aviation, technology, integration and operations conference, AIAA-2005-7322
Kerzhanovic V, Yavrounian A, Hall J et al (2008) Prototype design and testing of a Venus long duration, high altitude balloon. Adv Space Res 42:1648–1655
Kremnev RS et al (1986a) VEGA balloon system and instrumentation. Science 231(4744):1408
Kremnev RS et al (1986b) The Vega balloons—a tool for studying atmosphere dynamics on Venus. Sov Astron Lett 12(1):7

Landis GA (2001) Exploring Venus by solar airplane, STAIF conference on space exploration technology, Albuquerque, NM
Landis GA (2003) Colonization of Venus, conference on human space exploration, space technology & applications international forum, Albuquerque, NM
Landis GA, LaMarre C, Colozza A (2003) Atmospheric flight on Venus: a conceptual design. J Spacecraft Rockets 40:672–677
Landis GA (2006) Robotic exploration of the surface and atmosphere of Venus, Acta Astronautica 59(570-406)
Landis GA (2010) Low-altitude exploration of the Venus atmosphere by balloon, 48th AIAA aerospace sciences meeting, Orlando, FL
Landis GA, Schmidt G, Oleson S (2011) HERRO missions to Mars and Venus using telerobotic surface exploration from orbit, AIAA Space Conference & Exposition, Long Beach, CA
Landis GA, Colozza A et al (2013) A wind powered rover for a low-cost Venus mission, 51st AIAA aerospace sciences meeting, Grapewine, TX
Landis GA (2012) Report on Zephyr Rover. http://www.nasa.gov/directorates/spacetech/niac/2012_phase_I_fellows_landis.html (May 2014)
Landis G et al (2013) Venus landsailer: a new approach to exploring our neighbor planet presentation. NASA Glenn, Cleveland, OH
LASP SAGE mission website. http://lasp.colorado.edu/sage/ (May 2014)
Levine JS et al (2003) Science from a Mars airplane: the aerial regional scale environmental survey (ARES) of Mars, Paper AIAA 2003-6576, Second AIAA unmanned unlimited systems, technologies, and operations conference, San Diego, CA
Lorenz R et al (2011) A stand-alone long-lived Venus lander mission concept, 8th international planetary probe workshop, Portsmouth, VA
McRonald AD (1999) A light-weight inflatable hypersonic drag device for Venus entry, AAS/AIAA astrodynamics specialist conference, Girdwood, AK
NASA (2008) Robotic exploration of Venus promotional video. http://www.youtube.com/watch?v=oet63vzBvkg (2014)
NASA (2012) News release on Zephyr Rover. http://www.nasa.gov/directorates/spacetech/home/feature_windsurfing.html#.U6XrkV4WfZs (May 2014)
NASA (2013a) Dynamic penetrator experiment data sheet. http://nssdc.gsfc.nasa.gov/nmc/experimentDisplay.do?id=1981-110D-10 (May 2014)
NASA (2013b) Vega balloon data sheet. http://nssdc.gsfc.nasa.gov/nmc/spacecraftDisplay.do?id=1984-128F (May 2014)
Ozdemir K, Haeuplik-Meusburger S, Gruber P, Imhof B, Antonietta M, Waclavicek R (2007) Deployable structures for a human lunar base, Elsevier
Ozdemir K (2009) A methodical approach to the transfer and the integration of design knowledge from terrestrial extreme environment structure designs to inhabited space structure design concepts, PhD Thesis, Vienna University of Technology
Pellegrino S (ed) (2002) Deployable structures. Springer, CISM International Center for Mechanical Sciences
Sagdeev RZ et al (1986) The VEGA Venus balloon experiment. Science 231(4744):1407
Sagdeev RZ, Moroz VI (1986) Project vega first stage: missions to Venus. Sov Astron Lett 12(1)
Sagdeev RZ et al (1986) The VEGA balloon experiments. Sov Astron Lett 12(1, 3)
Smith BP et al (2010) A historical review of inflatable aerodynamic decelerator technology development, Aerospace conference 2010, IEEE
Space studies board of National Research Council NRC (2003) New frontiers in the solar system, an integrated exploration strategy, Technical report, Washington, DC
Squyres S et al (2011) Visions and voyages for Planetary Science in the Decade 2013-2022 (Decadal Survey). The National Academics Press, Washington, DC
Sweetster T et al (2003) Venus sample return missions—a range of science, a range of costs. Elsevier
Titov DV et al (2007) Venus express—the first European mission to Venus. Planet Space Sci 55:1636–1652

University of Cambridge, Structures Group Website, Department of Engineering. http://www-civ.eng.cam.ac.uk/dsl/ (May 2014)
Venera-D mission official website (2014). http://venera-d.cosmos.ru/
Venkatapathy E et al (2011) Going beyond rigid aeroshells: enabling Venus in-situ science missions with deployables, 8th international planetary probe workshop, Portsmouth, VA
Zacny K, Bar-Cohen Y (2009) Drilling in extreme environments. Wiley-Vch Verlag, Weinheim
Zadunaisky PE et al (1961) Experimental and theoretical results on the orbit of echo 1, SAO Special Report Issue 61
Zasova LV et al (2007) Structure of the Venus atmosphere. Planet Space Sci 55:1712–1728

Chapter 16
A Systems Approach to the Exploration and Resource Utilization of Venus and Mercury

Dragoş Alexandru Păun

16.1 Introduction

Missions to Venus have been launched starting with the year 1961 and a total of 38 missions have been so far launched covering a range of mission scenarios including fly-by, orbiting and atmospheric and landing probes. Of those missions, 16 had experienced partial or total failure and a total of 14 missions, mainly launched by the USSR, have been constructed with the aim of In-Situ studies of the Venusian atmosphere and surface (Fig. 16.1). Following a trial and error approach, initial missions have gathered increasing amounts of information on the environmental conditions of the planet's atmosphere and surface. The iterative approach of the Soviet probes has proven to be very successful and culminated with the Vega 1 and 2 missions that each included both a lander and a balloon in addition to the orbiting mother craft.

A total of 5 NASA missions also concentrated on Venus, with the Mariner 2, 5 and 10 performing fly-by type manoeuvers around the planet and culminating with the success of the Pioneer Venus 1 and 2 missions which were composed of an orbiter and a dispensing platform which released a total of 4 probes (one large and three small identical probes) (Bienstock 2003). These historical missions have shed light on how extreme the Venusian environment is, and have gathered invaluable experience in designing systems that would perform science operations even for extended periods of time.

The purpose of the present chapter is to give an overview of the state of the art technologies that could be involved in future exploration missions to the inner solar system and to exemplify one such mission by proposing a system architecture able

D.A. Păun (✉)
Airbus Defence and Space, Bremen, Germany
e-mail: alexandru.paun@yahoo.co.uk

V. Badescu and K. Zacny (eds.), *Inner Solar System*,
DOI 10.1007/978-3-319-19569-8_16

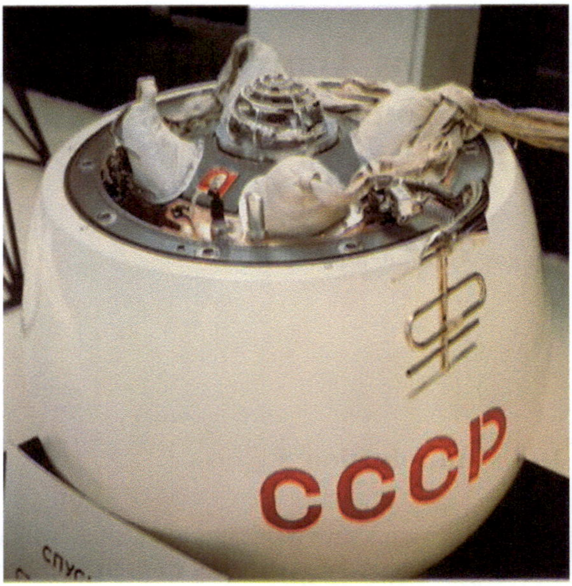

Fig. 16.1 Venera 4 descent capsule (Image credit: NASA)

to maximise the scientific return while utilizing very successful heritage technologies.

In addition, a short overview of what can be easily considered as being resources of Venus is given in the following section.

16.2 Venusian Resources

16.2.1 Venus Gravity Assist Manoeuvers

The manoeuvers generally performed by exploration spacecraft drive their mass and generally mission cost significantly. As such, both missions dispatched to the inner solar system (e.g. the future Solar Orbiter mission) as well as missions dispatched towards the outer solar system (e.g. Cassini, Voyager) make use of gravity assist manoeuvers in order to keep their mission cost down. This type of approach requires substantial planning and a significant increase in mission time but is an enabling factor for a number of missions.

A very good description of planetary fly-by manoeuvers can be found in Curtis (2014). Here, a very good theoretical description is presented including an example of a Venus fly-by. It is important to note that Venus has a relatively large synodic period of 1.6 years and that more complex missions requiring multiple approaches become increasingly more complex to design. Table 16.1 presents a short summary of past and future missions that utilized one or multiple fly-by manoeuvers by Venus.

Table 16.1 Past and future missions to use Venus gravity assist manoeuvres

Mission	Year	No. of Venus gravity assist manoeuvres	Mission target	Mission success
Mariner 10	1973	1	Mercury	Yes
Galileo	1989	1	Jupiter and its Moons	Yes
Cassini	1997	2	Saturn and its Moons	Yes
Messenger	2004	2	Mercury	Yes
BepiColombo	2016*	2	Mercury	N/A
Solar Orbiter	2017*	5	Sun	N/A
Solar Probe+	2018*	7	Sun	N/A

*-planned

16.2.2 Venus Trailing Orbit as Vantage Point for NEO Observations

Near Earth Objects (NEOs) are present in very large numbers in our Solar System. Current detection efforts have yielded impressive results with large numbers of new discoveries coming from both earth based observatories and space based assets. However, the need to dramatically increase the rate of discovery is pushing for novel concepts such as the Sentinel mission of the B612 Foundation now being developed at Ball Aerospace. The Sentinel mission scenario foresees an infrared space based telescope orbiting the Sun in what is being described as a Venus-like orbit offering the spacecraft a unique vantage point from where to detect objects otherwise very difficult to detect. This type of observations can only be performed by observatories orbiting in the inner Solar System thus being able to more easily identify objects in the IR spectrum. It is as such possible to avoid the potential destructive consequences of events such as the recent Chelyabinsk event that saw the atmospheric entry of a 20 m diameter NEO which remained undetected until the time of entry and caused significant material damage.

16.3 Venusian Exploration Systems

The following section aims at proposing a top level architecture that would ensure the achievement of several highly important scientific topics aimed at the Venusian atmosphere and geology as well as investigating the state of the art technologies presently available.

At the time this chapter is being written, the ESA Venus Express mission is drawing to a close having successfully performed an aero-braking manoeuvre with the purpose of studying the outer layers of the Venusian atmosphere. As a consequence of this and also of the JAXA Akatsuki mission's failure to achieve orbit in December 2010, any new mission to Venus involving In Situ research must employ a two segment approach with an orbiter providing continuous science operations as well as serving as a relay between Earth and the atmospheric/surface exploration probe.

Hall et al. (2009) presents a very complex and well-studied architecture that involves both of the aforementioned segments. The Venus Flagship Design Reference Mission (DRM) approaches the mission design much in the spirit of the Pioneer Venus probes, describing a dual launch architecture using an Atlas V 551 via a Type IV transfer orbit. However, given the recent advances in launch technology such as the success of the Falcon 9 and Angara launchers, it is conceivable that in the near future, launchers such as the Falcon 9 Heavy with superior lift capabilities w.r.t. the Atlas V launch vehicle will become a reality. Such a capability would allow for launching of both the Orbiting and In Situ Segments at the same time, greatly reducing both cost and system complexity.

16.3.1 Mission Statement

The mission statement of the proposed Venus mission takes into consideration the two segments necessary for maximising the scientific return of such a mission:

> To explore in sufficient detail the Venusian Atmosphere and its Geology by performing Remote Sensing and In-Situ Measurements.

The chosen name for such a hypothetical mission is VeGE which stands from Venus Global Explorer.

16.3.2 Mission Objectives

Limaye et al. (2011) and Esposito et al. (2007) have clearly defined the scientific goals for the future missions to Venus and a short summary of their classification is here given.

Three main goals have been identified for a comprehensive future exploration of Venus according to Limaye et al. (2011):

- 1st Goal - Identify the Origin and Evolution of Venus
- 2nd Goal - Identify the Processes that have shaped and are still shaping the planet
- 3rd Goal - Identify the greenhouse processes, search for the presence of water and characterise how the Venus interior, atmosphere and surface are interacting

These goals were identified as put forward in the frame of the Planetary Science Decadal Survey devised by US and international scientists with the purpose of driving future planetary missions being built by NASA in collaboration with international organizations.

So far, the NASA Solar System Exploration Roadmap (2006) considered two Venus missions, the Flagship Class Venus Mobile Explorer and the New Frontiers Class Venus In-Situ Explorer, however, the difficult environment in which future missions to Venus will have to operate, w.r.t. to the milder Martian environment for example, means that these missions could be considered only past the 2020–2030 timeframe. It is however fortunate that due to the position of Venus and its previously mentioned synodic period, transit times to the planet will be short and launch opportunities will repeat at almost one and a half year intervals providing numerous launch opportunities.

Esposito et al. (2007) summarizes in a very clear way the scientific issues that need to be addressed and also offers certain directions to be followed. As such, a prime candidate for an In situ mission would be represented by the Ishtar Terra highland region which was previously not explored by any of the Venera or Pioneer Venus landers. It can be noted from Fig. 16.2 that the Ishtar Terra region can be found at latitudes of above 60° north and that this will mainly drive the inclination of the necessary orbiting platform. Considering these aspects, for the purpose of the hypothetical VeGE mission described in this chapter, a polar orbit will be considered which would also allow for imaging coverage of the entire planet surface improving on the existing data set (Fig. 16.3).

Considering the information presented in this section, the following summary of objectives for the VeGE mission is presented:

- To provide high resolution remote sensing of Venus' entire surface and internal processes
- To develop and mature novel re-entry technologies aimed at reducing the high-g and high heat flux environment during Venus atmospheric entry
- To provide an in depth view of the Venusian atmosphere at different altitudes through the use of novel technologies for long duration atmospheric flight and atmospheric sampling and analysis
- To develop technologies and techniques enabling an extended Venus surface exploration
- To provide detailed information of the Venusian surface and sub-surface through the use of novel instrumentation that will improve on the collection of samples, analysis speed and accuracy for geological and petrological analysis

16.3.3 Mission Scenario

The presented scenario proposes an architecture focused on maximising scientific return by utilizing as much as possible heritage technologies in the areas such as thermal control while investigating novel concepts for areas such as entry, descend and landing (EDL) technologies, structures and sample acquisition (Fig. 16.4).

In order to maximize scientific return, a lander could also release one or several balloons during its descent without having to severely increase the mission

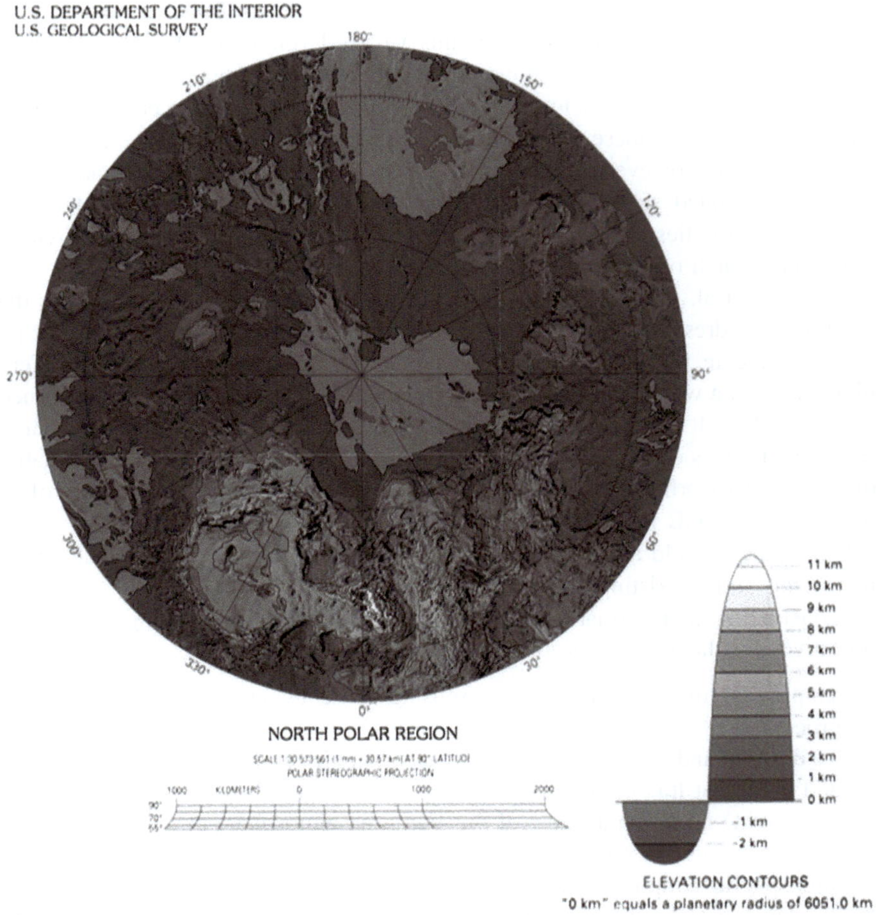

Fig. 16.2 Map of North Polar region of Venus with the Ishtar Terra region clearly visible (Image was extracted from a topographic map of Venus by USGS; Image credit: USGS)

complexity. Several balloon concepts have been proposed and several have also been flown successfully (Vega 1 & 2 balloons + Chassefière et al. 2006 proposal from ESA). Balloons have the advantage of being able to cover large distances in the atmosphere and sample the atmosphere at different locations also measuring wind speed and direction.

Although the orbiting segment poses a series of distinct technological challenges, significant past experience has been gained in constructing spacecraft dispatched towards the inner solar system such as the Venus Express and Magellan spacecraft. As such, the orbiting segment will represent a relatively easy and economic endeavour. Orbiting segment would have an altitude with a low point between of around 200–400 km around Venus at periapsis and a maximum of

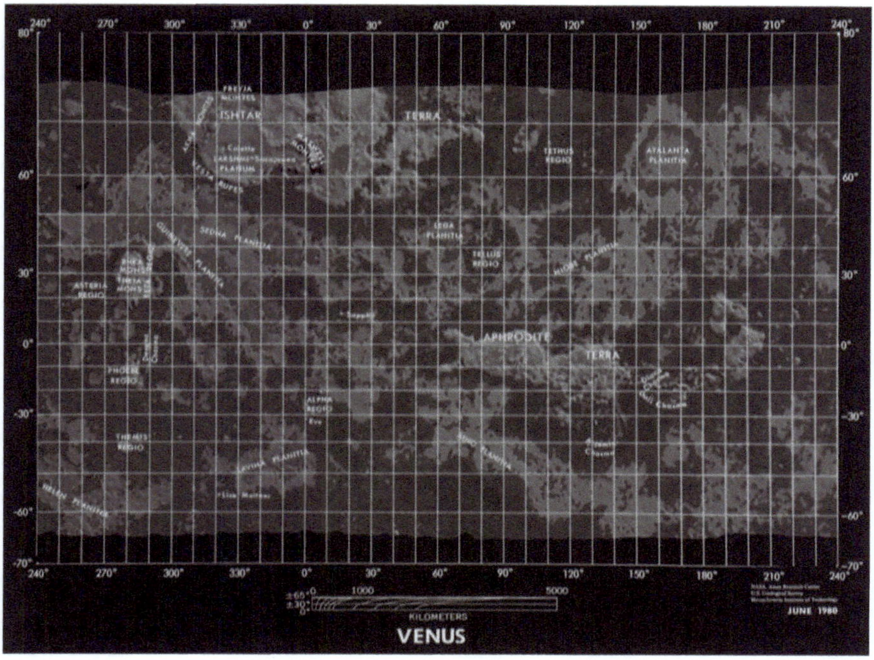

Fig. 16.3 Map of Venus (Image credit: NASA Ames Research Center/USGS)

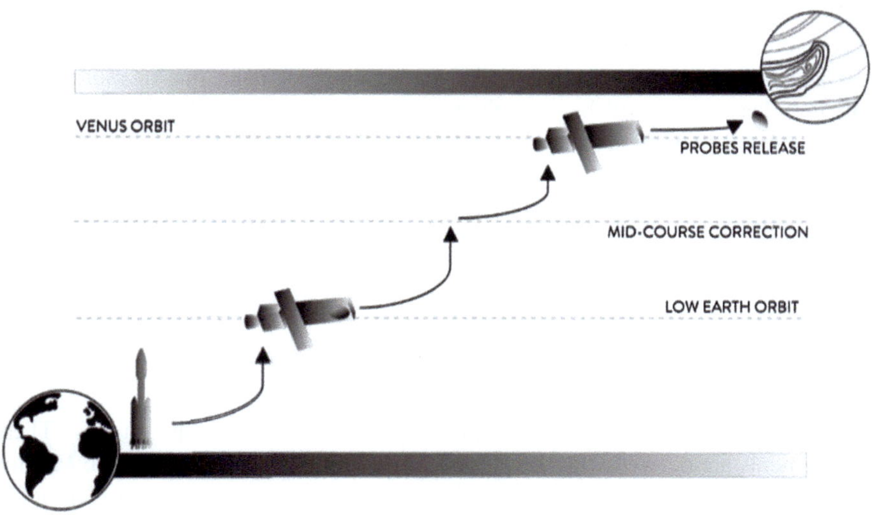

Fig. 16.4 Proposed mission scenario of the VeGE mission

Fig. 16.5 ESA Venus Express (Image credit: ESA)

2000 km at apoapsis orbiting the planet in a polar plane in order to ensure near full coverage of the planet (Fig. 16.5).

Once orbit of Venus is achieved by the orbiting segment, one or more probes will be released and will begin their descent. Upon re-entry, the heat shield will release such as in previous missions and data collection will begin. It is at this phase that any balloons foreseen for the mission shall be released. Upon landing, all instrumentation shall function simultaneously due to the short time of the mission on surface.

16.3.4 Top Level Requirements and Functional Analysis

The VeGE mission will have to satisfy specific top level requirements. These requirements are derived with the purpose of ensuring the fulfilment of the Mission Statement and Objectives in an iterative manner that takes into consideration a functional breakdown and analysis of the system.

Due to the usual complexity of a requirements document specific to exploration missions, only selected top level requirements shall be presented in the present section.

The top level requirements presented in this section are grouped in three categories:

- Common Top Level Requirements are used to define general architecture requirements
- Venus Orbiting Segment Top Level Requirements are used to define top level requirements specific to the orbiting segment
- Venus In Situ Segment Top Level Requirements (containing two sub-segments) are used to define top level requirements specific to the In Situ segment

16.3.5 Common Top Level Requirements

- R-1 The VeGE mission architecture shall provide for remote and in situ measurements of the Venusian Atmosphere, Surface and Sub-Surface for an extended period of time
- R-2 The VeGE mission architecture shall provide for communication services between the architecture elements and Earth
- R-3 The VeGE mission architecture elements shall be able to communicate with both DSN and ESTRACK networks
- R-4 The VeGE mission architecture elements shall withstand launch, transit and entry environment

16.3.6 VeGE Orbiting Segment

- R-1 The VeGE Orbiting Segment shall allow for mechanical, power, thermal and data interfaces for the VeGE In Situ segment
- R-2 The VeGE Orbiting Segment shall allow for separation and entry trajectory insertion of the VeGE In Situ segment
- R-3 The VeGE Orbiting Segment shall perform remote measurements
- R-4 The VeGE Orbiting Segment shall survive for a nominal period of 5 years in Venus orbit
- R-5 The VeGE Orbiting Segment shall survive for a nominal period of 10 years in Venus orbit
- COM-1 The VeGE Orbiting Segment shall provide for communications with Earth
- COM-2 The VeGE Orbiting Segment shall provide for communications between Earth and In Situ Segment

16.3.7 *VeGE In-Situ Segment*

16.3.7.1 VeGE Atmospheric Sub-Segment

- R-1 The VeGE Atmospheric Sub-Segment shall perform in situ measurements of the Venusian Atmosphere
- R-2 The VeGE Atmospheric Sub-Segment shall survive for a nominal period of TBD in the Venusian atmosphere

16.3.7.2 VeGE Surface Sub-Segment

- R-1 The VeGE Surface Sub-Segment shall survive for a nominal period of one Earth day on the Venusian Surface
- R-2 The VeGE Surface Sub-Segment shall survive for an extended period of one Earth week on the Venusian Surface
- R-3 The VeGE Surface Sub-Segment shall perform detailed Analysis of the Venusian atmosphere during the descent and landed phases
- R-4 The VeGE Surface Sub-Segment shall perform drilling and core extraction
- R-5 The VeGE Surface Sub-Segment shall perform detailed analysis of the Venusian soil
- R-6 The VeGE Surface Sub-Segment shall perform seismic measurements for it entire surface lifetime

16.3.8 *Sub-System Specific Aspects*

Due to the fact that the orbiting segment of the mission will be built using already tried and true technologies, this section will focus only on the atmospheric/surface segment and on moreover on the critical subsystems and technological choices that will drive the specific design of this segment. It is clear that enabling the required mission life on the surface of Venus would necessitate significant advances in all of the described areas and/or new strategies to be implemented and as such, following a comprehensive literature study several solutions are proposed.

16.3.8.1 Entry, Descent and Landing (EDLS)

A critical component of the atmospheric/surface segment lies with the atmospheric entry strategy. A thorough analysis of past Venus entry systems as well as alternative entry methods was presented by Dutta et al. (2012). Their work shows that previous missions have sustained loads in excess of 100 g's and total peak heat fluxes of around 1000 W/cm^2.

Dutta et al. (2012) presents various alternative entry systems such as inflatable systems and mechanically deployable systems that have the potential to reduce launch profile as well as decreasing mass and reducing the overall thermal load and acceleration upon re-entry. The recent successful test of the NASA Low-Density Supersonic Decelerator (LDSD) have raised the TRL of inflatable entry technology to 6.

16.3.8.2 Thermal Control Sub-System (TCS)

The TCS subsystem shall have to take into consideration the distinct phases starting from launch, transfer and Venus orbital/atmospheric/surface thermal environments. Each of these mission phases imposes specific thermal requirements. A short description of each environment is given herein.

16.3.8.3 Thermal Environments

Venus Transfer Environment
As previously mentioned in the Mission Scenario section of the present chapter, a direct transfer between Earth and Venus is assumed for the VeGE mission, hence direct solar radiation will be the predominant source of environmental heat. Depending on the departure date, the solar constant at Earth departure varies between 1420 W/m^2 in the hot case and 1360 W/m^2 in the cold case, Larson and Wertz (2006). Table 16.2 presents the solar constant at Venus for both cold and hot cases.

Venus Orbit Environment
Venus's orbital environment is influenced as stated by Gilmore (2002), by its predominant cloud cover which is responsible for the high albedo, low planetary IR and other complex effects.

Venus Atmospheric/Surface Environment
The atmosphere of Venus is largely composed of CO_2 and at meal elevation, the average temperature is of 463.5 K and the surface temperature varies slightly with latitude (Esposito et al. 2007).

Table 16.2 Venus orbital environments, Gilmore (2002)

	Perihelion	Aphelion	Mean
Direct solar (W/m^2)	2759	2650	2614
Albedo	0.8 ± 0.02	0.8 ± 0.02	0.8 ± 0.02
Planetary IR (W/m^2)	153	153	153

16.3.8.4 Thermal Requirements and Available Technical Solutions

Typical electronic components must be usually kept within a range of −15 °C to +50 °C (Fortescue et al. 2003), with specially designed electronics able to withstand higher temperatures (sometimes up to +300 °C) but at significantly higher cost and complexity. Moreover, the reduced mission life implies that, while on the planet, all available subsystems (e.g. scientific payload, C & DH) will have to function simultaneously, thus both utilizing peak power and rejecting peak heat.

Previous missions relied solely on a passive approach whereas, the entry probe was protected during re-entry by a heat shield, while during parachute phase and on the surface, the design goal has been to minimise thermal transfer between Venus' surface environment and the spacecraft allowing for sufficient time to provide for measurements to be taken and their transmission to the relaying spacecraft to be completed. This approach limited the capability of the landers to operate for more than 120 min and would not be feasible for long duration science missions (Fig. 16.6).

Hall et al. (2009) presents an evaluation of several refrigeration techniques that could be utilized and ultimately proposes a two stage system that would both cool the spacecraft and the power source. Refrigeration techniques however require large amounts of power which in itself requires more mass and heat rejection capabilities. Dyson and Bruder (2010) presents a summary of Venus lander architectures with different power generation and cooling strategies while expanding on the concept of

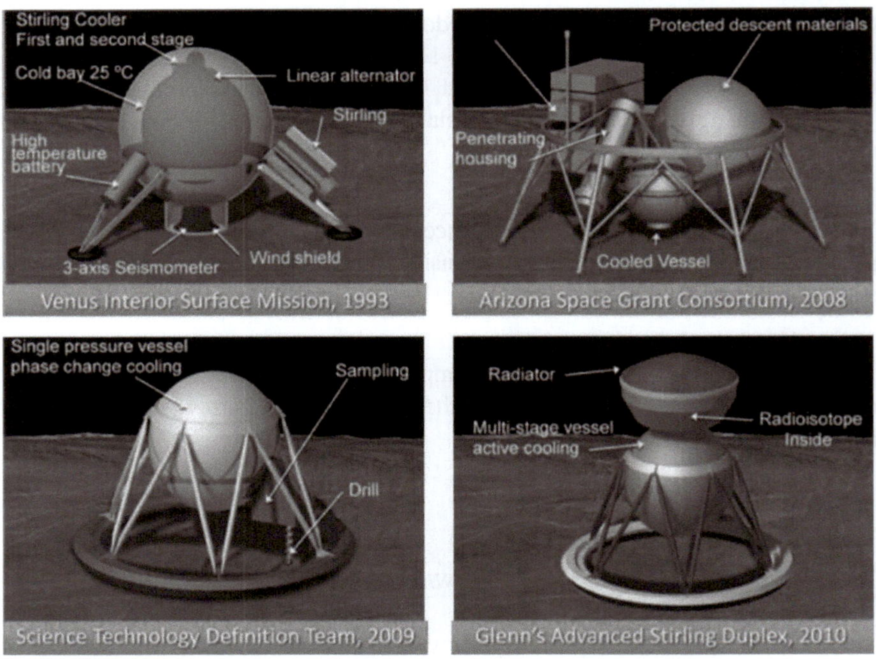

Fig. 16.6 Lander concepts (Image credit: Dyson and Bruder 2010)

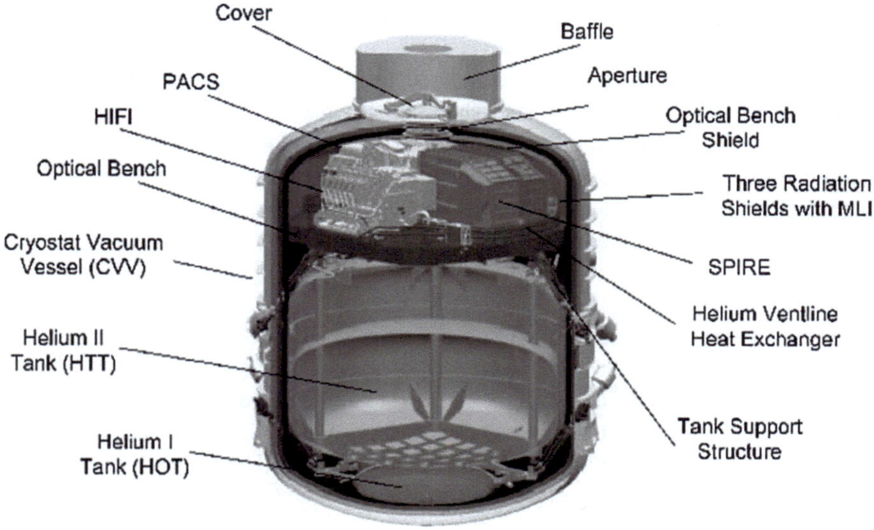

Fig. 16.7 Cryostat used during the ESA Herschel mission (Image credit: ESA)

Hall et al. (2009) and proposing a long term refrigeration system based on the Advanced Stirling Radioisotope Generator (ASRG). Such a system would enable missions in excess of several months due to the high power generation capability and the coupling of power generation and refrigeration. However, the ASRG program has been cancelled by NASA in 2014 and considering the added complexity, power needs and mechanical vibration levels coming from the mechanical apparatus a strategy involving phase change cooling would be desirable for a mission with the duration and scope of the proposed VeGE mission.

Several research spacecraft have already used cryostats to cool special electronics to very low temperatures, such as the case of the European Space Agency's Herschel mission. The Herschel mission made use of a cryostat using superfluid helium to cool its instruments (Pilbratt 2008) (Fig. 16.7). The maturity of such technology as opposed to the previously discussed strategies would highly recommend it for use in a short/medium length Venus mission.

16.3.9 Mechanical Sub-system

Given the previously described environmental conditions specific to Venus, it is evident that the high temperature and corrosive environment will substantially degrade the material performance of structural materials and useful life. For example, the Ti6Al4 V material, typically used in high strength, high temperature environments loses around 60 % of its room temperature allowable values when subjected to temperatures in excess of 500 degrees Celsius.

The construction of the main capsule body will have to take into consideration the high atmospheric pressure of over 92 bars as well as atmospheric entry and landing loads. Since pressure is considered to be the design driver on the planet surface, a mainly spherical pressure shell construction is implied.

Figure 16.8 exemplifies the internal construction of the NASA Pioneer Venus 2 (Large Probe). As it can be observed, internal components have been carefully placed in order to avoid as much as possible contact with the pressure shell and this minimize conductive heating.

16.3.9.1 Power Generation

Due to Venus' heavy cloud cover, solar power can be excluded from the onset as a viable power source for the atmospheric/surface sub segments. Traditionally, batteries have been used on all Venus landers and balloons. The Venus Vega missions have used silver-zinc batteries (Bienstock 2003) and these batteries have the advantage of good power density (around 2130 Wh/kg) and are ideal for the VeGE mission's duration. The NASA MMRTG currently being used on the Curiosity rover has a mass of around 45 kg and comparable power generation capability. However, as stated above, a Radioisotope Thermoelectric Generator would contribute significantly to the Thermal Control Subsystem mass while adding approximatively 10 kg of mass (the evaluation has been performed by considering the values offered by Hall et al. (2009).

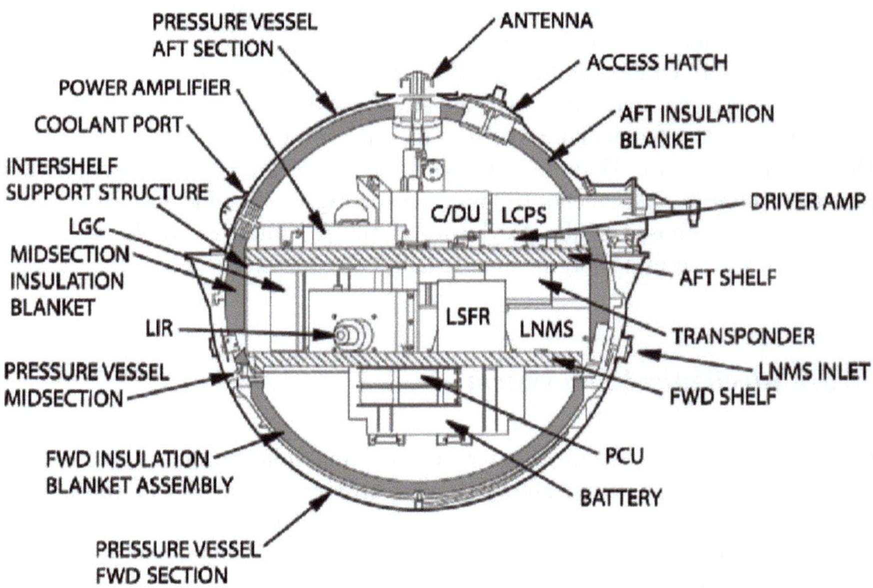

Fig. 16.8 Pioneer Venus 2 - large probe (Image credit: Bienstock 2003)

16.4 Mercury Exploration Systems

It is clearly visible from Table 16.3 that Mercury has not presented a large historical interest to planetary programmes. This fact is also related to the relative difficulty of obtaining and maintaining orbit around the planet given its proximity to the Sun as well as to the challenging thermal environment around the planet. However, the recent Messenger (Mercury Surface Environment Geochemistry and Ranging) mission has been the first mission to achieve orbit around Mercury and has since been continuously mapping the surface of the planet. The follow up mission, designed to complement the Messenger mission is the European Space Agency BepiColombo mission (Fig. 16.9). Hauck et al. (2010) proposes a landing mission to the planet and postulates that such a mission would answer several open questions.

Table 16.3 Past and future missions to Mercury

Mission	Year	No. of Venus gravity assist manoeuvres	Mission target	Mission success
Mariner 10	1973	1	Mercury	Yes
Messenger	2004	2	Mercury	Yes
BepiColombo	2016*	2	Mercury	N/A

* - planned

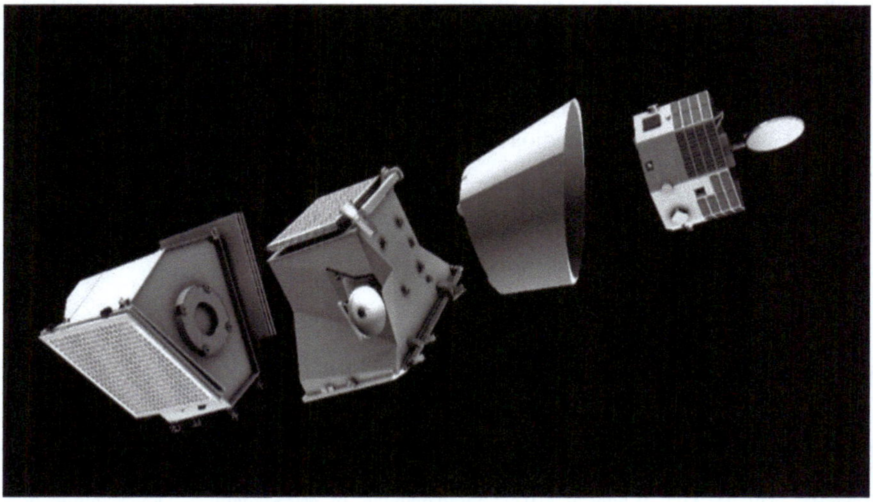

Fig. 16.9 BepiColombo spacecraft description (Image credit: ESA)

16.5 Conclusion

Although a significant number of missions have already visited Venus, a considerable amount of questions remain unanswered due to the severe environment and difficulty in gathering scientific measurements. This chapter shows that a considerable amount of work has been performed by teams around the world and that the concept of medium to long term science being performed at Venus is plausible. However, several key technologies such as power generation and thermal control need to be advanced beyond TRL4-5 and the costs and complexity of missions aiming to land on Venus or studying its atmosphere remain high.

The inner planets represent one of the keys to understanding the birth and evolution of our Solar System and their study should attract renewed efforts.

Glossary

Aerobraking Menoeuver	are manoeuvres that allow a spacecraft to use the upper atmosphere of a planet in order to reduce velocity through drag
ASRG	Advanced Stirling Radioisotope Generator
C & DH	Communication and Data Handling
Gravity Assist Manoeuver	are manoeuvres that are used to add momentum to a spacecraft by making use of its relative movement around a body and that body's gravity
RTG	a Radioisotope Thermoelectric Generator represents a type of power generation device that converts the heat generated by a radioactive material into electric power
TRL	the Technology Readiness Level reflects the level of maturity of a particular space technology and the degree of risk it implies

References

Baines KH, Atreya SK, Chutjian A (2007) Exploring Venus with balloons: science objectives and mission architectures. In: 5th international planetary probe workshop, Bordeaux

Bate RR, Mueller DD, White JE (1971) Fundamentals of astrodynamics. Dover Publications, INC., New York

Bienstock BJ (2003) Pioneer Venus and Galileo entry probe heritage. In: International workshop on planetary probe atmospheric entry and descent trajectory analysis and science, Lisbon, Portugal pp 6–9 October 2003

Curtis HD (2014) Orbital mechanics for engineering students, 3rd edn. Elsevier
Dutta S, Smith B, Prabhu D, Venkatapathy E (2012) Mission sizing and trade studies for low ballistic coefficient entry systems to Venus, IEEE
Dyson RW, Bruder GA (2010) Progress towards the development of a long-lived Venus lander duplex system, NASA Glenn Research Center
ECSS-E-ST-10C (2009) Space engineering: system engineering general requirements. Noordwijk, The Netherlands
Esposito LW, Stofan ER, Cravens TE (2007) Exploring Venus as a terrestrial planet, book. American Geophysical Union
Fortescue P, Stark J, Swinerd G (2003) Spacecraft systems engineering, 3rd edn. Wiley, Book
Gilmore DG (2002) Spacecraft thermal control handbook—volume I: fundamental technologies, 2nd edn. The Aerospace Press/American Institute of Aeronautics and Astronautics
Hall JL, et al. (2009) Venus flagship mission study: report of the Venus science and technology definition team, JPL
Hauck SA, Eng DA, Tahu GJ (2010) Mercury lander mission concept study, NASA
Landis GA (2010) Low-altitude exploration of the venus atmosphere by balloon. AIAA Aerospace Sciences Meeting, Olando
Larson WJ, Wertz JR (2006) Space mission analysis and design, 3rd edn. Space Technology Library and Microcosm Press, El Segundo
Limaye S, Smrekar S, VEXAG Executive Committee (2011) Venus white paper for planetary sciences decadal survey inner-planets panel
NASA (2006) Solar system exploration roadmap
Nash WA (1995) Hydrostatically loaded structures—the structural mechanics. Pergamon, Analysis and Design of Powered Sumbersibles
Pilbratt GL (2008) Herschel mission overview and key programmes, European Space Agency—ESTEC
Sergeyevsky AB, Yin NH (1983) Interplanetary mission design handbook, volume I, part 1—Earth to Venus ballistic mission opportunities, 1991–2005, Book, JPL, Pasadena

Chapter 17
Artificial Magnetic Field for Venus

Alexander A. Bolonkin

17.1 Introduction

The Earth's magnetic field is shaped roughly as a magnetic dipole, with the poles currently located proximate to the planet's geographic poles. At the equator of the magnetic field, the magnetic field strength at the planet's surface is 3.05×10^{-5} T, with global magnetic dipole moment of 7.91×10^{15} T m^3. According to dynamo theory, the field is generated within the molten outer core region where heat creates convection motions of conducting materials, generating electric currents. These in turn produce the Earth's magnetic field. The convection movements in the core are chaotic; the magnetic poles drift and periodically change alignment. This causes field reversals at irregular intervals averaging a few times every million years. The most recent reversal occurred approximately 700,000 years ago. The field forms the magnetosphere, which deflects particles composing the solar wind. The sunward edge of the bow shock is located at about 13 times the radius of the Earth. The collision between the magnetic field and the solar wind forms the Van Allen radiation belts, a pair of concentric, torus-shaped regions of energetic charged particles. When the plasma enters the Earth's atmosphere at the magnetic poles, it forms the aurora.

The Earth's magnetic field and the ozone layer blocked harmful solar radiation, and permitted formerly ocean-confined life to move safely to land. The physical properties of the Earth, as well as its geological history and orbit, have allowed life to persist.

In 1967, Venera 4 found the Venusian magnetic field to be much weaker than that of Earth. This magnetic field is induced by an interaction between the ionosphere and the solar wind, rather than by an internal geodynamo situated in the core like the one inside Earth. Venus's small induced magnetosphere provides

A.A. Bolonkin (✉)
C&R, Brooklyn, NY, USA
e-mail: aBolonkin@juno.com; aBolonkin@gmail.com

V. Badescu and K. Zacny (eds.), *Inner Solar System*,
DOI 10.1007/978-3-319-19569-8_17

negligible protection to the atmosphere against cosmic radiation. This radiation may result in cloud-to-cloud lightning discharges.

The lack of an intrinsic magnetic field at Venus was surprising given it is similar to Earth in size, and was assumed also to contain a dynamo at its core. A dynamo requires three things: a conducting liquid, rotation, and convection. The core is thought to be electrically conductive and, although its rotation is often thought to be too slow, simulations show it is adequate to produce a geodynamo. This implies the dynamo is missing because of a lack of convection in the Venusian core. On Earth, convection occurs in the liquid outer layer of the core because the bottom of the liquid layer is much hotter than the top. On Venus, a global resurfacing event may have shut down plate tectonics and led to a reduced heat flux through the crust. This caused the mantle temperature to increase, thereby reducing the heat flux out of the core. As a result, no internal geodynamo is available to drive a magnetic field. Instead, the heat energy from the core is being used to reheat the crust.

One possibility is that Venus has no solid inner core, or that its core is not currently cooling, so that the entire liquid part of the core is at approximately the same temperature. Another possibility is that its core has already completely solidified. The state of the core is highly dependent on the concentration of sulfur, which is unknown at present.

The weak magnetosphere around Venus means that the solar wind is interacting directly with its outer atmosphere. Here, ions of hydrogen and oxygen are being created by the dissociation of neutral molecules from ultraviolet radiation. The solar wind then supplies energy that gives some of these ions sufficient velocity to escape Venus's gravity field. This erosion process results in a steady loss of low-mass hydrogen, helium, and oxygen ions, whereas higher-mass molecules, such as carbon dioxide, are more likely to be retained. Atmospheric erosion by the solar wind probably led to the loss of most of Venus's water during the first billion years after it formed. The erosion has increased the ratio of higher-mass deuterium to lower-mass hydrogen in the upper atmosphere by 150 times compared to the ratio in the lower atmosphere.

In this chapter a method of creating artificial magnetic field on Venus is proposed. It will allow protection of Venus surface from harmful solar wind impact.

17.2 Explanation of Planet Magnetism. Dynamo Theory

In geophysics, the *dynamo theory* proposes a mechanism by which a celestial body such as Earth or a star such as our Sun generates a magnetic field. The dynamo theory describes the process through which a rotating, converting, and electrically conducting fluid can maintain a magnetic field over astronomical time scales.

Dynamo theory describes the process through which a rotating, converting, and electrically conducting fluid acts to maintain a magnetic field. This theory is used to explain the presence of anomalously long-lived magnetic fields in astrophysical bodies. The conductive fluid in the geodynamo is liquid iron in the outer core, and in the solar dynamo is ionized gas at the tachocline. Dynamo theory of astrophysical bodies uses magnetohydrodynamic equations to investigate how the fluid can continuously regenerate the magnetic field. It was once believed that the dipole, which comprises much of the Earth's magnetic field and is misaligned along the planet's rotation axis by 11.3°, was caused by permanent magnetization of the materials in the earth.

There are three requisites for a dynamo to operate:

- An electrically conductive fluid medium
- Kinetic energy provided by planetary rotation
- An internal energy source to drive convective motions within the fluid.

In the case of the Earth, the magnetic field is possibly induced and constantly maintained by the convection of liquid iron in the outer core. A requirement for the induction of field is a rotating fluid. Rotation in the outer core is supplied by the Coriolis effect caused by Earth's rotation. The Coriolis force tends to organize fluid motions and electric currents into columns (also see Taylor columns) aligned with the rotation axis.

Tidal forces between celestial orbiting bodies causes friction that heats up the interiors of these orbiting bodies. This is known as tidal heating, and it helps create the supposed liquid interior criteria, providing that this interior is conductive, that is required to produce a dynamo. For example, Saturn's satellite Enceladus and Jupiter's moon Io have enough tidal heating to liquefy their inner cores, even if a moon is not conductive enough to support a dynamo. Mercury, despite its small size, has a magnetic field, because it has an apparent conductive liquid core created by its iron composition and friction resulting from its highly elliptical orbit. It is theorized that the Earth's Moon once had a magnetic field, based on evidence from magnetized lunar rocks, due to its short-lived closer distance to Earth creating tidal heating. An orbit and rotation of a planet helps provide a liquid core, and supplements kinetic energy that supports a dynamo action (Fig. 17.1).

17.2.1 Ferromagnetism

Ferromagnetism is the basic mechanism by which certain well-known materials (such as iron) form permanent magnets, or are attracted to magnets. In physics, several different types of magnetism are distinguished. Ferromagnetism (including ferrimagnetism) is the strongest type; it is the only type that creates forces strong enough to be felt, and is responsible for the common phenomena of magnetism encountered in everyday life. Other substances respond weakly to magnetic fields with two other types of magnetism, paramagnetic and diamagnetism, but the forces are so weak that they can only be detected by sensitive laboratory instruments. The attraction between a magnet and ferromagnetic material is "the quality of magnetism first apparent to the ancient world, and to us today", especially in the form of the navigational compass.

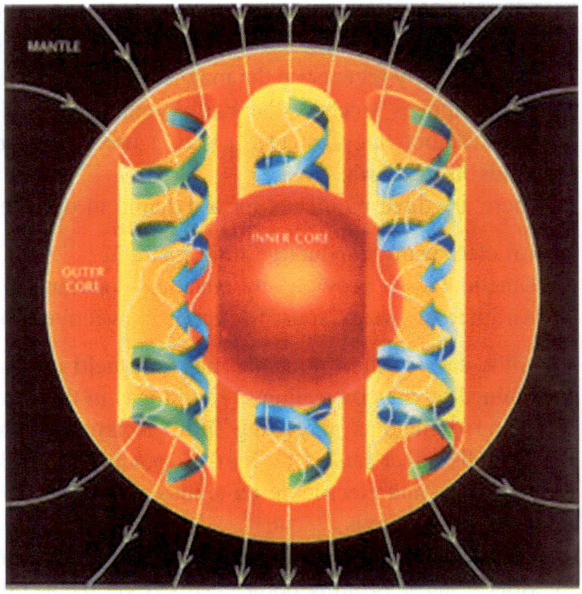

Fig. 17.1 Illustration of the dynamo mechanism that creates the Earth's magnetic field: convection currents of magma in the Earth's outer core, driven by heat flow from the inner core, organized into rolls by the Coriolis force, creates circulating electric currents, which generate the magnetic field

Permanent magnets (materials that can be magnetized by an external magnetic field and remain magnetized after the external field is removed) are either ferromagnetic or ferromagnetic, as are other materials that are noticeably attracted to them. Only a few substances are ferromagnetic. The common ones are iron, nickel, cobalt and most of their alloys, some compounds of rare earth metals, and a few naturally-occurring minerals such as lodestone.

As the temperature increases, thermal motion, or entropy, competes with the ferromagnetic tendency for dipoles to align. When the temperature rises beyond a certain point, called the *Curie temperature* (Table 17.1), there is a second-order phase transition and the system can no longer maintain a spontaneous magnetization, although it still responds paramagnetically to an external field. Below that temperature, there is a spontaneous symmetry breaking and magnetic moments become aligned with their neighbors. The Curie temperature itself is a critical point, where the magnetic susceptibility is theoretically infinite and, although there is no net magnetization, domain-like spin correlations fluctuate at all length scales.

Table 17.1 Curie temperatures for some crystalline ferromagnetic (*ferrimagnetic) materials

Material	Curie temp. (K)	Material	Curie temp. (K)	Material	Curie temp. (K)
Co	1388	$CuOFe_2O_3^*$	728	$MnOFe_2O_3^*$	573
Fe	1043	$MgOFe_2O_3^*$	713	$Y_3Fe_5O_{12}^*$	560
$Fe_2O_3^*$	948	MnBi	630	CrO_2	386
$FeOFe_2O_3^*$	858	Ni	627	MnAs	318
$NiOFe_2O_3^*$	858	MnSb	587	Gd	292

Table 17.2 Chemical composition of the Earth crust

Compound	Formula	Composition	
		Continental (%)	Oceanic (%)
Silica	SiO_2	60.2	48.6
Alumina	Al_2O_3	15.2	16.5
Lime	CaO	5.5	12.3
Magnesia	MgO	3.1	6.8
Iron(II) oxide	FeO	3.8	6.2
Sodium oxide	Na_2O	3.0	2.6
Potassium oxide	K_2O	2.8	0.4
Iron(III) oxide	Fe_2O_3	2.5	2.3

Table 17.3 Magnetic properties of iron and some its combination (Kalashnikov 1985, p. 229)

Material	Initial permeability	Maximal permeability	Maximal inductance (T)	Coercivity (A/m)
Iron	200	5000	2.15	80
Iron-silicate alloy	600	10,000	2	16
Iron-silicate alloy anneal in H	1500	40,000	2	8

The mass of the Earth is approximately 5.98×10^{24} kg. It is composed mostly of iron (32.1 %), oxygen (30.1 %), silicon (15.1 %), magnesium (13.9 %), sulfur (2.9 %), nickel (1.8 %), calcium (1.5 %), and aluminium (1.4 %); with the remaining 1.2 % consisting of trace amounts of other elements (Table 17.2). Due to mass segregation, the core region is believed to be primarily composed of iron (88.8 %), with smaller amounts of nickel (5.8 %), sulfur (4.5 %), and less than 1 % trace elements.

The iron, its alloys and chemical combinations significantly increases the magnetic field. For example, the iron has initial permeability 200, iron-silicon alloy has maximum permeability 5000. More detail data in Table 17.3.

The planet crust has a lot of the iron and iron chemical connection which can drastically increase the artificial magnetic field. For example, the average Earth magnetic intensity is 27.1 A/m. But at city Kursk (Russia) (where are located famous deposits of iron ore) the magnetic intensity reaches 160 A/m.

17.3 Main Ideas

17.3.1 Superconductivity Satellite Belt

The simplest method to create the artificial magnetic field is a superconductivity satellite belt (Fig. 17.2a). If the belt is delivered from another planet (for example, belt is delivered from the Earth to Venus), this method saves the rocket fuel because there no braking manoeuver necessary at the implantation planet. Cable can be delivered in segments and assembled in section by section in satellite orbit. The macro-problem may be in the protection of the implanted cable from constantly impinging solar radiation and in charging the orbiting cable ring by an electric currency.

17.3.2 Superconductive Ground Belts

The second method is shown on Fig. 17.2b. In this method two rings are located on planet poles. Here rings are less and assembled on the planet's surface. If superconductive ring is produced in situ at the host planet this method is cheaper and the artificial ring can be located near the poles where atmospheric temperature is less than at the equator. Charging of ring by energy in planet surface is simpler problem than charging the ring in outer space.

17.3.3 Protection of the Superconductive Belt from Heating

Protection of the superconductivity cable from the solar and planet radiations is shown in Fig. 17.3. These are screens developed in works (Bolonkin 2008, 2011).

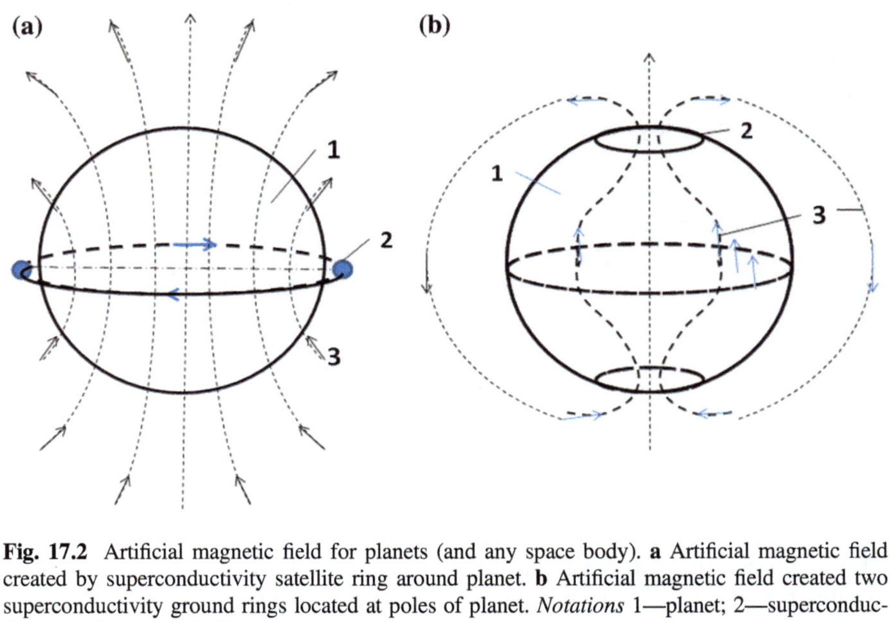

Fig. 17.2 Artificial magnetic field for planets (and any space body). **a** Artificial magnetic field created by superconductivity satellite ring around planet. **b** Artificial magnetic field created two superconductivity ground rings located at poles of planet. *Notations* 1—planet; 2—superconductive ring; 3—magnetic lines

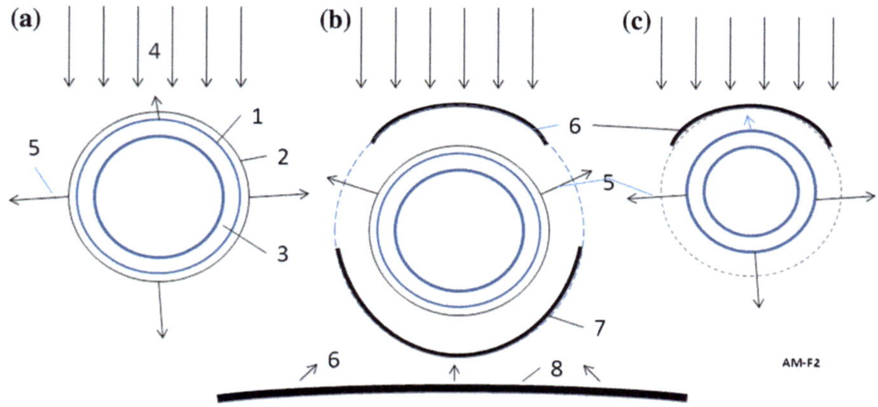

Fig. 17.3 Protection of superconductive cable (ring) from solar and planet radiation. ***a*** – Protection by heat conductive transparence film with control transparence (reflectivity) in different parts of the solar spectrum. ***b*** – Additional protection by screens, reflectors from a solar radiation and planet radiation. ***c*** – Additional protection by screens from the solar radiation. *Notations* 1 - superconductive layer; 2 - a film with a control transparence (reflectivity) layer; 3 - thermal conductivity of pipe; 4 - solar radiation; 5 - tube heat radiation; 6–7 - heat screens from solar and planet radiations; 8 - planet surface; 9 - planet radiation

17.4 Theory and Estimation of the Magnetic Intensity

When the electric currency flows into wire, one produces magnetic field. Intensity H of this magnetic field in the given point A from the electric currency $\boldsymbol{i}$ and the small element of wire $\boldsymbol{dl}$ (Fig. 17.4a) may be computed by the equation

$$dH = \frac{1}{4\pi}\frac{i \sin v}{r^2} dl \tag{17.1}$$

where v is angle between the direction of the currency vector (the element wire) and direction from given element to point A.

If we integrate along whole electric circle, we get the full magnetic intensity in given point of the ring. Unfortunately it is possible analytically only in simple cases. Two cases suitable for integration are shown in Fig. 17.4b, c. That is ring and straight wire. In case Fig. 17.4b (ring) we get for point A located in axis

$$\begin{aligned} dH &= \frac{1}{4\pi}\frac{i \sin v}{r^2} dl = \frac{1}{4\pi}\frac{iR}{r^3} dl, \\ H &= \frac{1}{4\pi}\frac{iR}{r^3}\int dl = \frac{1}{4\pi}\frac{iR}{r^3} 2\pi R = \frac{iS}{2\pi r^3} = \frac{iR^2}{2r^3} \end{aligned} \tag{17.2}$$

where S is area of circle (ring). Magnitude $iS = p_{m,}$ where p_m is magnetic moment. In final we get

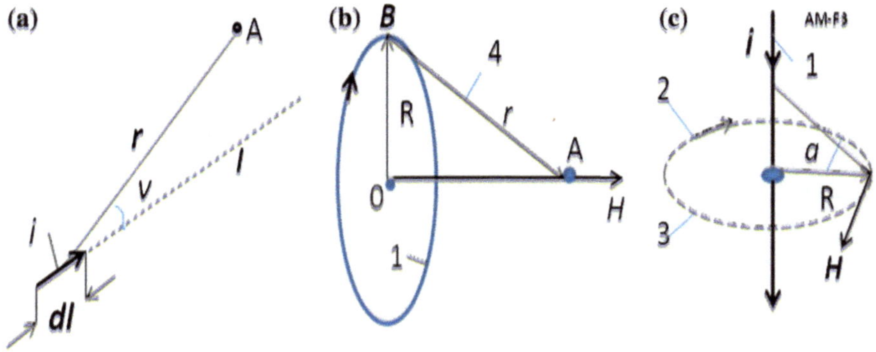

Fig. 17.4 Magnetic intensity: ***a*** – from element *dl* electric currency; ***b*** – from electric ring; ***c*** – from straight electric wire. *Notation* 1 – electric wire; 2 – (*H*) – magnetic intensity; 3 – magnetic line; 4 – distance from ring to point A; *R* – radius of the electric ring; *r* – distance BA

$$H = \frac{iS}{2\pi r^3} = \frac{p_m}{2\pi r^3}, \quad \text{where} \quad p_m = iS \tag{17.3}$$

In center of our ring we have

$$H = \frac{i}{2R} \tag{17.4}$$

If OA = R, $r = (2)^{0.5}R$, than magnetic intensity in point A is

$$H = \frac{i}{2^{2.5}R} \tag{17.4a}$$

In case 2 (Fig. 17.4c, straight wire) we have

$$\frac{\sin v}{r}dl = \frac{\cos\alpha}{r}dl = \frac{ds}{r} = d\alpha, \quad r = \frac{R}{\cos\alpha}$$
$$dH = \frac{1}{4\pi}\frac{i \cdot \sin v}{r^2}dl = \frac{i}{4\pi}\cos\alpha dl, H = \frac{i}{4\pi R}\int_{-\pi/2}^{+\pi/2}\cos\alpha d\alpha = \frac{i}{2\pi R} \tag{17.5}$$

Energy W is requested by ring Fig. 17.4b is

$$W = \frac{Li^2}{2} \tag{17.6}$$

where L is inductance of the electric circle, Hm.

Inductance of electric solenoid is about

$$L \approx \mu_0 \frac{N^2 S}{l} \tag{17.4}$$

where $\mu_o = 4\pi 10^{-7}$ is permeability constant; N–number of turns; $S = \pi R^2$ - cross-section area of solenoid, m^2; l–length of the solenoid, m. Assume $N = 1$, $l = 1$ for the ring (Fig. 17.4b) inductance is about

$$L \approx \mu_0 S \approx \mu_0 \pi R^2 \tag{17.8}$$

17.4.1 Short Information About Superconductivity

The magnetic ring 2 in Fig. 17.2 may be only superconductive. Let us to consider some properties of superconductive material.

There are hundreds of new superconductive materials (type 2) having critical temperature 70–120 K and more. *Some of the superconductible materials* are presented in Table 17.4. The widely used $YBa_2Cu_3O_7$ has mass density 7 g/cm^3.

The last decisions are: Critical temperature is 176 K, up to 183 K. Nanotube has critical temperature of 12–15 K. Some organic matters have a temperature of up to 15 K. Polypropylene, for example, is normally an insulator. In 1985, however, researchers at the Russian Academy of Sciences discovered that as an oxidized thin-film, polypropylene have a conductivity 10^5–10^6 that is higher than the best refined metals.

Boiling temperature of liquid nitrogen is 77.3 K, air 81 K, oxygen 90.2 K, hydrogen 20.4 K, helium 4.2 K (Bolonkin 2013). Specific density of liquid air is 920 kg/m^3, nitrogen 804 kg/m^3; evaporation heat is liquid air is 213 kJ/kg, nitrogen 199 kJ/kg (AIP 2003). Unfortunately, most superconductive material is not strong and needs a strong covering for structural support.

Table 17.4 Transition temperature T_c and upper critical field $B = H_{c2}(0)$ of some examined superconductors (AIP 2003, p. 752)

Crystal	T_c (K)	H_{c2} (T)
$La_{2-x}Sr_xCuO_4$	38	≥80
$YBa_2Cu_3O_7$	92	≥150
$Bi_2Sr_2Ca_2Cu_3O_{10}$	110	≥250
$TlBa_2Ca_2Cu_3O_9$	110	≥100
$Tl_2Ba_2Ca_2Cu_3O_{10}$	125	≥150
$HgBa_2Ca_2Cu_3O_8$	133	≥150

17.4.2 Estimation the Magnetic Ring for Venus

Let us to estimate the minimal Magnetic Ring for Venues. Assume the superconductivity ring magnetic ring is located at altitude h = 350 km over the Venus surface and has the electric currency i = 100 millions amperes. One will produces the next magnetic intensity on Venus surface at equator:

$$H = \frac{i}{2\pi h} = \frac{10^8}{6.28 \cdot 350 \cdot 10^3} = 45.5\,\mathrm{A/m}$$

That is more than the Earth magnetic intensity at the Earth equator (27.1 A/m). If we calculate the magnetic intensity at the Venus Pole we get

$$H = \frac{i}{4\sqrt{2}R} = \frac{10^8}{5.66 \cdot 6.4 \cdot 10^8} = 2.76\,\mathrm{A/m}$$

That is less than the Earth geomagnetic intensity at the Earth North Pole (52.5 A/m).

Let us take the safety (for superconductivity) magnetic intensity B = 100 T. Than minimal radius of superconductivity tube (wire) is:

$$r_w = \frac{\mu_0 i}{2\pi B} = \frac{4\pi 10^{-7} 10^8}{2\pi 100} = 0.2\,\mathrm{m}$$

Assume the tube and its protection is the thin inflatable film with the thin superconductivity layer filled the gas from the Venus atmosphere. One has the specific mass m = 0.1 kg/m. That the mass of Venus ring will be:

$$L_w = 2\pi R = 6.28 \cdot 6.4 \cdot 10^6 \approx 4 \cdot 10^7\,\mathrm{m},$$
$$M = mL = 0.1 \cdot 4 \cdot 10^7 = 4 \cdot 10^6\,\mathrm{kg} = 4000\,\mathrm{tons}$$

where L_w is length of ring. The inductance of Venus ring is:

$$L = \mu_0 S = \mu_0 \pi R^2 = 4\pi 10^{-7} (6.4 \cdot 10^6)^2 = 1.6 \cdot 10^8\,\mathrm{Hm}$$

The energy is needed for charging this ring is

$$E = 0.5 \cdot Li^2 = 0.5 \cdot 1.6 \cdot 10^8 10^{16} = 8 \cdot 10^{23}\,\mathrm{J}$$

This energy is very high, in 4–5 order over energy of the Earth magnetic field. If we make the assumption of the electric currency as about one million ampere we decrease the request energy by 10,000 times.

Some scientists claim that 1 % of the Earth's magnetic field is enough for protecting our planet from the high energy particles impacting it. If it is true the request energy is decreased by 10^8 times.

17.5 Conclusion

In future, it is possible to protect Venus from the solar wind and some high energy space particles by the artificially generated magnetic fields.

Acknowledgments The author wishes to acknowledge Richard B. Cathcart for correcting the English and for offering useful advices and suggestions.

References

AIP (2003) Physics desk references, 3rd edn. Springer

Bolonkin AA (2008) Femtotechnology and innovative projects. Lulu Publisher, USA. http://viXra.org/abs/1309.0191

Bolonkin AA (2011) Femtotechnology and revolutionary projects. LAB Lambert Academic Publisher. http://archive.org/details/FemtotechnologiesAndRevolutionaryProjects_873

Bolonkin AA (2013) Innovations and new technologies. Scribd, 30/7/2013. 309 pp. http://viXra.org/abs/1307.0169, https://archive.org/details/Book5InnovationsAndNewTechnologiesv2102014

Kalashnikov SG (1985) Electricity. Moscow, Nauka (in Russian)

Nachshekin VV (1969) Technical thermodynamic and head transferring. Moscow, Visshaya Shcola (in Russian)

Chapter 18
Business Modalities of the Inner Solar System: Planets with Potential?

Mike H. Ryan and Ida Kutschera

18.1 Introduction

The idea of considering business opportunities within the inner solar system is clearly speculative. However, the danger with most speculative opportunities is not being ahead of your time, but failing to recognize that virtually all such possibilities may be realized in time. The inner solar system is inherently more challenging in terms of prospective business activity than the Moon, Mars, or even asteroids. The primary inhabitants of the inner solar system, Venus and Mercury, are particularly inhospitable to human life and their utility as business locations or sources of business resources within the bounds of current technology is limited. However, humans are a very industrious and creative species whose engineering and commercial endeavors have repeatedly exceeded what was often initially viewed as sheer fantasy.

18.2 What We Know About Mercury (NASA 2014a)

Mercury's eccentric orbit takes the small planet as close as 47 million km (29 million miles) and as far as 70 million km (43 million miles) from the sun. Temperatures on Mercury's surface can reach 430 degrees Celsius (800 degrees Fahrenheit). Nighttime temperatures on the surface can drop to −180 degrees Celsius (−290 degrees Fahrenheit) as Mercury has no atmosphere to retain the heat.

M.H. Ryan (✉) · I. Kutschera
Bellarmine University, Louisville, Kentucky, USA
e-mail: mryan@bellarmine.edu

I. Kutschera
e-mail: ikutschera@bellarmine.edu

V. Badescu and K. Zacny (eds.), *Inner Solar System*,
DOI 10.1007/978-3-319-19569-8_18

Mercury possesses a thin exosphere made up of atoms blasted off the surface by the solar wind and striking micrometeoroids instead of an atmosphere. Its magnetic field has just one percent the strength of Earth's magnetic field. Mercury's surface features are close to those of Earth's Moon in that both are scarred by large numbers of impact craters. Images also reveal a surface that is shaped by extensive volcanism. Mercury has a remarkably high abundance of the volatile elements sulfur and potassium. Earth-based radar observations detected that Mercury may have water ice at its north and south poles inside deep craters. These materials appear to be present only in regions of permanent shadow, consistent with the idea that these permanently-shadowed areas are cold enough to preserve water ice, despite the extreme high temperatures experienced on sunlit areas of the planet. Mercury does not initially appear to have the type of appeal that might make space-based business possible, much less practical. Lack of immediate business appeal might also suggest that Mercury will require closer examination to discern business opportunities. Initial surveys of territory often miss crucial features especially if you do not know what to look for. For example, until the utility of oil was determined, where it bubbled to the surface was often considered an indication that the land was useless. Large parcels of land across the southwest United States, Africa, and the Middle East were viewed as worthless or nearly so until oil became a valuable commodity. Mercury may not be the next "oil patch" but it may very well be much more valuable than current knowledge and experience suggest.

18.3 What We Know About Venus (NASA 2014b)

Venus and Earth are similar in size, mass, density, composition, and gravity. However, the similarities end at this point. Venus is covered by a thick, rapidly spinning atmosphere, creating a scorched world with temperatures hot enough to melt lead and a surface pressure 90 times that of Earth. NASA's Magellan mission during the early 1990s used radar to image 98 percent of the surface, and the Galileo spacecraft (launched in 1989) used infrared mapping to view both the surface and mid-level cloud structure as it passed by Venus in 1990 on the way to Jupiter. Infrared surface images by the European Space Agency's Venus Express in 2010 provided evidence for volcanism within the past several hundred thousand years. Venus' atmosphere consists mainly of carbon dioxide combined with clouds of sulfuric acid droplets. The thick atmosphere traps the sun's heat producing a greenhouse effect resulting in surface temperatures higher than 470 degrees Celsius (880 degrees Fahrenheit). The few probes that have landed on Venus have not survived longer than two hours in the intense heat. Sulfur compounds are abundant in Venus' clouds and cause significant surface weathering and erosion when combined with the moving atmosphere. The top level of cloud layers move rapidly around the planet completing their rotation every four Earth days. These clouds appear to be driven by hurricane-force winds traveling at roughly 360 km (224 miles) per hour. What factors create the violence of Venus weather continues as topic of scientific investigation. Venus has no global magnetic field even though its core iron content is similar to that of Earth but rotates too slowly to generate the type of magnetic field that Earth has. Venus is similar to Mercury in its lack of innate business appeal. Neither planet appears to have any obvious potential as a source of resources for space-based business or as a general location of interest to business. As has often been the circumstance historically, perhaps initial appearances do not tell the entire story either? In the case of Mercury and Venus, an examination of these planets will not guarantee their business viability but should reveal their business potential.

18.4 The Inner Solar System: Limits of Commerce

The patterns of commercial development are not always clear in the beginning and opportunities frequently evolve over time rather than appear fully developed. Examples abound of business opportunities that were never imagined at the beginning of an era but appeared none-the-less. Whether or not the inner solar system, Mercury and/or Venus, have business potential is, at best, a forecasting exercise that requires speculation, good guesses and foresight. Nothing in the descriptions of either planet suggests that they are massive business opportunities that would attract entrepreneurs, investors or coalitions of companies interested in exploiting resources. Still, the patterns of future development and opportunity are not always evident in the short term. Examining the possibilities of the inner solar system over a longer time horizon within the context of other opportunities, or even a general human expansion outward toward Mars is worthwhile. Business opportunities develop in interesting ways and often in improbable places. For example, various proposals to return humans to the Moon have been accompanied by a variety of lunar business proposals ranging from power production to tourism (Ryan and Kutschera 2007).

18.5 Unanticipated Business Opportunities

Unanticipated Resources: It is not unexpected on a planet which is 71 percent covered by water that its inhabitants would seek to explore beneath the water's surface (USGS 2014). Efforts to explore, travel and use the areas below the surface of the oceans have long been a preoccupation for humanity. However, commercial resource development below the ocean surface is a relatively recent accomplishment. The first successful off-shore petroleum drilling occurred in 1897 (American Oil & Gas Historical Society 2014a). However, it was not until 1947 that the commercial opportunities for off-shore drilling encouraged companies to attempt drilling beyond the sight of land. "Drilling equipment specifically designed for off-shore drilling did not exist. Off-shore exploration was extraordinarily speculative and risky, and an offshore dry hole could be financially catastrophic for oil firms involved (American Oil & Gas Historical Society 2014b)." The oil industry eventually pursued off-shore oil drilling opportunities and overcame the technical, organizational and political obstacles.

New techniques for construction of off-shore facilities, transportation and storage of petroleum, and transportation and housing of off-shore oil field workers were developed since the first crude rigs were deployed; large scale commercial operations are now the norm. What began as a highly speculative and financially risky venture has developed and grown into an industry supplying a significant percentage of the world's oil supplies. As the Deepwater Horizon disaster (2010) in the Gulf of Mexico illustrated, oil exploration is still risky. Regardless, "more than 5,000 off-shore oil and natural gas platforms operate in the Gulf of Mexico around the clock, seven-days a week. It is the largest artificial reef system in the world (American Oil & Gas Historical Society 2014c)." And, it would not have been possible to predict that that the top three off-shore oil fields in the world would reside in the Persian Gulf. The sheer size and scope of these operations were unimagined when the first haltering steps to drill for oil from below the ocean surface were taken. Commercial opportunities to use ocean-based resources still are considered in their infancy as the majority of such activities have occurred in the relatively shallow depths of the continental shelves. We may know more about the Moon than we do about the expansive terrain below the surface in the oceans of the Earth. It is not unreasonable to extend this idea and suggest that potential inner solar system business opportunities may not be as unimaginable as generally claimed.

Opportunities of Distance and Location: Location, distance, and time have always been variables of importance to business. Examples of extreme risk taking to obtain the benefits of specific locations, reduce time between destinations, and conquer previously unavailable opportunities have long been part of business. The development of fast sailing ships whose extreme form lead to the 1850s Clippers is representative of how business influences technology development (Engle and Lott 1975). The California gold rush in the 19th century pushed ship design to the limits as the price of goods delivered to California made it possible for builders and owners to generate extraordinary profits. Design elements in which speed took precedence over other factors to move high value cargo from port to port are good examples of a business response to time and distance (Chapelle 1967). Perhaps more appropriately to the issues of business opportunities within the inner solar system is the experience of Pan American Airlines (PanAm) and its Pacific operations. PanAm created and developed a novel approach involving technology (aircraft design) and optimal base construction to overcome the barriers of distance and improve the time it took to travel across the Pacific Ocean to points in the Far East. Since aircraft of the time lacked the range and the Pacific routes lacked the airfields to make land-based trans-Pacific flight practical, PanAm decided to deal with these limitations by a combination of novel aircraft technology and a dedicated program of constructing intermediate bases for these aircraft to use (Allen 2000). That these bases used the most plentiful resource available in the Pacific, i.e. water, as their airfields only made the entire operation more ingenious.

Base construction required sending self-sufficient ships and crews to numerous Pacific islands to build infrastructure to support a flying boat network. Facilities for maintaining and refueling aircraft, residential facilities for support personnel, in-transit passengers, and aircraft crews were all built where no infrastructure existed for flying boat service. Areas within protected anchorages were cleared of coral and other obstructions to make landing of large flying boats practical and safe. PanAm had previously become the leader in commercial use of flying boats by creating viable passenger, freight, and mail routes in the Caribbean, South and Central America, and across the Atlantic. The Pacific routes were a more significant challenge requiring equipment, techniques and supporting facilities not readily available. PanAm and Boeing collaborated to design and build the Boeing 314 flying boat, generally regarded as among the best of its type (Allen 2000, pp. 82–83). A combination of technology and original thinking allowed Pan-American Airways Corporation to build and operate a unique and original transportation infrastructure. PanAm was able to secure routes and access to the Far East market by building a series of immediate stops for its flying boats. The chain of flying boat stations would ultimately be supplanted when longer-range land-based aircraft were developed. Still, PanAm successfully demonstrated with its flying boats the technology and organization that made long-distance commercial flight viable, undoubtedly leading to future investments in long-range commercial aviation. The transportation infrastructure created by PanAm would have been impossible without the appropriate aircraft technology and the corresponding business organization working in concert. Each was necessary for the overall success of the business enterprise in an environment that was technologically, financially, and operationally challenging. These are exactly the sort of hurdles that commercial use of inner solar system resources will face and that space-based entrepreneurs will have to surmount in the 21st century.

As exploratory hurdles are overcome and Earth-based enterprises move outward away from planet Earth, a variety of bases and work platforms will be required. The planet Venus may yet represent another location where an intermediate station might make sense. The position of the planets at any given moment determines the relative ease of a specific orbital path between them. Even assuming that space engine technology will make even more rapid transits between the various outposts built within the solar system, the distances remain daunting and dangerous. Creating various way stations for emergency use, resupply or just to make movement of goods more efficient will harken back to the age of the Flying Clippers when such staging points were also crucial to economic and operational success. Travelers on their way to Mars may see a way station located proximate to Venus as a welcome respite on a long voyage, and shippers may see such a point as a place for sending goods and materials to other locations. Orbital Lagrange points might also come into play as locations for stations of various types and functions. And while advances in propulsion technology could supplant such resting points by making direct trips to places such as Mars significantly faster, way stations would still be critical in any number of possible emergencies. Consequently, it may make sense to think of Venus as a primary way station, transfer point or life boat station until such time as technology makes those locations unnecessary. It might not even be the development of new technology that results in changes but potentially the development of new business models that could make the use of existing solar system infrastructure obsolete.

Opportunities Derived from New Business Models: How a business creates value is a crucial issue for all businesses. New ways of looking at things provide insights that suggest alternate business operations, new methods of value creation, new processes, or new applications of old technologies. A new approach radically may change a business proposition evaluation from little or no opportunity to a good business opportunity. The new business model frequently provides new ways to generate revenue, reduce cost, expand into new areas or otherwise improve the opportunity to do business. SpaceX represents operations that generate new opportunities because of their novel approaches to business. The company is one of many that have attempted to make access to space less expensive. SpaceX is the first such firm to put its low cost access model to a serious commercial test. The price for putting an object in low earth orbit (LEO) prior to SpaceX was around $10,000 per pound, even after 50 years of space operations. SpaceX sought to reduce this cost to less than half by using simple proven designs with a focus on reliability and strict control of operating costs. The resulting business model could operate well below the cost of previous or current rocket providers. SpaceX has become the first private company to provide cargo delivery service to the International Space Station (ISS), and SpaceX continues to pursue other avenues for resetting the space business model. Its pursuit of a reusable orbital spacecraft may well become one of the enduring stories of the 21st century, if successful (Seedhouse 2013).

18.6 Implications of Historical Examples

The coming and going of businesses has become increasingly noteworthy. Firms replete with history such as Sears and Kodak found themselves unable to compete and compelled to fundamentally change or go bankrupt. Other firms, such as Radio Shack, coupled to large brick and mortar operations, found changing customer preferences made their operations unstainable financially. Alternatively, firms such as Amazon developed imaginative operations that have become businesses in themselves capable of generating revenue, supporting employees, and providing a profit or cover their respective costs. Not since the dot com boom have so many unique enterprises been proposed. The more outrageous an undertaking is described as being, the more likely it is that some expert will proclaim that it is impossible and therefore unlikely to provide any commercial opportunity what-so-ever. The pages of history are brimming with such examples. Most experts of their day found that their assertions of impossibility frequently failed to discourage those who have a different view of reality. This is not to suggest that unheralded enterprises are not difficult. Almost all start up enterprises are difficult. But where one person sees only difficulty, another sees opportunity. Opportunities trump difficulties with the development of new scientific understandings, new technology, or just a new way of looking at the problem. The noted political philosopher, Machiavelli, observed that no greater difficulty exists than attempting to introduce a new order of things (Machiavelli 2012). Postulating opportunities within the inner solar system that await the farsighted business person is speculative, but it is not without a reasonable basis. Many new inner solar system business enterprises may require only extensions of what we know now and can accomplish.

What the Inner Solar System Has in Terms of Potential: One obvious problem with business opportunities on Mercury and Venus is the fact that inhabiting either planet is not practical at the present time. There are no existing technologies that we are aware of that are collectively waiting in the wings, economically viable and/or currently ready for deployment that would make living on either planet possible. It is possible to envision explorers establishing scientific outposts to explore these two planets with proper preparation, transportation, and some innovative breakthroughs. However, the long list of "*ifs and buts*" makes any business scenario read more like science fiction than a business plan. The potential for both planets is initially more likely to be related to location than resource utilization. By comparison, the Moon provides both location and potential resources that could be exploited using current technology in many cases. This does not take into account the legal issues involved, only that mining on, building with lunar materials, and getting to and from the moon are things we know can be done. Therefore the primary advantage of exploring both inner system planets is at present in their proximity to the Sun. Other advantages may be determined once adequate time and attention has been devoted to what is possible at those locations.

Close Proximity to the Sun: Solar Power Production: One possibility for exploiting the proximity of Mercury and Venus to the sun is they could provide a suitable location for collecting solar energy that could be transmitted (Flournoy 2012). Current technology does not yet support power transmission on this scale but the potential for collecting energy at those locations does exist and could be quantified. Quantification is an important consideration because it provides a baseline for potential revenue critical to any viable business model. The situation is analogous to the current inability to exploit some known energy resources on earth because of costs or technological limitations. Examples include extensive use of geothermal energy or wave-based power systems here on Earth as alternatives to fossil fuel or as supplements to wind or solar systems. While the technology works, it is not always clear exactly how practical or cost effective it might be. So although the distances are daunting, and the technological challenges seemingly insurmountable, the potential still exists for solar-based power systems proximate to Venus or Mercury (Jones and Baghchehsara 2013). When does a potential resource become valuable? Or more interestingly, does an untapped resource have no value? The realistic interpretation is that untapped potential is simply a resource that might be of indeterminate value at some future date. The problem in dealing with potential well beyond current business practice is that while value, as in the case of solar power production, can be estimated, the inability to exploit the resource within a current time frame pretty much makes the entire venture highly speculative. Even so, the possibility of creating unique and interesting enterprises will remain as will the value of the resource even if it appears unlikely to be exploited within the near future.

Location, Location, Location: Location is important for planets, much as location can be a critical consideration for many other real estate transactions. Inter-planetary trade is a long-established science fiction concept. Trade between planets would operate much like international trade has operated for thousands of years. Space-based enterprises will mirror terrestrial predecessors in their need to move raw materials, finished goods or people from one point to another.

The pattern of exchange was based historically on the need for specific things in specific places. The differences in needs, relative scarcity, and comparative importance result in price differences that justify transportation costs. Transportation costs are secondary to the need for some cargo because of the cargo's inherent value. Supplying off-planet outposts may be critically important for years until the outposts become self-sufficient. Outposts may be able to develop reciprocal trade when the outposts themselves have developed sufficient surplus resources that can be traded for supplies/goods from other locations. Balance of trade discussions will move to a large and previously unknown scale as patterns of trade expand across the solar system.

Port or Way Station on the Way to Somewhere Else: One of the enduring features of space is the distance between places. It is a very long way between planets and few roadside parks, way stations, or amusement parks exist to break the monotony of long distance travel. More critically, no rescue stations, emergency airfields, or the equivalent of a coast guard exist in case of serious emergencies. Short-term and long-term travel for the foreseeable future requires a mindset of *"if you might need it, you have to take it with you."* Virtually all proposed space missions from earth have a fatalistic quality in that not much can be done if things go wrong. The numbers of ships lost in the earth-bound exploration, colonization, or trade may be incalculable. Efforts to reduce risk and minimize loss of life have engendered Coast Guards, satellite tracking, and global communications. And ships and lives are lost when things go wrong, such as the loss of Malaysian Air 370 in 2014.

Technologically sophisticated space operations may share more with the risky early voyages of exploration than comparatively straightforward international airline travel. And, even with airline travel strange things have been known to occur. A variety of approaches have been suggested to deal with possible catastrophe on long space voyages. Self-sufficiency may partially compensate for the absence of rescue. Self-sufficiency solutions may include traveling in multiple vessels or including large scale onboard 3-D printing capability on board to obviate the need for large parts inventories (Ryan and Kutschera 2013). These solutions would address initial missions. Longer-term solutions may include development of intermediate destinations for ongoing support and supply where assistance might be obtained.

Direct flights are not always practical to the most often identified destinations of either Mars or various asteroids. More economical orbital trajectories might utilize the gravity of Venus and perhaps Mercury to accelerate or decelerate vehicles transiting to other destinations (Thomson 1986).

> "Starting with Mariner 10, which flew by Venus on its way to Mercury, there have been a series of missions which have taken advantage of the so called 'gravitational slingshot' effect to visit other targets (Wiesel 2010, p. 47)."

Improvements in space transportation technology, including propulsion, radiation shielding and/or operational reliability, could either increase the need for an orbiting Venus location or reduce it. With greater speed or enhanced radiation shielding or both, trips could be made without the need of way stations. It is also possible that those same capabilities would increase the need for way stations much in the way that regional airports make flying to larger hubs more efficient. Either way, having transportation stops within the inner solar system may well make business sense. Delivery of supplies and other goods by way of Venus may be reasonable depending on the respective locations of Earth and Mars. It also may make sense for a facility to be moved into an orbit of Venus that could serve as an emergency port or repair depot should a Mars mission encounter difficulties. A parallel may be the United States Coast Guard maintained weather stations that were placed off the U.S. east and west coasts to provide weather information and rescue points for ships and aircraft. Most were replaced by automated buoys that monitor the weather, and the weather stations' secondary utility as "rescue stations" became largely unnecessary as aircraft reliability increased. Four-engine aircraft were initially required for trans-oceanic travel because of safety and reliability concerns. Twin-engine aircraft, including the Boeing 777 and Airbus 330, are now used for intercontinental travel. Similar improvements from initial offerings are likely to be seen in space transportation systems. In much the same manner, it might be expected that as space transportation technology matures, that the infrastructure which once served well would be replaced by other more robust facilities or perhaps fewer facilities as long-haul space transport makes such facilities obsolete. Regardless, an inner solar system facility could have both business utility or transportation importance as more crewed vessels leave earth for points outward from the Sun.

Prospective Resource or Fuel Depot: Reasonable business or operational arguments based on efficiency, safety, and frequency may support a fuel depot orbiting Venus. Much as ships are unable to carry sufficient fuel or expendables for indefinite periods, space craft will face the same limitations. The ability to resupply or to alter trajectories by replenishing fuel or consumables (as in the case of crewed vehicles) may well be advantageous to any number of public or private ventures. The need for such depots ultimately is a question of whether or not sufficient trips to Venus or by way of Venus to other places represent effective resource utilization. Venus' atmosphere is primarily carbon dioxide and may be a useful resource if and when accessed. Technology to remotely distill constituent elements from carbon dioxide for a multitude of uses by automated processes may be needed given the difficulties of atmospheric pressure, temperature and other barriers. A corresponding case might be made for Mercury if materials on or near its surface could be mined and transported for use elsewhere. This assumes that the materials involved could not be obtained less expensively or with less difficulty than other options. These approaches are not presently feasible but their current lack of feasibility does not undermine their possible utility that Mercury or Venus or both might eventually have business potential.

A Port-Based Model of Commerce for the Inner Solar System: Possible use of Mercury and Venus as ports may prevail in the absence of specific commercial viability for Mercury/Venus-derived resources. Current technology does not suggest construction of port facilities on either planet as a likely and reasonable scenario. However, the inner planets' locations could still be advantageous locations for orbiting facilities. These facilities could be engineered in lunar or earth orbit and moved to orbit either Mercury or Venus. These port facilities also may serve as outposts to study the planets they orbit or as ports for transiting spacecraft. The capability of having potential support for missions at points beyond Earth, Earth orbit, or even lunar orbit would greatly extend mission potential and business potential. It would also improve safety and reliability as crewed vessels move further into the solar system. The potential advantages of multiple locations to repair, restock or refuel would become a factor of increasing importance as the number of trips to and from Earth and its vicinity increase.

18.7 A Timeline for Inner Solar System Opportunities

Creating timelines for business operations involving either Mercury or Venus is a combination of extrapolating current capability combined with "good" guesses as to the state of tomorrow's technology. It will remain so until an individual or an organization sees the value or necessity of investing in activity related to those two planets. Predicting the future is a pursuit fraught with error even at its best. Many factors can impact the timing, manner or form of forecasted activities. It would have been difficult in 1969 when men from Earth first walked on the Moon to envision that those initial steps would not be followed up by more permanent facilities. A return lunar visit or permanent facilities are still years if not decades away more than forty years after explorers ventured to the Moon. Serious consideration of exploiting inner solar system business potential could easily be another fifty years away if the ultra-slow expansion into the solar system continues. Three factors may be critical in the Mercury and Mars business development discussion.

The first is if the Earth's population or specific nations on the planet become serious regarding Mars exploration. A manned Martian trip could become more than wishful thinking if discoveries from the various unmanned probes encourage additional exploration. Occupying or colonizing Mars is unlikely if Mars exploration follows the lunar exploration pattern with only a few trips to make quick surveys or to plant the flag. Mercury and/or Venus are unlikely to rise to importance for any opportunities previously discussed until Mars is explored. The second condition would be if permanent Martian facilities are constructed, maintained and expanded over time. In that case, the potential for Mercury and Venus increase as possible transit points. The use of these transit points may evolve into other uses over time as other compelling opportunities are found or created. Once constructed expensive facilities often find new uses; recycling can be less expensive, less troublesome and more efficient than building a new facility. Orbiting facilities would undoubtedly be evaluated in the same manner. The third condition that might motivate serious interest in both Mercury and Venus is if the Moon were to become a construction materials source. The Moon's surface contains large amounts of aluminum and the Moon could become a practical source of refined building materials for orbital manufacturing (Ryan and Kutschera 2012). The Moon has no atmosphere, and finished goods or materials could be transported from the surface with comparative ease. The availability of aluminum components and structural elements could provide an economically justifiable basis for constructing orbiting facilities or solar power satellites located tangentially to the Moon, Earth, Mercury or Venus. A wide range of Moon-based "business related" activities might begin to make sense without the gravity penalty of launching structural elements from Earth. The ability to build structures – especially large structures – in lunar orbit, and then relocate these structures to other planetary orbits including Mercury or Venus, could alter the calculus of what makes sense and where.

18.8 Points of Departure for Business Within the Inner Solar System

Not everything that is possible actually happens. Nor do the events that take place always follow predicted times, locations, or methods. However, broad patterns of historical commercial activity provide foundations for what may happen in the future. The planets of the inner solar system are locations no less than China or the Americas, locations Europeans once viewed as places shrouded in mystery whose potential was often dismissed. The problems of transportation, support and development for a commercial model of the inner solar system are complex, difficult and hard to imagine but not impossible. The implications of what is suggested above have the echo of science fiction but are not outside the realm of current or near-term engineering technology. Possibilities for both Mercury and Venus currently exceed the likely investment envelope for virtually all private sector enterprises. Mercury and Venus may well be untapped opportunities for a considerable period in the absence of an extraordinary financial return or an overall improvement in space technology that would significantly reduce the cost, dramatically improve access, or provide a stronger rationale. Time is not particularly relevant to whether or not business potential exists. The issue is whether the precedent exists to suggest that at some undetermined time Mercury, Venus, or both could have business potential.

That precedent exists. Commercial utility does not always correspond to commercial success. Many reasons for the business potential of both Mercury and Venus relate to their utility to another objective, such as Mars exploration, and not to business factors, such as revenue or possible profit. Such factors, while critical to ultimate business success, are difficult to determine in the context of places so far removed from conventional business opportunities. Future business opportunities on or near Mercury and Venus may emerge as technology, capability, and vision permits. Identifying these opportunities remains a challenge for the future.

References

Allen R (2000) The PAN AM Clipper: the history of Pan American's flying boats 1931 to 1946. Barnes and Noble, Inc., by arrangement with Amber Books Ltd., New York

American Oil & Gas Historical Society (2014a) http://aoghs.org/offshore-oil-and-gas-history/offshore-oil-history/. Accessed 11 May 2014

American Oil & Gas Historical Society (2014b) http://aoghs.org/offshore-oil-and-gas-history/offshore-oil-history/. Accessed 11 May 2014

American Oil & Gas Historical Society (2014c) http://aoghs.org/offshore-oil-and-gas-history/offshore-oil-history/. Accessed 11 May 2014

Chapelle HI (1967) The search for speed under sail: 1700–1855. W. W. Norton & Company Inc, New York

Engle E, Lott AS (1975) America's maritime heritage. United States Naval Institute, Annapolis, pp 140–145

Flournoy DM (2012) Solar power satellites. Springer, New York

Jones DR, Baghchehsara A (2013) Electric space: space-based solar power technologies and applications, North Charleston. CreateSpace Independent Publishing Platform

Machiavelli N (2012) The prince, the project gutenberg ebook of the prince (trans: Marriott WK). Release date: February 11, 2006 [EBook #1232], last updated: (November 5, 2012)

NASA (2014a) http://solarsystem.nasagov/planets/profile.cfm?Object=Mercury&Display=Overview Long. Accessed 10 June 2014

NASA (2014b) http://solarsystem.nasa.gov/planets/profile.cfm?Object=Venus&Display=Overview Long. Accessed 10 June 2014

Ryan MH, Kutschera I (2007) Lunar-based enterprise infrastructure—hidden keys for long-term business success. Space Policy 23:44–52

Ryan MH, Kutschera I (2012) Fundamentals of modern lunar management: private sector considerations. In: Badescu V (ed) Moon-prospective energy and material resources. Springer, Heidelberg, pp 703–724

Ryan MH, Kutschera I (2013) The case for asteroids: commercial concerns and considerations. Badescu V (ed) Asteroids—prospective energy and material resources. Springer, Heidelberg, pp 645–657

Seedhouse E (2013) SpaceX: making commercial spaceflight a reality. Springer Science + Business Media, New York

Thomson WT (1986) Introduction to space dynamics. Dover Publications, New York

USGS (2014) http://water.usgs.gov/edu/earthhowmuch.html. Accessed 11 May 2014

Wiesel WE (2010) Modern orbit determination. Aphelion Press, Beavercreek

Chapter 19
Economic Development of Mercury: A Comparison with Mars Colonization

Alexander A. Bolonkin

19.1 Introduction

Let us review the known conditions on Mercury, and shortly on Mars, because we will compare them.

Mercury has the most eccentric orbit of all the planets; its eccentricity is 0.21 with its distance from the Sun ranging from 46,000,000 to 70,000,000 km. It takes 87.969 earth days to complete an orbit. This varying distance to the Sun, combined with a 3:2 spin–orbit resonance of the planet's rotation around its axis, result in complex variations of the surface temperature. This resonance makes a single day on Mercury last exactly two Mercury years, or about 176 Earth days. Due to Mercury's 3:2 spin–orbital resonance, a solar day (the length between two meridian transits of the Sun) lasts about 176 Earth days. A sidereal day (the period of rotation) lasts about 58.7 Earth days.

Mercury's surface is very similar in appearance to that of the Moon, showing extensive mare-like plains and heavy cratering, indicating that it has been geologically inactive for billions of years (Fig. 19.1). Because our knowledge of Mercury's geology has been based only on the 1975 Mariner flyby and terrestrial observations, it is the least understood of the terrestrial planets. As data from the recent MESSENGER flyby is processed, this knowledge will increase. For example, an unusual crater with radiating troughs has been discovered that scientists called "the spider". It later received the name Apollodorus.

Albedo features are areas of markedly different reflectivity, as seen by telescopic observation. Mercury possesses dorsa (also called "wrinkle-ridges"), Moon-like highlands, montes (mountains), planitiae (plains), rupes (escarpments), and valles (valleys).

A.A. Bolonkin (✉)
C&R, Brooklyn, NY, USA
e-mail: aBolonkin@juno.com; aBolonkin@gmail.com

V. Badescu and K. Zacny (eds.), *Inner Solar System*,
DOI 10.1007/978-3-319-19569-8_19

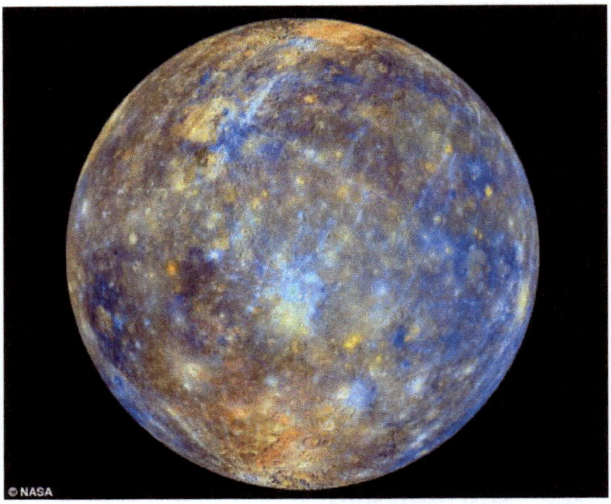

Fig. 19.1 Mercury

Mercury was heavily bombarded by comets and asteroids during and shortly following its formation 4.6 billion years ago, as well as during a possibly separate subsequent episode called the late heavy bombardment that came to an end 3.8 billion years ago. During this period of intense crater formation, the planet received impacts over its entire surface, facilitated by the lack of any atmosphere to slow impactors down. During this time the planet was volcanically active; basins such as the Caloris Basin were filled by magma, producing smooth plains similar to the maria found on the Moon. Data from the October 2008 flyby of MESSENGER gave researchers a greater appreciation for the jumbled nature of Mercury's surface. Mercury's surface is more heterogeneous than either Mars's or the Moon's, both of which contain significant stretches of similar geology, such as maria and plateaus.

The surface temperature of Mercury ranges from 100 K to 700 K at the most extreme places: 0°N, 0°W, or 180°W. It never rises above 180 K at the poles, due to the absence of an atmosphere and a steep temperature gradient between the equator and the poles. The sub-solar point reaches about 700 K during perihelion (0°W or 180°W), but only 550 K at aphelion (90° or 270°W). On the dark side of the planet, temperatures average 110 K. The intensity of sunlight on Mercury's surface ranges between 4.59 and 10.61 times the solar constant (1,370 $W{\cdot}m^{-2}$).

Although the daylight temperature at the surface of Mercury is generally extremely high, observations strongly suggest that ice (frozen water) exists on Mercury. The floors of deep craters at the poles are never exposed to direct sunlight, and temperatures there remain below 102 K; far lower than the global average. Water ice strongly reflects radar, and observations by the 70 m Goldstone telescope and the VLA in the early 1990s revealed that there are patches of very high radar reflection near the poles. Although ice was not the only possible cause of these reflective regions, astronomers believed it was the most likely. The icy regions are believed to contain about 10^{14}–10^{15} kg of ice, and may be covered by a layer of regolith that significantly inhibits sublimation. By comparison, the Antarctic ice sheet of Earth has a mass of about 4×10^{18} kg, and Mars's south polar cap contains about 10^{16} kg of water. The origin of the ice on Mercury is not yet known, but the two most likely sources are from out gassing of water from the planet's interior or deposition by impacts of comets. On November 29, 2012, NASA confirmed that images from MESSENGER had detected that craters at the north pole contained water ice. Sean C. Solomon was quoted in *the New York Times* as estimating the volume of the ice as large enough to "encase Washington, D.C., in a frozen block two and a half miles deep".

Mercury is too small and hot for its gravity to retain any significant atmosphere over long periods of time; it does have a "tenuous surface-bounded exosphere" containing hydrogen, helium, oxygen, sodium, calcium, potassium and others.

This exosphere is not stable—atoms are continuously lost and replenished from a variety of sources. Hydrogen and helium atoms probably come from the solar wind, diffusing into Mercury's magnetosphere before later escaping back into space. Radioactive decay of elements within Mercury's crust is another source of helium, as well as sodium and potassium. MESSENGER found high proportions of calcium, helium, hydroxide, magnesium, oxygen, potassium, silicon and sodium. Water vapor is present, released by a combination of processes such as: comets striking its surface, sputtering creating water out of hydrogen from the solar wind and oxygen from rock, and sublimation from reservoirs of water ice in the permanently shadowed polar craters. The detection of high amounts of water-related ions like O^+, OH^-, and H_2O^+ was a surprise. Because of the quantities of these ions that were detected in Mercury's space environment, scientists surmise that these molecules were blasted from the surface or exosphere by the solar wind. Sodium, potassium and calcium were discovered in the atmosphere during the 1980–1990s, and are believed to result primarily from the vaporization of surface rocks, struck by micrometeorite impacts. In 2008 magnesium was discovered by MESSENGER probe. Studies indicate that, at times, sodium emissions are localized at points that correspond to the planet's magnetic poles. This would indicate an interaction between the magnetosphere and the planet's surface.

Despite its small size and slow 59-day-long rotation, Mercury has a significant, and apparently global, magnetic field. According to measurements taken by Mariner 10, it is about 1.1 % as strong as Earth's. The magnetic field strength at the Mercurian equator is about 300 nT. Like that of Earth, Mercury's magnetic field is dipolar. Unlike Earth, Mercury's poles are nearly aligned with the planet's spin axis. Measurements from both the Mariner 10 and MESSENGER space probes have indicated that the strength and shape of the magnetic field are stable. Mercury's magnetic field is strong enough to deflect the solar wind around the planet, creating a magnetosphere. The planet's magnetosphere, though small enough to fit within Earth, is strong enough to trap solar wind plasma.

Mars is the fourth planet from our Sun and the second smallest planet in this Solar System, after Mercury. Named for the Roman's god of war, it is often described as the "Red Planet" because the iron oxide prevalent on its crustal surface gives it a reddish appearance. Mars is a terrestrial planet with a thin atmosphere, having surface features reminiscent both of the impact craters of the Moon and the volcanoes, valleys, deserts, and polar ice-caps of Earth. The rotational period and seasonal cycles of Mars are likewise similar to those of Earth, as is the tilt that produces the seasons. Mars is the site of Olympus Mons, the second highest known mountain within our Solar System (the tallest on a planet), and of Valles Marineris, one of the largest canyons. The smooth Borealis basin in the northern hemisphere covers 40 % of the planet and may be a giant impact feature. Mars has two known moons, Phobos and Deimos, which are small and irregularly shaped. These may be captured asteroids, similar to 5261 Eureka, a Martian trojan asteroid.

Until the first successful Mars flyby in 1965 by the USA's *Mariner 4*, many speculated about the presence of liquid water on the planet's surface. This was based on observed periodic variations in light and dark patches, particularly in the polar latitudes, which appeared to be seas and continents; long, dark striations were interpreted by some as irrigation channels for liquid water. These straight line features were later explained as optical illusions, though geological evidence gathered by unmanned missions suggest that Mars once had large-scale water coverage on its surface. In 2005, radar data revealed the presence of large quantities of water ice at the poles and at mid-latitudes. The Mars rover *Spirit* sampled chemical compounds containing water molecules in March 2007. The *Phoenix* lander directly sampled water ice in shallow Martian soil on July 31, 2008.

As of 2014, Mars hosted five functioning spacecraft: three in orbit – the *Mars Odyssey*, *Mars Express*, and *Mars Reconnaissance Orbiter* – and two on its surface – Mars Exploration Rover *Opportunity* and the Mars Science Laboratory *Curiosity*. Defunct spacecraft on the surface include MER-A *Spirit* and several other inert landers and rovers such as the *Phoenix* lander, which completed its mission in 2008. Observations by the *Mars Reconnaissance Orbiter* have revealed possible flowing water during the warmest months on Mars. In 2013, NASA's Curiosity rover discovered that Mars' soil contains between 1.5 % and 3 % water by mass (about 33 liters per cubic meter, albeit attached to other compounds and, thus, not freely accessible in puddles or pools).

19.2 Colonization of Mercury and Comparison with Mars

There is a good reason for colonizing another planet, which is to avoid extinction if the Earth is hit by a 10 km or larger asteroid, as has happened repeatedly in the Earth's geological history. Colonization of Mercury, to this author, appears to be a very real and practical technical and social possibility. Mercury has been suggested as a possible target for space colonization of the inner Solar System, along with the Moon (and Lagrange points), Mars (including its small satellites: Phobos and Deimos), and the asteroid (Ceres).

19.2.1 Human Settlement on Mercury

Typical human settlement for Mercury and other planets is presented in Fig. 19.2.

19.2.2 Heating of the Crust of Mercury

The surface temperature of a no-atmosphere planet can be estimated by:

$$(1-a)E = \varepsilon C_s \left(\frac{T}{100}\right)^4 \tag{19.1}$$

where a is the rock-surface planet's albedo; E is energy of solar radiation, W/m^2; ε is average blackness coefficient of the planet surface; C_s = 5.67 is radiation constant, W/(m^2·K^4); T is temperature in K. Mercury has $a \approx 0.106$; E_{max} = 14,446 W/m^2; $\varepsilon \approx (1 - a)$. Substitute these values into Eq. (19.1) we get T_{max} = 710 K ≈ 700 K.

Average temperature in depth of Mercury may be estimated by equations:

$$q_1 = \lambda_1 \Delta T_1 \tau_1 / \delta_1 , \quad q_2 = \lambda_2 \Delta T_2 \tau_2 / \delta_2 , \tag{19.2}$$

where q is head energy throw 1 m^2, J/m^2; λ is heat transfer co-efficient, W/mK; ΔT is temperature difference, K; τ is time, sec.; δ is thickness of Mercury crush, m. Sub index "$_1$" treads to the heat flow from surface to crush depth in day time, sub index "$_2$" treads to the heat flow from crush depth to surface in night time. For Mercury

$$q_1 = q_2 , \quad \lambda_1 = \lambda_2 , \quad \tau_1 = \tau_2 , \quad \delta_1 = \delta_2 , \quad \text{hence} \quad \Delta T_1 = \Delta T_2 , \tag{19.3}$$

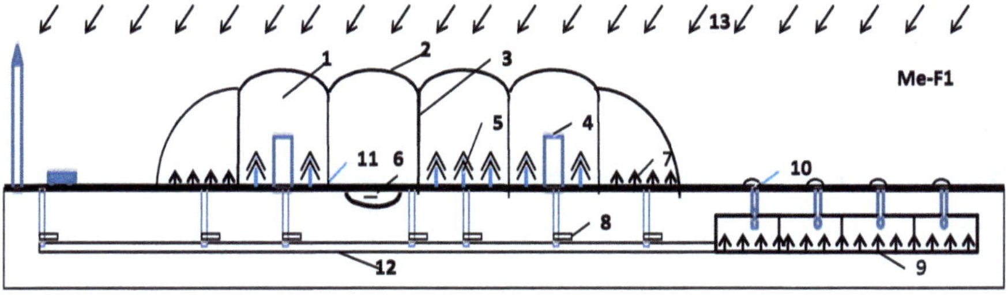

Fig. 19.2 Typical human settlement for Mercury. *Notations*: 1 is an area covered by thin double film with the control reflectivity and clarity, 2 is thin double film with the control reflectivity and clarity, 3 is internal walls which is separated internal volume in hermetic sections, 4 is house, 5 is forest, 6 is water pool, 7 is agriculture area, 8 is underground shelter in period strong solar radiation, 9 is underground shelter, 10 is solar collector, 11 is hermetic door in section, 12 is underground connection tunnel, 13 is solar radiation

But

$$\Delta T_1 = T_{\max} - \Delta T_a\,, \quad \Delta T_2 = T_{\min} + \Delta T_a\,, \tag{19.4}$$

where ΔT_a is stable average difference temperature in crust depth, K. From Eq. (19.4) we get

$$\Delta T_a = 0.5(T_{\max} - T_{\min}) \approx 0.5(700 - 100) = 300\,\mathrm{K}\,. \tag{19.5}$$

For Mercury in equator the average temperature in crush depth is

$$T_a = T_{\max} - \Delta T_a \approx 700 - 300 = 400\,\mathrm{K}\,. \tag{19.6}$$

In middle latitude of Mars the average temperature in crust depth is about 300 K. That is acceptable for underground human home, a "subterranean" shelter as it were.

The next question is: At what crust depth does this temperature is appear? Equations for this estimation are below:

$$E_a = \frac{\lambda \Delta T_a}{\delta}\,, \quad E_a \approx c_1 \varepsilon E_{\max}\,, \quad \delta = \frac{\lambda \Delta T_a}{E_a}\,, \tag{19.7}$$

where E_a is the energy flow from the Sun's radiation, W/m^{2}s; λ is heat transfer coefficient, W/mK; δ is thickness of of crust, m; $c_1 \approx 0.01 \div 1$ is production co-efficient latitude, average daytime, average day energy, etc. The typical heat coefficients are presented in Table 19.1.

Example. If we take $E_a = 2300$ W/m^2, $\lambda = 5$ (Mercury's crust contains metals) and $\Delta T_a = 300$ K, we get the depth for constant temperature 0.65 m.

19.2.3 Mercury Is Similar to Earth's Nearby Moon

Like the Earth's Moon, Mercury does not have any significant atmosphere. It is close to the Sun and rotates slowly with a very small tilt of its axis. The lack of an

Table 19.1 Typical heat coefficients [1, p. 451]

Matter	Density ρ kg/m^3	Heat resistance λ W/mK	Heat capacity c kJ/kg K	Thermal conductivity b 10^6, m^2/s
Concrete	2300	1.279	1.13	0.622
Brick	1800	0.768	0.879	–
Sand (dry)	1500	0.326	0.785	2.74
Steel	7900	45	0.461	14.7
Aluminum	2670	204	0.921	86.7
Glass	2500	0.744	0.67	0.444
Air (dry)	1.293	0.0244	1.005	18.8

atmosphere also means the lack of heat transfer by convection from any nearby hot objects. However, thermal protection of colonists and sensitive equipment would still require shielding from the intense solar radiation that reaches Mercury's surface and from the infrared radiation of any very hot region of Mercury's surface and insulation from heat conducted through the ground from any very hot surrounding ground. Hot areas can be avoided by moving to remain in the neighborhood of the terminator. Because of the similarity to Earth's moon, any colonization of Mercury might be performed with the same general technology, approach and equipment as a colonization of the Moon. Unlike the Moon however, Mercury has the additional advantage of a magnetic field protecting it from cosmic rays and solar storms, and a larger surface gravity of about 0.38 g, nearly equal to that of Mars.

19.2.4 Ice Present Within Polar Craters

Researchers at Arecibo observatory have found high radar-reflective areas near Mercury's poles in permanently shadowed crater bottoms, and this material was recently proven to be nearly pure water ice using neutron spectrometry data from the Mercury MESSENGER spacecraft, as described in Thermal Stability of Volatiles in the North Polar Region of Mercury and Evidence for Water Ice Near Mercury's North Pole from MESSENGER Neutron Spectrometer Measurements. Many of the craters are 1–2 km deep (Fig. 19.3). The wavelength dependence of radar signals indicates a minimum of several meters of water ice over a total area of 10^{10} m^2, resulting in 10^{14}–10^{15} kg of water. A more recent estimate using MESSENGER data is 0.5–20 m of water ice over 5.65×10^{10} m^2, resulting in 2.1×10^{13}–1.4×10^{15} kg of water. Laser altimeter reflections indicate that most of the ice is covered by a dark material which is thought to be a layer of hydrocarbons about 10 cm thick (amounting to $\sim 5.65 \times 10^9$ m^3), as described in Bright and Dark Polar Deposits on Mercury: Evidence for Surface Volatiles.

Due primarily to its proximity to the Sun, the surface of Mercury can reach 700 K (427 °C, 800 °F) near the equator, hot enough to melt lead. However, temperatures at the polar regions are much colder, less than 273 K (0 °C, 32 °F). There is little

Fig. 19.3 Crater on Mercury

doubt that there are considerable deposits of ice and perhaps other volatiles in the shadowed regions of polar craters. The polar areas do not experience the extreme daily variation in temperature seen on more equatorial areas of Mercury's surface. For these reasons there would be less difficulty in maintaining human colony enclosing structures in the polar regions than elsewhere on Mercury.

19.2.5 Agriculture

Agriculture would be possible with 2.1×10^{13}–1.4×10^{15} kg of water covered by 5.65×10^9 m^3 of carbon-rich hydrocarbons. Crops would provide food and oxygen, and consume the carbon dioxide we exhale. All human habitation and agriculture would be underground to avoid temperature extremes, ionizing radiation, and the loss of oxygen, water and carbon dioxide to the surface. Filtered light could be used for crops, but it is likely that rapidly growing crop varieties could be developed which would take advantage of the high light intensity and the long Mercury day, where sunrise to sunset lasts for 88 Earth days.

19.2.6 Solar Energy

Being the closest planet to the Sun, Mercury has vast amounts of solar power available. Its solar constant is 6.3–14.5 kW/m^2, on average 6.5 times that of the Earth-Moon system. Table 19.2 shows the amount of solar radiation received inner planets in the Solar System.

A very important advantage for Mercury is the high solar light intensity, which is stronger than on Earth by a factor of 10.6 at perihelion and 4.6 at aphelion. This strong sunlight intensity would provide virtually unlimited power via electronic solar arrays, and the resulting vertical temperature gradients of ~200° C/m would provide even more electrical power via thermal solar arrays. With such an "unlimited", and inexpensive power source, almost anything needed for human survival could be produced.

Because the tilt of its axis of rotation relative to its orbit is so low, approximately 0.01 degrees, there is also the possibility of so-called peaks of eternal light, similar to those of the Moon—high points located at the poles of the planet that are

Table 19.2 Radiation received from Sun

Planet	Distance (AU)		Solar radiation (W/m^2)	
	Perihelion	Aphelion	Maximum	Minimum
Mercury	0.3075	0.4667	14,446	6,272
Venus	0.7184	0.7282	2,647	2,576
Earth	0.9833	1.017	1,413	1,321

continuously radiated by the Sun. Even if they do not exist, it is possible that they could be constructed artificially.

Surface temperature of the planets with absent or minimal atmospheres, and temperature of space bodies surface, may be computed by equation:

$$T = 100\left(\frac{(1-r)E}{C_s}\right)^{0.25}, \qquad (19.8)$$

where T is surface temperature, K; E is energy flow from Sun, W/m^{2}s; r is coefficient reflectivity; $C_s = 5.67$ is constant. Result of computation is presented in Fig. 19.4.

The coefficient of reflectivity the different materials of thin-film covered by layer of matter is shown in Table 19.3. Film can have a control clarity (reflectivity, control temperature).

19.2.7 Mineral Resources

X-ray spectrometry for Si, Mg, Al, S, Ca, Ti, Fe, Cl, Cr and Mn, gamma-ray spectrometry for K, Th and U and gamma-ray spectrometry for Al, Ca, S, Fe and Na from Mercury MESSENGER shows the average composition of Mercury's soil compared to Earth (Table 19.4).

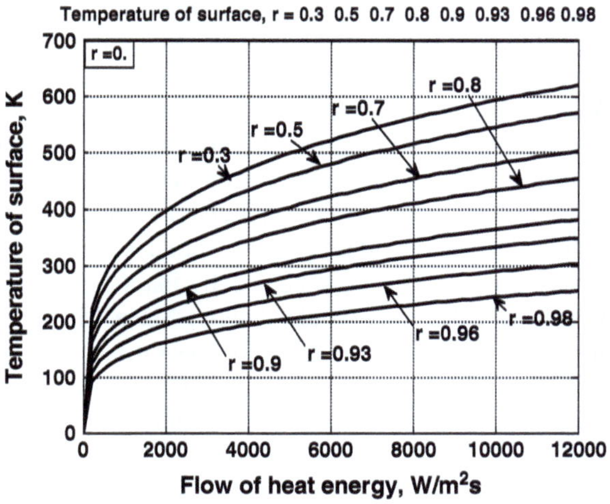

Fig. 19.4 Temperature of the surface planet (without atmosphere) and space body via the solar radiation energy flows for different reflectance r of body and albedo planet

Table 19.3 Average reflectance of material or polished metal layer [2, p. 465]

Matter	Reflectance r	Matter	Reflectance r	Matter	Reflectance r
Aluminum	0.94–0.96	Gold	0.97–0.98	Steel	0.2
Copper	0.98	White lacquer	0.80–0.95	Brick	0.07–0.12
Silver	0.97–0.98	Asbestos board	0.96	Glass	0.06–0.09

Table 19.4 Comparison element concentration of Mercury and Earth. Average element concentration (%weight)

Planet\element	O	Si	Mg	Al	S	Ca	Ti	Fe	Cl
Mercury (tens of cm depth)	42.3	24.6	12.5	7.1	2.3	5.9	0.2	1.9	<0.2
Earth (continental crust)	47	28	2.5	8	0.04	4	0.5	5	0.02
Planet\element	Cr	Mn	Na	K	Th	U	C	N	H
Mercury (tens of cm depth)	<0.5	<0.5	2.9	0.1	0.00002	0.00001	?	?	?
Earth (continental crust)	0.02	0.1	2	2	0.0007	0.0002	0.1	0.002	0.1

The consistency of the X-ray and gamma-ray spectrometry data indicates that the compositions above apply down to depths of tens of centimeters, and that the soil is homogeneous to this depth. It is not known for sure that there is enough of certain elements to grow crops, particularly the volatile elements C and N which cannot be measured by these methods. However, the fact that Mercury's K/Th ratio is higher than that of the Earth strongly suggests that Mercury's volatile elements have not been boiled away at some point in the planet's history. The S abundance supports this conclusion. The K abundance also supports this conclusion, being similar to the abundance on Earth as a whole. However, K abundance is low compared to that in the Earth's continental crust, which might be a problem for plant growth.

It is also possible that some of the elements are locked up in minerals which cannot be metabolized by plants. The MESSENGER probe is currently gathering ultraviolet and infrared spectrometry data and other data about the soil and the high radar-reflective areas, so some of these questions might be answered soon.

There are predictions that Mercury's soil may contain large amounts of helium-3, which could become an important source of clean nuclear fusion energy on Earth and a driver for the future economy of the Solar System. However, Mercury's magnetic field could have prevented helium-3 from reaching the surface.

Mercury is also theorized to have a crust rich in iron and magnesium silicates, with the highest concentrations of many valuable minerals of any surface in the Solar System, in highly concentrated ores. Some geologists have suggested this will make Mercury an ideal place to build solar sails, which could be launched in spacecraft as folded up "chunks" by mass driver from Mercury's surface. Once in space the solar sails would deploy. Since Mercury's solar constant is 6.5 times higher than Earth's, energy for the mass driver should be easy to come by, and solar sails near Mercury would have 6.5 times the thrust they do near Earth. This could make Mercury an ideal place to acquire materials useful in building hardware to send to (and to terraform) Venus.

19.2.8 Gravity

Mercury is larger than the Moon, with a diameter of 4,879 km versus 3,476 km, and has a higher density due to its large iron core. As a result, gravity on the surface of

Mercury is 0.377 g, more than twice that of the Moon (0.1654 g) and very close to the surface gravity on Mars. Since there is evidence of human health problems associated with extended exposure to low gravity, from this point of view, Mercury might be more attractive for long-term human habitation than the Moon.

The gravity on Mercury is 38 % that of Earth, which is strong enough to avoid the reduction in bone mass that occurs in very low gravity and weightless environments. There are no temperature variations over periods longer than the Mercury day (like Earth's seasons), which avoids the need for heating/cooling equipment within the 22 ± 1°C underground.

19.2.9 *Comparison of Colonization Mercury and Mars*

Mars automatically comes to mind when discussing planetary colonization, and manned missions to Mars have been the long term focus of US space exploration plans since 2004. But despite all the hype, Mars is really a poor prospect for colonization. The solar light intensity on Mars is 0.43 that of Earth, which makes solar power and agriculture much less practical than on Mercury. The gravity of Mars is 38 % of Earth, essentially equal to Mercury. The magnetic field of Mars is 0.1 % of Earth, and its atmosphere density is 2 % that of Earth, so protection from ionizing radiation would require underground habitation, the same as on Mercury. The average equatorial surface temperature of Mars is about −45 °C (−50 °F), which would be the constant temperature underground. And of course the temperature gets much lower away from the equator. Such low temperatures can be withstood by machines such as the Spirit, Opportunity and Curiosity Mars rovers, but not by people. Human habitation of Mars would be problematic because of the very low temperatures, limited solar power capacity, and a biological history which precludes oil, gas and coal deposits. Human habitation would probably be impossible without nuclear power. Uranium mining and nuclear plants would be very challenging to build, operate and maintain in an airless, cold environment. Also, concentrated uranium ore deposits are probably less common than on Earth because they depend on sedimentary and hydrothermal processes which are more prevalent on Earth. The other planets, moons and asteroids have even worse drawbacks than Mars.

19.2.9.1 Disadvantages

The first thought about Mercury is that it would have very high temperatures and no water, because the equatorial surface temperature range between −183 °C and 427 °C as the planet rotates. But an analysis of temperature vs. latitude and depth shows that the temperature is nearly constant at room temperature (22 ± 1 °C) in underground rings circling the planet's poles, and deeper than 0.7 m below the surface.

The lack of any substantial atmosphere, close proximity to the Sun and long solar days (176 Earth days) would all lead to significant challenges for any future human settlement.

Mercury is also deep in the Sun's gravitational potential well, requiring a larger velocity change (delta V) to travel to and from Mercury than is needed for other planets, although, in the past, gravity assist orbits using Venus have been used to reach Mercury. However, entering orbit around Mercury and landing on the surface would take 6 years with current propulsion methods. Solar sails and mass drivers may assist in transportation in the future, but are not viable options at present.

A Mercury colony would be a challenging and costly effort for sure. The voyage to Mercury (in two ends, to and from) might take 6.5 years like the MESSENGER trip because of the large velocity change involved, and the spacecraft would require heavy shielding against ionizing radiation. Much planning and preparation would be needed to ensure that the colony could get through the first weeks, months, and years, with little or no re-supply from Earth.

19.2.10 Reasons to Colonizing Mercury Instead of Other Planets

Asteroid impacts of 5 km diameter or greater occur roughly once every 10 million years, and those of 10 km or greater occur roughly once every 100 million years. In the past 540 million years there have been 5 extinction events where more than 50 % of the Earth's species were killed off, including the Permian-Triassic extinction where 90 % of the species were lost. Most scientists think that some of these were caused by asteroid impacts. A well proven example is the Chicxulub impact which resulted from a 10 km asteroid impact at the Cretaceous-Tertiary boundary 65 million years ago and caused the extinction of 70 % of the Earth's species, including the dinosaurs. Even larger impacts have occurred at earlier times, of which only a few are known because their impact craters get erased by the Earth's geological processes over time. It is thought that a 20 km or larger asteroid would cause the extinction of all higher order animals and plants, leaving only micro-organisms. While the likelihood of such an event is very small in any given year, it could happen at any time, and it is almost guaranteed to happen eventually.

Given the facts above, it appears that the focus of USA's space exploration plans should be shifted from Mars to Mercury.

The main motivation for investigating Mercury is its potential for hosting a self-sustaining human colony, which would protect humanity from extinction in the event of a catastrophic asteroid impact. A second motivation is simply to increase our scientific understanding of the solar system. It is very unlikely that Mercury could ever be a practical source of minerals or energy to be transported back to Earth, or that Mercury would ever have any other Earth-serving economic value. But surely preservation of the human species and scientific curiosity are better motivations than economic benefit. Humans are part of a universe where time is measured in billions of years. We need to take a long term view, and consider the future of the human species in the next thousand, million and billion years, not just the next 10 or 100 years.

19.3 Discussion and Conclusion

Mercury is more suitable planet for human colonization than any other planets of our solar system including Mars. Mercury is more appropriate for active agriculture and solar energy, for flights to other planets. The high temperature of Mercury surface is easy to decrease by covering it surface by the film having control reflectivity or a white dust. The intensive solar light promises the rich harvests and transferring the CO_2 to oxygen needed for nutrition and breathing of people. On Earth the hydroponics agriculture gives in 2–3 times more harvests than conventional agriculture. On Mercury the control constant solar light, control pressure, CO_2 environment and closed loop water cycle (as described in [3, Chap.1] can create the conditions generally comfortable for human settlement. Especially if the idea of author [4, Chap.19.4, p. 335] about living in outer space without the special spacesuit became reality in the future [5].

We must study the development of the plants in permanent strong light in rich CO_2, rare atmosphere and hydroponic surface. It is expected that, like the Moon, Mercury's surface contains tritium, a fuel for nuclear reactor. Mercury's surface must contain more tritium than Moon surface because Mercury is located closer to Sun. The flight to Mercury requests on 30 % more Δ-speed (energy) that flight to Mars. But flight time is only 105 days in one way (in Mars 260 days). That save a food and oxygen need for crew. The colonization of Mercury is more perspective for humanity than colonization of Mars. We should pay more attention to research of Mercury.

Acknowledgments The author wishes to acknowledge Richard B. Cathcart (USA) for correcting the English and for offering some useful advice and suggestions.

References

1. Bolonkin AA (2012) Re-entry space apparatus to Earth, General Science Journal, #5289, http://www.gsjournal.net/Science-Journals/Research%20Papers-Engineering%20(Applied)/Download/5289, http://archive.org/details/ReentryOfSpaceCraftToEarthAtmosphere, http://intellectualarchive.com #1381, http://viXra.org/abs/1212.0003
2. Naschekin VV (1969) Technical thermodynamic and thermotransfer. Moscow, High School, p 351 (in Russian)
3. Bolonkin AA, Cathcart RB (2007) Macroprojects. Environment and Technology, NOVA, p 536, Chapter 1. http://www.archive.org/details/Macro-projectsEnvironmentsAndTechnologies, http://viXra.org/abs/1309.0192, https://independent.academia.edu/AlexanderBolonkin/Papers
4. Bolonkin AA (2008) Non rocket space launch and flight. Elsevier, p 488. https://archive.org/details/Non-rocketSpaceLaunchAndFlightv.3, http://vixra.org/abs/1504.0011 v4
5. Bolonkin AA (2009) Man in outer space without a special space suit. Am J Eng Appl Sci 2 (4):573–579. http://viXra.org/abs/1309.0199, http://www.archive.org/details/LiveOfHumanityInOuterSpaceWithoutSpaceSuite

Chapter 20
Terraforming Mercury

Kenneth Roy

20.1 Introduction

What role will Mercury play once Humanity becomes a space faring race and establishes a civilization that spans the solar system? It could become the industrial center of such a civilization because of its light gravity, material resources, and plentiful solar energy that can be concentrated to achieve very high temperatures or converted into almost unlimited quantities of electrical energy. One can envision robotic factories turning out space ships and components needed to assemble vast space settlements.

But might our descendants have something else in mind? While Mercury is a small planet, its surface area is still roughly half the total land area of Earth. If we could terraform Mercury, we would increase the amount of land area available for humanity to live on by 50 % (somewhat less if we insist on having oceans). But is it possible to terraform Mercury? This chapter will attempt to answer that question.

Martyn J. Fogg defines terraforming as "*…a process of planetary engineering, specifically directed at enhancing the capacity of an extra-terrestrial planetary environment to support life. The ultimate in terraforming would be to create an unconstrained planetary biosphere emulating all the functions of the biosphere of the Earth—one that would be fully habitable for human beings*" (Fogg 1995). Planetary engineering is only possible for a spacefaring civilization able to utilize energy and material resources on a scale well beyond anything available today. Barring super advanced or "magic" technology, such projects will take many centuries or even millennia.

One criticism of the idea of terraforming is based on the ethical question, "What right do we have to play 'God' and alter the fate of a planet?" Here it is important to

K. Roy (✉)
Tennessee Valley Interstellar Workshop, Chattanooga, TN, USA
e-mail: kiroy@tviw.us

V. Badescu and K. Zacny (eds.), *Inner Solar System*,
DOI 10.1007/978-3-319-19569-8_20

make the distinction between a living world and a lifeless planet. A living world has life that may evolve over time into something intelligent, something that is deserving of a chance to exist. No such argument applies to a lifeless planet. As far as is known, the planet Mercury has been lifeless since its creation and will remain lifeless even as the sun enters its red giant phase and engulfs Mercury, the Earth and maybe Mars some five billion years from now. One could argue that it is unethical not to bring life to such a planet, if the capability to do so exists.

A second criticism is based on the idea that a space faring civilization can build space settlements in quantity and doesn't really need to terraform planets. This argument has some merit. But several open questions need to be answered before this argument can be fully accepted.

(1) Can centrifugal acceleration resulting from rotation serve as an adequate substitute for gravity?
(2) Can space settlements adequately address radiation protection issues?
(3) Can human beings psychologically adapt to life in a space settlement, or is there something about living on a planet that is vital for long term health and sanity?
(4) Can a small contained ecology function successfully in the long term (several millennia)? And even if space settlements prove to be viable, that does not rule out the possibility of terraformed planets as a possible home for some members of a space faring civilization.

But, can the planet Mercury be considered as a candidate for terraforming? If Mercury can be terraformed, then many other marginal planets orbiting other stars can also be made into living worlds based on Earth life. It's good to have options.

20.2 A Question of Gravity

The planet Mercury has a surface gravity of 3.7 m/s^2 (about 38 % that of Earth and very close to the surface gravity of Mars). Humans need gravity to function. Ignoring for a moment the question of atmosphere and temperature, is Mercury's gravity enough for humans to survive and even thrive?

The International Space Station and earlier space flight activities in Earth orbit have provided some opportunity to study human biology in a zero-g environment. However, there is effectively no experience with humans in a low-g (as opposed to a zero-g) environment. The time that astronauts have spent on Earth's moon is too limited to draw any conclusions. Low-g environments cannot be created on Earth for more than a few minutes at a time. There are many open questions relating to how much gravity human beings need to survive, grow, and reproduce. Developing children seem to need gravity similar to that of Earth to develop bone and muscle, and balance abilities. They would probably do just fine in a gravity that is 80 % or 90 % that of Earth. But what about 38 % that of Earth?

For a human colony to be successful, it must allow for human reproduction. However, we know very little about human reproduction in low-g environments. Nothing is known about human embryogenesis under such conditions. Studies with vertebrate animals seem to indicate that there are critical periods of development where gravity is important for proper embryogenesis (Wakayama et al. 2009; Horn and Gabriel 2011). How much gravity is needed for a successful pregnancy is unknown.

To some extent the same questions apply to plants. Experiments on the International Space Station indicate that plants do reasonably well in micro gravity and should do well in low gravity environments (Paul et al. 2012). How well animals and insects would do in such an environment is an open question.

It is unknown if Mercury's gravity is adequate for a successful human colony due to embryogenesis and child development considerations. Adult humans can probably do well in a gravity that is only 38 % that of Earth, although that too is an unsettled issue.

It is possible to envision rotating structures on the surface of Mercury (and Mars) that combine centrifugal acceleration and the planet's natural gravity to provide Earth-normal acceleration for pregnant women and children. It is also possible to modify, or engineer, humans, animals, insects, and plants to thrive in low-g environments. It is even possible that Earth life is adaptable enough that low-g environments pose no real obstacle to survival. The issue of low-g environments (in this case, 38 % that of Earth) must remain an open question subject to future research, but, at this point, is not an automatic show-stopper.

20.3 The Problem of Atmosphere

Mercury currently has a very thin atmosphere consisting of trace amounts of hydrogen and helium. The atmospheric pressure is about a trillionth of the atmospheric pressure on Earth. A terraformed Mercury would have an atmosphere consisting of 21 % oxygen, 78 % nitrogen, and 1 % argon at one atmosphere of pressure at the surface. Other mixtures and pressures are certainly possible, but it is assumed that what is required needs to be almost identical to Earth's atmosphere with respect to composition and pressure. Argon is useless to known biological systems and could probably be omitted. Mercury has little nitrogen, and any oxygen is locked up as oxides in the crust. Gases for the atmosphere will need to be imported from elsewhere. Mercury may have some oxides in its crust that could yield oxygen, and that could reduce, to some extent, the amount of material needed to be imported. CO_2 is available in quantity at Venus. If CO_2 is imported, the carbon would be used to enrich the soil and the oxygen would be released to form the atmosphere. Nitrogen is available in the outer solar system (and also from Venus) and could be imported from the moons of the gas giants.

How much gas must be imported? The mass of the atmosphere on a world can be approximated by using the ideal gas shown below:

$$PV = nR_uT \quad \Rightarrow \quad n = \frac{PV}{R_uT} \tag{20.1}$$

where: n = number of moles of gas, P = Pressure of gas, (Pa), V = Volume of gas, (m^3), R_u = Universal Gas Constant, (8.21E-5 atm-m^3/mole K), T = Temperature of gas, (K).

A spherical element of atmosphere has a constant pressure and a volume that can be described as:

$$dV = 4\pi(W + x)^2 dx \tag{20.2}$$

where: x = height above world's surface, (m), W= the radius of the planet, (m).

An atmosphere on a planet has a density that is highest at the surface and then decreases with distance above the surface. This barometric effect is caused by gravity and is described by Eq. (20.3):

$$p = p_o e^{-\frac{x}{H}} \tag{20.3}$$

where: p = atmospheric pressure at elevation x, p_o = atmospheric pressure at the surface of the planet, H = scale height in meters, (m).

The scale height "H" is the vertical distance over which the pressure of the atmosphere changes by a factor of e = 2.7182 (decreasing upward). Every time the altitude is increased by one scale height the pressure is reduced by 62.2%. It can be calculated by Eq. (20.4):

$$H = \frac{kT}{Mg} \tag{20.4}$$

where: k = Boltzmann constant = $1.38 \cdot 10^{-23}$ J/K; T = mean atmospheric temperature, (K), M = mean molecular mass of atmosphere, (kg), g = acceleration due to gravity on planetary surface, (m s^{-2}).

Equations (20.1) to (20.4) can then be combined into the following:

$$dn = p_o e^{\frac{-x}{H}} \frac{4\pi(W + x)^2}{R_uT} dx \tag{20.5}$$

Integration from x = 0 to x = a yields:

$$\begin{aligned} n &= \int_0^a p_o e^{\frac{-x}{H}} \frac{4\pi(W + x)^2}{R_uT} dx \\ &= \frac{4_o}{R_uT}\left(-e^{\frac{-a}{H}}(2H(a + W) + (a + W)^2\right. \\ &\qquad \left. + 2H^2) + 2H^2 + 2HW + W^2\right) \end{aligned} \tag{20.6}$$

Equation (20.6) is based on the ideal gas law and the barometric relationships listed in Eqs. (20.3) and (20.4), both of which are approximations, so Eq. (20.6) is also an approximation. However, it seems to give reasonable results provided that the scale height (H) is not excessive. Built into the equation is the assumption that the atmosphere is at a constant temperature, which is somewhat problematic. However, Eq. (20.6) can give a rough approximation of the quantities of gas necessary to make a world habitable.

To provide enough gas on Mercury to yield one atmosphere of pressure at the surface at a temperature of 288 K (15 °C) requires approximately $2.1 \cdot 10^{15}$ metric tons of gas of which, by weight, 75 % is nitrogen, 23 % is oxygen and 1 % is argon. This is about 40 % of the mass of Earth's atmosphere.

This requires $1.6 \cdot 10^{15}$ metric tons of nitrogen that must be imported from the outer solar system. If CO_2 is imported from Venus, then $6.6 \cdot 10^{14}$ metric tons are required. Argon would probably come from the outer solar system as well. Only $2.7 \cdot 10^{13}$ metric tons are required. Atmospheric CO_2 (plants need this in the air, assume normal levels 0.03 %) will also be required and an additional $9.0 \cdot 10^{11}$ metric tons will need to be imported for this.

Aside from the massive quantities of gas required, Mercury's low gravity and proximity to the sun present two challenges to maintaining this atmosphere. Planets lose atmosphere to space in three ways:

(1) Gas molecules at the top of the atmosphere sometimes acquire enough velocity to escape. The lower the planet's gravity and the hotter the top layer of atmosphere, the more likely this is to happen. Cold gas giants are very good at keeping their atmosphere; small planets like Mercury close to their sun are very poor at retaining their atmospheres.

(2) The solar wind interacts with the top of the atmosphere and strips off gas molecules through collusion and electric and magnetic interactions between ions. If a planet has a strong magnetic field, this field deflects the solar wind away from the top of the atmosphere and protects the gases from this interaction. Earth has a strong magnetic field that protects our atmosphere, while Mars and Mercury have a very slight magnetic field. Mercury being closer to the Sun would also experience a more intense solar wind than planets farther out.

(3) Atmospheric turbulence sometimes results in energetic upwells that in effect throw gases into space.

Even if an atmosphere could be imported to Mercury, it would be stripped away by the above mechanisms. It is difficult to calculate how long it would take before life on the planet's surface became untenable due to low pressure. The question is somewhat moot due to the fact that life on Mercury's surface would be untenable even with an Earth type atmosphere due to temperature considerations.

The solar wind is a continuous stream of plasma released from the upper atmosphere of the Sun. It consists of mostly electrons and protons but some alpha particles are also present. This solar wind varies in intensity over time. The average speed of the solar wind is approximately 400 km/s. Mercury's weak magnetic field offers some protection but some of the more energetic particles would reach the surface, even with an atmosphere, resulting in radiation damage. In addition, much more energetic bursts of plasma, called coronal mass ejections (CMEs), occasionally hit the planet. CMEs are caused by release of magnetic energy at the Sun and are often called "solar storms" or "space storms."

Mercury rotates slowly. To an observer standing on the surface, a single day on Mercury would take about 176 Earth days. With no atmosphere, the average night side temperature drops to around 100 K. The average day side reaches about 700 K. Assuming an atmosphere capable of evenly distributing heat around the planet with only limited greenhouse gases then the typical temperature would be about 400 K (127 °C or 260 °F). Mercury's current albedo varies across its surface between 0.08 to 0.25 with an average of about 0.12. The addition of an atmosphere would change the albedo, increasing it due to cloud cover. With enough cloud cover, it might be possible to drop the average temperature to something less than the boiling point of water. However, an atmosphere also reduces the ability of a planet to radiate heat into space. An Earth-like atmosphere would contain the greenhouse gas CO_2. This greenhouse gas will tend to restrict the ability of the planet to radiate heat to space, raising the temperature.

Positioning solar sail craft at the Mercury-Sun L1 point offers the ability to reduce the solar flux hitting Mercury, thus reducing the surface temperatures to reasonable levels. L1 points are not stable and, because of the high eccentricity of Mercury's orbit, this is especially true for Mercury. These solar sail craft could do the job but would have to actively work to maintain their position. Thus some intelligence and maneuvering ability are required on the part of each craft. They would have to interact with each other to avoid collisions and remain on station. If enough such craft were positioned somewhat sunward of the L1 point (to counteract the thrust resulting from light pressure) they could reduce the solar flux at Mercury by almost any desired amount. To reduce the solar flux at Mercury to something resembling the solar constant at Earth, some 16 to 17 million craft (each with an area of one square kilometer) would be needed. This number is approximate because of the difficulty of calculating radiation reductions due to the phenomena of solar limb darkening (where the center of the sun supplies more radiation than the outer edges) and the somewhat variable diameter of the sun. This approach has the advantage of shielding, to some extent, the new atmosphere from the solar wind. The solar sails could be designed to convert some of the intercepted solar radiation to energy, which would be beamed back to Mercury as microwaves and converted to electricity on the surface.

These intelligent and maneuverable solar sails would be exposed to intense solar radiation, high temperatures, vacuum, the constant solar wind, occasional CMEs, and the other hazards of space. As old units fail, they would have to be replaced. To an advanced space faring civilization, these constraints may not be too difficult to overcome.

This, however, does not address the eventual loss of the atmosphere to space due to Mercury's low gravity. The possibility of a long term, habitable, uncontained atmosphere on Mercury begins to look more and more problematic. Is there an alternative?

20.4 Contained Atmospheres

A contained atmosphere would address many of the issues listed above. Another big advantage of a contained atmosphere is that the amount of atmosphere that is necessary to provide a planet with an Earth-normal atmosphere at the surface is greatly reduced when compared to an uncontained atmosphere. If the roof/ceiling is three kilometers above the surface, about equal to the tallest mountains on Mercury (the rim wall of the Calorus Basin), then only 12.4 % of the uncontained atmospheric mass is required. If the roof/ceiling is reduced to one kilometer it is only 4.3 % as much, but at this altitude several mountains would be projecting through the roof.

Two approaches have been proposed in the terraforming literature to provide for a planetary contained atmosphere.

Paraterraforming was suggested in 1992 by Richard Taylor (Taylor 1992). The basic idea is to construct a structure capable of holding a gas-tight transparent roof one to three kilometers above the surface. The outer walls would also be required to contain the pressure. This structure must support this gas-tight roof both while un-pressurized during construction and after being pressurized with an atmosphere. The advantage is that this structure can be built incrementally, starting with a large building and adding to it until it covers the planet. This world spanning structure is sometimes called a world house.

This world house must be sturdy enough to withstand large stresses resulting from atmospheric gases pushing on large areas, occasional seismic events, and interior meteorology processes. The roof is supported/restrained by high-rise like structures and cables. Like a high-rise building on Earth, it will need maintenance and periodic replacement. Unlike a high-rise building on Earth, these structures will spend most of their existence in tension and the foundation for such structures would pose interesting engineering challenges.

Taylor's original proposal was for the planet Mars. Trying to apply it to Mercury adds the complexity of cooling the contained atmosphere and its containing structure while on the day side and heating it (and lighting it) while on the night side. The exposed exterior will undergo significant thermal stress as it transitions from the day side to the night side.

The gas-tight transparent roof/ceiling would also be the primary radiation shielding for people and equipment under it. Even three kilometers isn't enough to permit natural weather patterns but is high enough that some weather would result. It could be fairly luxurious with gardens and open spaces and life inside would be similar to living in a large shopping mall. It would be an engineered environment depending on climate control machinery.

The second contained atmosphere concept, shell worlds, was published in 2009 by Kenneth Roy, Robert Kennedy, and David Fields (Roy et al. 2009). This approach uses the pressure of the contained atmosphere to support a containment shell above the surface of the planet. No structures connect this shell with the planet's surface. Such a shell is subject to two primary stresses. The first is compression within the shell caused by gravity. The second is tension caused by the pressure of the contained atmosphere acting on the underside of the shell. By careful design, the net stress within the shell can be close to zero.

Assuming a spherical thin shell with a mass in the center, the compressive stress within the shell resulting from gravitational attraction between the shell and the central planet can be calculated by Eq. (20.7):

$$C = \frac{GM\rho s}{2rd} \tag{20.7}$$

where: C = gravitationally-induced compressive stress in a shell, [N m^{-2}], G = universal gravitational constant, $6.67 \cdot 10^{-11}$, [N m^2 kg^{-2}], M = mass of central planet, [kg], ρ = average density of shell, [kg / m^3], s = total shell thickness, [m], r = radius of shell, [m], d = thickness of load-bearing portion of shell, [m].

If the shell is composed of a single material that is load-bearing, then the s and d terms cancel out. Assuming that the atmospheric pressure above the shell is negligible, then the tensile stress created from atmospheric pressure on the underside of the shell can be calculated from:

$$T = \frac{ar}{2d} \tag{20.8}$$

where: T = atmospheric pressure-induced tension in a shell, [N m^{-2}], a = atmospheric pressure at the underside of the shell, [N m^{-2}], r = radius of shell, [m].

Setting the two terms equal to each other, assuming that the atmosphere has Earth normal pressures and temperatures on the surface of Mercury, assuming densities of the shell, calculating the pressure on the underside of the shell (a function of altitude per Eqs. (20.3) and (20.4)), knowing the mass and diameter of Mercury, it is then possible to solve for various values of s at different altitudes above the surface of the planet. Using Eq. (20.6) it is possible to calculate how much atmosphere is required relative to an uncontained atmosphere. Using Eq. (20.3) it is possible to calculate the pressure on the underside of the shell. The results are shown in Table 20.1.

Although the steel shell assumed above is fairly thick, it is under very low stress. The actual shell could be made of several materials with the load bearing portion being fairly small.

If we select a shell height of ten kilometers then a shell at this altitude could consist of fifty centimeters of steel to carry any stresses, either compression or

Table 20.1 Shell parameters for the planet Mercury

Height of shell above surface (km)	Shell thickness (steel[a]) (m)	% of atmosphere under shell, relative to uncontained	Atmospheric pressure at shell (atm)
3	3.04	12.4 %	0.87
5	2.789	19.8%	0.80
10	2.23	35.6%	0.64
20	1.44	58.7%	0.41

[a]The shell is assumed to be solid steel having a density of 7900 kg/m^3

tensile, topped with nine meters of dry soil (assumed density of 1500 kg/m^3). Other combinations are possible. Air-tight Kevlar® covered with steel plates, covered with lead sheets, covered with crushed rock, covered with solar cells is another possibility. Sometimes the dead mass can be replaced with vacuum-loving, radiation tolerant industrial facilities on the surface of the shell.

Ten kilometers above the surface is a reasonable height to consider. It is well above any Mercurial mountains. On Earth that is the altitude at which most airliners cruise. On a Mercury with Earth normal pressures and temperatures at the surface, the underside of the shell would see pressure that is 64 % of Earth-normal. This is similar to pressures at Pikes Peak on Earth and represents an altitude that humans can function at fairly comfortably. Maintenance of equipment and utility lines hung on the underside of the shell will be required on a regular basis, and there is no reason occupied structures couldn't be hung from the shell (the ever changing view would be spectacular).

Other heights are possible and the actual height selected is a design choice and will probably be a compromise between safety, cost, esthetics, radiation protection, and the desired weather patterns.

But is a shell supported by air pressure, and subject to gravity from a central body, stable? If the shell moves relative to the central world and collides with it, the results would be exciting but tragic for any colonists and a complete loss for any investors. Newton's shell theorem indicates that the gravitational interaction between the shell and the central body provides no restoring force. But gravity also results in the barometric effect described in Eqs. (20.3) and (20.4). As long as the central body is large enough that atmospheric pressure varies faster than gravity with changes in altitude, a restoring force is present that will resist any movement of the shell relative to the central world. The greater the displacement, the greater the restoring force on the shell. Mercury is large enough for this to apply and thanks to the unusually high density, the gradient is stronger than on most other planets or moons.

The surface of Mercury under an opaque shell would be perpetually shaded from the sun. Lights hung from the bottom of the shell could provide Earth-normal lighting. Lighting technology is advancing rapidly today. By the time the Mercury terraformers start ordering their fixtures, highly efficient, long duration lights tailored to almost any wavelength or combinations of wavelengths should be available.

Artificial lighting offers the terraformers the ability to specify the lighting intensity, cycling, and wavelength completely independent of the actual sun's location. The day/night cycle could be adjusted to be exactly 24 h. Agricultural areas could get as much light at the proper wavelength as they desired 24 h a day. Light over beaches could include enough UV light to ensure sunburns. Lighting would be a design choice.

The lighting bill will be high for a shell world. Just to duplicate the Earth's solar constant for Mercury through artificial lighting would take some 71,000 TW (Earth's solar constant of 1361 W/m^2 reduced by 30 %, due to the Earth's albedo, multiplied over the projected area of Mercury visible to the sun). This number can be greatly reduced by providing only the lighting necessary. Much of the light falling on Earth is wasted. The chlorophyll-A and chlorophyll-B molecules each have absorption peaks in the blue-violet and orange-red regions of the spectrum. Plants only use about 20 % of the light that impinges on them.

Assume that the lighting needs can be reduced by 80 % to 14,200 TW by only providing enough light at the wavelength and intensity needed for plants and for human vision. Luckily, Mercury has vast amounts of solar energy available. Its solar constant varies (due to a highly elliptical orbit) between 6.3–14.5 kW/m^2, on average 6.5 times that of Earth. Assuming that 25 % of the surface of the shell is covered in solar cells having an efficiency of 20 %, then at aphelion the total energy production can be calculate to be approximately 117,800 terawatts. By comparison, the actual US electricity generation rate in 2012 was less than half a terawatt. With more efficient solar cells, and recognizing that the planet is usually much closer to the sun, the power can be much higher.

Tidal forces on the shell must be considered. The portion nearest the sun will experience a greater pull than the farther portion, resulting in the spherical shell being pulled into a spheroid shape. Tidal forces also retard rotation of bodies. If the shell is constructed with no rotation with respect to the sun (the shell would be tidally locked to the Sun with the same side always facing the Sun) then several issues can be addressed. The tendency to be pulled into a spheroid shape can be countered by increasing the mass of the shell at the sunward and anti-sunward points. This allows the gravitational pull of Mercury to pull the spheroid back into a spherical shape. The sunward portion of the shell is where the solar cells are positioned. The albedo of the shell facing the sun and not screened by solar cells can be maximized, reducing the amount of sunlight absorbed. The far side of the shell, the portion that will never see the sun, can be optimized to radiate heat, maybe even incorporating active cooling systems. Hot atmospheric gases could be run through ducts near the surface of the shell where they would lose heat and then be dumped as cold air and snow onto Mercury's surface, setting up an atmospheric circulation. Mercury itself will continue to rotate under the shell, allowing the surface to experience "seasons". The Mercurial "year" would then become 176 Earth days long. Ocean tides would vary with this Mercurial "year" and not the 24-hour day/night cycle.

The question of a detailed heat calculation given such a scenario is very complex and depends on the albedo of the sunward facing shell, energy usage within the shell, the radiative characteristics and cooling ability of the dark side of the shell, and the internal temperature of Mercury itself. The weather and air movements under the shell along with ocean currents will play a significant part in heat redistribution.

The emissivity of a given material (usually indicated by ε) is the ability of a surface to emit radiation. The value of ε is the ratio of energy emitted compared to a black body at the same temperature (it is a dimensionless value that ranges from 0 to 1). The fraction of radiation absorbed by a material is known as its absorptance (usually indicated by α). The value of α is also a dimensionless value that ranges from 0 to 1 with 0 indication a perfect reflector and 1 indicating a black body that absorbs all radiation impinging on it. For a real material, both values vary with wavelength. The sun's energy distribution is approximately 7% in the UV region, 46 % in the visible light region and 47 % in the short IR region. Values of α (for sunlight) and ε (for long wave IR) for various materials are available in several references. It is possible to find materials that are highly reflective to sunlight and yet have a high emissivity relative to long wave IR. One such material is a magnesium oxide-aluminum oxide white paint. It has an α value of 0.09 and an ε value of 0.88 (Gilmore 2002). This material is very good at reflecting sunlight (91 %) and is also very good at radiating heat into the 3-kelvin cold of space.

If a reference, conductive metal sphere is coated with this material and is placed into an orbit identical to Mercury's, then the temperature of that sphere can be calculated using the following equation:

$$t^4 = \frac{\alpha S_c}{4\sigma\epsilon} \tag{20.9}$$

where: t = temperature of metal sphere, (K), α = absorptance of solar radiation for the coating, (dimensionless), S_c = solar constant at sphere, (W m^{-2}), σ = Stefan-Bolzmann constant, ($5.669 \cdot 10^{-8}$ W m^{-2} K^{-4}), ε = emissivity of coating for long IR radiation, (dimensionless).

At Mercury's mean distance from the sun, the solar constant is 9126 W / m^2 and this yields a temperature of 252 K (−20 °C). At Mercury's perihelion the solar constant is 14,462 W/m^2 yielding a temperature of 283 K (10 °C). At Mercury's aphelion the solar constant is 6278 W/m^2 and this yields a temperature of 229 K (−43 °C).

It is reasonable to assume that a future space faring civilization will have access to materials with similar or better values, materials that will withstand the vacuum and radiation of this environment for long periods of time. A world house or shell world is not a simple metal sphere, and the problem begs for a more detailed analysis. But it seems reasonable to state that a world house or shell world as described above could achieve reasonable temperatures that would allow for an Earth-like environment to exist within the contained atmosphere. A world house would have to have large and complex heat transfer systems to move heat from the "day" side and radiate it to the "night" side. On a shell world, oceans may be required to stabilize internal temperatures under the shell, but oceans are desirable unto themselves and support the principle to import as many Earth life forms as possible.

A world house would have a very thick roof to provide radiation protection and reduce the tension in the support structures. It is probably not a good idea to make it transparent, and it too would likely utilize a solid roof and artificial lighting.

With a shell world, the shell and the kilometers of atmosphere provide significant radiation protection. It should be able to withstand the solar wind, CMEs, cosmic radiation, and even gamma ray bursts. Gamma ray bursts are intense, short duration, bursts of high energy gamma rays. On Earth they would destroy the ozone layer allowing intense UV light to reach the surface, destroying most life. A world house or shell world doesn't have to worry about an ozone layer, they have something far better: solid matter.

20.5 Oceans

The presence of liquid water is a fundamental requirement for Earth life. On a terraformed planet liquid water must be plentiful. Because of the solar flux variations imposed on Mercury by its orbital eccentricity ($e = 0.2057$) a large body of water is needed to help mitigate temperature extremes and aid in heat transport. Oceans are also needed to facilitate precipitation of (evaporated) salt-free water over land. This water then flows back into the ocean carrying salts via lakes, rivers and streams. How much water is needed is an open question. Some minimal quantity of water is required as water vapor in the atmosphere and as water absorbed in the land to maintain a moist soil, necessary for life, before water begins to accumulate into seas and oceans. Establishing oceans in a world house is possible but difficult. There, water would probably be in the forms of fountains, pools, and reservoirs. On a terraformed world with an uncontained atmosphere and on a shell world, oceans are probably essential.

Some ice may be present on Mercury. There are crater floors at both poles that are never illuminated by the Sun and water-ice may have survived there. This water was probably delivered from impacts by comets (Beech 2009). Some ice may be present, but it is probably well short of what is needed to create an ocean. It will have to be imported. A water molecule is composed of one oxygen atom attached to two hydrogen atoms. Oxygen is available from Venus, but the hydrogen may have to be imported from the outer gas giants.

Water covers 71 % of Earth's surface with an average depth of over 3,500 m. It is the habitat of over 200,000 known species. The oceans on Mercury will, of necessity, be far more modest. To begin to understand the level of effort required the assumption is made that 33 % of Mercury will be covered with oceans to an average depth of 300 m. This requires 7.4×10^{15} metric tons of water. If the oxygen and hydrogen are imported separately and then combined at Mercury, then 6.6×10^{15} metric tons of oxygen and 8.3×10^{14} metric tons of hydrogen are required.

To put this in perspective, assume that the future terraformers have access to a standard space tanker capable of moving a sphere of liquid gases. Assume that this sphere is 2 km in radius and insulated adequately to maintain the cargo in its liquid state when delivered to Mercury. The oxygen coming from Venus requires 172,000 such vessels. The hydrogen coming from the gas giants require 350,000 such vessels (because liquid hydrogen has a very low density).

To supply the atmosphere (assuming a shell 10 km above the surface) an additional 21,000 tankers with liquid nitrogen, 4500 oxygen tankers and 205 argon tankers are required. The total tanker fleet (assuming no reuse) is thus slightly less than 573,000 ships. This can be reduced by producing the oxygen on site and/or reducing the size and depth of the ocean. It will be a massive effort, quite impossible today, but the technology and energy resources available to a future space faring civilization may be adequate for the task.

Part of the terraforming effort will involve design of the new Mercurial surface to create ocean basins, continents, islands, and mountains. Implementation of this design will involve moving vast amounts of surface material, probably before the shell is constructed. The tidal forces on Mercury caused by the Sun are approximately fifteen times as strong as on Earth. The resulting ocean tides on Mercury could be fifteen times as high and the design of the ocean basin will need to reflect this reality and mitigate the effects.

Once the shell is erected and the ocean warmed, it will take centuries or even millennia of an active hydrological cycle with rains washing over the land and then carrying salts and other minerals into the oceans before the waters are ready to receive the first life from Earth.

20.6 The Biosphere

Martyn Fogg defines the Biosphere to be the contents of the volume enclosing a planet, extending as far up into the atmosphere and as far down into the crust, in which life can exist (Fogg 1995). The only biosphere that is known of at this time is the one on Earth. It is sufficiently large and complex, with redundancies and feedback loops, that it has endured for millions of years without intelligent control. The creation of biospheres around other worlds is the ultimate justification for terraforming, because it provides not just humans but all of Earth's life another home.

Although Humanity seems intent on conducting experiments with Earth's biosphere, it has no real experience in creating a biosphere from scratch. The above sections have described how Earth-normal temperatures, atmospheric composition and pressure, lighting, and oceans might be established on Mercury. The task of introducing Earth life into this environment and creating a biosphere that could support not just humans but all of Earth's life is difficult to envision. The soil will at first be sterile and maybe even toxic. Once the hydrological cycle has been established, fresh water will rain down on this sterile soil, washing the salts and other toxic compounds into the ocean. Bringing life to this sterile soil is the next logical step.

On Earth, single-cell organisms live in soil with populations ranging up to 3 billion in a single gram of soil. This includes bacteria and fungi. Fungi have no chlorophyll and, like animals, require a chemical source of energy as well as organic materials to get carbon for growth and development. Many depend on dead plants for this and are invaluable in recycling dead matter. But without plants, how might a population of fungi be established? Certain bacteria (which manufacture their own carbohydrate supply without photosynthesis) transform ammonium (produced by the decomposition of proteins) into nitrates, which are available to growing plants who convert them into proteins. How are they to be established in the soil prior to plants? Understanding soil biology is critical to a successful terraforming project. This area requires additional research.

Once the soil food web can support plants, they will be introduced. Then insects and animals will be carefully introduced until an entire Earth-type ecology has been established. A decision will need to be made at this point whether to include mosquitoes, fleas, and other parasites. Exactly how essential are these creatures to the ecosphere? Ecosystem construction will be a difficult and challenging process and could take many centuries. But once this difficult task is complete, most, if not all, of the life that evolved and survived on Earth, humans included, could have a new home.

20.7 Conclusion

Will Mercury be terraformed at some point in the future? It will require a tremendous investment of time, material and energy. The project will take many centuries or even millennia to complete. To a future space faring civilization, rich beyond anything we can envision, the costs in materials and energy could be trivial. Perhaps their planning horizon will be much longer than ours and a project taking millennia would be acceptable. Provided that the light gravity isn't a show stopper, a terraformed Mercury, with a third of its surface covered in ocean, would increase the land area available to humanity by a third. If the shell-world approach is chosen, the biosphere should be large enough that it could endure for thousands or hundreds of thousands of years without much human management. Life from Earth would have a second home. A once dead planet would experience life.

The choice to terraform Mercury or to use it for some other purpose will belong to a future space faring civilization, one far richer and more powerful than ours, and hopefully wiser. But terraforming appears to be one option.

References

Beech M (2009) Terraforming: the creating of habitable worlds. Springer, New York
Fogg MJ (1995) Terraforming: engineering planetary environments. Society of Automotive Engineers Inc, Warrendale
Gilmore DG (ed) (2002) Spacecraft thermal control handbook: volume 1 fundamental technologies, 2nd edn, American Institute of Aeronautics and Astronautics, The Aerospace Press, El Segundo CA
Horn ER, Gabriel M (2011) Gravity-related critical periods in vestibular and tail development of Xenopuslaevis. J Exp Zool Ecol Genet Physiol 315:505–511
Paul A, Amalfitano CE, Ferl RJ (2012) Plant growth strategies are remodeled by spaceflight. BMW Plant Biol 12:232
Roy KI, Kennedy RG, Fields DE (2009) Shell Worlds: an approach to terraforming moons, small planets and plutoids. J Br Interplanet Soc 62:32–38
Taylor RLS (1992) Paraterraforming: the Worldhouse Concept. J Br Interplanet Soc 45:341–352
Wakayama S, Kawahara Y, Li C, Yamagata K, Yuge L, Wakayama T (2009) Detrimental effects of microgravity on mouse preimplantation development in vitro. PLoS ONE 4:6753

Chapter 21
Terraforming Mercury and Venus

Alexander A. Bolonkin

21.1 Introduction

In previous essays [1–4] author showed the basis of the Universe: Time, Matter, Charge, Distance (dimensions), Volume, is energy. Energy may be positive and negative. All particles are only different forms; collections of energy units. Author offered new artificial forms of matter having gigantic strength, heat resistance (millions times more than steel) and other awesome properties [5–6]. He also offered method for conversion any matters in energy and using than in aerospace [6–10].

The concept of negative energy explains the main riddle of the Universe: From where did the vacuum take a huge amount of energy for creation of our Universe? If we include the equal virtual pair positive and negative energy in vacuum, the total (sum) energy in vacuum equals zero. It may help to explain the dark energy and dark matter in the current model of the Universe. If there exist two Universes (positive and negative), embedded one (positive) Universe into the other (negative) Universe. As shown in [11] the negative Universe is full identical to our positive Universe because the choice of sign is arbitrary. Negative energy generates negative matter. But the relationship of negative matter to positive matter is different. Positive matter attracts positive matter but repels negative matter and the negative matter attracts the negative matter but powerfully repels the positive matter. They fly away one from other with a speed close to light speed. If they are connected, the positive and negative matters are annihilated (the sum of energy becomes zero).

This assumption about the existence of negative energy may give big possibilities in human technology. We can produce required matter in a needed amount

A.A. Bolonkin (✉)
C&R, Brooklyn, NY, USA
e-mail: aBolonkin@juno.com; aBolonkin@gmail.com

V. Badescu and K. Zacny (eds.), *Inner Solar System*,
DOI 10.1007/978-3-319-19569-8_21

(including a new planetary atmosphere) in needed place, annihilate the positive matter (for example, the bad-for-humans atmosphere of Venus) and create large space colonization.

21.2 Basic Idea

Vacuum energy is an underlying background energy that exists in space throughout the entire Universe. One contribution to the vacuum energy may be from virtual particles which are thought to be particle pairs that blink into existence and then annihilate in a time span too short to observe yet. They are expected to do this everywhere—that is, throughout the known Universe. Their behavior is codified in Heisenberg's energy–time uncertainty principle. Still, the exact effect of such fleeting bits of energy is difficult to quantify.

The effects of vacuum energy can be experimentally observed in various phenomena such as spontaneous emission, the Casimir effect and the Lamb shift, and are thought to influence the behavior of the Universe on cosmological scales. Using the upper limit of the cosmological constant, the vacuum energy in a cubic meter of free space has been estimated to be 10^{-9} J (10^{-2} ergs). However, in both Quantum Electrodynamics (QED) and Stochastic Electrodynamics (SED), consistency with the principle of Lorentz covariance and with the magnitude of the Planck constant requires it to have a much larger value of 10^{113} J/m^3. This huge discrepancy is known as the vacuum catastrophe. Vacuum energy can also be thought of in terms of virtual particles (also known as vacuum fluctuations) which are created and destroyed out of the vacuum. These particles are always created out of the vacuum in particle-antiparticle pairs, which in most cases, shortly annihilates each other and disappear. However, these particles and antiparticles may interact with others before disappearing, a process which can be mapped using Feynman diagrams. Note that this method of computing vacuum energy is mathematically equivalent to having a quantum harmonic oscillator at each point and, therefore, suffers the same renormalization problems. Additional contributions to the vacuum energy come from spontaneous symmetry breaking in quantum field theory.

Other predictions are harder to verify. Vacuum fluctuations are always created as particle–antiparticle pairs. The creation of these virtual particles near the event horizon of a black hole has been hypothesized by physicist Stephen Hawking to be a mechanism for the eventual "evaporation" of black holes. The net energy of the Universe remains zero so long as the particle pairs annihilate each other within Planck time. If one of the pair is pulled into the black hole before this, then the other particle becomes "real" and energy/mass is essentially radiated into space from the black hole.

In 1973, Edward Tryon proposed the zero-energy universe hypothesis: that the Universe may be a large-scale quantum-mechanical vacuum fluctuation where positive mass-energy is balanced by negative gravitational potential energy. During the 1980 s, there were many attempts to relate the fields that generate the vacuum energy to specific fields that were predicted by attempts at a S. Hawking Grand unification theory (2010) and to use observations of the Universe to confirm one or another version. However, the exact nature of the particles (or fields) that generates vacuum energy, with a density such as that required by inflation theory, remains a mystery. Main problem with Edward Tryon and Crand unification theory is the gigantic difference of the energy between the positive mass-energy and negative gravitational potential energy.

Contrary to Edward Tryonand, in Hawking's theory, the author believes the vacuum has the zero energy and simultaneously produces equal amount of the positive and negative energy. The positive and negative matter repels one other and the part of the negative matter may fly with light speed in space or in the negative Universe. That also means that, at the same time that our (positive) Universe was created, the same negative Universe which does not have interaction with our Universe was created. We cannot view the negative Universe. Contact between positive and negative matter requires high speed (returning the energy was received in removing) and produces full annihilation with zero-point energy. The zero-annihilation is opposed by the conventional annihilation of a positive particle-antiparticle which produces the maximum of energy.

Ratio between positive and negative energies and positive and negative matters are described by equations:

$$E_+ = m_+ c^2, \quad E_- = -m_- c^2, \quad E_+ + E_- = 0, \tag{21.1}$$

where E_+ is positive energy, J; E_- is negative energy, J; m_+ is positive matter, kg; m_- is negative matter, kg; $c = 3 \times 10^8$ m/s is the light speed.

21.3 Creating Earth-Like Atmospheres

For colonization of Mercury and Venus it is desirable that the planet has the atmosphere similar to Earth's atmosphere (pressure and composition). There cannot be any wide colonization of other planets if people have to live in special hermetic settlements and walk out of them only when wearing a special space suit. It will be possible only if men can be in space without special space suit [11] or the biological civilization will be changed by E-exists [12] (electronic civilization). We can compute the needed positive mass and energy for creating of new atmosphere. The need equations are:

$$p = mg, \quad S = 4\pi R^2, \quad M = pS/g, \quad E = Mc^2, \tag{21.2}$$

where p is surface pressure of atmosphere, N/m^2; m is specific mass of atmosphere on 1 m^2, kg/m^2; S is planet surface, m^2; R is radius of planet, m; M is total mass of planet atmosphere, kg; g is gravity of planets, m/s^2; E is energy need for creating the planet atmosphere, J. Result of computation is presented in Table 21.1.

Table 21.1 Mass and energy is needed for creating the artificial Earth atmosphere on Mercury and Venus

Planets and moon	Planet gravity g, m/s^2	Planet Radius R × 10^6 m	Mass of atmosphere, M × 10^{18} kg	Desire energy E × 10^{18} J
Mercury	3.72	2.43	2.69	2.42
Venus	8.69	6.05	5.3	4.77

Table 21.2 Mass of water and energy is needed for creating the artificial ponds in planets for $k = 0.1$ and $h = 5$ m

Planets	Mass of water, M × 10^{14} kg	Need energy, E × 10^{18} J
Mercury	0.74	0.733
Venus	4.6	5.94

If planet has a bad atmosphere (as does Venus) the old atmosphere may be annihilated by a negative matter and the Earth-like atmosphere can then be created. The excess of positive matter may be used for getting an agriculture soil and as construction material.

21.3.1 Creating Water on Mercury and Venus

For comfortable living conditions the humanity needs ponds, lakes, small rivers and possibly, small sea.

Assume the water surface cover k relative part of a planet surface S and has an average deep h meters. The needed mass of water may be estimated by equations:

$$M = k\gamma Sh, \quad E = Mc^2, \tag{21.2a}$$

where M is planet mass of water, kg; γ = 1000 kg/m^3 is specific mass of water, kg/m^3; h is average deep of water ponds, m. Result of estimations for $k = 0.1$ and $h = 5$ m is presented in Table 21.2.

21.4 Creating the Day Time (Angle Rotary of Planet)

It is important for people to have an alternation in the duration of a day and night (24 h). If we have unlimited energy, we can change the angle speed of planets and natural satellites. For computation of the desire energy we use the following equations:

Table 21.3 Desire energy for change the angle speed

Planet	Planet radius R_p, 10^6 m	Mass, 10^{24} kg	Angle inertia J, 10^{36} kg·m^2	Period, Earth days	Need energy E, 10^{25} J	ΔE
Mercury	2.43	0.33	0.773	59	0.1314	≈E
Venus	6.05	4.87	71.3	247	12.12	≈E

$$\omega = \frac{1}{2\pi T}, \quad J = \frac{2}{5} m R^2, \quad E_0 = \frac{J\omega^2}{2}, \quad \Delta E = \frac{J(\omega - \omega_p)^2}{2}, \tag{21.3}$$

where ω is desire angle speed of planet, rad/sec; *T* is time of one revolution, sec; *J* is angle inertial moment, kg · m^2; *m* is mass of planet, kg; *R* is radius planet, m; *E* is rotation energy of planet, J; Δ*E* is energy requested for change the angle speed of planet or satellite, J; ω_p is angle speed of planet, rad/s. Results of computation of need energy are presented in Table 21.3.

21.5 Creating of Magnetic Field

Earth's magnetic field, also known as the *geomagnetic field*, is the magnetic field that extends from the Earth's interior to where it meets the solar wind, a stream of charged particles emanating from the Sun. Its magnitude at the Earth's surface ranges from 25 to 65 μT (0.25–0.65 Gauss). The magnetosphere is the region above the ionosphere and extends several tens of thousands of kilometers into space, protecting the Earth from the charged particles of the solar wind and cosmic rays that would otherwise strip away the upper atmosphere, including the ozone layer that protects the Earth from harmful ultraviolet radiation.

Earth's magnetic field serves to deflect most of the solar wind, whose charged particles would otherwise strip away the ozone layer that protects the Earth from harmful ultraviolet radiation. One stripping mechanism is for gas to be caught in bubbles of magnetic field, which are ripped off by solar winds. Calculations of the loss of carbon dioxide from the atmosphere of Mars, resulting from scavenging of ions by the solar wind, indicate that the dissipation of the magnetic field of Mars caused a near-total loss of its atmosphere.

Earth's magnetic field, predominantly dipolar at its surface, is distorted further out by the solar wind. This is a stream of charged particles leaving the Sun's corona and accelerating to a speed of 200–1000 km/s. They carry with them a magnetic field, the interplanetary magnetic field (IMF). As well as deflecting the solar wind, the Earth's magnetic field deflects cosmic rays, high-energy charged particles that are mostly from outside the Solar system. (Many cosmic rays are kept out of the Solar system by the Sun's magnetosphere, or heliosphere.) By contrast, astronauts on the Moon risk exposure to radiation. Anyone who had been on the Moon's surface during a particularly violent solar eruption in 2005 would have received a lethal dose.

Magnetic field is important for humanity planets. One protects the humanity from ultraviolet radiation and decreasing the leak atmosphere in space. The magnetic field is about 25 ÷ 65 A/m near the Earth surface. The energy of outer Earth magnetic field is about 4 ÷ 8 × 10^{18} J. The similar magnetic field must have any human planet located near the Earth orbit. The creation the artificial magnetic field is described and computed in author article [10].

21.6 Relocation Mercury and Venus on Earth Orbit

The planets are located in different orbits (Table 21.4). All orbits (except Earth's) have an excess or a lack of solar energy. We have only one best (Earth's) orbit.

The offered concept of negative energy theoretically gives the possibility to receive a gigantic energy needed for the planet relocation. Let us to estimate the energy needed for planet relocation to Earth's orbit.

For planet relocations from one circle to other circle by Hohmann transfer we need two impulses. Desired Delta Speeds, time of transfer and specific energy may be computed by equations:

$$a=\frac{r_1+r_2}{2},\quad \Delta V_1=\sqrt{\frac{\mu}{r_1}}\left(\sqrt{\frac{r_2}{a}}-1\right),\quad \Delta V_2=\sqrt{\frac{\mu}{r_2}}\left(1-\sqrt{\frac{r_1}{a}}\right),$$
$$\Delta V=\Delta V_1+\Delta V_2, \qquad (21.4)$$
$$t=\pi\sqrt{\frac{a^3}{\mu}},\quad E_1=\frac{\mu}{2a},\quad E=ME_1,\quad \Delta E=M(E_{1,1}-E_{1,2}),$$

where a is average distance (average radius of planet orbits), m; r_1 is radius of the first orbit, m; r_2 is the radius of the second orbit, m; ΔV_1 is delta speed of the first impulse, m/s; ΔV_2 is delta speed of the second impulse, m/s; ΔV is total delta speed, m/s; $\mu = 1.328 \times 10^{20}$ is solar constant, t is transfer time, sec (or days or years); E_1 is energy of 1 kg planet mass in given orbit, J; M is mass of planet, kg; E is full energy of planet in given orbit, J; ΔE is relocation energy, J; $E_{1,I}$ is energy of planet in desire position, J.

Result of computations for Mercury and Venus are present in Table 21.5 and Figs. 21.1, 21.2 and 21.3. In those figures computations for Mars are also shown, as a reference.

Table 21.4 Data on Mercury and Venus

Space body	Distance from Sun, 10^{10} m	Mass, 10^{24} kg	Gravitation constant μm³/s²	Circulars speed, V_O km/s	Escape speed, V_e km/s	Orbital speed, V_{or} km/s	Gravity constant g, m/s²	Planet radius R_p, 10^6 m
Mercury	5.79	0.33	$2.2 \cdot 10^{13}$	3	4.25	47.87	3.72	2.43
Venus	10.8	4.87	$3.2 \cdot 10^{14}$	7.25	10.36	35.02	8.69	6.05

Table 21.5 Delta Speeds, time of transfer, specific energy and total energy of planets

Planet	$r \times 10^{10}$ m	$a \times 10^{10}$ m	$\Delta V_1 \times 10^3$ m/s	$\Delta V_2 \times 10^3$ m/s	$\Delta V \times 10^3$ m/s	t, day	$E_1 \times 10^8$ J/kg	$M \times 10^{18}$ kg	$E \times 10^{27}$ J
Mercury	5.79	10.37	7.5	9.5	17	105.5	6.4	2.69	1.72
Venus	10.5	12.88	2.86	2,74	5.6	145.8	5.15	5.3	2.73

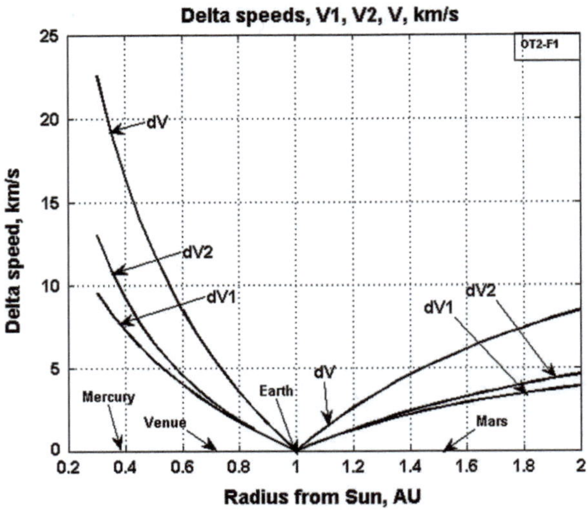

Fig. 21.1 Desire delta speeds for relocation of planets the Earth's group (Mercury, Venus, Earth, Mars) via radius from Sun in astronomical units, AU = 150 million km; dV1 is the first impulse, dV2 is the second impulse, dV = dV1 + dV2 km/s

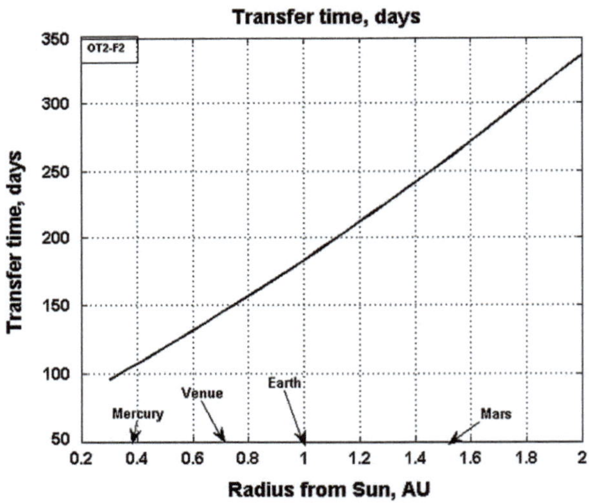

Fig. 21.2 Transfer time (in days) for relocation of planets the Earth's group (Mercury, Venus, Mars) via planet radius from Sun in astronomical units, AU = 150 million km

Example. Let us to estimate the desire energy for relocation 1 kg from Mercury orbit to Earth orbit (or back). In Mercury's orbit 1 kg mass has $E_{1,1} = 6.4 \times 10^8$ J, in Earth's orbit one has $E_{1,2} = 4.4 \times 10^8$ J (see also Fig. 21.3). Consequently, the transfer energy is $\Delta E = E_{1,1} - E_{1,2} = 2 \times 10^8$ J.

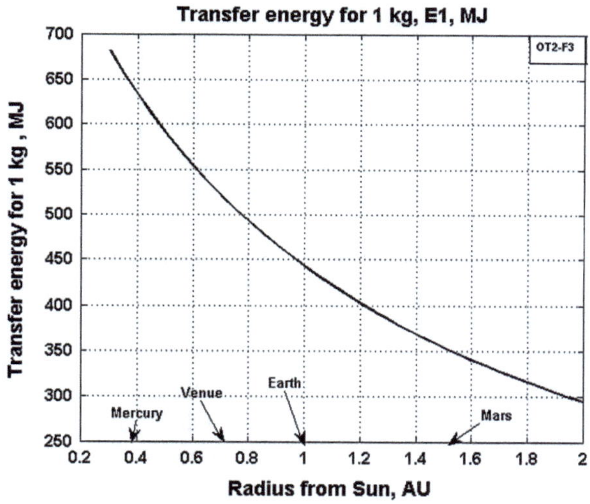

Fig. 21.3 Energy (in million J) 1 kg planet the Earth's group (Mercury, Venus, Mars) via radius of planet orbit from Sun in astronomical units, AU = 150 million km

21.7 Colonization of Mercury and Venus

Space colonization (also called *space settlement* or *extraterrestrial colonization*) is permanent human habitation that is not on Earth. Many arguments have been made for space colonization. The two most common are survival of human civilization and the biosphere from possible disasters (natural or man-made), and the huge resources in space for expansion of human society.

As of right now building new space colonies on Mercury and Venus present a number of huge challenges, both technological and economic. Space settlements would have to provide for all the material needs of hundreds or thousands of humans in an environment out in space that is very hostile to human life. They would involve technologies, such as controlled ecological life support systems, that are yet to be developed in any meaningful way. They would also have to deal with the unknown issue of how humans would behave and thrive in such places over a long period of time. Because of the huge cost of sending anything from the surface of the Earth into orbit (roughly $20,000 USD per kilogram) a space colony would be a massively expensive proposition.

No space colonies have been built so far, nor are there any timetables for building one by any large-scale organization (either government or private sector). However, there have been many proposals, speculations, and designs for space settlements that have been made, and there are a considerable number of space colonization advocates and groups. Several famous scientists, such as Freeman Dyson, have come out in favor of space settlement.

Also on the technological front, there is an ongoing progress in making access to space cheaper, and in creating automated manufacturing and construction techniques. This could in the future lead to widespread space tourism, which could be a stepping stone to space colonization.

21.8 Discussion

The reader can ask: where is the evidence that the negative energy and negative matter exist? That can only be proved in the future. The author gives examples when scientists made the assumption which received weak proof after long time. Some of them did not have any proof up to a recent time, however their theories are widely used now.

Examples: In 1974, S. Hawking assumed the black hole has radiation. All attempts done over 40 years found this radiation to be not true. But most scientists still believe this radiation exists and research it. In 2014, S. Hawking received 3 millions prime for his assumption. In 1927 the hypotheses about creating the Universe from Big Bang about 14 billion years ago was offered. No strong proofs this theory. But many scientists believe in it and develop it. Perturbation theory believes the vacuum has virtual particles (Dirac sea). This theory assumes that the Casimir force supports this theory. But other scientists explain the Casimir forces by other factors. Some theories contain conditions which can never be validated. For example, quark. This theory believes that if we try to separate two quarks, they convert to other particles. Strings Theory believes: there are many other space dimensions, which are impossible to see.

Any theory may be useful if one simplify current theory, explain the obscure pheromones, prompts for new experiments. Assumption of positive and negative energy allows to develop not only the theory positive and negative matter, but to build the simple Universe from simplest units of energy.

Let us shortly consider that the propulsion system uses the positive and negative matter. Ever since Newton first formulated his theory of gravity, there have been at least three conceptually distinct quantities called mass: inertial mass, "active" gravitational mass (that is, the source of the gravitational field), and "passive" gravitational mass (that is, the mass that is evident from the force produced in a gravitational field).

In 1957, Hermann Bondi suggested in a paper in Reviews of Modern Physics that mass might be negative as well as positive [13]. Although no particles are known to have negative mass, physicists (primarily Hermann Bondi and Robert L. Forward) [14] have been able to describe some of the anticipated properties such particles may have. Assuming that all three concepts of mass are equivalent, the gravitational interactions between masses of arbitrary sign can be explored.

For two positive masses, nothing changes and there is a pull on each other causing an attraction. Two negative masses would produce a pull on one another, but would repel because of their negative inertial masses. For different signs there is a push that repels the positive mass but attracts the negative mass. Bondi pointed out that two objects of equal and opposite masses would produce a constant acceleration of the system towards the positive-mass object. Forward used the properties of negative-mass matter to create the diametric drive, a design for spacecraft propulsion using negative mass that requires no energy input and no reaction mass to achieve arbitrarily high acceleration.

The diametric drive was a speculative proposal for an "engine" which would create a non-conservative gravitational field with non-zero curl. It was argued that in such circumstances, the side of the field which creates more force on the spacecraft will accelerate the spacecraft in the direction of the force. One idea for realizing this concept involved hypothetical particles with negative mass, originally proposed by Robert Forward and Jamie Woodward. If one were to construct a block of negative mass, and then attach it to a normal "positive" mass, the negative mass would fall towards the positive as does any mass toward any other. On the other hand, the negative mass would generate "negative gravity", and thus the positive mass (the spaceship itself generally) would fall away from the negative mass. If arranged properly, the distance between the two would not change, while they continued to accelerate forever.

The negative mass propulsion offered in this article is in principal different from Forward's engine. Forward believed that inertial, "active" and "passive" masses have different properties (an assumption not so far not supported by experiments). He designs from them a rigid structure which produces a thrust despite the rigid linkage between them. His engine is a reactionless drive.

A reactionless drive (also known by many other names, including as an inertial propulsion engine, a reactionless thruster, a reactionless engine, a bootstrap drive or an inertia drive) is a fictional or theorized method of propulsion wherein thrust is generated without any need for an outside force or net momentum exchange to produce linear motion. The name comes from Newton's Third Law of Motion, which is usually expressed as, "[f]or every action, there is an equal and opposite reaction". Such a drive would necessarily violate laws of classical physics, the conservation of momentum and the conservation of energy. In spite of their physical impossibility, devices are a staple of science fiction, particularly for space propulsion. Devices and methods are still being proposed as working technologies only now they are based on the real or imagined principles from modern physics.

The author's theory asserts that the inertial, active and passive positive matter are one positive matter (supported by experiments) and inertial, active and passive negative matter are also one (same) negative matter, which has but a single difference in properties (other than mathematical sign): that negative matter repels positive matter. The Forward drive violates Newton's Third Law of Motion, violates laws of classical physics, the conservation of momentum and the conservation of energy. The suggested engine does not do this (the total energy and mass are zero!). The Forward engine is a reactionless drive, the offered engine works as a conventional reactive engine, sending away the negative mass with speed close to light speed. The Forward drive has constant mass in during the entire flight. The offered drive increases its' positive mass in the flight and can use it for further increasing the speed, decreasing the flight time or creating an artificial space body i.e. tapping the matter influx for space construction.

21.9 Conclusion

The current physics believes that vacuum can produce energy and Universes. Author assumes: the basis of any Universe is energy. Energy may be positive or negative. Positive energy produces our positive matter, negative energy produces negative matter. Using this effect, the author offers the terraforming of Mercury and Venus, making them suitable for people, for humanity. That include: the production of Earth-like atmosphere, water, magnetic field on Mercury and Venus, or change their angle speed, and relocate them to Earth orbit. If author's assumption about the existing of the negative energy is tested, the probability of negative matter existing and exhibiting repulsion from our positive matter is very high.

Acknowledgments The author wishes to acknowledge Richard Cathcart for correcting the English and offering useful advice and suggestions.

References

1. Bolonkin AA (2012a) Universe (part 1). Relations between time, matter, volume, distance, and energy. J Energy Storage Convers 3:141–154. http://viXra.org/abs/1207.0075, http://intellectualarchive.com #139, http://archive.org/details/Universe.RelationsBetweenTimeMatterVolumeDistanceAndEnergy
2. Bolonkin AA (2009a) Universe (part 2): rolling of space (volume, distance), time, and matter into a point. http://www.scribd.com/doc/120693979
3. Bolonkin AA (2012b) Remarks about Universe (part 1–2). Int J Adv Eng Appl .1:62–67. http://viXra.org/abs/1309.0196, http://fragrancejournals.com/wp-content/uploads/2013/03/IJAEA-1-3-10.pdf
4. Bolonkin AA (2010) Article Universe (part 3). Relations between charge, time, matter, volume, distance, and energy. Gen Sci J (9 January 2014). The General Science Journal #1192. IJAEA, GSJ, http://www.gsjournal.net/Science-Journals/Research%20Papers-Mechanics%20/%20Electrodynamics/Download/5245, http://viXra.org/abs/1401.0075, http://www.scribd.com/doc/197830994, http://www.IntellectualArchive.com/, Reference #1192, https://archive.org/details/universepart3.RelationsBetweenChargeTimeMatterVolumeDistance, http://www.IntellectualArchive.com/getfile.php?file=gwBJfgtbOeS&orig_file=Article Universe3 after Friedlander 01 09 14.doc
5. Bolonkin AA (2012c) Femtotechnology. AB-matter. Properties, stability, possibility production and applications. Glob Sci J (24 March 2014). http://fragrancejournals.com/wp-content/uploads/2013/03/IJAEA-3-1-3.pdf, http://www.gsjournal.net/Science-Journals/Research%20Papers-Quantum%20Theory%20/%20Particle%20Physics/Download/5244, http://viXra.org/abs/1312.0017, https://archive.org/details/StabilityAndProductionSuper-strongAb-matter, http://www.IntellectualArchive.com/, Reference #1178
6. Bolonkin AA (2011) Femtotechnologies and revolutionary projects. Lambert, USA, p 538. http://viXra.org/abs/1309.0191, http://www.archive.org/details/FemtotechnologiesAndRevolutionaryProjects, https://independent.academia.edu/AlexanderBolonkin/Papers
7. Bolonkin AA (2012d) Converting of any matter to nuclear energy by AB-generator and aerospace. J Energy Storage Convers 3:43–69. http://www.archive.org/details/ConvertingOfAnyMatterToNuclearEnergyByAbgeneratorAndAerospace

8. Bolonkin AA (2014) Method for interstellar flight. GSJ, 9 August 2014. http://viXra.org/abs/1408.0055, https://archive.org/details/, http://intellectualarchive.com #1312, http://gsjournal.net/Science-Journals-Papers/Author/1481/Alexander,%20Bolonkin
9. Bolonkin AA (2012e) Space wing electro relativistic AB-ship. I J N N A, 4(2):13–19 (January–June 2012). http://www.archive.org/details/SpaceWingElectroRelativisticAb-ship
10. Bolonkin AA (2015) Artificial magnetic field for Venus and other planets (This volume)
11. Bolonkin AA (2009b) Man in outer space without a special space suit. Am J Eng Appl Sci 2 (4):573–579. http://viXra.org/abs/1309.0199
12. Bolonkin AA (2010) Universe, human immortality and future human evaluation. Scribd., 1p 24. http://viXra.org/abs/1207.0020, http://www.archive.org/details/UniverseHumanImmortalityAndFutureHumanEvaluation
13. Bondi H (1957) Negative mass in general relativity. Rev Mod Phys 29(3):423
14. Landis G (1991) Comments on negative mass propulsion. J Propul Power 7(2):304

Chapter 22
Cloud Ten

Magnus Larsson and Alex Kaiser

22.1 Introduction: Hypatia Revisited

> I think there is a strong humanitarian argument for making life multi-planetary in order to safeguard the existence of humanity in the event that something catastrophic were to happen, in which case being poor or having a disease would be irrelevant, because humanity would be extinct [1].

The topic of this chapter is not an easy one. Not only is space architecture possibly humanity's greatest architectural challenge. Of the terrestrial sites that could potentially allow for the expansion of human life in space—planets Mercury, Venus, and Mars, together with the possibility of a lunar colony and the development of bodies positioned within the asteroid belt—Venus is certainly not the easiest target [2]. But it might well be the most promising one. Indeed, while our growing knowledge of the extremely hostile surface environment on Venus has made many theorists and commentators shift focus away from Earth's "sister planet" towards the potential colonization of the Moon and Mars instead, a deeper investigation of the unique opportunities offered by a future Venus habitat makes for a compelling, if elaborate (by necessity, given the circumstances) argument.

So why attempt to venture into outer space in the first place? Why go through the trouble of designing speculative structures for inhospitable environments that are today absolutely hostile to human life? Why envisage projects such as this, which present huge technological and financial challenges? Why allow ourselves to dream

M. Larsson (✉) · A. Kaiser
Ordinary Ltd., London, UK
e-mail: magnus@ordinary.london

A. Kaiser
e-mail: alex@ordinary.london

V. Badescu and K. Zacny (eds.), *Inner Solar System*,
DOI 10.1007/978-3-319-19569-8_22

about the construction of space colonies that are based on as-yet-undeveloped automated manufacturing and construction techniques, and that cannot even be satisfactorily proven to support human life in the long term?

Because, as theoretical physicist and cosmologist Steven Hawking famously told the Telegraph newspaper in 2001, it might be our only chance of survival as a species [3]. Recently, entrepreneur, engineer, inventor and investor Elon Musk has echoed the same opinion [4]. The argument is based on the scenario that a natural or man-made planetary-scale disaster would force us to crawl off our home planet just as our primitive progenitors crawled out of the ocean [5]. That dystopic sequence of events has a more utopian corresponding analogue: that the vast resources in space would allow the human race to expand beyond its current limits in an eventually sustainable fashion, an argument made even more pertinent given the current combination of a growing population and an unsustainable energy consumption on planet Earth [6].

The brightest natural object in the night sky after the Moon, Venus is the second planet from the Sun, and has probably been known almost since the dawn of history [7]. This neighbour of the Earth is the only planet in the solar system to be named after a female figure – the Roman goddess of love, beauty, fertility and desire (qualities borrowed from her mythical Greek counterpart, Aphrodite; to the Babylonians before them, her incarnation was called Ishtar—the goddess of womanhood and love. The northern highland "continent" on planet Venus is called Ishtar Terra, and is about the size of Australia; the southern continent is called Aphrodite Terra, at roughly the size of South America). Venus the godess was supposedly born of sea foam; the Roman people viewed her as their divine mother through her mythological son Aeneas, the first true hero of Rome. Venus has an ambivalence to her that is etymologically suggested by the relationship of the root *venes-* with the Latin *venenum* (poison): a godess of prostitutes (Latin *venus* means sexual love/desire) that simultaneously turns the hearts of men and women from sexual vice to virtue [8].

This multifaceted character is to an extent carried across to planet Venus as well – depending on your point of view, this heavenly body can be an interstellar inferno or a potential paradise or, failing that, perhaps the closest we can get to a paradise in the aftermath of an extinction of Earth. Either way, while Venus might be considered to be Earth's twin, the two planets are certainly not identical twins. If understanding the site is a main component of the architectural process, then this is indeed quite a site to comprehend. It is true that Venus is similar to Earth in terms of some rather important planetary parameters (size, mass, bulk composition, solar distance, presence of an atmosphere), but in terms of others, equally important ones, it is substantially different (absence of an intrinsic magnetic field, presence of sulfuric acid clouds, large atmospheric mass, carbon dioxide composition of the atmosphere, lack of water, probably a non-existent carbon cycle and organic life, and a very high surface pressure and temperature) [9].

That certainly sounds like a reasonably deterring list of obstacles to overcome before being able to welcome the first space settlers to a habitat on Venus. Unlike Earth, Venus lacks any detectable magnetic field. Its average surface temperature is about 735 K (462 °C), which is not only much higher than that on Earth, but actually makes it by far the hottest planet in the entire solar system, despite Mercury being closer to the Sun. Indeed, the heat is intense enough to have earned Venus its epithet as a *hellish* planet: walking across its surface equals moving through a temperature higher than that used to achieve sterilization. Winds are slow but, due to the high density at the surface, exert a significant amount of force, and transport both dust and small stones across the surface [10]. Venusian winds move at up to 60 times the speed of the planet's rotation, whereas Earth's fastest winds are only 10–20 % of the rotation speed [11]. Despite the planet's slow rotation (slower than any other planet in the solar system), the transfer of heat by these winds in the lower atmosphere, coupled with thermal inertia, means the temperature of the Venusian surface does not vary significantly between the night and day sides. It is always diabolically hot.

Venus's atmosphere is composed predominantly of CO_2, and its average surface atmospheric pressure is much higher than ours (about 93 bar), whereas the pressure at the planet's surface is about 92 times that at Earth's surface – equivalent to the pressure at a depth of nearly 1 km under Earth's oceans [12]. This dense carbon dioxide atmosphere makes conditions on the Venusian surface differ radically from those on Earth – the mass of the Venus atmosphere is 96.5 % carbon dioxide, with most of the remaining 3.5 % being nitrogen, generating a runaway greenhouse effect, the strongest in the solar system [13]. The entire planet is shrouded by an opaque layer of highly reflective clouds, made not of particles of liquid or solid water as on Earth, but of sulfuric acid, dense enough to prevent its surface from being seen from space in visible light. As if that wasn't awkward enough, a year on Venus (its orbital period) is 225 Earth days, while its rotation (in a retrograde manner) is 243 days – the combination makes the solar day on Venus around 116.75 Earth days long, which in turn means any given place on the planet except the poles is illuminated by the Sun for 58.37 Earth days, and then remains in darkness for another 58.37 days [14]. That's a long working day (and a long night's journey into day) by any standards, in particular if the workplace happens to be devoid of oxygen and set in a dry desertscape pierced with wrinkled slabs of volcanic rock.

But then there's the good news. Venus orbits the Sun at a mean distance of 108,000,000 km, some 0.72 of the mean Earth–Sun distance. This means the Sun illuminates Venus with an intensity that is only a factor of 1.9 higher than that with which it illuminates Earth. The radius of Venus is 6,051.8 km (making for a diameter only 650 km less than the Earth's) and its mass is 4.87×10^{24} kg (0.814 of Earth's) [15]. It is likely that the planets share a similar internal structure, featuring core (the Venusian one being at least partially liquid), mantle, and crust [16]. But far more important than such similarities are the conditions at approximately 55 km above the fiendish surface of Venus. At that altitude, the atmospheric pressure and temperature turn out to be quite close to those at the surface of the Earth: 30 °C and 0.5 bar, respectively. (However, the predominantly CO_2-filled atmosphere is mixed with small amounts of sulfuric acid aerosols, noble gases, and other constituents; the cloud layer extends across an atmospheric layer between ~45 and ~70 km, and the winds in this superrotating atmosphere reach mean velocities between ~7.5 and ~60 m/s) [17]. Furthermore, Earth air (predominantly a mixture of nitrogen and oxygen) would be a lifting gas in the primarily carbon dioxide-composed Venusian atmosphere. This has led to a number of proposals for "floating cities" (essentially superpressure balloon structures) in the Venusian atmosphere, of which the one presented in a Russian magazine in 1971 might have been the first [18] and the one presented here the latest addition.

Perhaps the foremost among them is American scientist (at the NASA John Glenn Research Center) and science fiction writer Geoffrey A. Landis's 2010 short story, The Sultan of the Clouds. The storyline is based on a premise that Landis – who in his dayjob works on NASA missions to Mars, Venus, and the solar corona – has backed up with a series of scientific papers [19]), in which human settlements float inside huge balloons at a level of the atmosphere high enough above the surface for the atmospheric temperature to be roughly Earth-like.

Landis's narrative traces the history of the "fabled floating cities" through the journey to Venus by its two protagonists, invited by Carlos Fernando Delacroix Ortega de la Jolla y Nordwald-Gruenbaum, the son of Udo Nordwald, "tyrant and patriarch of the Nordwald industrial empire," head of one of the 20 families who own the future solar system, who "made his reputation by colonizing what was casually called the solar system's Hell planet. Venus". Upon arrival, the first vehicle the narrator encounters is a dirigible-like transfer barque that is over a kilometre long, and yet deceivingly simple in its construction: "…no more than a thin skin over a hollow shell made of vacuum-foamed titanium surrounding a vast empty chamber. It was designed not to land, but to float in the atmosphere, and to float it required a huge volume and almost no weight". Piloted by a yellow-robed Buddhist monk, the barque takes them through the atmosphere and up above the…

> "Clouds. A hundred and fifty million square kilometers of clouds, a billion cubic kilometers of clouds. In the ocean of clouds the floating cities of Venus are not limited, like terrestrial cities, to two dimensions only, but can float up and down at the whim of the city masters, higher into the bright cold sunlight, downward to the edges of the hot murky depths" [20].

And then onwards to the floating city of Hypatia, shimmering high above Venus:

> The pilot-monk rolled the barque back, and then pointed, forward and slightly to the right. "There. See it?"
>
> I didn't know what to see. "What?"
> "There."
> I saw it now, a tiny point glistening in the distance. "What is it?"
> "Hypatia. The jewel of the clouds."
> As we coasted closer, the city grew. It was an odd sight. The city was a dome, or rather, a dozen glistening domes melted haphazardly together, each one faceted with a million panels of glass. The domes were huge, the smallest nearly a kilometer across, and as the barque glided across the sky the facets caught the sunlight and sparkled with reflected light. Below the domes, a slender pencil of rough black stretched down toward the cloudbase like taffy, delicate as spun glass, terminating in an absurdly tiny bulb of rock that seemed far too small to counterbalance the domes.

> "Beautiful, you think, yes? Like the wonderful jellyfishes of your blue planet's oceans. Can you believe that half a million people live there?"
> The pilot brought us around the city in a grand sweep, showing off, not even bothering to talk. Inside the transparent domes, chains of lakes glittered in green ribbons between boulevards and delicate pavilions. At last he slowed to a stop, and then slowly leaked atmosphere into the vacuum vessel that provided the buoyancy. The barque settled down gradually, wallowing from side to side now that the stability given by its forward momentum was gone. Now it floated slightly lower than the counterweight. The counterweight no longer looked small, but loomed above us, a rock the size of Gibraltar. Tiny fliers affixed tow ropes to hardpoints on the surface of the barque, and slowly we were winched into a hard-dock.
>
> "Welcome to Venus," said the monk [21].

And then, in a passage that, at less than ten sentences, reads as a model of brevity, the author summarizes the entire premise of his storyline:

> The surface of Venus is a place of crushing pressure and hellish temperature. Rise above it, though, and the pressure eases, the temperature cools. Fifty kilometers above the surface, at the base of the clouds, the temperature is tropical, and the pressure the same as Earth normal. Twenty kilometers above that, the air is thin and polar cold.
>
> Drifting between these two levels are the ten thousand floating cities of Venus.
> A balloon filled with oxygen and nitrogen will float in the heavy air of Venus, and balloons were exactly what the fabled domed cities were. Geodetic structures with struts of sintered graphite and skin of transparent polycarbonate synthesized from the atmosphere of Venus itself, each kilometer-diameter dome easily lifted a hundred thousand tons of city.
> Even the clouds cooperated. The thin haze of the upper cloud deck served to filter the sunlight so that the intensity of the Sun here was little more than the Earth's solar constant [22].

Landis's Venus-based story was nominated for the 2011 Hugo Awards in the "Best novella" category. Understandably, his employer has expressed an interest in the ideas presented in the short story, and – whether there is a relation or not (Landis does not appear to be officially on the team), NASA have subsequently presented their plans for exploratory missions to the "exciting destination for both further scientific study and future human exploration" that is the atmosphere of Venus. They plan on carry these out using a "lighter-than-air vehicle [that] can carry either a host of instruments and probes, or a habitat and ascent vehicle for a crew of two astronauts to explore Venus for up to a month. The mission requires less time to complete than a crewed Mars mission, and the environment at 50 km is relatively benign, with similar pressure, density, gravity, and radiation protection to the surface of Earth. A recent internal NASA study of a High Altitude Venus Operational Concept (HAVOC) led to the development of an evolutionary program for the exploration of Venus, with focus on the mission architecture and vehicle concept for a 30 day crewed mission into Venus's atmosphere" [23].

The space architecture scheme presented in this chapter sits somewhere in between Landis's visionary space urbanism and NASA's more pragmatic mission description. In brief, we propose to launch into the Earth-like part of the Venus atmosphere a capsule structure that contains everything needed to grow – in situ, in space – the materials needed to create the skeleton frame and membrane skin for an expandable/deployable structure that increases in size through an adaptation of the famous sequential geometries called the Jitterbug Transformation, invented by neo-futuristic architect and systems theorist Richard Buckminster Fuller. Following several of these transformations, the initial cell (capsule) divides to create two cells, and so on. The cells then self organise and attach to each other using algorithms borrowed from the field of evolutionary logics, in order to create a habitat floating high above the Venus surface. Not until we have reached this stage do we begin building up the interior atmosphere while adding natural biosystems and implementing different harvesting processes, before allowing the first settlers move in. In order to explain this process, we will first cover the material, geometrical, and cellular organisation of the scheme, before going through the project phases in chronological order, and finally conclude with some remarks on how the scheme might be improved and enhanced through future studies.

Many challenges are of course left to overcome before the Cloud Ten habitat and its methods become a viable alternative to existing, more well-documented technologies. We understand that what we suggest is easily dismissed as improbable, implausible, and possibly impossible. We appreciate that in all likelihood, humanity's colonization of Venus is not a feat scheduled for the next decade or two. The proposal is an easy target for criticism: not only is what we suggest perhaps a far-fetched process to begin with; we are the first to admit that many crucial moments within the scheme remain unresolved. But that does not mean we shouldn't present these ideas as an outline for others (or ourselves at a later date) to expand and elaborate on. As we have pointed out elsewhere [24] architecture is all about opportunities, all acts of design inevitably and inherently about change, and any architectural project – in particular any space architecture project – therefore radically optimistic: we build because we believe in making the world, on this planet as well as others, a better place. As US engineer, professor, physicist, and inventor Robert Goddard (credited with building and successfully launching the world's first liquid-fueled rocket in 1926) put it: "Nobody knows what is impossible, for the dream of yesterday is the hope of today and the reality of tomorrow" [25].

22.2 From ISRU to ISMG

> Because we humans are big and clever enough to produce and utilize antibiotics and disinfectants, it is easy to convince ourselves that we have banished bacteria to the fringes of existence. Don't you believe it. Bacteria may not build cities or have interesting social lives, but they will be here when the Sun explodes. This is their planet, and we are on it only because they allow us to be [26].

In space exploration, and ultimately space colonization, the concept of in situ resource utilisation (ISRU) – in simple terms, using resources that can be found at a present location – has been suggested as a way to drastically reduce the amount of payload that must be launched from Earth in order to explore a given planetary body. Despite not yet having received any practical application over and above the common use of spacecraft solar panels, NASA describes the idea of using resources found on or manufactured from other astronomical objects as making it possible to "enable the affordable establishment of extraterrestrial exploration and operations by minimizing the materials carried from Earth" [27].

Numerous studies have indicated the potential benefits of using extraterrestrial resources for both robotic and human space exploration missions. These benefits typically result in smaller spacecraft and reduced launch requirements, both of which reduce mission costs [28]. The idea of using natural resources from heavenly bodies such as for instance asteroids is older than the Space Age. Already in 1903, Konstantin Tsiolkovskii included the "exploitation of asteroids" as one of his fourteen points for the conquest of space in *The Exploration of Cosmic Space by Means of Reaction Motors* [29]. NASA, under the Exploration Technology Development Program, has made significant investments in the development of Space resource utilization technologies as a part of the In Situ Resource Utilization (ISRU) project. This has led to a shift away from academic towards applied research, with real-life tests of built machinery including a system capable of producing oxygen from Lunar soils (or terrestrial equivalents) [30]. In the 1990s and early 2000s, NASA performed significant research at the technology and system level aimed at developing ISRU concepts and hardware to make propellant from carbon dioxide in the atmosphere of Mars, and in 2004, research into the development of ISRU strategies for extraction of oxygen from lunar regolith was begun [31].

It has been pointed out that the Apollo program, and most of NASA's human exploration studies following this (which tend to build on their predecessor) was neither a sustainable nor an affordable scheme, and that these sustainability issues connected to the existing approach can essentially be divided into three main areas: (1) a focus on upfront cost and mass at the expense of life cycle analyses and long-term cost implications, (2) a lack of interaction between sub systems that are individually designed, and (3) a single application-based technology optimisation and development that lacks coordination with other parts of the architecture [32]. While this is probably correct to an extent, we would argue that at least an additional three areas could and should be added to these three, namely (4) the underutilised potential to conceive of innovative ways to systematically *grow*, as opposed to *extract*, resources in space for both extraterrestrial and terrestrial use (this is in line with current thinking about sustainable material usage on Earth [33] and we see no reason why that concept couldn't be extended to a space context), (5) the exhaustive study of geometric-spatial concepts and relationships, in particular with regards to folded/deployable structures, and (6) the strategic appropriation of ideas from concilient fields such as evolutionary logics in search of amplified optimisation strategies. The Cloud Ten project employs all of these three additional and potentially innovative strategies, and we will now attempt to explain them in successive sections, beginning with the material considerations.

As mentioned, the proposal presented in this chapter is for the design of a sustained human settlement floating high above the poisonous acid clouds that shroud the fiendish surface of Venus (Fig. 22.1). Any such scheme will by necessity have to place particular emphasis on addressing material challenges and strategies. This is arguably the case with all acts of space architecture, but perhaps more so when that architecture is to be located on Venus, a planet equipped with an atmosphere and a surface that appear to make it a perfectly Hadean environment, devoid of the life-supporting material supplies available on the surface of, for instance, Mars, the planet most hotly tipped as humanity's next long-term outpost in the extraterrestrial universe [34].

Potential space resources that can be utilised in situ include "water, solar wind implanted volatiles (hydrogen, helium, carbon, nitrogen, etc.), vast quantities of

Fig. 22.1 The proposed design is for a sustained human settlement floating high above the poisonous acid clouds that shroud the fiendish surface of Venus

metals and minerals, atmospheric constituents, abundant solar energy, regions of permanent light and darkness, the vacuum and zero-gravity of space itself, and even trash and waste from human crew activities. Suitable processing can transform these raw resources into useful materials and products" [35]. In our view, one potential processing technology that might well be more suitable than all others is being more or less consistently overlooked, and that is the opportunity to harness microbial action to *generate* materials at macroscopic scales in situ. Fundamentally, there are three ways of providing the necessary material supplies for space architecture schemes. Building materials can be brought from Earth (at great expense and effort), extracted locally where the architecture is to be constructed (ISRU), or somehow produced in space (a concept for which we will use the neological acronym ISMG, in situ materials generation).

Why is ISMG a promising alternative? Let's begin by considering the alternatives. Bringing building materials into space is no longer as expensive as it has been: in 2011, the SpaceX Falcon Heavy broke the long-sought 1,000 US dollars a pound (0.453592 kg) to orbit price barrier as the company published details about its capacity to put 53 metric tons in low Earth orbit at a launch cost, published on the SpaceX website, of 100 million US dollars per launch, equalling a payload price of 1.9 million USD per ton. Bringing the same payload into low Earth orbit using the comparable Delta 4 Heavy rocket would be almost exactly ten times more expensive. Clearly, the private initiatives are currently performing better than their state-run counterparts when it comes to lowering manufacturing and operational costs while achieving high efficiency performance in flight [36].

But 1.9 million USD per ton is still not exactly cheap. While a Buzz Lightyear action figure orbited the Earth for 468 days as a passenger aboard the International Space Station, and astronaut Alan Shepard brought a 6-iron golf club with him to the Moon [37] the costs associated with sending anything up in space is clearly prohibitive, and adds a new dimension to Buckminster Fuller's famous question "How much does your building weigh?" – for space architects, that query certainly needs to be dealt with from the very outset of the design process. But while it is certainly a major obstacle, cost is not the only reason why bringing supplies from Earth is an unsustainable and ultimately unrealistic option: resupply missions are expensive and exclusively relying on them may put crews at risk. While several units could be interlocked in situ, prefabricated volumes would have to be imported in the payload bay of the space shuttle, imposing boundaries on the scale that could be attained. While the performance of building materials and structures assembled using Earth materials may be unknown in its final context, allowing the use of local resources means working with what's already there – if the materials survived to date, they are likely to continue performing well even after they are reorganised into a human habitat.

Which brings us to the ISRU option. ISRU has been heralded as a groundbreaking paradigm shift in the context of designing human colonies in outer space, and at first glance, the reasoning appears to be perfectly obvious: using the local indigenous resources on a heavenly body in order to reduce the amount of material that must be brought from Earth makes a lot of sense. But it doesn't automatically follow that ISRU is always the preferable option. As Donald Rapp has pointed out, if, but only if, the savings "resulting from a reduction of resources brought from Earth outweigh the cost of prospecting, developing, testing, validating in situ, and implementing ISRU in missions, it follows that ISRU will have a favorable benefit/cost ratio" [38]. The virtues of ISRU are relative, says Rapp, and while "many ISRU advocates within NASA seem to take it on faith that the benefit/cost ratio is always favorable for ISRU, my analysis indicates that this is not always so" [39].

While it is certainly possible that ISRU materials can save substantial amounts of mass brought into space, it should be remembered that such concepts come with significant costs in prospecting for resources and validating the potentials for ISRU. Resources need to be identified, extracted, processed, and stored before they can be utilised. A lot of preparatory work that has to happen before an efficient ISRU system can be set up, but even following such an undertaking, relying on the extraction of existing raw materials from often inhospitable sites appears to be a potentially short-sighted strategy. It is for instance not immediately obvious how a successful ISRU operation could be established on the surface of a planet like Venus. Certainly, some lessons learnt on our own planet could be applied to other heavenly bodies as well: harmful effects and the large impact on the topography, vegetation, and water resources of both underground and surface mining (including strip mining, open-pit mining and mountaintop removal mining) [40] not to mention the current debate on the environmental dangers of "fracking" (hydraulic fracturing) [41] are obvious calls for caution. There are definitely interesting options to explore within the area of ISRU (including NASA's Electron Beam Melting machine and a range of proposals for interplanetary sintering machines) [42], but erosion, emissions, loss of biodiversity, toxic contamination, extreme pollution, and so on – do we really wish to export such consequences to space?

There are no guarantees that ISMG would not come with adverse effects as well – indeed, issues are likely to arise from any attempt to erect architectural structures anywhere in the universe – but the "softer approach" at its heart seems to offer a less aggressive way of obtaining the necessary constituents. Essential requirements to support human life include water, food, air, and shelter. If we want to create a functioning extraterrestrial habitat capable of sustaining a substantial population, we also need at the very least energy and spacecraft propellants. To make our Venus habitat a sustainable colony, just as on Earth, as many of these resources as possible need to be renewable. As the version of ISMG that we wish to develop and promote aims to be fully renewable, this is where the concept potentially excels.

In March 2012, we had a chapter published in which we suggested that *bacillith*, a novel conceptual material that we invented as the result of a process that used microbially-induced calcite precipitation by *Bacillus pasteurii* with lunar regolith, could be used as a building material on the Moon [43]. At the time, we positioned this process within the conceptual realms of ISRU; today we would have defined it as a cross between ISRU and ISMG. The proposal seems to have been a timely one: half a year after the publication, in October 2012, New Scientist reported that professor Lynn Rothschild at NASA's Synthetic Biology (SynBio) initiative at the

Ames Research Center was busy working on space applications for engineered microbes [44] and in January 2013 the architecture firm Foster + Partners made public their concept for a lunar habitat based on a highly similar idea. At least two important papers have since been published that further discuss the potential of using advances within synthetic biology to support human habitation in outer space [45].

We want to explore methods that allow us to grow materials in space, including the possibility to harness the power of microbial activity in the creation of new building materials for a variety of space contexts and applications. We argue that more energy, time and resources should be invested in investigating the potential benefits of such a research trajectory, which would be analogue to current discussions on Earth, where renewable materials are singled out as one potential way of increasing the sustainability credentials of the building sector [46]. For the present project, our material palette is composed of a space-grown grass and a bacterial skin – bamboo and kombucha – both to be produced above Venus using ISMG strategies.

As we have seen, in *The Sultan of the Clouds* Geoffrey Landis describes his science fiction city as being composed of “geodetic structures with struts of sintered graphite and skin of transparent polycarbonate synthesized from the atmosphere of Venus itself” [47]. This is a perfectly reasonable but fairly aggressive and machining-intensive material strategy, and we have opted for a more sustainable and renewable option, which, as with Landis’s precedent, employs a skin-and-bone model based on a geometry associated with Buckminster Fuller.

In the context of human spaceflight, plants can be grown and used for food, as a way of refreshing the atmosphere, to scrub carbon dioxide and return oxygen, adjust humidity, and provide a psychological benefit to human crews [48]. A less-discussed option is to use plants to grow building materials. A range of plants have already been grown in space, including maize, flax, rice, tulips, onions, peas, radishes, lettuce, wheat, garlic, cucumbers, parsley, and dill. Space stations such as ISS and Mir have been used to carry out experiments with plant growth, and in an article on the greenhouse aboard the latter, Dr Tania Ivanova showed that in principle, a factor such as weightlessness is not necessarily an obstacle to the growing of space plants [49]. Certainly, soil is also overrated: as Nobel laureate Richard Feynman explains in a famous video interview, trees are essentially made from the atmosphere:

> The substance of a tree is carbon, and where does that come from? It comes from the air. It’s carbon dioxide from the air. (T)he carbon dioxide in the air goes into the tree, and it changes it, kicking out the oxygen and pushing the oxygen away from the carbon, and leaving the carbon substance with water. Water comes out of the ground, you see. Only how did it get in there? It came out of the air, didn’t it? It came down from the sky. So in fact most of the tree, almost all of the tree, is out of the (…) air [50].

This is of course true for a grass like bamboo as well. Once there is water and trace elements, we can grow bamboo, and most of its mass will come directly from the atmosphere. While plenty of research needs to be carried out to make the process work, an automatic aeroponics growth system could be designed to grow bamboo in space (aeroponics is the process of growing plants in an air or mist environment without the use of soil or an aggregate medium). While the issue of supplying the bamboo with water in the essentially waterless context of Venus (trace amounts of water have been found in the planet's thick atmosphere) would need to be resolved, an ultrasonic (as opposed to the traditional piezometrical or centrifugal) atomiser could be used to atomise a nutrient solution to efficient ultrafine particles, and a temperature-humidity control system be implemented to control water-fertiliser and temperature factors in the bamboo root growth environment. Precedents for such a design already exist, but further experimental design research, outside the scope of this study, is needed to achieve a prototypical design [51]. Additional challenges in the context of Venus include how to efficiently protect the growing seedlings from the surrounding sulfuric acid environment while allowing for an optimal enclosed growth envelope, how to adjust for the possible (beneficial or detrimental) effects of prolonged solar exposure due to the extended day/night cycle, and how to negotiate the atmospheric conditions. We never said it would be easy.

Why bamboo? Because it is not only easily structurally strong enough for us to be able to construct the lightweight habitats needed for Venus, but also possesses two other properties that makes it an excellent plant for this kind of interstellar cultivation: the combination of low water demands and exceptional growth rates. Very little bamboo is irrigated and there is sound evidence that the water-use efficiency of bamboo is twice that of trees, making the species particularly able to handle harsh weather conditions including droughts, floods and high temperatures [52]. This sturdiness makes bamboo a distinctly well-suited candidate for space harvesting.

While there are many types and qualities of bamboo – this tribe of flowering perennial evergreen plants in the *Poaceae* family includes more than a thousand species of woody grasses in more than a hundred genera – most of them are in many ways superior in strength and resilience to other natural, fibrous building materials, including wood. This structural strength and hardness arises from the heavily lignified tracheids and fibres associated with the vascular bundles: the thick-walled fibres of bamboo contain not only cellulose and lignin, but also up to five percent silica in the form of silicon dioxide (SiO_2), making them very hard [53]. Interestingly for our intended application, the internodal regions of the stem in bamboos are hollow, with the vascular bundles in the cross section being scattered throughout the stem rather than arranged cylindrically, and the dicotyledonous woody xylem is absent: as there is no secondary growth wood, the stems of even large bamboos are columnar rather than tapering [54].

Giant bamboos are the largest members of the grass family, and bamboos are some of the fastest-growing plants in the world [55]. At a growth rate of up to 900 mm/day or 0.00003 km/h, they hold the Guiness World Record for fastest-growing plant [56]. Future research will have to show whether this growth rate can be increased further in a space setting. Over and above our intended primary use as a building material, bamboo is also an exceptionally diverse commodity that can be used alone or together with other materials to produce a wide range of products that could support human life in space, including food, clothing, beer, alcohol, wind turbine blades, and deodorant.

The bamboo members will constitute the structural part of our design (Fig. 22.2) and also be used (eventually) for floors, partition walls, and other surfaces within the habitat. To achieve an enclosed volume, however, we need to pair these structural elements (bones) with membranes (skins). This is achieved using our second material: kombucha. This lightly effervescent and often mildly alcoholic fermented drink of sweetened black and/or green tea may not be familiar to a Western audience (in particular as the word was first recorded in 1995 and has an uncertain etymology) [57], but in Eastern Europe and Asia – in countries such as China, Japan, Korea, Indonesia, and Russia – it is a traditionally consumed beverage, often claimed to have several (scientifically unproven) beneficial effects on health.

Kombucha is produced by fermenting the tea using a symbiotic colony of bacteria and yeast ("scoby") that contains a combination of *Acetobacter* (a genus of acetic acid bacteria) and one or more yeasts – usually *Saccharomyces cerevisiae, Brettanomyces bruxellensis, Candida stellata, Schizosaccharomyces pombe, Torulaspora delbrueckii,* or *Zygosaccharomyces bailii* – which form a zoogleal mat. The bacterial component of the culture comprises several species, almost always including *Gluconacetobacter xylinus* (formerly *Acetobacter xylinum*), which ferments the alcohol(s) produced by the yeast(s) into acetic acid, increasing the acidity while limiting the kombucha's alcoholic content; alcohol production by the yeast(s) also contributes to the creation of acetic acid by the bacteria. Sucrose is broken down into fructose and glucose, which the bacteria and yeast convert into gluconic acid and acetic acid, respectively [58]. Important to our plan for future space cultivation, kombucha is also relatively easy to maintain as a culture outside of

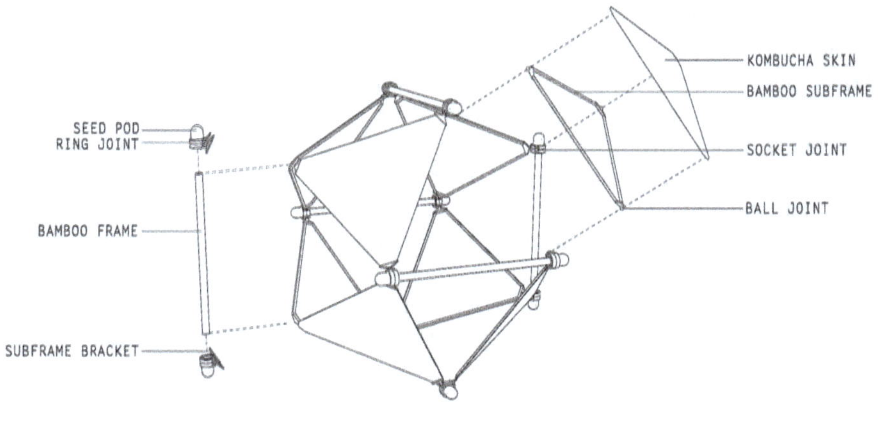

Fig. 22.2 Exploded sketch model of a dipolygonal bamboo/kombucha structure showing seedpod joints

sterile conditions, as its acidity and mild alcoholic element resists contamination by most airborne molds or bacterial spores – it inhibits growth of harmful microorganisms such as *Escherichia coli, Salmonella enteritidis, Salmonella typhimurium,* and *Shigella sonnei* [59].

So far so good, but why are we discussing microbial tea, and won't we contaminate space by bringing terrestrial bacteria into the Venus atmosphere? Let us try to answer the latter question first. Interplanetary contamination is certainly an important topic of debate, and one that is not to be taken lightly: the discussion has lasted at least since the International Astronautical Federation VII Congress in Rome in 1956 [60] and is still far from resolved. In 1959, the Committee on Contamination by Extraterrestrial Exploration (CETEX) recommended that interplanetary spacecraft be sterilised, but that the "need for sterilization is only temporary. Mars and possibly Venus need to remain uncontaminated only until study by manned ships becomes possible" [61]. If our in-space materials growth process forward contaminates (transfers viable organisms from Earth to) the Venus environment, we run the risk of endangering resources of value for future human missions. Having said that, the forward contamination argument has met with criticism [62] and at least three primary reasons have been used to argue the case for a softening of the present rules: (1) If the end goal is to find new habitats for humans, careful protection against forward contamination makes little sense, as humans will eventually be present, and humans can't be sterilised, (2) we have already carried bacteria out into space for four decades (contamination has most likely already happened), and (3) meteorites from Earth also pose a contamination threat, and have done for some four billion years (ditto) [63].

The Committee on Space Research (COSPAR) develops recommendations for avoiding interplanetary contamination, and positions Venus in its Category II of five mission groups, which comprises "Any mission to locations of significant interest for chemical evolution and the origin of life, but only a remote chance that spacecraft-borne contamination could compromise investigations" [64]. Astrobiologist Christopher McKay, on the other hand, has argued that space explorations should be biologically reversible [65]. While arguing either side of the case is outside the scope of this chapter, using bacteria to grow materials in space should in theory pose less of a challenge in terms of potential contamination issues than sending human astronauts to space, as it is possible to sterilise remote controlled/robotic growth machines (there are trillions upon trillions of microbes living on and in the human body). We acknowledge the potential contamination issue, but leave that to be discussed by the specialists.

So why microbial tea? Because kombucha can be used as a starter culture (carbon source) in the production of bacterial cellulose (Fig. 22.3).

The *Gluconacetobacter xylinus* appears to be what produces this bacterial cellulose, which has been shown to be possibly created using several bacterial genera including *Acetobacter, Agrobacterium, Gluconacetobacter, Rhizobium,* and *Sarcina. Gluconacetobacter sp.* are known to have the highest cellulose production capacity [66]. Kombucha fermentation is brought about by the symbiosis of yeast species and acetic acid bacteria [67] and the process is reasonably short: in one

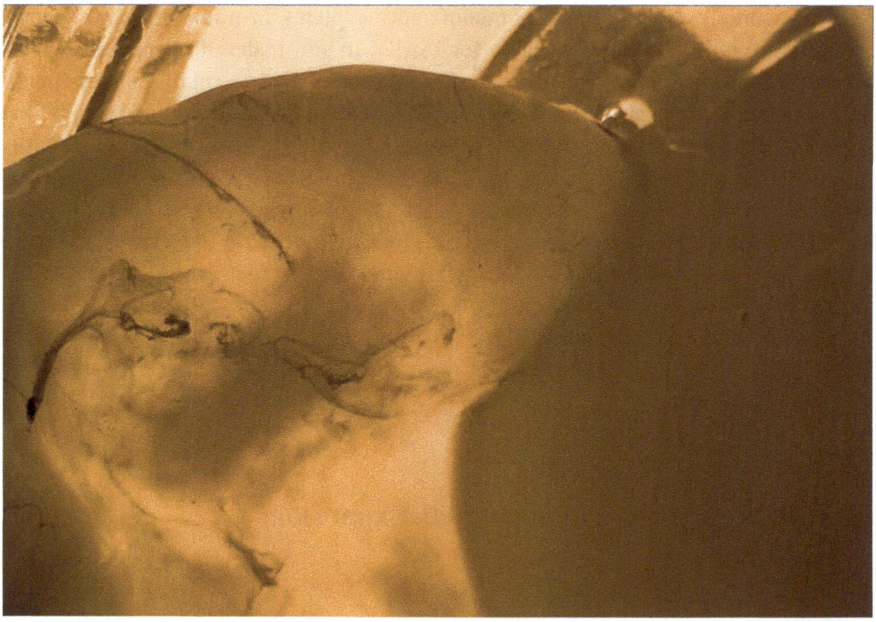

Fig. 22.3 Kombucha skin grown at Ordinary Labs, the research facility attached to the authors' practice

study, the bacteria and yeasts in kombucha promoted microbial growth for the first six days of fermentation, after which they steadily declined [68]. (It is generally agreed that the maximum production of bacterial cellulose from a tea fungus takes a minimum of 7–8 days, and that variables such as the pH of the substrate, and the effect of surface area, volume, and depth on bacterial cellulose production also have an impact on the final product) [69]. One research team concluded that the results and material obtained were "comparable with the commercial microcrystalline cellulose, indicating that the process developed (…) can be explored industrially on a pilot scale" [70]. Interestingly, the same study discusses how sucrose "at a concentration of 90 g/L produced the highest yield (66.7 %) of cellulose, and increasing the sucrose concentration from 110 to 250 g/L produced a gradual decrease in the yield" [71].

A bacterial skin that could potentially be grown – and grown again – quickly in situ in space at an industrial scale by feeding it sugar? Needless to say, the *Gluconacetobacter*-produced cellulose will have to be studied further before it can be used as one half of our material palette, but it does offer a promising membrane counterpart to the bamboo structure. In order to control the growth of both bacterial skin and bamboo, we will need to rely on engineered microbes, in all likelihood (given the conditions in the Venusian atmosphere) extremophiles, that will have to be singled out amongst the approximately five nonillion (5×10^{30}) bacteria on Earth [72]. Equipped with remote controllable qualities, effectively trained to become workers in our materials manufacturing plants in miniature, and sent out into space: could this be a roadmap for ISMG, in situ materials generation, above Venus? Many details remain to be thought through, many experiments conducted and question marks straightened out, but the material strategy outlined above does initiate a discussion about alternative ways – instead of or in conjunction with a more traditional ISRU-based approach – through which we can produce the raw materials needed to support human life in interstellar environments. Those materials truly come into their own when they are allowed to give birth to exactly the kind of lightweight, cellular, balloon-like structures that could float 50 km above Venus, structures synonymous with their own geometrical efficiency and surprising capabilities for volumetric expansion. Let us move on to how a unique set of transformations between particular polyhedra – dipolygonids – can make for a plethora of architectural constructs.

22.3 Geometries: Dipolygonidal Expansions

> In the jitterbug, we have a sizeless, nuclear, omnidirectionally pulsing model. The vector-equilibrium jitterbug is a conceptual system independent of size, ergo cosmically generalizable [73].

It should come as no surprise, given the title of this chapter and its eponymous topic, that we are fascinated, as have so many been before us, with the ideas and inventions of Richard Buckminster Fuller (1895–1983). We're not alone: Albert Einstein called Fuller's ideas about energy significant advancements, and no lesser luminaries than Marshall McLuhan and Steve Jobs were fans [74]. Few people are still ahead of their time more than 30 years after their death, but in the case of "Bucky" (the name he himself preferred) – visionary futurist, scientist, inventor, theorist, architect and mathematician – this actually appears to be essentially true. In his lifetime, Fuller published more than 30 books. He coined or popularised several terms and structures, including "synergetic" and the geodesic dome. He was the second president of Mensa from 1974 to 1983, and graced the cover of Time Magazine in 1964 [75].

The title of our project, Cloud Ten, is of course a play on Fuller's famous Cloud Nine proposal for airborne habitats created from giant geodesic spheres, which were to be made to levitate by slightly heating the air inside the spheres above the ambient temperature [76]. The geodesic sphere is a structure made from triangular components. It is a geometry with the interesting property of becoming stronger the larger it gets, due to the change in stress distribution across its surface: as the spherical form expands, the volume it encloses increases much faster than the mass of the enclosing structure. The mass of a mile-wide geodesic sphere, Fuller argued, would be negligible compared to the mass of the air it would hold. Heat the air inside this sphere by just a degree above the ambient temperature of its surroundings, and you will not just make it airborne, but allow it to lift a considerable mass. Airborne towns, or Cloud Nines, for thousands of people could be built this way, and be manoeuvred to migrate in response to climatic and environmental changes [77]. This is very much in line with our vision for how a ballon-like habitat floating in the atmosphere above Venus could be arranged and re-arranged to adjust for environmental changes (this will be further discussed in the next part).

But we shamelessly borrow more than just the idea for an updated Cloud Nine: in 1948, Fuller first proposed the "Jitterbug" transformation as a geometrical construct that "…oscillates, expanding and contracting over tetrahedrons, octahedrons, icosahedrons, to again end with the cuboctahedron" [78]. In a 2004 paper [79] Joseph D. Clinton sketches a brief history of the transformation's research and development: in the early 1950s, the Jitterbug concept was expanded upon by Duncan Stuart (whose work started while collaborating with Fuller on the latter's "energetic/synergetic geometry" [80] and which was independently published in 1963) to include face-, edge-, and vertex-connected transformations of the regular and semi-regular polyhedra [81]. Also in the early 1960s, Ronald D. Resch – seemingly always one step ahead of or behind Fuller in terms of geometry and origami-inspired structures – was independently studying these transformation concepts, [82] while under NASA sponsorship, Clinton himself further expanded on the system proposed by Stuart, before making the first dual-face polyhedral transformation model in 1965; his final research report was published in 1971 [83]. Verheyen published work on the transformation in 1989 and 1996 [84] followed by a private Internet publication by Gray in 1991 [85]. In recent years, many other individuals including Wohlhart, Röschel, and Kiper have also contributed notably to the research on Jitterbug-like transformations.

Fuller's Jitterbug structure is a set of eight identical regular triangles connected by their vertices. This construct is capable of performing a symmetrical expansion motion, transforming from an octahedron to a cuboctahedron, followed by an equally symmetrical impansion motion, moving back to the original octahedron (Fig. 22.4). In Clinton's 1971 report on these expanding rigid structures, the author observes that each triangle is subject to a translation-rotation along its symmetry

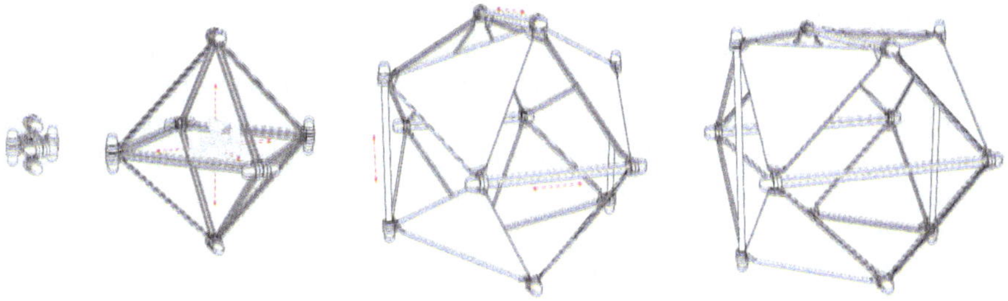

Fig. 22.4 The initial three Jitterbug transformations

axis. Explains Verheyen: "When starting from the position in the octahedron, these axes are the four triangular symmetry axes of the octahedron. When describing cylinders about the triangles along the axes, each vertex common to two triangles moves along the intersecting curve of the two cylinders" [86].

In his extensive and meticulously illustrated paper, Verheyen expands on what he calls the "Fuller-Stuart-Clinton transformers," positioning them within a larger geometrical definition aimed at providing a complete classification and a new definition for a set of transformable types, which he names dipolygonids [87]. Having thoroughly explained in a polyhedronist's terms exactly how the motions work, the author lists a large number of possible dipolygonids expansions, including the "Verheyen's Vampire". While an exhaustive account of the details are not within the realm of this chapter, during the reading it soon became clear to us that the traditional Jitterbug transformation is forced to impend simply because it is fixed at its vertices, and that the shackles of the clockwise and counterclockwise symmetry relationships of the original Jitterbug transformation can be broken if we allow the expansion to continue into new dipolygonid polyhedra. The reason why this trajectory appears to not have been thoroughly investigated in the past is perhaps that such a transformation, if carried out using a physical model, needs to have additional material added for following geometries to emerge, which would typically call for a new model to be made, rather than for the initial model to grow into a second one [88]. However, since we are growing our structure in space, we can allow for this material to be added organically, "from within," as the structure evolves and expands. A physical sketch model was produced to test this idea of "internal expansion", and as was to be expected this showed the promise of allowing one volume transform into another through such an organic material addition (Fig. 22.5).

While several precedents for built structures made from living trees (as opposed to cut and sawn timber) exist – including the Ficus Hut on Okinawa by Arborsmith Studios; Oliver Storz, Ferdinand Ludwig and Hannes Schwertfeger's Botany Buildings; the Baubotanical Tower by Ferdinand Ludwig; The Ground Beneath by Finnish artist Jaakko Pernu; Patrick Dougherty's The Nest Houses project, and many others – we have not been able to find examples of an expandable structure that allows for a bacterial skin to be grown at the pace of its volumetric expansion, and so to the best of our knowledge, this is the first time such a material strategy to support a geometric transformation of this sort is proposed.

In practical terms, the scheme as we envisage it would probably have to rely on advances to come in the field of GMO (genetically modified organisms) engineering: ideally, we should be able to achieve the mechanics involved in the dipolygonids expansion through the growth of the bamboo alone, with the bacterial skins accruing at the same pace, so that the membranes "fold up" from within the structure as the bamboo plants grow. This would effectively make for a structural module that not only carries out its own construction and material formation through photosynthetic growth, but also develops its formal characteristics through the biochemical conversion of carbon dioxide and water into simple sugars that

Fig. 22.5 Physical sketch model testing the idea of "internal expansion," allowing one volume to transform into another through an organic material addition

drive the volumetric expansion. If the germplast of the bamboo could be programmed to change its metabolism and adjust for a particular growth rate to within a small enough tolerance, and the bacterially-induced growth of the kombucha skins could be programmed to correspond to that growth rate (while adjusting for the volumetric expansion of the structure), then a synergetic system could be achieved that would yield an expanded module in a defined period of time. While a successful implementation of such a technology might still be some years away, GM crops have in recent years been taken up faster than any other agricultural technology since the plow was introduced some 8,000 years ago [89] and a number of recent experiments show that controlling growth rate in this way is not an unlikely possibility in the near future.

One such experiment was carried out by Israeli biotech firm FuturaGene, whose trial plantations of thousands of GM eucalyptus and poplar trees on 100-hectare plots in Israel, China and outside São Paulo in Brazil have produced quite interesting results. According to the company, a gene taken from the common, fast-growing *Arabidopsis* weed has been used to alter the structure of plant cell walls to stimulate the natural growth process, allowing a modified eucalyptus tree to grow by some five metres a year, and with 20–30 % more mass than normal eucalyptus: at five and a half years old these trees were 27 m high [90]. Admittedly, that may still not be an exact enough growth rate control mechanism for the process we outline above, but as a precedent scheme it does show the promise for increasing – and at least to an extent regulating – the growth rate of a species that can typically grow at a rate of up to 1500–2000 mm/year (as compared to the bamboo rate already mentioned above: up to a theoretical 328–500 mm/year). Furthermore, there is no reason why we would have to limit our toolbox for achieveing a controlled growth rate to just genetic engineering: other technologies that could be of interest in this context include adjustments of temperature, fertiliser, and light. One example of the latter is the magenta LEDs that line the floor-to-ceiling cultivation racks at the Mirai company's facilities in Tagajo, northeastern Japan. These utilise wavelengths conducive to photosynthesis and cell division to mimic an accelerated day/night cycle, speeding up plant growth while using 40 % less electricity than comparable fluorescent grow lights. Mirai claims that the technology helps the company achieve impressive results: lettuce allegedly grows 2.5 times faster, has beta-carotene levels that are 10 times higher and vitamin C levels 1.8 times higher, consumes 1 % of the water, and occupies about 2 % of the land compared to its counterpart grown outside in soil [91].

The geometry of the self-deploying dipolygonid's expansion modules depend upon connections that allow for the necessary rotation of the bacterial membranes whilst securing the positions of the bamboo members vis-à-vis each other. While the detailed design of these joints are beyond the scope of this chapter, the properties and conditions for wholly representative Jitterbug-like linkages that might provide the starting point for that design process are rigorously mathematically described by Gökhan Kiper in his 2011 PhD thesis Design Methods for Planar and Spatial Deployable Structures, which includes descriptions of how extra joints can be added to the triangular links (making, for instance, the polygonal links of an octahedron become regular hexagons, which is one possible design trajectory) [92]. The connections would need to provide an intelligent and adaptive "buffer zone" that could stabilise the geometry in situations when the growth of the individual bamboo members differ too much, which could perhaps be achieved with a muscle-type actuator based on fluidic principles [93]. It is the materials themselves that drive the volumetric expansion, but the joints provide the trajectories necessary for the structure to deploy according to plan. At times, these trajectories follow the dipolygonal rotations, while at other times, they combine with proportional elongations of the subframes to allow for the continuous transformation of cellular units into their respective iterative states (Fig. 22.6).

Japanese architect Kengo Kuma once wrote that if the act of designing were suddenly to be redefined as "selecting particles," we would be "thrown into a new world" [94]. The Cloud Ten project goes at least one step further: we are not only making a novel building from novel materials that utilise microbial manufacturing and genetically modified and controlled growth of grass, but designing a self-deployable system that grows on its own accord, guided by relatively minuscule connection points. We like to think Fuller would have enjoyed that idea.

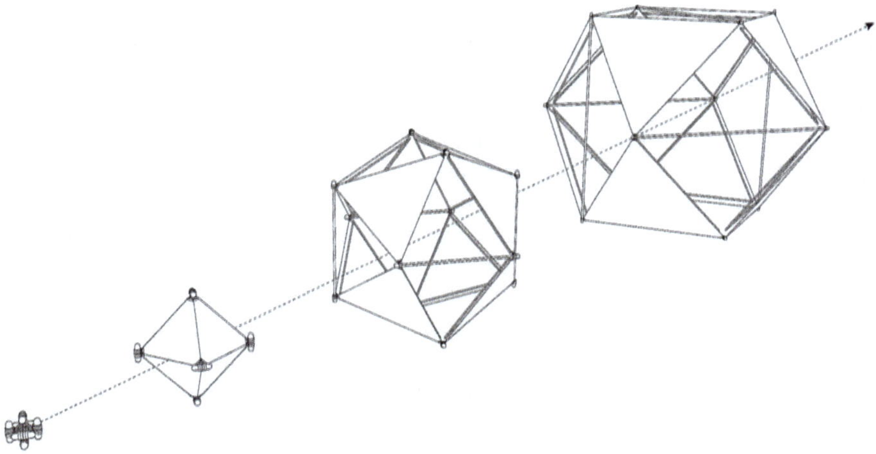

Fig. 22.6 Dipolygonal expansions combine with proportional elongations of the subframes to allow for the continuous transformation of cellular units into their respective iterative states

22.4 Cells: Evolutionary Logics

> …robustness is complementary to innovation: Any network that can evolve new features and forms among a vast array of alternatives must necessarily be robust against small changes, because it almost certainly has an alternative on hand that performs equally well. This realization offers an antidote to an excessively deterministic view of genes: Exactly which genes you have may not matter so much (within reason), because the job they do is more a property of the network in which they are embedded [95].

Our dipolygonidal expansions produce volumes up to a certain scale, but for these to be sufficiently and efficiently supportive of human habitation, the geometries will have to be combined into larger structures, moving from the building scale through the city scale and beyond, towards the interstellar mega city, echoing Landis's vision of a Hypatia in which "each kilometer-diameter dome easily lifted a hundred thousand tons of city" [96].

Such aggregations of spaces (or "modules," or "cells") can employ various organisational principles and different modes of connectivity to construct the overall structure's performance, allowing for different densities and arrangements within the assembly. As selected parameters can be allowed to modulate both individual cells and the overall conglomeration of modules, highly specific programmatic needs and functional requirements – as well as specific contextual conditions brought about by the harsh atmosphere of Venus – can all be achieved. The final form can be computer controlled to vary according to underlying (live or set) data in order to respond to a wide variety of design constraints in order to obtain performative gains for instance in terms of scale, program, construction, orientation, direction, shelter, protection, solar radiation, and so on. The cell and the system simultaneously become architecture and structure, node and network, part and whole.

As structural, programmatic, spatial, and economic conditions change, the habitat's layout can change with them, using basic principles of cellular aggregation together with low-energy strategies for disconnecting existing arrangements and reconnecting into new ones. If adopting a decentralised procedure, the process could resemble the way ants collectively transport irregular-shaped objects with seemingly simple coordination [97] or the way a sponge will reassemble following a catastrophic event: "Disassemble the cells of a sponge, (by passing them through a sieve, for instance)," writes Bill Bryson, "then dump them into a solution, and they will find their way back together and build themselves into a sponge again. You can do this to them over and over, and they will doggedly reassemble because, like you and me and every other living thing, they have one overwhelming impulse: to continue to be" [98].

So what should the underlying logic for these agglomerations of volumetric modules be? In what way should the cells acquire order and structure? How might we go about finding a principle that guides the accumulation of spatial units and the collective rearrangement of cells with respect to each other? To answer such questions, we investigated three conceptual constructs that could help arrange the modules: those of Swarming, Stigmergy, and the Compromise.

22.4.1 Swarming

The word swarm comes from Old English *swearm*; it is related to German *Schwarm* and has an etymology that can be traced back to Germanic origins, and probably also to the base of Sanskrit *svarati* ("it sounds"), which makes sense to anyone who's heard a swarm of bees fly by. There are many examples of biological systems that harness the collaboration of vast numbers of limited individuals, a plethora of simple agents that self-organise locally to produce highly complex and robust global multi-agent behaviours. From multicellular organisms and bacterial aggregates to social insects (such as ants and bees), shoals of fish and swarms of birds, the kind of collective intelligence within such groups enables them to reliably achieve complex tasks. The clustering of beetle larvae, weaver ant colonies, the synchronous night flashing of male fireflies, and the pheromone-based interactions of termites as they construct their intricate mounds are other examples of such self-organising biological mechanisms [99].

A few examples of schemes from the world of architecture that in one way or another build on ideas of swarm behaviour spring to mind: Carlo Ratti's *FlyFire* project at the MIT; Swiss duo Gramazio & Kohler's swarm of robotic bricklayers in their *Flight Assembled Architecture* scheme; the *Swarm Construction* project by Neri Oxman & Markus Kayser, also at the MIT (aimed at building architectural structures autonomously); Skylar Tibbits's *Fluid Crystallization* scheme (which investigates hierarchical and non-deterministic self-assembly with large numbers of parts in a fluid medium), and so on [100]. At Harvard University, the Self-Organizing Systems Research Group is interested in "self-organizing multi-agent systems, where large numbers of simple agents cooperate to produce complex and robust global behavior" [101]. The swarm paradigm has been applied to a broad range of studies, giving rise to new concepts within fields such as theoretical biology, economics, and philosophy: not only might we, as multi-cellular organisms, be composed of swarms – so might everything from the economy to the human mind. Humanity's fascination for swarms is of course often given other cultural expressions aside from architecture: the tormented protagonist's reoccuring visions of menacing murmurations (or perhaps actual murmurations?) in the 2011 Jeff Nichols film *Take Shelter* is but one recent example of this [102].

Architecture can be viewed as being fundamentally an exercise in designing for swarm intelligence. A door is essentially "an on-off switch in the membrane, the movement of stuff is structured as to flow through that door. Doors are open or closed (or half open and half-closed), the spaces are switched on or off, or sort of switched on or off. The membranes are semi-permeable envelopes around a certain quantized volume of space. The semi-permeable membranes let through people, light, heat, cold, small animals, air, radiation, information, food, water, gas, waste, molecules, wind, sun, moist, materials, cars, shopping bags, television programmes, waves, books, paper. A wide range of different materials is coming in through the membranes, another wide range of materials is leaving the space somewhat later. Some things come in through explicit holes, others come in by diffusion, by radiation, by transmission, or are carried by other messengers. Much of it is carried by people, coming in and going out. People are information carriers, they run in, about and out the house. The information they carry out of the house is of different content then the information – in whatever disguise – they take out of their house. The information content and some material properties of incoming information is changed inside the space. This space can be considered as a content transformer, it digests the incoming material/information. Taken to the extreme all material is a form of information, and taken even further all information is a form of computation. Thus space computes information" [103].

For Cloud Ten to be a valid proposal, we need to find a way of not only having our spaces compute information, but also of having that information guide the positioning of the modules with respect to each other as well as to the overall collective performance, which in turn calls for some method of physical movement to be added as a feature within the structure. Swarm behaviour is the collective motion of a large number of self-propelled entities [104] and so the topic of self organisation and self assembly of swarms is interesting as the structure we envisage would fundamentally be a swarm of agglutinated modules. In the traditional sense, self-assembly is the autonomous organisation of components into patterns or structures entirely without human intervention; a field that originated in the study of molecules in organic chemistry and essentially is about a soup of particles spontaneously arranging themselves into a coherent structure. Such processes are common throughout nature and technology, involving components that range in scale from the molecular (for instance crystals) to the planetary (like entire weather systems) [105]. As a strategy (as opposed to a singular chemical process), it is applicable at all scales. One definition limits the term to describe "processes that involve pre-existing components (separate or distinct parts of a disordered structure), are reversible, and can be controlled by proper design of the components" [106].

The self assembly of swarms and other similar phenomena can be broadly divided into two tiers: passive and active. In passive self assembly, "particles interact according to geometry or surface chemistry and tend toward a thermodynamic equilibrium at which the system is assembled. For example, phospholipids stick to each other along hydrophobic regions to form membranes. In active self-assembly, the particles can somehow decide in what interactions to partake. For example, proteins in cells may undergo conformational switching that changes the outcomes of their subsequent interactions in very specific ways" [107]. We like to think of the cells or modules that make up Cloud Ten as being self assembling in the sense that they act "on their own accord" following evaluations of algorithm-based rules that change with the incoming data. While we (humans) install the system, we do not intervene in the ongoing reorganisation of the cells: this is left to the "organism" that is the building. In this respect, they would constitute an example of active self assembly in a swarm manner.

22.4.2 Stigmergy

We might be able to sharpen our swarm intelligence tool by focusing on a specific type of self organisation, which has become a key concept in the field of swarm intelligence. In 1959, French biologist and zoologist Pierre-Paul Grassé introduced the neologism *stigmergy* to refer to termite behaviour [108]. Grassé's own definition of stigmergy was "stimulation of workers by the performance they have achieved". Derived from the Greek words *stigma* (mark, sign, goad) and *ergon* (work, action), the concept is based on the notion that an agent's actions leave imprints on the environment, signs that all agents within the system sense and that determine and incite their subsequent actions [109].

Stigmergy is thus a mechanism of indirect coordination between agents or actions [110] guiding additional work (following Grassé, Edward O. Wilson introduced the term *sematectonic* communication to describe a more general non-work-specific incitation) [111]. Stigmergy is the phenomenon of indirect communication mediated by modifications of the environment [112]. It is an act of "invisible writing" [113] an "incitement to work by the products of work" [114]. In short, the fundamental principle of stigmergy can be viewed as an attempt to answer the question "how can the seemingly uncommunicative and certainly very simple creatures we know as termites be responsible for the epic feats of organisation and creativity that is the termite mound?"

The answer lies at least partly in pheromones: the natural system of social insects achieves remarkable feats through exploiting the single principle that performing complex tasks quickly and reliably might be an undertaking better suited to many simple agents rather than one or a few complex agents, and trail pheromones – semiochemicals secreted from the body of one individual to impact the behaviour of another – is an important medium through which they communicate with each other. In nature, construction teams are common (just think of hornero couples and families of beavers building away), but the scale of the undertaking becomes truly impressive when considering social insects, where millions of tiny individuals might collectively produce functional global structures, without BIM software or even orthographic projections to help guide their assembly. Could this power-of-the-locally-toiling-many be harnessed as a strategy for arranging our modules in space? Could the previous behaviour of the structure as modules are added produce a certain behaviour as the structure is further advanced?

When building their nest, termites make small mud balls and deposit them probabilistically (initially in a random fashion) on the substrate. Having thus modified their local environment, they then impregnate each mud ball with a minute quantity of a particular pheromone. The probability of a mud ball being deposited at a given location increases with the sensed presence of other mud balls as well as the sensed concentration of pheromone. Initial random placements increase the probability that other termites will put their loads at the same place. Thus small columns are formed, and pheromone drifting from neighbouring columns cause the tops of the columns to be built with a bias towards the neighbouring columns, so that eventually they meet to form arches, the basic compressive building unit. Other stigmergic processes that involve water vapour and carbon dioxide concentrations as well as modulation by the queen's presence produces the internally complex structure of the mound, including efficient solutions for the circulation of termites and air, the programming of spaces (such as food stores and brood nurseries), and other architectural features contained within one of the largest, and arguably most well engineered, non-excavated structures built by any creature except man [115].

In the context of Cloud Ten, a highly interesting paper was published in 1994, in which a series of experiments were presented that had a group of mobile robots gather 81 randomly distributed objects and cluster them into one pile using stigmergy to coordinate the agents' movements [116]. The experiments extend some basic principles of stigmergy to the logic of robots, and finds a fit between stigmergy and behaviour-based robotics. States the authors: "Conventional robots are too slow to cope with an environment containing other moving robots, and too expensive for anyone to be able to experiment with large numbers of them; behaviour-based robots cope well with unstructured dynamic environments and are cheap. We might expect the biological principle of stigmergy to fit better with the biologically inspired architectures of behaviour-based robots than with the alien computational paradigm of conventional robotics" [117]. For their experiments, they designed and built a robot that could move small numbers of objects and that was more likely to leave them in locations where other objects had previously been left. Several such robots were then left to push circular pucks across a table following stigmergically programmed instructions. The researchers were able to show that the robots were able to perform a potentially useful task through stigmergic control and coordination.

While clearly able to produce impressive feats of large-scale engineering in the natural world, as a concept, stigmergy appears to be a particularly good (and comparatively cheap) method for controlling and coordinating the positioning of modules in a *local* context, in particular in a situation where communication might be restricted, where sensor capability and working memory might be limited, where mechanics are best left as simple as possible, and where some tolerance in terms of errors and non-flawless behaviour is welcome. A collection of modules that perform smaller arranging tasks – multi-agent, redundant systems with decentralised control mechanisms – as opposed to larger, multi-functional, and more complex systems, allow for an efficient parallel construction process that doesn't rely on key individuals: the failure of one module need not obstruct the entire system from succeeding [118]. However, while this strategy appears to outperform that of traditional robots (which typically rely on centralised control mechanism that are easy to design and allow for efficient routing of all parts of the system), as we want our habitat to optimise for environmental parameters that constitute a much wider context, this decentralised procedure is likely to function at the *global* level only up to a point. We probably need an underlying, primary system that feeds the local stigmergy units with global information turned into instructions through a series of algorithms and processes that allows for environmental data to inform the overall structure. Enter the conception of the compromise as constructive design tool.

22.4.3 The Compromise

The idea of the compromise as organisational principle perhaps calls for a slightly more comprehensive background sketch. A speculative line can be drawn from the assembly of the oldest of the Chinese classics, the ancient divination text *I Ching,* around 3,000 years ago [119] to the near future of not just space architecture, but the entire discipline of architecture. As with most conceptual arguments, as it connects the dots this theoretical line passes by the meditations of Georg William Friedrich Hegel (1770–1831) [120]. Hegel's philosophy is based on a dialectical way of thinking: every situation inherently holds it own negation; energetic resistance hides within every status quo; wars give rise to anti-war movements; to live is also to die.

Historically, this idea of a reality based on dualities has been used to describe the universe as a place in which every yin has its yang (the I Ching), where change is the only constant (Lao-Tzu, China, around 2,500 years ago), and where all change comes through the struggle of opposites (Heraclitus, Greece, at roughly the same time). Other cultures that have used similar concepts include the Aztects, the Lakotas in North America, the Dogon people of Mali, and many others. At least in part because of Aristotle's formal logic, which argued that things have to be either/or rather than both/and, the idea of everything being made of opposites essentially died out in Western thought. This Aristotelian view was also supported by the most important Catholic medieval philosopher and theologian, Thomas Aquinas (1225–1274), and therefore turned into the official word of the church throughout the Middle Ages.

This traditional dialectic way of thought, made popular in the West by Plato in his Socratic dialogues, was revived by Hegel, who, while not himself using the thesis-antithesis-synthesis formula so often attributed to him (the terminology of which originated with Kant and was later elaborated by Fichte) [121] often based his arguments on a similar three-valued logical model defined by the terms abstract-negative-concrete. According to this model, as a thesis is too abstract it deserves its antithesis, and having absorbed this negative it completes the triad by becoming concrete.

What (Kant and) Hegel did, in effect, was to reintroduce the idea of dialectics to a world that had failed to recognise that change comes through the conflict between opposing forces. While Western philosophy had mystified or even denied change, Eastern philosophy recognises change as the movement of opposites, but generally views this movement as cyclical, without forward motion, without evolution. Using ancient Greek philosophy (from which the word *dialectics* was taken) as his starting point, Hegel wrote that "dialectic. . . is no novelty in philosophy" [122] before essentially explaining the hypothesis of evolution before it even existed as a concept. Indeed, German philosopher Walter Kaufmann has pointed out that Hegel's famous first book, The Phenomenology of Spirit (1807), came out "over half a century before Darwin published his Origin of Species and impressed the idea of evolution on almost everybody's mind" [123].

How might notions of evolution inform space architecture? Through the introduction of processes that use evolutionary logic to mediate between creative conflicts. A creative conflict might be described as the opposite of the politically correct sociological term for conflict, "misunderstanding". If there is a conflict (and we usually know one when we see it), we understand certain things about the parties' desires, positions, stances, attitudes, agendas and so forth. It is not that we don't see that the trajectories do not align: on the contrary, we see precisely that they diverge. In our method for using creative conflicts, there is no misunderstanding, but a highly effective combination of an informed understanding of the different desires within the system and a method for allowing those driving forces to yield optimal equilibria. As one recent quote from popular culture had it: "You want change? Then learn to compromise." [124].

That line could be further amplified thus: if you want productive change within your architectural design – and what is design really if not a process devised to create productive change, from a blank sheet of paper through to a finished object or construct – then you need to teach the system that you base your design upon the art of the constructive compromise. As we have pointed out elsewhere [125] architecture is a conditional field of constantly varying parameters. We design compromises between those parameters. Perhaps it follows logically that the most successful architects are those that manage the inevitable compromise the best. So how might we go about this task?

One way would be to investigate the world of evolutionary computation:

"If we allow several competing objectives (genes) to participate in the creation of successful compromised positions (genomes), we might find interesting and novel material, spatial, and programmatic organisations between the controlled and the serendipitous. We might allow gradients of objectives to inform different (or indeed the same) parts of our buildings, pit material properties against circulatory efficiency, structural considerations against financial implications. It might well be that the most interesting works to come out of such a paradigm will arise from the most unexpected combinations of objectives: what is the outcome of a balancing act between atomic structure and programmatic diversification? Energy generation and speed of construction? Cost and happiness?" [126].

If form follows compromise, then form follows fitness – though in biological evolution, fitness is a tricky concept that is often very difficult to explain (though being fit is clearly not to do with being the strongest or bravest or fastest or most vicious). As David Rutten points out, the reason there are no flying dogs "isn't that evolution hasn't gotten around to making any yet, it is that the dog lifestyle is supremely incompatible with flying and the sacrifices required to equip a dog with flight would certainly detract more from the overall fitness than flight would add to it. Fitness is the result of a million conflicting forces. Evolutionary Fitness is the ultimate compromise" [127].

This insistence on finding "ultimate compromises" is quite anatheme to the tradition of engineering and space architecture based on the customary problem-solving paradigm. This strategy, which has been researched extensively at least since the 1960s (when the American Nobel laureate, political scientist, economist, sociologist, psychologist, and computer scientist Herbert Simon and his colleagues began investigating human subjects solving difficult problems [128]) normally seeks to find *one singular* solution to *one singular* problem: it poses that challenges need to be broken down into smaller portions (problems) that can be dealt with individually and with relative ease.

But optimisation of the kind we need to have Cloud Ten perform as efficiently as possible in the confines of the atmospheric conditions of Venus doesn't lend itself very well to this sort of subtractive simplification: to paraphrase Donald Rumsfeld, there are too many known knowns, known unknowns, and unknown unknowns to have the structure operate on a problem-by-problem basis. Once you've dissected the challenge into a number of problems, the initial situation has already changed unretractably. Optimisation, in short, is not so much about problem solving as about a continuous search for responses.

While an engineer might calculate the stresses in a structure based on certain loads (that is: analysing, determining the response of a specified system to its combination of input parameters), design is the process of defining a system (for instance selecting and specifying the particular dimensions of structural members and their locations to allow the structure to withstand the specified load). The process of design is about finding the minimum or maximum of some characteristics (the objective functions), while staying within the requirements (design constraints) posed by the context of the brief at hand. Optimisation, then, is the process of searching for the minimum or maximum of our objective functions while satisfying all the required design constraints. So long as we can quantify the objective functions (preferably through a system that allows us to give them individual weightings), contemporary computer technologies based on evolutionary logics allow us to analyse alternative designs with unsurpassed speed and accuracy, in order to rationally search through a wide range of potential solutions to find the optimal compromise between objective functions within the design constraints [129].

This also removes some weight off the shoulders of the individual objectives (the genes in our biological metaphor): as Philip Ball points out, the genes operate within the network in which they are embedded, and their collective performance – essentially the resulting global compromise in the system – gives it an inherent robustness "complementary to innovation" [130]. The resulting performance isn't so much a series of individual compromises (that is, compromises used as "solutions" to "problems") as a result of Rutten's "million conflicting forces". But if we're not looking for individual compromises but rather for something closer to a compromising force that rises like an emergent behaviour within the structure, how can this be visualised?

US geneticist Sewall Green Wright (1889–1988) carried out influential work on evolutionary theory and introduced, in 1932, the idea of visualising the distribution of fitness values as a kind of landscape in order to better understand the relationship between genotypes and reproductive success. Wright's model allows the relationship between genotype or phenotype and fitness to be modelled as a fitness surface or evolutionary landscape. Mean population fitness is the "height" of the landscape (and so plotted on the Z axis) against horizontal (X,Y) axes representing continuous phenotypic traits. The height at each point thus represents the corresponding organism's fitness. Highly similar genotypes are said to be "close" to each other, while those that are very different are "far" from each other. Natural selection leads to a population climbing the nearest peak, while genetic drift causes random wandering. The combination of all possible genotypes, their degree of similarity, and their related fitness values becomes the fitness landscape, upon which the population can follow various mutational paths through its evolution [131].

Fitness landscapes offer helpful visualisations of our field of potential compromises. They don't, however, give us a final design "solution". Rather, they produce a series of iterative suggestions that move towards the best-possible compromises given the circumstances. We can then use algorithms (solve equations) to bring us closer to an optimal, "final" situation (the best – or least bad – compromise between given objectives at the time we switch off the optimisation process). In practical terms, optimisation is a delicate area: while evolutionary solvers (the family of software applications used to implement evolutionary logics as organisational strategies) are used to optimise for certain qualities, the solver itself knows nothing about what an optimal solution to the situation in question might be. At best, genetic, or evolutionary, algorithms might find a good solution (or suggestion) to a reasonably well-scaled model [132]. The evolutionary solver has no way of determining whether a given solution is optimal – it only assesses whether it is better than the previous solutions found. It follows that the solver never knows when to stop its search: as designers, we need to provide instructions for this, as left to its own devices, the inherently-optimistic solver will simply keep feeding back new suggestions through a near-endless stream of potential vectors and locations within its search space of possibilities. The traditional single-problem solution does not exist. The search must go on.

22.4.4 Cellular Aggregate Models

The cellular aggregate models we created as an experiment to illustrate this chapter are based on a combination of all three conceptual constructs outlined above to arrange the modules. These are initially treated as part of a swarm, brought together in space after the growth process and the dipolygonidal expansions have produced the volumes and geometries needed for the habitat. Three simple rules borrowed from the world of swarming intelligence – cohesion, separation, and alignment – produces our initial configuration. We use the Locust plug-in for the Rhino Grasshopper environment to simulate this swarm behaviour [133] based on the famous 1986 algorithm for boids by Craig Reynolds [134] which might begin to illustrate the way our cells could come together above Venus in a self-propelled fashion without colliding (Fig. 22.7).

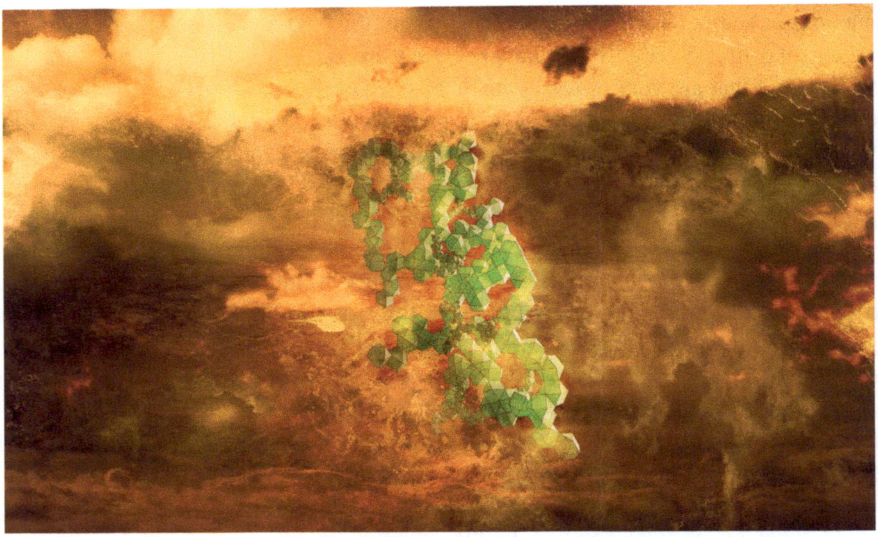

Fig. 22.7 A boid-based algorithm is used to make cells come together above Venus in a self-propelled fashion without colliding

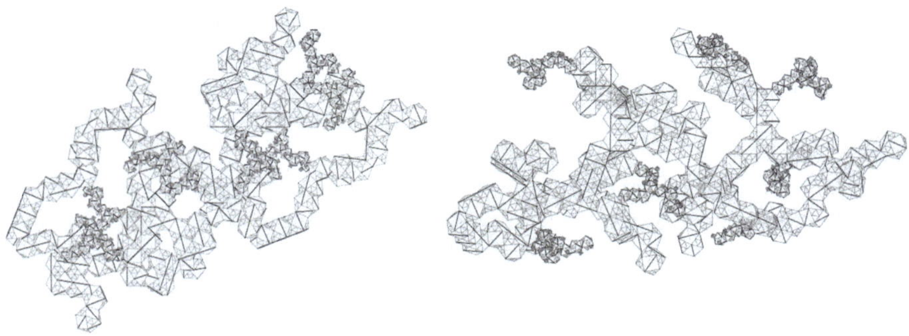

Fig. 22.8 Plan/elevation. The habitat supports human habitation as a mega city in outer space – as strange, beautiful, and unexpected as any terrestrial planned city

We then allowed a combined evolutionary solver/stigmergy optimisation to rearrange the initial swarm into first branching patterns (by feeding the evolutionary solver instructions that mimic stigmergy: essentially a diffusion-based pathfinding exercise that forces the cells to cluster influenced by previous arrangements) and then optimal configurations based on a series of desires (or conflicts) that include surface exposure to atmosphere, sun radiation, light values, circulation path lengths, amount of structure needed, and overshadowing.

More design constraints can easily be added to this model; the point here is not so much to achieve a "perfect" configuration of cells, but rather to illustrate the validity of this three-tiered approach. The end result is a habitat that supports human habitation as a mega city in outer space, as strange, beautiful, and unexpected as any terrestrial planned city (Fig. 22.8).

22.4.5 The Ten Phases of Cloud Ten

> [The cloud tops of Venus] is arguably the most hospitable region for humanity in our solar system, outside of Earth itself [135].

Whether one agrees with his political agenda or not, activist writer George Monbiot clearly adopts an applaudable attitude when he defines the process of fulfilling his self-imposed mission to be that of seeking to "suggest nothing that cannot be achieved with our own resources, starting from our current circumstances" before complaining that "(t)oo many of the schemes some members of (the 'Global Justice Movement') have put forward appear to be designed for implementation by the people of another time or another planet" [136]. While the strange, beautiful, and unexpected Cloud Ten scheme is not to be implemented by people *of* another planet, it is to be implemented *above* another planet, and we firmly believe the only way to get there is precisely to suggest something that probably cannot be achieved with our current knowledge and technologies, but that probably could if in the near future those were expanded and built upon so as to make a new model possible that, as Buckminster Fuller suggested, "makes the existing model obsolete" [137].

Actor, director and stand-up artist Woody Allen once quipped that when you do comedy, "you're not sitting at the grown-ups' table, you're sitting at the children's table" [138]. The same might be said of "doing speculation" in science; it's not *quite* solid science, and yet the act of speculating – proposing suppositions, postulating theorems, formulating hypotheses – is an absolutely crucial and essential part of not only science, but critical thinking in any field, as well as (at least in an ideal scenario) the design process at large. When it comes to the topic of space architecture in particular, the "what-if" position clearly needs to be invoked as often as, if not more often than, the traditional "what-is" perspective. Just as our evolutionary logic-based optimisation algorithms will inevitably spend a non-trivial amount of time working on producing iterations that will never actually be used – essentially doing the opposite of a hard-nosed optimisation process whereby productivity is increased by reducing the number of running processes as much as possible – our speculative designs for the Cloud Ten habitat might very well end up becoming a series of attempts at answering strange questions, and a list of design research results from investigating unexpected trajectories [139]. And yet even if that were the case, we would still have achieved something: not just a utopian vision for a mission that could potentially become possible in the distant future, but a critical blueprint sketch for how to get from present-day thinking (in the realm of human space inhabitation primarily centered on missions to the Moon and Mars, and only secondarily to the potentially much more promising option of mining the asteroid belt for building materials [140] to what at first appears to be a virtually-impossible pursuit, but which, once probed through the process of speculative design, reveals itself to be a tenable, if optimistic, proposal.

Let us recap, then, on the particulars of the project as presented above, list some advantages and disadvantages of the scheme, and then run through the ten phases that would make Cloud Ten go from imaginative idea to finalised habitat. It might be prudent, before expanding on this topic, to note that while there is a distinct lack of recent peer-reviewed articles on this subject, it is quite actively discussed on different web forums; we are beholden to our precursors and appreciate their inventive endeavours to illuminate these essentially academically unchartered territories.

As we have seen, the surface of Venus is far too hot and its atmosphere far too dense for Earth life to prosper. However, on Venus, our terrestrial air mixture of 78.09 % nitrogen, 20.95 % oxygen, 0.93 % argon, 0.039 % carbon dioxide, and small amounts of other gases becomes a lifting gas, with roughly half the lifting power of helium on Earth. If we fill our habitat with normal air, it will float high in the Venus atmosphere, where the pressure is the same as at Earth sea level (1 bar). This pressure will be the same on the interior of the habitat as on the exterior: there is no need for heavy engineering to withstand high pressures – our habitat can (and should) be of a light construction. While we have seen that the surface of Venus is harsh in the extreme for most forms of Earth life (possibly excluding some extremophile bacteria), temperatures at this region of the Venusian atmosphere will be quite beneficial for our purposes, at just above 0 °C. It is probably one of the most hospitable territories for human life in our solar system, outside of Earth itself [141]. This is our main advantage over other locations in space, and also over other architectural structures outside our atmosphere, all of which (to date) have had to be massively engineered to withstand exceptional pressures, and therefore essentially limited to cylindrical or spherical configurations with very little fenestration. Our habitat, on the contrary, hold the promise of many more and different options for architectural design, with slimmer and lighter elements and larger surfaces holding fenestration. Humanity, including the architectural and engineering professions, has very limited experience of constructing high-pressure space habitats. Under Earth-like circumstances, however, our knowledge when it comes to achieving high-quality spatial enclosures – including dirigibles and inflatables, deployable volumes, dymaxion-type domes, and other skin-and-bones-based membrane structures – is immense.

We propose to construct these lightweight structures – a modular system in which cellular volumes combine in accordance with reasoning borrowed from the world of swarm intelligence and evolutionary logics – out of space-grown grass (bamboo) and bacterial skins (kombucha). This ISMG (in situ materials generation) concept goes well beyond today's discussions about payload costs and even ISRU (in situ resource utilisation), which tends to focus more on extraction and mining of materials than on renewable technologies. The ISMG concept is particularly relevant for the Venusian climate, where the atmosphere contains three bars of nitrogen compared with 0.78 bars of nitrogen for Earth. Generally speaking, nitrogen is in short supply in the solar system (with the rare exception of nitrogen-rich meteorites), but the Venus location will not run out of nitrogen for bamboo and other plants, as well as for human breathing. By contrast, access to water is a major issue (we will expand on this in the section on disadvantages below), but if for a moment we suspend disbelief and assume that a solution for this can be found, then we have everything needed (water, CO_2 – also available on Venus – and trace elements, which if not mined from space could be brought from Earth at a relatively low cost) to grow materials in space.

Sending people and supplies to Cloud Ten would also be relatively easy and similar to Earth re-entry – aeroshells equipped with parachutes and balloons that hold lifting gas could be made to slow down and level out close to the habitat. If enclosed in titanium spheres with the interior adjusted to the right buoyancy, supplies would not necessarily even need balloons or parachutes. Fired on a trajectory to take them into the Venus atmosphere at the right point, such spheres would survive atmospheric entry intact all the way to our Cloud Ten and take Amazon's idea of drone deliveries to the next level [142]. It is also easy to get to Venus – there are more opportunities than there are for going to Mars, for instance. The transit time is shorter (five months) and the fuel requirements lower compared to Mars – while it is possible to travel to Venus on a minimum energy path every

1.6 years, we would have to wait more than two years to carry out an equivalent mission to Mars [143]. Final descent into Cloud Ten would be via parachutes and deployable balloons/dirigibles, eliminating the need for costly landing mechanisms and processes. (Getting back to Earth from Venus is harder; this will be covered in the disadvantages section below).

So long as the nitrogen and oxygen inside our bamboo-and-kombucha structure doesn't diffuse out into the Venus atmosphere, Cloud Ten will float at just the right level for an atmospheric pressure level equal to Earth normal (though the habitat would be configured so as to have a slightly higher internal pressure in order to tighten the kombucha skins around the perimeter of the structure). At this ideal pressure, there would also be no need to wear pressurised spacesuits inside the habitat (when venturing outside the kombucha screens, a breathing apparatus combined with acid protection gear would be needed). As the kombucha dries, it hardens: while more research is needed, it is safe to assume that this process could be altered to achieve different levels of opacity, elasticity, and so on to allow for the construction of membranes that either let light through or protect against the sun's rays. At this upper-cloud level of the Venusian atmosphere, the temperature stays much the same throughout the day and the year: we would not have to worry too much about the implications of cracking due to expansion and contraction based on large differences in temperature. (In one of his scientific papers, Geoffrey A. Landis illustrates the atmospheric pressure on Venus as a function of altitude above the surface, pointing out that the "altitude where terrestrial aircraft operate, between sea level and 24 km, corresponds to atmospheric pressure from 1 bar to 30 millibar. On Venus, this pressure range is found from 50 to 75 km above the surface. At these flight altitudes, the temperature varies from 80 °C at 45 km, decreasing to –10 to –35 °C at 60 km") [144].

Due to the super-rotation of the atmosphere (which carries the entire atmosphere around the planet every four days), the Venus day is equivalent to four Earth days at the level of Cloud Ten. While not an optimal 24 h, this is close enough to be reasonable for human habitation. Gravity is roughly the same as on Earth; micro-meteorites is not as big a hazard as for other proposed space habitats (and as opposed to Mars, the atmosphere would protect against larger meteorites as well); if the habitat gets punctured, it can be repaired without the risk of explosive decompression associated with structures at other (more pressurised) locations; and health threats from cosmic rays would also be substantially smaller than elsewhere – with a radiation protection of RP100 (that is, 100 % of Earth's protection) this is one of the few places in the solar system outside of our own atmosphere where we can remain protected from cosmic radiation and solar radiation.

So those are the main advantages of creating a Cloud Ten habitat above Venus. What about the disadvantages? First and foremost, the entire structure would need to be acid resistant in order to resist the concentrated sulfuric acid in the Venusian clouds. At a high concentration, sulfuric acid can cause very serious damage to materials and humans upon contact: in the latter case, the harm is not even limited to carcinogenic effects or chemical burns via hydrolysis, but is also likely to cause secondary thermal burns through dehydration, while irreversible damage might result if the acid is swallowed, and permanent blindness occur if it is splashed onto eyes [145]. In theory, materials such as carbon steel, stainless steel, tantalum, zirconium, cast high silicon iron, lead, tungsten, glass, or plastics such as Teflon, PVC, and ETFE could be used to protect the habitat against the acid, but for obvious reasons, most of them do not lend themselves to protecting our structure, and this area needs more research in order to find a solution that works to shield the exterior of the Cloud Ten habitat, as well as the ISMG seedpods (further explained below).

The second main disadvantage is to do with the lack of water. While a reasonable amount of water vapour exists, at the cloud-top levels where Cloud Ten is situated, this is bound up in the concentrated sulfuric acid. The pH is less than 0: attempting to drink or have something grow from this water is equivalent to trying the same with battery acid. Again, more research is needed, but one possibility has been suggested: to use solar furnaces to separate the sulfuric acid from the water by heating the Venusian cloud droplets with concentrated solar energy [146]. In the long term, another means of getting water to Venus would be to use gravity assists to send water (ice) to Venus from elsewhere in the solar systems, again using aeroshells equipped with parachutes [147].

Other disadvantages include high UV radiation from the sun, possible planetary protection issues (including the potential dangers of introducing terrestrial bacteria to an extraterrestrial environment as discussed in the previous section on materials), and the issues associated with returning humans to Earth from Venus. This latter difficulty is far from trivial: while the suspension of return-flight rockets could do away with the need for building launch pads, hard-to-get fuel would be needed. Again, there might be a way to separate the hydrogen in the sulfuric acid, or it could perhaps be sent as hydrogen feedstock [148]. Aside from somewhere to launch a rocket from and fuel to move it back to Earth, we also need the rocket itself, which would probably in turn need to be divided into two vehicles, similar to JP Aerospace's Airship To Orbit concept: a large-distance spaceship and a staging post holding airships designed to fly in the upper Venusian atmosphere (with today's technology and materials, an airship can't fly straight to orbit because if made strong enough to survive the relatively turbulent lower atmosphere, it would be too heavy to lift payloads to space) [149]. The engineering of a return mission process needs serious consideration and is beyond the scope of this chapter.

To produce an outline for how the schedule for constructing Cloud Ten might be constituted, we have subdivided the process into the following ten phases:

(1) Preparation phase

The creation of Cloud Ten needs substantial amounts of research and development of the technologies necessary to support the engineering and construction of the habitat. This would include (but be far from limited to) unresolved difficulties associated with the in situ growing of materials – how to shield these from the acidic environment, how to control and synchronise the growth rates of the bamboo and kombucha, how to choose and engineer the microbes for maximum yield and other properties, and so on – and also the implementation of strategies for how to produce water for the habitat. The habitat needs to be designed in detail, including the programming of the algorithms that control its cellular aggregation (as well as the creative conflicts that yield the final compromise configuration). Programmatic strategies would also need to be developed for supporting life in four-day cycles. Energy generation, food supplies, fuel production for return missions, and many other details would need careful consideration and planning.

(2) ISMG seedpod phase

Following such preparatory arrangements, the ISMG seedpods would be designed and constructed. These would essentially be the jitterbug joints (the connections between the individual bamboo elements), filled with bamboo seeds and bacterial colonies. As described above, these joints perform several functions at once, and need to be meticulously engineered so as to not only support the growth of both bamboo "bones" and kombucha "skins," but also allow for the necessary tolerances for the growing/expanding/deploying geometry, while also holding all sensors and communication equipment to allow for Earth monitoring and cellular aggregation manoeuvering, as well as means for propelling the final volumes into position. If the seeds and bacteria are sent into space in a dormant state, these seedpods could perhaps be enclosed in titanium spheres similar to the ones outlined above, and fired on trajectories that take them to the correct positions within the Venus atmosphere. The spheres might then be opened remotely (or using a timed release mechanism) before the growth is initiated. As mentioned above, during both this and the third phase, the structure needs to be coated in some way to withstand the chemical pressure from the sulfuric environment; further research is needed to conceive of plausible ways through which this might be achieved.

(3) Jitterbug growth phase

One thing that makes bamboo distinctly different from all trees is that their individual stems (or culms) emerge from the ground at their full diameter, and then grow to their full height in a single growing season of three to four months. During these months, new shoots grow vertically into culms without any branching out, until the majority of the mature height is reached. It is only then that the branches extend from the nodes and the leafing out occurs. During the second year, the pulpy wall of each culm slowly hardens, and in the third year, the culm hardens further until the shoot is considered a fully mature culm. Over the next two to five years (depending on species), fungus forms outside the culm, eventually (five to eight years later, depending on species and climate) causing it to collapse and decay. In a normal, terrestrial setting, this means culms are ready for harvest – and use in construction – after a mere three to seven years [150]. As mentioned above, genetic modification might assist with controlling this process and potentially altering it to become more applicable to our proposed scenario.

If arranged and supported correctly, the bamboo elements could grow from seed (inside one seedpod) to culm (into the opening of another seedpod). When three bamboo elements stretch in that way between three seedpod jitterbug joints, we achieve a triangle that grows (expands) in height (outwards) but not in diameter. If we are able to make this happen, then the continuous growth of the bamboo would push the seedpods away from each other. Connecting several such triangles – and possibly other geometries – would produce a dipolygonidal expansion that grows

Fig. 22.9 Cloud Ten, exterior perspective of fully expanded state

into its fully expanded state, through the volumetric iterations that we have designed (Fig. 22.9).

In tandem with the growth of the bamboo, the bacterial skin begins to grow between the members. This growth needs to become faster and faster as the bacterial skin increases in size due to the expansion of the form and its escalating surface area, and of course it needs to accelerate in a way that corresponds to the growth of the bamboo. This calls for either highly engineered bacteria or programmable microbes that we can instruct to change production rate – yet another interesting research area calling for further development.

(4) Cell swarm phase

At their full expansion, the growth of the dipolygonidal expansion volumes stops and we are left with a swarm of in situ-grown modules orbiting Venus at the top of the clouds, where the wind speed reaches up to 95 m/s and the super-rotation makes them circle the planet every four Earth days. Using the three basic rules of swarm intelligence (cohesion, separation, and alignment) we now bring these initial "cells" together into an ordered swarm, self propelled into positions relative to each other while avoiding damaging collisions.

(5) Stigmergy/evolutionary logics phase

The swarm is then rearranged in accordance with live data from the site, which is fed into our combined evolutionary solver/stigmergy optimisation system. The algorithms rearrange the initial swarm into stigmergy-based branching patterns which are then configured into first local clusters and then a globally optimised form based on evolutionary conflicts/compromise relationships as discussed above. Locking mechanisms would need to be incorporated within the modules, probably again as part of the ISMG seedpod joint structure, to allow for the cells to attach to each other while withstanding the forces from the surrounding winds and atmosphere, as well as mechanical stresses due to thrust, drag, and torque brought about by the trajectories of the volumes. Sensors and/or other communication devices would guide the cells as they cluster together into combined volumes.

(6) Interior phase

With the modules securely locked in place, the interior configuration can begin. Firstly, the interior volumes would be connected through the subtraction of double surfaces at the points where the initial volumes have been intersected. Secondly, ground datums would be established and covered with bamboo surfaces. This construction process needs to be designed in detail, but we envisage it as being based on a combination of an interior (so sheltered from the surrounding atmosphere) bamboo nursery and a robotic workforce. The robots would of course need to be introduced once the dipolygonidal expansions are fully developed, and we have not been able to find examples of robots being used in this way in an extraterrestrial setting, but terrestrial precedents for pruning robots exist, which could presumably be used as starting points for space-based equivalents [151].

(7) Atmospheric control phase

Once the interior has been configurated and the exterior checked for leaks and repaired where necessary to create a hermetically sealed shell, the interior atmosphere can begin to be controlled through a system of sensors and remote-controlled devices to achieve, in particular, proper oxygen levels. This was famously an issue at the Biosphere 2 Earth systems science research facility in Oracle, Arizona (constructed between 1987 and 1991). For its first mission, the oxygen within this facility began at 20.9 % and then fell at a steady pace to be down at 14.5 % 16 months later, resulting in sleep apnea and fatigue amongst inhabitants and leading to the medical team administering injections of oxygen. Daily (and seasonal) fluctuation of carbon dioxide dynamics were also high (typically 600 ppm) due to the strong drawdown during sunlight hours by plant photosynthesis, followed by a similar rise during the nighttime when system respiration dominated; similar, but more extreme, effects might be expected for the day/night cycle above Venus [152]. All such issues would of course need to be tackled before the habitat could welcome its first inhabitants.

(8) Inhabitation phase

When we have a sealed habitat with a functioning interior atmosphere, and not before then, we allow a team of technicians and scientists to carry out all necessary tests before welcoming the first settlers to Cloud Ten.

A combined team of agriculturalists and construction workers would join forces to create a materials factory that allows for fabrication of further elements and the erection of partition walls together with the construction of circulation paths, programmed spaces, and so on, and to further develop the interior of the habitat. Atmosphere-derived CO_2 polyurethane foam could perhaps be used for insulation, internal walls filling, and so on. Within our new spatial envelope, we could begin to grow more and other bioplastics. And of course slower-growing materials, such as wood, could also be grown once the issue of harvesting water has been overcome.

Perhaps the combination of bamboo and microbes could be further explored and integrated into a novel 3d-printing technology, whereby the bamboo fibres are used as filament materials for the construction of new elements that can be both (first) grown and (then) printed in situ in space, but again, such an idea is outside the scope of this chapter.

(9) Ecosystems phase

The next step in the development of Cloud Ten is to turn the habitat into a closed ecological system that uses biomes to create a varied set of contiguous areas with similar climatic conditions as on Earth, creating typical ecosystems that help sustain human life in space. The Biosphere 2 project might again function as a source of inspiration: this project contained "a 1,900 m^2 rainforest, an 850 m^2 ocean with a coral reef, a 450 m^2 mangrove wetlands, a 1,300 m^2 savannah grassland, a 1,400 m^2 fog desert, a 2,500 m^2 agricultural system, a human habitat, and a below-ground infrastructure. Heating and cooling water circulated through independent piping systems and passive solar input through the glass space frame panels covering most of the facility, and electrical power was supplied into Biosphere 2 from an onsite natural gas energy center" [153]. Such systems would of course need to be diligently adjusted for the properties of the site, but could equally be devised so as to utilise the unique qualities of the Venusian environment to produce effects and yields currently beyond our imagination.

(10) Permanent living phase

In the final phase, actions are taken and systems implemented that allow for a controlled and sustainable exploration of the habitat and its environments, including missions to harvest the surface of Venus for commodities. This would not necessarily have to be a very complex operation: airships could be remote controlled from the habitat and sent down to deploy nets or grabs to collect boulders or sand or other elements back up to the floating structure. This is also the phase to introduce proper long-time energy production facilities, storage alternatives for electricity and fuel, and so on. From this phase onwards, Cloud Ten becomes a true alternative to terrestrial living, a human outpost in outer space, a second solar system haven to safeguard the future existence of humanity in the unthinkable event of complete human eradication on our planet.

Coda: Paradoxical Territories

Even beginning to pursue the Cloud Ten project in any shape or form demands a certain capability for stretching one's imagination. We appreciate this is a scheme that might be exceptionally hard to achieve (as is arguably any future transformation worth investing time and effort into), and a terribly difficult struggle (all worthwhile projects are); one that might be received with disdain, intellectual opposition, snide remarks, and ridicule (as have most important ideas at one time or another). It might be all of those things, but what it won't be is insignificant or affirmative to the status quo. Cloud Ten challenges narrow assumptions of what is possible and not, and offers a challenging-yet-plausible roadmap towards a Venus habitat delineated in the tradition of critical design schemes, as expressed by a range of speculative contingents and productions, from the Italian Radical Design

Fig. 22.10 Cloud Ten is conceivable conjecture; science fact rather than science fiction

movement of the 1970s [154] through to Anthony Dunne's 1999 minor design theory classic, Hertzian Tales [155]. A list of further studies that need to be carried out and challenges that need to be overcome before disbelief in the suspended city can be conclusively suspended would run long, but it wouldn't include unsurpassable obstacles. Cloud Ten, in short, is conceivable conjecture. It is science fact rather than science fiction (Fig. 22.10).

At Ordinary Ltd, the architecture and design studio that the authors co-founded and run from our headquarters in east London, we use exactly such sharp speculation tools in our attempts to pierce the mundane and find the marvellous.

Our projects begin and end with notions of materiality, ideas about the particular and peculiar material flows that make up our buildings and cities, whether they happen to be situated on Earth or elsewhere in the solar system (or beyond). The outcome of these speculations is generally focused more on exploration than explanation, intellectual excursions ending more often than not in paradoxical territories: humanity may need to use its most advanced futurologist technologies to construct an alternative dwelling place elsewhere in space – and the way to get there might hinge on growing a plant that evolved from prehistoric grasses between thirty and forty million years ago. One definition of the word paradox (from Greek *paradoxon* – "contrary opinion," *para* meaning "beyond," *doxa* "opinion") is "a seemingly absurd or contradictory statement or proposition which when investigated or explained may prove to be well founded or true" [156].

Canadian philosopher of communication theory and public intellectual Marshall McLuhan gave that notion an optimistic slant when writing that paradox "is the technique for seizing the conflicting aspects of any problem. Paradox coalesces or telescopes various facets of a complex process in a single instant" [157]. In 1994, author Charles Handy had a book published in which he identifies nine global paradoxes in an unsettling critical assessment of the contemporary world. The fifth paradox on Handy's list is that of riches: economic growth depends upon more people wanting more things. But increasingly, the things people want most (clean air, safe environment) are collective and cannot be bought by individuals at any price. And because there is no customer, organisations cannot produce them [158]. From both of those standpoints, Cloud Ten is the best of paradoxes: a cellular habitat at the intersection of many conflicting aspects – an interstellar lightweight megastructure, an enormous city made from grass and bacteria, volumetrically expanded in an exponential fashion from within, folded into shape through a geometric dance in space, self-assembled as an incredibly intelligent swarm, soaring high in the acid clouds above the most fiendish planet we know – that indeed represents precisely such a single instant that coalesces varying strands of a mindboggingly intricate process, while at the same time having the potential to become a true sanctuary in a post-terrestrial universe, offering the clean air and safe environment that we struggle to find on Earth. Now we just need to find a customer (or organisation) that can make it happen.

Acknowledgments It is fitting for an editor so passionate about human endeavours in outer space that Viorel Badescu is a superhumanly patient editor. This time around his forbearance was tested far beyond the call of duty, and we are immensely grateful for his graceful capacity to tolerate the innumerable delays on our part as the manuscript was (slowly) prepared. Our heartfelt thanks go to Svetla Yordanova, and Camille Lot, who offered exceptional assistance throughout the production of this chapter, and a particular word of appreciation is due to Jack Munro, who made the geometry happen. Thank you, as always, to Madelaine Levy and Thidaa Roberts for their ceaseless support. Emmett Larsson Levy provided excellent distractions and diversions. The Cloud Ten project is dedicated to Mike Skinner (1974–2014), a true Bucky fan with whom we would have loved to discuss its propositions had he not been suddenly taken away from us, far too young, as we wrote this chapter. Rest in peace, Mike.

References

1. Musk E (2014) Interviewed by Ross Andersson for the article Exodus. Aeon Magazine, 30 Sept 2014. http://www.aeon.co/magazine/technology/the-elon-musk-interview-on-mars/
2. Cf. Hender M (2009) A permanent habitat for the colonization of Mars. Masters thesis, University of Adelaide School of Mechanical Engineering, Aug 2009, in which the author makes a compelling argument for Mars being the best option. His prime reason is perhaps that the issues associated with a Mars habitat still potentially make for an easier life than that elsewhere in space, for instance on a space station, due to its similar gravity, day/night cycle, and plentiful local surface resources
3. The danger is that either by accident or design, we create a virus that destroys us. I don't think the human race will survive the next thousand years, unless we spread into space. Roger Highfield, Colonies in space may be only hope, says Hawking. Article in The Telegraph, 16 Oct 2001. http://www.telegraph.co.uk/news/uknews/1359562/Colonies-in-space-may-be-only-hope-says-Hawking.html
4. Musk/Andersson, ibid

5. For more reasons, see Pass J, Dudley-Rowley M (2006) The cultural imperative to colonize space: an astrosociological perspective. Space 2006, San Jose, California, AIAA 2006-7488, 19–21 Sept 2006
6. Global human population growth amounts to around 75 million annually, or 1.1 % per year. The global population has grown from 1 billion in 1800 to 7 billion in 2012. See Population Reference Bureau, 2013 World Population Factsheet (www.pbr.org). Human energy consumption in the United States appears to have risen by around 2.9 % a year between 1650 and 2009—see Tom Murphy, Galactic-Scale Energy. Blog post, 12 July 2011. http://physics.ucsd.edu/do-the-math/2011/07/galactic-scale-energy/. It is perhaps prudent to note that harbouring illusions of space as the only answer to our collision course of unprecedented growth on a finite planet is immature at best and downright fatal at worst
7. Colin L (1983) Basic facts about Venus. In: Hunten DM et al. (eds) Venus. University of Arizona Press, Tucson, p 19
8. Staples A (1998) From Good Goddess to vestal virgins: sex and category in Roman religion. Routledge, pp 12, 15–16, 24–26, 149–150
9. Basilevsky AT, Head JW (2003) The surface of Venus. Rep Prog Phys 66:1699–1734. PII: S0034-4885(03)08878-X), p 1699
10. Ibid., p 1701
11. Normile D (2010) Mission to probe Venus's curious winds and test solar sail for propulsion. Science 328(5979):677
12. For an initiated discussion about the Venusian atmosphere, see Lewis JS, Prinn RG (1982) Planets and their atmospheres: origins and evolution. Academic Press, London, pp 128–197
13. "Atmosphere of Venus" in David Darling, The Encyclopedia of Science (website). http://www.daviddarling.info/encyclopedia/V/Venusatmos.html. It should be noted that John Ackerman has presented a different viewpoint in his paper An Alternate View of Venus. Ackerman maintains that "the surface temperature is maintained at 450 °C by the evaporation of raining sulfur," that "it is the great mass of sulfur suspended in the lower atmosphere (Hadesphere) which produces the high surface pressure, not CO_2," and that "that the dominant gas in the lower atmosphere is S_8, not CO_2, and there is no 'runaway greenhouse effect'"
14. Colin L, ibid., p 18
15. Basilevsky AT, Head JW, ibid., p 1701
16. Faure G, Mensing TM (2007) Introduction to planetary science: the geological perspective. Springer eBook collection, Springer, p 201
17. Balint TS, Kerzhanovich VV, Hall JL, Baines KH, Stephens SK (2010) Four aspects of a venus balloon mission concept. In: 41st lunar and planetary science conference
18. Technica Molodezhi TM—9 1971. Available in facsimile format online: http://www.secretprojects.co.uk/forum/index.php?topic=14334.0#!
19. Including Landis GA, Colozza A, LaMarre CM (2002) Atmospheric flight on venus. NASA/TM-2002-211467, AIAA-2002-0819, June 2002; Landis GA (2001) Exploring Venus by solar airplane. In: Presented at the STAIF conference on space exploration technology, NM, Albuquerque, 11–15 Feb 2001; Landis GA (2003) Astrobiology: the case for Venus. NASA/TM-2003-212310, July 2003; Landis GA (2010) Low-altitude exploration of the Venus atmosphere by balloon. Paper 2010-628, 48th AIAA Aerospace Sciences Meeting, FL, Orlando, 6–9 Jan 2010
20. Landis GA (2010) Sultan of the clouds. Asimov's Science Fiction, Sept 2010
21. Ibid
22. Ibid
23. Havoc, NASA website: http://sacd.larc.nasa.gov/branches/space-mission-analysis-branch-smab/smab-projects/havoc/#sthash.3JnsSjrS.dpuf
24. Larsson M, Kaiser A (2012) Moon Dune: Bacillithic Cratertecture. In: Badescu V (ed) Moon—prospective energy and material resources. Springer, pp 661–702
25. Goddard G (1963) In the introduction to E.M. Emme, The history of rocket technology. Wayne State University Press in cooperation with the Society for the History of Technology, Michigan, Detroit

26. Bryson B (2003) A short history of nearly everything. Black Swan, Uxbridge, pp 302–304
27. Ames Technology Capabilities and Facilities, In-Situ Resource Utilization (website, NASA/Ames Research Center, last updated 2008). http://www.nasa.gov/centers/ames/research/technology-onepagers/in-situ_resource_Utiliza14.html
28. In Situ Resource Utilization Technical Interchange Meeting, Lunar and Planetary Institute, Houston Texas, 4–5 Feb 1997. Introductory notes. http://www.lpi.usra.edu/meetings/isru97/isru97.1st.html
29. Mazanek DD, Merrill RG, Brophy JR, Mueller RP Asteroid redirect mission concept: a bold approach for utilizing space resources. In: 65th international astronautical congress, Toronto, Canada. Copyright 2014 United States Government, as represented by the Administrator of the National Aeronautics and Space Administration, and Jet Propulsion Laboratory—California Institute of Technology, USA. IAC-14-D3.1.8, p 1
30. Larson WE, Sanders GB, Sacksteder KR, NASA's In-Situ Resource Utilization project: current accomplishments and exciting future plans. American Institute of Aeronautics and Astronautics
31. Larson W, Sanders G, Hyatt M (2011) ISRU—From concept to reality: NASA accomplishments and future plans. In: AIA a space 2011 conference and exposition, Long Beach, California, AIAA 2011-7114, 27–29 Sept 2011
32. Ibid
33. Larsson M, Kaiser A (2013) Invincible cities for the materiomic age. In: Rassia STh, Pardalos PM (eds) Cities for smart environmental and energy futures: impacts on architecture and technology. Springer Science & Business Media, 15 Aug 2013
34. Proposals for Mars exploration and colonisation projects abound. For a discussion about the use of ISRU in this context, cf. James G, Chamitoff G, Barker D, Resource utilization and site selection for a self-sufficient Martian outpost. NASA/Lyndon B Johnson Space Center, April 1998. NASA/TM-98-206538
35. Sanders GB, Dr. Duke M (chairs) (2005) NASA In-Situ Resource Utilization (ISRU) Capability Roadmap: Final Report, 19 May 2005
36. Vogler A, Vittori A (2008) Space architecture for the mother ship: bringing it home. In: Howe AS, Sherwood B (eds) Out of this world: the new field of space architecture, AIAA Reston, VA, USA, 2009, ISBN 978-1563479823)
37. Dove LL, 10 Offbeat Things Humans Have Launched Into Space (HowStuffWorks website article). http://science.howstuffworks.com/10-offbeat-things-humans-have-launched-into-space.htm
38. Rapp D, The problems with lunar ISRU. The Space Review, 5 Sept 2006. http://www.thespacereview.com/article/697/1
39. Ibid
40. Cf. Palmer MA et al (2010) Mountaintop mining consequences. Science 327:148
41. Cf. Broomfield M, Support to the identification of potential risks for the environment and human health arising from hydrocarbons operations involving hydraulic fracturing in Europe (Report, 2012-08-10) (17c). European Commission. pp vi–xvi. ED57281
42. Wingo D (2008) To ISRU or Not to ISRU, This is the Dumbest Question. Spaceref, 18 May 2008. http://www.spaceref.com/news/viewnews.html?id=1290
43. Larsson and Kaiser, ibid
44. Coghlan A (2012) Build a Mars base with a box of engineered bugs. New Scientist, 4 Oct 2012. http://www.newscientist.com/article/mg21628855.100-build-a-mars-base-with-a-box-of-engineered-bugs.html#.VPCXR1OsXG8
45. The papers we refer to are Menezes AA, Cumbers J, Hogan JA, Arkin AP (2015) Towards synthetic biological approaches to resource utilization on space missions. Interface/The Royal Society, April 2015, doi:10.1098/rsif.2014.0715, 5 Nov 2014, and Montague M, McArthur IV GH, Cockell CS, Held J, Marshall W, Sherman LA, Wang N, Nicholson WL, Tarjan DR, Cumbers J (2012) The role of synthetic biology for In Situ Resource Utilization (ISRU). Astrobiology 12(12). doi:10.1089/ast.2012.0829

46. Cf. Kim J-J, Rigdon B, Graves J (1998) Qualities, use, and examples of sustainable building materials. Compendium, National Pollution Prevention Center for Higher Education, 1998, p 9: "…a material is only considered a renewable or sustainable resource if it can be grown at a rate that meets or exceeds the rate of human consumption"
47. Landis GA, Sultan of the clouds, ibid
48. Ivanova T, Sapunova S, Kostov P, Dandolov I (2001) First successful space seed-to-seed plant growth experiment in the SVET-2 space greenhouse in 1997. Space Research Institute, Bulgarian Academy of Sciences. http://www.space.bas.bg/astro/Aerosp16/tania1.pdf
49. Ivanova T (2002) Space farming on Mars: greenhouse aboard mir shows plants can thrive in space. 21st Century, summer 2002
50. BBC TV series Fun to Imagine, interview with Richard Feynman (1983). www.bbc.co.uk/archive/feynman/
51. Liu J, Zhang Y (2013) An automatic aeroponics growth system for bamboo seedling and root observation. Appl Mech Mater 307:97–102
52. Cf. Dierick D, Hölscher D, Schwendenmann L (2010) Water use characteristics of a bamboo species (Bambusa blumeana) in the Philippines. Agric For Meteorol 150(12):1568–1578
53. Armstrong WP (2015) Bamboo: Remarkable giant grasses. http://waynesword.palomar.edu/ecoph39.htm. Accessed 28 Feb 2015
54. Wilson CL, Loomis WE, Croasdale HT (1962) Botany: third edition. Holt, Rinehart & Winston, Austin
55. Farrelly D (1984) The book of bamboo. Sierra Club Books. ISBN 0-87156-825-X
56. The entry is at http://www.guinnessworldrecords.com/world-records/fastest-growing-plant/
57. Algeo J, Algeo A (1997) Among the new words. Am Speech 72(2):183–97. doi:10.2307/455789
58. Sreeramulu G, Zhu Y, Knol W (2000) Kombucha fermentation and its antimicrobial activity. J Agric Food Chem 48(6):2589–2594. doi:10.1021/jf991333m
59. Sreeramulu G, Zhu Y, Knol W, ibid
60. NASA Office of Planetary Protection, Planetary Protection History. http://planetaryprotection.nasa.gov/history
61. Committee on Preventing the Forward Contamination of Mars, Space Studies Board, Division on Engineering and Physical Sciences, National Research Council, Preventing the Forward Contamination of Mars. National Academies Press, 22 March 2006
62. Cf. Fairén A and Schulze-Makuch D (2013) The over protection of Mars. Nat Geosci 6(7):510
63. Ibid
64. NASA Office of Planetary Protection, About The Categories. http://planetaryprotection.nasa.gov/about-categories/
65. McKay CP (2009) Biologically reversible exploration. Science 323(5915):718
66. Aydin YA, Aksoy ND (2009) Isolation of cellulose producing bacteria from wastes of vinegar fermentation. In: Proceedings of the world congress on engineering and computer science 2009, vol I, WCECS 2009, San Francisco, USA, 20–22 Oct 2009
67. Teoh AL, Heard G, Cox J (2004) Yeast ecology of kombucha fermentation. Int J Food Microbiol 95:119–126
68. Nguyen VT, Flanagan B, Gidley MJ, Dykes GA (2008) Characterization of cellulose production by a gluconacetobacter xylinus strain from kombucha (Current. Microbiology 57 (5):449–453. doi:10.1007/s00284-008-9228-3
69. Goh WN, Rosma A, Kaur B, Fazilah A, Karim AA, Rajeev B (2012) Microstructure and physical properties of microbial cellulose produced during fermentation of black tea broth (Kombucha). II. Int Food Res J 19(1):109–117
70. Goh WN, Rosma A, Kaur B, Fazilah A, Karim AA, Rajeev B (2012) Microstructure and physical properties of microbial cellulose produced during fermentation of black tea broth (Kombucha). II. Int Food Res J 19(2):153–158
71. Goh WN, Rosma A, Kaur B, Fazilah A, Karim AA, Rajeev B, Ibid., 19(1):111
72. Whitman W, Coleman D, Wiebe W (1998) Prokaryotes: the unseen majority. Proc Natl Acad Sci USA 95(12):6578–6583. doi:10.1073/pnas.95.12.6578.PMC33863.PMID9618454

73. Fuller RB (1979) Synergetics: explorations in the geometry of thinking. Macmillan, 1975, 1979, 460.08
74. The exchange with Einstein is revealed in Robert Snyder, Buckminster Fuller: An Autobiographical Monologue/Scenario (St. Martin's Press, New York. c1980), pp 68–69: "… young man, this chapter on Mrs. Murphy—you amaze me. I cannot myself conceive of anything I have done ever having the slightest practical application". It may be prudent to note that not everyone agreed—architect Philip Johnson, for instance, was a little less easily impressed: "Bucky Fuller was no architect and he kept pretending he was. It was very annoying. We all hated him because he really thought the profession was unnecessary". Interview with Philip Johnson in Kirk Simon & Karen Goodman, Buckminster Fuller: Thinking Out Loud (PBS "American Masters" tv documentary series, 1996). Transcript available online: http://www.thirteen.org/bucky/johnson.html
75. The Dymaxion American, TIME Magazine (cover story), vol 83, No 2, 10 Jan 1964
76. Baldwin J (1997) BuckyWorks: Buckminster Fuller's ideas for today. Wiley, p 190
77. Sieden LS (2000) Buckminster Fuller's Universe. Basic Books p 409
78. Krausse J, Lichtenstein C (eds) (1999) Your private sky. Lars Müller Publishers, Baden, Switzerland, pp 286
79. Clinton JD (2004) Intersecting cylinders and the Jitterbug. In: Proceedings, bridges: mathematical connections in art, music, and science, p 346
80. Private collection of Joseph D. Clinton—unpublished notes of Duncan Stuart. As referenced in Clinton, ibid
81. Stuart D (1963) Polyhedral and mosaic transformation, vol 12, No 1. The Student Publications of the School of Design, University of North Carolina, Raleigh, North Carolina
82. Resch RR (1973) The topological design of sculptural and architectural systems. In: FAIPS-conference proceedings, vol 42, AFJPS Press, NJ, Montvale, pp 643–650
83. Clinton JD (1971) Advanced structural geometry studies: part 11- a geometric transformation concept/or expanding rigid structures. NASA CR-173S, NASA, Washington D.C
84. Verheyen HF (1989) The complete set of jitterbug transformers and the analysis of their motion. Comput Math Appl 17(1–3):203–250, and Verheyen HF (1996) symmetry orbits. Birkhäuser, Massachusetts, Boston, respectively
85. Gray RW, The Jitterbug motion. http://www.rwlllY.Projects.comlrbfnotes/toc.html
86. Verheyen (1989), ibid., p 203
87. Verheyen, ibid
88. It would of course be quite possible to build a series of models that would illustrate such series of expansions, but these appear to be few and far between, over and above Verheyen's drawings. On the computer, simulating the end results of such additions of membranes is as easy as lofting surfaces between the curves in their new positions following each transformation
89. James C (2011) Global status of commercialized biotech/GM crops: 2011. ISAAA Brief No. 43. ISAAA, NY, Ithaca
90. Vidal J (2012) The GM tree plantations bred to satisfy the world's energy needs. The Guardian, Thursday 15 Nov 2012
91. Bird W (2014) In the pink: speeding up crop growth. Futurefood2050 website article, 23 Oct 2014. http://futurefood2050.com/in-the-pink-speeding-up-crop-growth/
92. Kiper G (2011) Design methods for planar and spatial deployable structures. PhD thesis, The Graduate School of Natural and Applied Sciences of Middle East Technical University
93. As outlined, for instance, in Pilkauskas K, Gaidys R, Lira C (2007) Adaptive structures based on smart stick concept for robotic applications. In: Proceedings of IMAC-XXV: conference & exposition on structural dynamics
94. Kuma K (2005) The relativity of materials, JA The Japan Architect 38/Summer 2000, reprinted in Luigi Alini, Kengo Kuma—Works and Projects. Electa Architecture, Milan, pp 222–223
95. Ball P (2015) The strange inevitability of evolution—good solutions to biology's problems are astonishingly plentiful. Nautilus, issue 020, chapter one, 8 Jan 2015. http://nautil.us/issue/20/creativity/the-strange-inevitability-of-evolution

96. Landis GA (2010), ibid
97. Rubenstein M, Cabrera A, Werfel J, Habibi G, McLurkin J, Nagpal R Collective transport of complex objects by simple robots: theory and experiments. In: Proceeding AAMAS '13 proceedings of the 2013 international conference on autonomous agents and multi-agent systems, pp 47–54
98. Bryson B, ibid
99. For a good overview of self-organised biological and social systems, see Camazine S, Deneubourg J-L, Franks NR, Sneyd J, Theraulaz G, Bonabeau E (2001) Self-organization in biological systems. Princeton University Press Princeton, New Jersey
100. Carlo Ratti/*FlyFire:* http://senseable.mit.edu/flyfire. Gramazio & Kohler/*Flight Assembled Architecture:* www.gramaziokohler.com/web/e/projekte/209.html. Neri Oxman & Markus Kayser/*Swarm Construction:* https://www.media.mit.edu/research/groups/mediated-matter. Skylar Tibbits/*Fluid Crystallization:* http://www.suckerpunchdaily.com/2013/07/17/fluid-crystallization
101. Cf. http://www.eecs.harvard.edu/ssr/
102. Nichols J (2011) Take shelter. Sony Pictures Classics
103. TU Delft, Swarm Architecture II—Space is a computation. http://www.bk.tudelft.nl/en/about-faculty/departments/architectural-engineering-and-technology/organisation/hyperbody/research/theory/swarm-architecture-ii/2006-swarm-architecture-ii/
104. O'Loan OJ, Evans MR (1998) Alternating steady state in one-dimensional flocking. J Phys A Math Gen 32(8):L99–L105
105. Whitesides GM, Grzybowski B (2002) Self-Assembly at all scales. Science 295:2418–2421
106. Ibid
107. Klavins E, Burden S, Napp N, Optimal rules for programmed stochastic self-assembly. Electrical engineering, University of Washington, Seattle, WA 98195
108. Grassé P-P (1959) La reconstruction du nid et les coordinations interindividuelles chez Bellicositermes natalensis et Cubitermes sp. La théorie de la stigmergie: Essai d'interprétation du comportement des termites constructeurs. Insectes Sociaux 6(1):41–83
109. Bonabeau E (1999) Editor's introduction: stigmergy. Special issue of Artificial Life on Stigmergy. vol 5, Issue 2 / Spring 1999, pp 95–96
110. Marsh L, Onof C (2007) Stigmergic epistemology, stigmergic cognition. Cognitive Systems Research
111. Wilson EO (1975/2000) Sociobiology: the new synthesis. Harvard University Press, p 186
112. Marsh L, Onof C, ibid
113. Gordon D, Collective intelligence in social insects (website article). http://ai-depot.com/Essay/SocialInsects.html
114. Beckers R, Holland OE, Deneubourg JL (1994) From local actions to global tasks: stigmergy and collective robotics. Artificial life IV, p 181
115. Ibid
116. Ibid., pp 181–189
117. Ibid., p 182
118. Petersen KH (2014) Collective construction by termite-inspired robots. PhD dissertation, Harvard University School of Engineering and Applied Sciences, Sept 2014, p 4
119. Originating in a Western Zhou divination text called the Changes of Zhou or Zhou yi, the I Ching can be said to contain references to events that played out as early as the 11th century BC. In his 1983 Ph.D. dissertation The Composition of the Zhouyi (Stanford University), Edward Shaughnessy concludes that the text was assembled in approximately its current form in the last quarter of the 9th century BC
120. Maurice Merleau-Ponty: "All the great philosophical ideas of the past century—the philosophies of Marx and Nietzsche, phenomenology, German existentialism, and psychoanalysis—had their beginnings in Hegel". Maurice Merleau-Ponty (trans. Herbert L. and Patricia Allen Dreyfus), Sense and Nonsense, Northwestern University Press, 1964, p 63. According to Michel Foucault, all contemporary philosophers may be

"doomed to find Hegel waiting patiently at the end of whatever road [they] travel". Andrew Bowie, Schelling and Modern European Philosophy, 2 Routledge London, 1993, p 2
121. So often, indeed, that it has become known as "Hegelian dialectic". Writes German philosopher Walter Kaufmann: "Fichte introduced into German philosophy the three-step of thesis, antithesis, and synthesis, using these three terms. Schelling took up this terminology. Hegel did not. He never once used these three terms together to designate three stages in an argument or account in any of his books. And they do not help us understand his Phenomenology, his Logic, or his philosophy of history; they impede any open-minded comprehension of what he does by forcing it into a scheme which was available to him and which he deliberately spurned… The mechanical formalism… Hegel derides expressly and at some length in the preface to the Phenomenology. Walter Kaufmann, Hegel: A Reinterpretation, 1966, Anchor Books, § 37, p 154. See also G. E. Mueller (June 1958), "The Hegel Legend of 'Thesis-Antithesis-Synthesis," 166ff
122. Hegel's Logic, Part One of the Encyclopaedia of the Philosophical Sciences, 1830, translated by J.N. Findlay, Oxford at the Clarendon Press, 1975, p 117
123. Kaufmann, ibid., p 132
124. The quote is by the character Claire Underwood (Robin Wright) in House of Cards (tv series, Netflix, Season 3/episode 6, 2015)
125. Larsson M (2014) Conflict and compromise: an evolutionary framework for the design of multi-storey timber buildings. Licentiate thesis, Luleå University of Technology
126. Ibid
127. Rutten D (2011) Evolutionary solvers: fitness functions. Blog post, I Eat Bugs for Breakfast, March 4, 2011. http://www.ieatbugsforbreakfast.wordpress.com/2011/03/04/fitnessfunctions/
128. Dunbar K (1998) Problem solving. In: Bechtel W, Graham G (eds) A companion to cognitive science. Blackwell, London, England, pp 289–298
129. Larsson, ibid
130. Ball, ibid
131. Cf. Provine W (1986) Sewall wright and evolutionary biology. University of Chicago Press, Chicago
132. Frontline systems, Basic Solver—Evolutionary Solving Method Stopping Conditions (website entry). http://www.solver.com/content/basic-solver-evolutionary-solving-methodstopping-conditions
133. For more information. http://www.cerver.org/research/locustswarm/
134. For more information. http://www.red3d.com/cwr/boids/
135. Walker R (2014) Will We Build Colonies That Float Over Venus Like Buckminster Fuller's "Cloud Nine"? (website article, science20.com, 12 Jan 2014). http://www.science20.com/robert_inventor/will_we_build_colonies_that_float_over_venus_like_buckminster_fullers_cloud_nine-127573. Parts of this description of the Cloud Ten project is particularly indebted to Walker's lucid explications in this highly readable article
136. Monbiot G (2003) The age of consent: a manifesto for a new world order. Flamingo/HarperCollins, p 6
137. As quoted in Quinn D (1999) Beyond Civilization: Humanity's Next Great Adventure. Harmony Books, New York, p 137
138. This was in a Newsweek cover story in 1978, as recounted by Pauline Kael in Taking It All In. Henry Holt & Co., New York, 1984
139. Nowviskie B (2014) Speculative computing & the centers to come. Notes for a short talk. http://nowviskie.org/2014/speculative-computing/
140. Cf. Walker R (2013) Asteroid Resources Could Create Space Habs For Trillions; Land Area Of A Thousand Earths (website article, science20.com, 17 July 2013). http://www.science20.com/robert_inventor/blog/asteoid_resources_could_create_space_habs_trillions_land_area_thousand_earths-116541. We plan to explore this option architecturally in the near future
141. Walker (2014), ibid
142. Landis GA (2010) Low-altitude exploration of the Venus atmosphere by balloon. Paper 2010-628, 48th AIAA aerospace sciences meeting, FL, Orlando, 6–9 Jan 2010, p 5

143. Foster C, Daniels M, Mission opportunities for human exploration of nearby planetary bodies. American Institute of Aeronautics and Astronautics. doi:10.2514/6.2010-8609. Conference: AIAA SPACE 2010 Conference & Exposition
144. Landis GA, Colozza A, Lamarre CM (2002) Atmospheric flight on Venus. NASA/TM-2002-211467, AIAA-2002-0819, June 2002, p 2
145. Pritchard JD (2007) Sulphuric acid toxicological overview. UK Health Protection Agency, vol 2, 2007. https://www.gov.uk/government/uploads/system/uploads/attachment_data/file/338272/hpa_sulphuric_acid_toxicological_overview_v2.pdf
146. Walker (2014), ibid
147. Ibid
148. Zubrin R (1996) The Promise of Mars. Ad Astra, May/June 1996. http://www.nss.org/settlement/mars/zubrin-promise.html. While presented as a technology for Mars, the idea translates to the context of Venus
149. United States Patent No. 7614586
150. Armstrong WP, Bamboo: Remarkable giant grasses, ibid
151. Cf. Ishigure Y, Hirai K, Kawasaki H (2013) A pruning robot with a power-saving chainsaw drive. In: 2013 IEEE international conference on mechatronics and automation, 4–7 Aug 2013, pp 1223–1228
152. Chronology of Biosphere 2 (website article). http://www.biospherics.org/biosphere2/chronology/
153. UA Science, Fast Facts: Stats on the Biosphere (web page). http://b2science.org/who/fact
154. Cf. Rossi C, Coles A (2013) EP vol 1, The Italian Avant-Garde: 1968–1976. Sternberg Press
155. Dunne A (2006) Hertzian tales. The Royal College of Art, 1999/The MIT Press
156. The New Shorter Oxford English Dictionary, 1993: 2093, 2095
157. McLuhan M (1972) Take today: the executive as dropout, p 106
158. Handy C (1994) The age of Paradox. Harvard Business School Press

Author Index

V. Badescu and K. Zacny (eds.), *Inner Solar System*,
DOI 10.1007/978-3-319-19569-8

Subject Index

V. Badescu and K. Zacny (eds.), *Inner Solar System*,
DOI 10.1007/978-3-319-19569-8

MIX
Papier aus verantwortungsvollen Quellen
Paper from responsible sources
FSC® C105338

If you have any concerns about our products,
you can contact us on
ProductSafety@springernature.com

In case Publisher is established outside the EU,
the EU authorized representative is:
Springer Nature Customer Service Center GmbH
Europaplatz 3, 69115 Heidelberg, Germany

Printed by Libri Plureos GmbH
in Hamburg, Germany